3521356175

AF598624

NO LONGER THE PROPERTY OF THE
UNIVERSITY OF LINCOLN LIBRARY

Methods in Cell Biology

VOLUME 108

Lipids

Series Editors

Leslie Wilson
Department of Molecular, Cellular and Developmental Biology
University of California
Santa Barbara, California

Paul Matsudaira
Department of Biological Sciences
National University of Singapore
Singapore

Methods in Cell Biology

VOLUME 108

Lipids

Edited by

Gilbert Di Paolo

Department of Pathology and Cell Biology,
Taub Institute for Research on Alzheimer's Disease and the Aging Brain,
Columbia University Medical Center, New York, USA

Markus R Wenk

Department of Biochemistry and Department of Biological Sciences,
Yong Loo Lin School of Medicine, National University of Singapore,
Singapore, Singapore

AMSTERDAM • BOSTON • HEIDELBERG • LONDON
NEW YORK • OXFORD • PARIS • SAN DIEGO
SAN FRANCISCO • SINGAPORE • SYDNEY • TOKYO
Academic Press is an imprint of Elsevier

Academic Press is an imprint of Elsevier
225 Wyman Street, Waltham, MA 02451, USA
525 B Street, Suite 1900, San Diego, CA 92101-4495, USA
32, Jamestown Road, London NW1 7BY, UK
Linacre House, Jordan Hill, Oxford OX2 8DP, UK

First edition 2012

Copyright © 2012 Elsevier Inc. All rights reserved

No part of this publication may be reproduced, stored in a retrieval system or transmitted in any form or by any means electronic, mechanical, photocopying, recording or otherwise without the prior written permission of the publisher

Permissions may be sought directly from Elsevier's Science & Technology Rights Department in Oxford, UK: phone (+44) (0) 1865 843830; fax (+44) (0) 1865 853333; email: permissions@elsevier.com. Alternatively you can submit your request online by visiting the Elsevier web site at http://elsevier.com/locate/permissions, and selecting *Obtaining permission to use Elsevier material*

Notice
No responsibility is assumed by the publisher for any injury and/or damage to persons or property as a matter of products liability, negligence or otherwise, or from any use or operation of any methods, products, instructions or ideas contained in the material herein. Because of rapid advances in the medical sciences, in particular, independent verification of diagnoses and drug dosages should be made

ISBN: 978-0-12-386487-1
ISSN: 0091-679X

For information on all Academic Press publications visit our website at elsevierdirect.com

Printed and bound in USA

12 13 14   10 9 8 7 6 5 4 3 2 1

Working together to grow libraries in developing countries

www.elsevier.com | www.bookaid.org | www.sabre.org

ELSEVIER   BOOK AID International   Sabre Foundation

CONTENTS

PART III—Imaging

CONTRIBUTORS

Numbers in parentheses indicate the pages on which the author's contributions begin.

Marie-France Bader (445), Institut des Neurosciences Cellulaires et Intégratives, CNRS UPR3212, Strasbourg, France

Nirmalya Bag (395), Department of Chemistry, National University of Singapore, Singapore

Vytas A. Bankaitis (249), Department of Cell & Developmental Biology, Lineberger Comprehensive Cancer Center, School of Medicine, University of North Carolina at Chapel Hill, Chapel Hill, North Carolina, USA

Christin Bissig (19), Department of Biochemistry, University of Geneva, Geneva, Switzerland

Kristian Bredies (345), Institute of Mathematics and Scientific Computing, University of Graz, Graz, Austria

John H. Brumell (173), Cell Biology Program, Hospital for Sick Children, Toronto, Ontario, Canada; Institute of Medical Science, University of Toronto, Ontario, Canada; Department of Molecular Genetics, University of Toronto, Ontario, Canada

Belle Chang-Ileto (187), Department of Pathology and Cell Biology, Taub Institute for Research on Alzheimer's Disease and the Aging Brain, Columbia University Medical Center, New York, USA

Ketpin Chong (319), Cubic Membrane Laboratory, Department of Physiology, Yong Loo Lin School of Medicine, National University of Singapore, Singapore

Matthias Corrotte (445), Institut des Neurosciences Cellulaires et Intégratives, CNRS UPR3212, Strasbourg, France

James M. Davison (249), Department of Cell & Developmental Biology, Lineberger Comprehensive Cancer Center, School of Medicine, University of North Carolina at Chapel Hill, Chapel Hill, North Carolina, USA

Pietro De Camilli (3), Howard Hughes Medical Institute, Department of Cell Biology, Program in Cellular Neuroscience, Neurodegeneration and Repair, Yale University School of Medicine, New Haven, CT 06520, USA

Yuru Deng (319), Cubic Membrane Laboratory, Department of Physiology, Yong Loo Lin School of Medicine, National University of Singapore, Singapore

Gilbert Di Paolo (187), Department of Pathology and Cell Biology, Taub Institute for Research on Alzheimer's Disease and the Aging Brain, Columbia University Medical Center, New York, USA

Guillaume Drin (47), Institut de Pharmacologie Moléculaire et Cellulaire, Université de Nice Sophia-Antipolis and CNRS, 660 route des lucioles, 06560 Valbonne, France

Weihua Fei (303), School of Biotechnology and Biomolecular Sciences, the University of New South Wales, Sydney, Australia

Samuel G. Frere (187), Department of Pathology and Cell Biology, Taub Institute for Research on Alzheimer's Disease and the Aging Brain, Columbia University Medical Center, New York, USA

Michael A. Frohman (131), Department of Pharmacology, Center for Developmental Genetics, Stony Brook University, Stony Brook, New York, USA

Ratna Ghosh (249), Department of Cell & Developmental Biology, Lineberger Comprehensive Cancer Center, School of Medicine, University of North Carolina at Chapel Hill, Chapel Hill, North Carolina, USA

Nancy J. Grant (445), Institut des Neurosciences Cellulaires et Intégratives, CNRS UPR3212, Strasbourg, France

Sergio Grinstein (173, 429), Program in Cell Biology, The Hospital for Sick Children, Toronto, Canada; Cell Biology Program, Hospital for Sick Children, Toronto, Ontario, Canada; Department of Biochemistry, University of Toronto, Ontario, Canada; Institute of Medical Science, University of Toronto, Ontario, Canada

Jean Gruenberg (19), Department of Biochemistry, University of Geneva, Geneva, Switzerland

Xue Li Guan (149), Department of Medical Parasitology and Infection Biology, Swiss Tropical and Public Health Institute, Basel, Switzerland; University of Basel, Basel, Switzerland

Kentaro Hanada (117), Department of Biochemistry and Cell Biology, National Institute of Infectious Disease, Shinjuku-ku, Tokyo, Japan

Volker Haucke (209), Institute of Chemistry and Biochemistry, Freie Universität Berlin, Takustraße, Berlin, Germany; Leibniz-Institut für Molekulare Pharmakologie (FMP), Robert-Rössle-Straße, Berlin, Germany

Huiyan Huang (131), Department of Pharmacology, Center for Developmental Genetics, Stony Brook University, Stony Brook, New York, USA

Shem Johnson (19), Department of Biochemistry, University of Geneva, Geneva, Switzerland

Anjali Jotwani (93), Department of Cell Biology, Yale School of Medicine, Connecticut, USA

Omar Julca-Zevallos (93), Department of Cell Biology, Yale School of Medicine, Connecticut, USA

Nawal Kassas (445), Institut des Neurosciences Cellulaires et Intégratives, CNRS UPR3212, Strasbourg, France

Thang Manh Khuong (227), VIB Center for the Biology of Disease, Leuven, Belgium; K.U. Leuven, Center for Human Genetics, Leuven, Belgium

Sepp D. Kohlwein (345), Institute of Molecular Biosciences, University of Graz, Graz, Austria; Institute of Mathematics and Scientific Computing, University of Graz, Graz, Austria

Rachel Kraut (395), School of Biological Sciences, Nanyang Technological University, Singapore

Michael Krauß (209), Institute of Chemistry and Biochemistry, Freie Universität Berlin, Takustraße, Berlin, Germany

Keigo Kumagai (117), Department of Biochemistry and Cell Biology, National Institute of Infectious Disease, Shinjuku-ku, Tokyo, Japan

Cécile Leduc (47), Laboratoire Photonique, Numerique et Nanosciences (LP2N), Institut d'optique Graduate School, Universite de Bordeaux and CNRS, Talence, Cedex, France

Jean-Baptiste Manneville (47), Unité Mixte de Recherche 144, CNRS, Institut Curie, 26 rue d'Ulm, 75248 Paris Cedex 05, France

Frederick R. Maxfield (367), Department of Biochemistry, Weill Cornell Medical College, New York, USA

Thomas J. Melia (93), Department of Cell Biology, Yale School of Medicine, Connecticut, USA

Isabelle Motta (93), Laboratoire de Physique Statistique, Ecole Normale Supérieure, Paris, France

Masahiro Nishijima (117), National Institute of Health Science, Setagaya-Ku, Tokyo, Japan

Diana N. Richerson (93), Department of Cell Biology, Yale School of Medicine, Connecticut, USA

Bernhard Roppenser (173), Cell Biology Program, Hospital for Sick Children, Toronto, Ontario, Canada

Helen Sarantis (429), Program in Cell Biology, The Hospital for Sick Children, Toronto, Canada

Jan R. Slabbaert (227), VIB Center for the Biology of Disease, Leuven, Belgium; K.U. Leuven, Center for Human Genetics, Leuven, Belgium

Benoit Sorre (47), Laboratory of Theoretical Condensed Matter Physics, The Rockefeller University, New York

Tamou Thahouly (445), Institut des Neurosciences Cellulaires et Intégratives, CNRS UPR3212, Strasbourg, France

Petra Tryoen-Tóth (445), Institut des Neurosciences Cellulaires et Intégratives, CNRS UPR3212, Strasbourg, France

Patrik Verstreken (227), VIB Center for the Biology of Disease, Leuven, Belgium; K.U. Leuven, Center for Human Genetics, Leuven, Belgium

Nicolas Vitale (445), Institut des Neurosciences Cellulaires et Intégratives, CNRS UPR3212, Strasbourg, France

Markus R. Wenk (149), Department of Medical Parasitology and Infection Biology, Swiss Tropical and Public Health Institute, Basel, Switzerland; University of Basel, Basel, Switzerland; Department of Biochemistry and Department of Biological Sciences, Yong Loo Lin School of Medicine, National University of Singapore, Singapore

Marnix Wieffer (209), Institute of Chemistry and Biochemistry, Freie Universität Berlin, Takustraße, Berlin, Germany

Thorsten Wohland (395), Department of Chemistry, National University of Singapore, Singapore

Heimo Wolinski (345), Institute of Molecular Biosciences, University of Graz, Graz, Austria

Thomas Wollert (73), Molecular Membrane and Organelle Biology, Max Planck Institute of Biochemistry, Martinsried, Germany

Min Wu (3), Howard Hughes Medical Institute, Department of Cell Biology, Program in Cellular Neuroscience, Neurodegeneration and Repair, Yale University School of Medicine, New Haven, CT 06520, USA; Present address: Department of Biological Sciences, Centre for BioImaging Sciences and Mechanobiology Institute, National University of Singapore, Singapore 117546

Daniel Wüstner (367), Department of Biochemistry and Molecular Biology, University of Southern Denmark, Odense M, Denmark

Hongyuan Yang (303), School of Biotechnology and Biomolecular Sciences, the University of New South Wales, Sydney, Australia

PREFACE

The idea for a "Methods in Cell Biology" volume entirely dedicated to lipids originated from the International Singapore Lipid Symposium (ISLS), a biennial event that brings together researchers with diverse interests in lipid biology. In fact, the generation of new connections across disciplines is a defining characteristic of ISLS in the context of novel technology developments and pioneering applications.

With encouragement and support from Dr. Paul Matsudaira, we have thus decided to invite several of the participating speakers as well as other leaders in the field to contribute a "Methods" chapter for this volume. The response has been enthusiastic. As a result, we have at hand a collection of chapters that cover broad aspects of lipid cell biology and biochemistry. We have ordered these chapters into three main categories, namely "membrane dynamics and reconstitution assays", "lipid metabolism and signaling", and "imaging".

Membrane trafficking has been a very intense area of cell biological research in the past two decades. Investigations of subcellular dynamics have made seminal contributions in the field and demonstrated the role of lipids for example in the budding and fusion of membranes from various organelles. Tremendous progress has also been made in the reconstitution of trafficking steps that are critical for various transport reactions in the cell. Additionally, membrane lipids and their metabolites are central components of signaling cascades in addition to their structural roles in large part due to their ability to recruit or activate effector proteins at the membrane-cytosol interface. Understanding the complex reactions that underlie *lipid signaling* requires integrated approaches to provide information in space and time. Reversible phosphorylation has been successfully studied in such a fashion in the case of phosphoinositides. Finally, a lot of progress was made in the development of tools and technologies allowing for the visualization of several classes of lipids and cell membranes in their native state through various state-of-the-art *imaging* approaches.

Of course, we cannot expect this volume to comprehensively cover all aspects and developments of these emerging fields. However, we believe that it is a unique contribution for investigators who are interested to advance knowledge at the interface of multiple disciplines. It also provides an overview of recently-developed methodologies and techniques for scientists who are new to the field of lipid biology and would like to put efforts into it.

Finally, we would like to thank the staff of Elsevier/Academic Press, especially Zoe Kruze and Shaun Gamble as well as Hong Huimin at the National University of Singapore for their enthusiastic support of our Methods in Cell Biology volume: Lipids.

Markus R. Wenk and Gilbert Di Paolo

PART I

Membrane Dynamics and Reconstitution Assays

CHAPTER 1

Supported Native Plasma Membranes as Platforms for the Reconstitution and Visualization of Endocytic Membrane Budding

Min Wu[*,†] **and Pietro De Camilli**[*]

[*]Howard Hughes Medical Institute, Department of Cell Biology, Program in Cellular Neuroscience, Neurodegeneration and Repair, Yale University School of Medicine, New Haven, CT 06520, USA

[†]Present address: Department of Biological Sciences, Centre for BioImaging Sciences and Mechanobiology Institute, National University of Singapore, Singapore 117546

METHODS IN CELL BIOLOGY, VOL 108
Copyright 2012, Elsevier Inc. All rights reserved.

0091-679X/10 $35.00
DOI 10.1016/B978-0-12-386487-1.00001-8

Abstract

Cell-free assays represent an important complement to studies in living cells for the elucidation of mechanisms underlying the dynamics of biological membranes, such as budding, fission, and fusion reactions. Here we describe a method for the reconstitution of endocytosis, the process through which cells internalize portions of the plasma membrane along with extracellular material, under conditions that allow the visualization of individual budding events with high spatial and temporal resolution. The method, which is based on the generation of planar plasma membrane sheets attached to a glass substrate and their subsequent incubation with a cytosolic extract, results in a very robust formation of endocytic buds, which upon appropriate conditions undergo fission. The synchronization of the endocytic events and the accessibility of the material to a variety of manipulations make this experimental system a powerful tool for the molecular dissection of endocytosis.

I. Introduction

Endocytosis plays general and fundamental functions in all eukaryotic cells. In specialized cells, endocytosis has been adapted for unique functions, such as the recycling of synaptic vesicle membranes in neurons. The accumulation of information on the molecular players involved in endocytosis, including their structures and interactions, has greatly advanced our knowledge of this process. However, a full mechanistic understanding of the endocytic membrane transport will require its reconstitution *in vitro*. Several individual steps of the endocytosis have been reconstituted with purified proteins and artificial lipid membranes (Farsad *et al.*, 2001; Ford *et al.*, 2001, 2002; Itoh *et al.*, 2005; Pucadyil and Schmid, 2008; Roux *et al.*, 2006; Takei *et al.*, 1999). These reductionist "minimal" systems are extremely powerful, but usually require supraphysiological levels of charged lipids or recombinant proteins. It remains to be understood how cooperativity between lipids of the bilayer and cytosolic proteins can be harnessed to bypass these requirements in physiological endocytic events. To elucidate the integration logic of the native machinery, *in vitro* reconstitutions of increasing complexity are required.

"Cell-free" reconstitutions of endocytic events under more physiologically relevant conditions have been reported. These include the use of cytosolic extracts and native plasma membranes, such as substrate-attached plasma membrane sheets (Gilbert *et al.*, 1997; Lin *et al.*, 1991; Moore *et al.*, 1987), plasma membrane fractions or perforated cells (Izumi *et al.*, 2004; Schmid and Smythe, 1991; Seaman *et al.*, 1993; Shi *et al.*, 1998; Smythe *et al.*, 1989; Takei *et al.*, 1995, 1996). Until recently, these cell-free systems have mostly relied on bulk biochemical readouts that have little spatial resolution, or morphological readouts using electron microscopy (EM) that are usually low-throughput. To overcome these limitations, we have recently developed a simple fluorescence imaging-based protocol for the reconstitution and visualization of endocytic events, using planar plasma membrane sheets and brain cytosolic extracts (Wu *et al.*, 2010). The method allows thousands of individual reactions to be monitored in parallel. In addition, it is possible to zoom in on individual endocytic event and characterize reaction intermediates at high spatial resolution.

Plasma membrane sheets obtained by sonicating adherent cells have been used previously in many studies to visualize the inner leaflet of the plasma membrane and associated structures (Heuser, 2000). For example, they have been employed to investigate plasma membrane organization (Lang *et al.*, 2001, 2002; Prior *et al.*, 2003) and dynamics (Sieber *et al.*, 2007), clathrin assembly (Moore *et al.*, 1987), exocytosis (Avery *et al.*, 2000; Crabb and Jackson, 1985), actin assembly (Collins *et al.*, 2011; Morone *et al.*, 2006; Rodal *et al.*, 2005; Yamada *et al.*, 2005) and cortical microtubule dynamics (Reilein *et al.*, 2005). We found that upon incubation with cytosolic extracts and nucleotides these sheets support robust formation of clathrin-coated buds and tubular endocytic intermediates. This experimental system has great potential for a systematic analysis of the *de novo* assembly of endocytic structures on native plasma membranes. In this chapter we describe in details the experimental setup for this simple assay, including how the membranes and cytosolic extracts are prepared. We will also review the methods best suited for visualization of the membrane budding and fission reactions.

II. Preparation of Membrane Sheets

The procedure for generating membrane sheets includes three steps performed on three different days:

Day 1: substrate coating;
Day 2: cell seeding;
Day 3 or 4: cell-free reaction.

A. Substrate Coating

1. Materials

- MatTek dishes (MatTek Corporation, p35g-1.5-10-c)
- HCl (1N)
- PBS (sigma, P4417-100TAB)
- Borate buffer (0.1 M, pH8.5)
- poly-d-lysine (sigma, P0899-10MG): make 100x stock (2 mg/mL in borate buffer) and store at $-80\ ^{\circ}C$
- Steriflip (Millipore, SCGP00525)
- MilliQ plus Water

2. Procedure

1. Wash MatTek 35 mm glass bottom dishes with 1N HCl. We usually use 250 μL to coat the glass area only, and incubate 15 min at room temperature.
2. Remove HCl and wash 3x with PBS. Air dry between washes.
3. Make freshly diluted solution of poly-d-lysine (0.02 mg/mL) from 100x stock in borate buffer. Filter using Steriflip. Incubate each dish with 250 μL poly-d-lysine for 30 min at 37 °C.
4. Wash 3x with water and keep the dishes in water overnight. We find this washing step helpful to reduce the clumping of the cells.

Note: We use poly-d-lysine coated glass substrates because this coating procedure is simple, reproducible, and yields membranes that can support invagination and budding. For long-term cell culture, glass surface modification based on covalent conjugations can be considered, as described elsewhere (Drees *et al.*, 2005). However, it is conceivable that such substrates may lead to tighter adhesion between the plasma membrane and the glass surface, compromising its ability to detach and thus to invaginate. It will be important to evaluate the impact of surface modification on the properties of the membranes and on the endocytic intermediates that originate from them.

B. Cell Culture

1. Choice of Cell Type

Membrane sheets have been routinely prepared from a wide variety of cell types, including fibroblast (Moore *et al.*, 1987), PC12 (Avery *et al.*, 2000), RBL-2H3 (Wilson *et al.*, 2000), T cells (Lillemeier *et al.*, 2006), MDCK (Yamada *et al.*, 2005), and erythrocytes (Jacobson and Branton, 1977). It should be noted that different cell types may need different glass coating conditions (Jacobson, 1977). The following procedures have been optimized for PTK2 cells, a commonly used adherent epithelial cell line.

2. Tissue Culture Conditions

PTK2 cells are maintained at 37 °C in 10% CO_2 in D-MEM (Invitrogen, 11965-118) supplemented with 10% (v/v) fetal bovine serum (Sigma F4135), 100 μg/mL penicillin/streptomycin (Invitrogen, 15140-122). PTK2 cells stably expressing palmitoylated and myristoylated GFP, that is, a plasma membrane targeted GFP (PM–GFP) (Chen and De Camilli, 2005) are maintained with the above medium also containing 0.5 mg/mL G418 (Invitrogen, 11811-031). Cells are passed twice a week, usually at 1:3 or 1:4 dilutions.

The best membrane sheets are obtained when cells form a confluent monolayer at the time of sonication. Towards this aim, we usually seed PTK2 cells in MatTek dishes that had been coated with 20 μg/mL poly-d-lysine the day before at a density of 0.5–1 $\times$ 10^6 cell/dish. Cells were grown for 24–48 h until they form a confluent monolayer.

C. Generating Membrane Sheets by Sonication

Surface-attached plasma membrane sheets are derived from adherent cells by a brief sonication pulse. We use probe sonication for this purpose (Fig. 1). Sonication is effective in generating plasma membrane sheets because less power is required to rupture the plasma membrane (mediated by the lysis tension of the membrane) than to detach the adherent cells from the substrates. Compared to the alternative "rip-off" method, which unroofs cells by depositing and then removing a glass slide over the cell monolayer (Sanan and Anderson, 1991), sonication yields a larger number of membrane sheets. The surface coverage of the membrane is also much higher (Fig. 2). However, the "rip-off" method allows the isolation of the apical plasma membrane, which remains attached to the glass surface used for the unroofing, in contrast to our method, which isolates basal membranes.

1. Materials and Equipment

- PBS
- Cytosolic buffer: 25 mM Hepes-NaOH (pH 7.4), 120 mM potassium glutamate, 20 mM potassium chloride, 2.5 mM magnesium acetate, 5 mM EGTA.

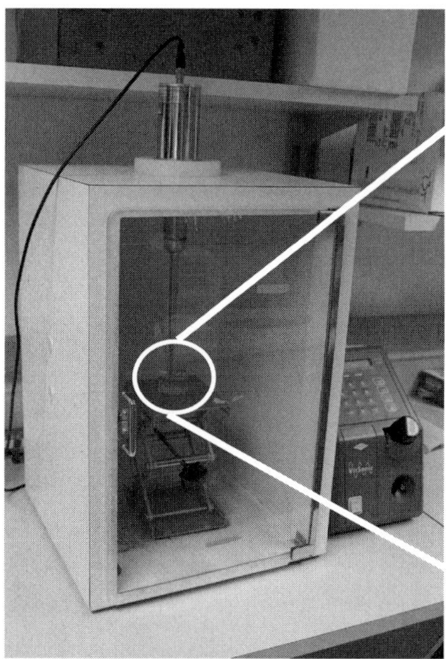

Fig. 1 Sonication setup. Photo credit: Ruben Fernandez-Busnadiego. (For color version of this figure, the reader is referred to the web version of this book.)

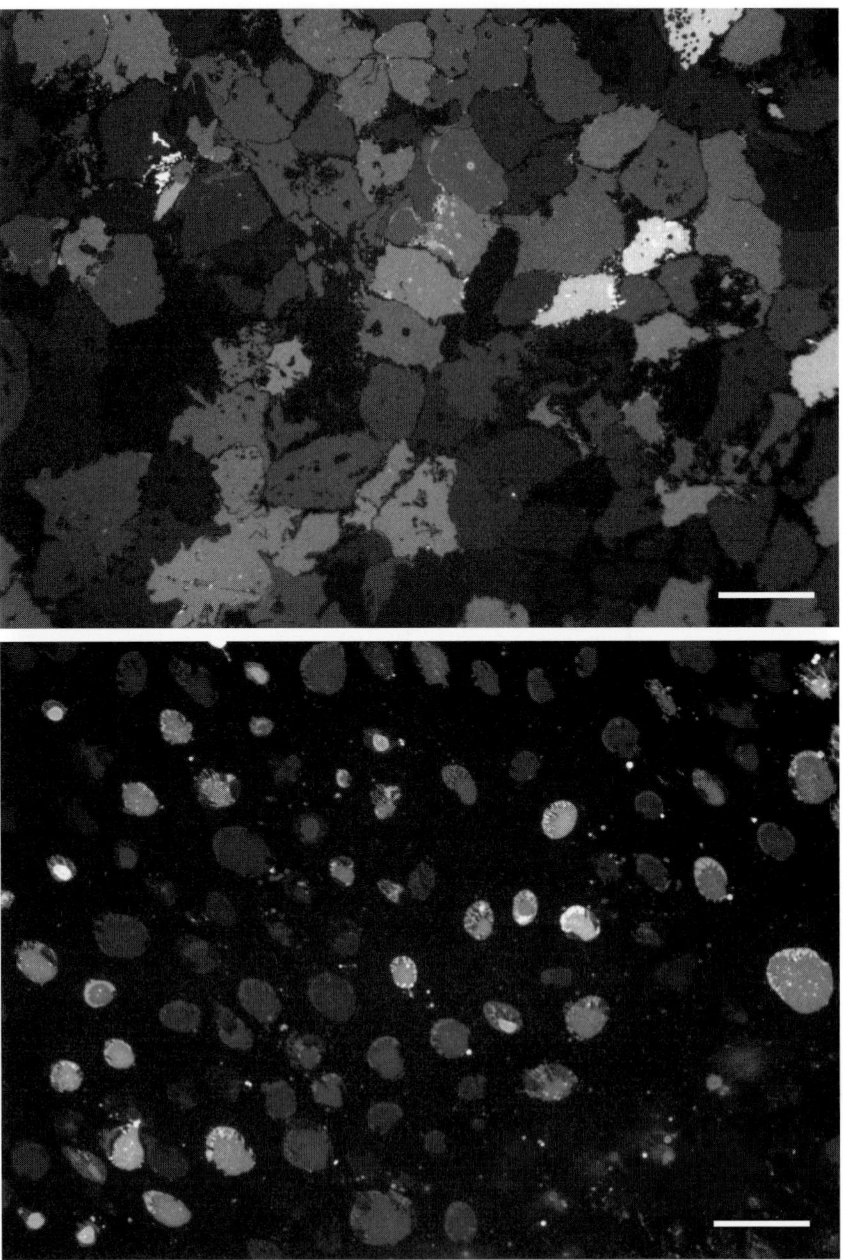

Fig. 2 Plasma membrane sheets generated by sonication (top) or the "rip-off" method (bottom). In both cases, membrane sheets are from PTK2 cells stably transfected with PM-GFP. Scale bar: 50 μm.

The buffers should be filtered, stored at 4 °C and supplement with 1 mM DTT before use.

- Crystallizing dish (VWR International, #89000-280)
- Probe sonicator with sonic enclosure (VirTis Ultrasonics)
- Tweezer

2. Procedure

1. Confirm at the microscope that cells in the MatTek dish have spread and formed a confluent monolayer.
2. Place the MatTek dish on ice. Discard the medium and wash 2x with ice cold PBS.
3. Fill the crystallizing dish with 50 mL cytosolic buffer pre-chilled on ice. The use of this additional container, which is taller than the MatTek dish, allows positioning the tip of sonicator further away from the cells.
4. Position the sonicator tip (1/8-inch microprobe) at 15 mm above the bottom of the crystallizing dish. The distance is arbitrarily determined but should be kept constant to ensure reproducibility.
5. The appropriate settings of the sonicator (% of the output power) should be determined by test experiments to ensure efficient cell lysis but preservation of basal membranes. If the power is too high, basal membranes will undergo fragmentation.
6. The optimal duration of the sonication depends on the desired area of disruption. Usually, for 1 s we obtain ~10 mm (diameter) of membrane sheets. For a smaller area, a shorter pulse will suffice; for a larger area, multiple pulses will be necessary. This should be empirically determined for individual cell lines and sonicator setting.
7. Remove PBS from the MatTek dish and immerse it into the crystallizing dish pre-filled with sonication buffer. Quickly apply the brief pulse of sonication. Remove the dish with the attached plasma membrane sheets using a tweezer and rinse it with cytosolic buffer by gentle pipetting. From this point on, keep samples with plenty of buffer at all times, as exposure to air will irreversibly damage the membranes.
8. Examine the membrane sheets at the microscope (see below).

3. Choice of Sonication Buffer

We obtain good yields of membrane sheets using either isotonic cytosolic buffer (Avery *et al.*, 2000) or hypotonic buffer (Lin *et al.*, 1991). It has been reported that water will also work (Bezrukov *et al.*, 2009). It should be noted that while all these conditions may be effective in generating membrane sheets, peripheral protein composition as well as the mechanical properties of the isolated membranes might differ. Biochemical and biophysical studies will be needed to characterize these potential differences.

4. Visualization of Membrane Sheets

Membrane sheets will appear blank compared to intact or half-broken cells under bright field microscopy. We routinely use a fluorescent lipid marker to examine the quality of membrane sheets in more details. Membrane sheets are incubated with the lipid dye (5 μM) right after sonication, at 37 °C for 5 min, and then rinsed to remove excess dye. Most of the commercial fluorescent lipids work. We prefer to use BODIPY®TR ceramide (Invitrogen B-34400) and NBD C6 sphingomyelin (Invitrogen N-3524), because they produce consistently a more homogeneous labeling than other dyes. To simplify the procedure and to shorten the time between cell rupture and the cell-free reaction, we also generated PTK2 cells stably expressing PM-GFP. No obvious differences were observed between the original PTK2 cells and this fluorescent PTK2 cell line, or between PTK2 cells expressing different levels of PM-GFP.

III. Preparation of Brain Extract

A. Materials and Equipments

- Fresh adult mouse brains (6–8 weeks old)
- Washing buffer: 25 mM Tris-HCl (pH 7.4), 320 mM sucrose
- Homogenization buffer: 25 mM Tris-HCl (pH 8.0), 500 mM KCl, 250 mM sucrose, 2 mM EGTA, 1 mM DTT, supplemented with protease inhibitor cocktail mix (Roche, 11836170001) just before use.
- Potter tissue homogenizer.
- Tabletop ultracentrifuge (Beckman Optima TLX) and rotor (TLA 100.3)
- Cytosolic buffer: 25 mM Hepes-NaOH (pH 7.4), 120 mM potassium glutamate, 20 mM potassium chloride, 2.5 mM magnesium acetate, 5 mM EGTA, filtered and stored at 4 °C; supplemented with 1 mM DTT before use.
- Disposable column PD-10 (Amersham Biosciences, 17-0851-01)
- Bio-Rad Protein Assay (Bio-Rad, 500-0006)

B. Procedure

To prepare a brain cytosolic extract, a simplified protocol based on published procedures (Malhotra *et al.*, 1989) was devised. All procedures are carried out on ice or at 4 °C.

1. Rinse four freshly dissected adult mouse brains (wet weight: ~ 1.5 g) in washing buffer.
2. Homogenize brains in 2 mL homogenization buffer, using a 5 mL Potter tissue grinder operated at 2000 rpm, with 20 passes.
3. Centrifuge the homogenate at 160,000 g for 2 h.
4. Collect the supernatant without disturbing the pellet.

5. Exchange the medium of the supernatant to cytosolic buffer on the PD-10 column.
6. Some precipitation will occur upon buffer exchange, due to polymerization of the cytoskeleton upon nucleotide removal (Jahraus *et al.*, 2001; Kurzchalia *et al.*, 1992). This process can be inhibited by ATP addition (Traub *et al.*, 1993). Spin again at 160,000 g for 20 min to clarify the cytosolic extract.
7. Determine the protein concentration of the cytosolic extract using the Bio-Rad Protein Assay with BSA as a standard. The concentration is usually ~6–8 mg/mL.
8. Aliquots of the cytosolic extract are flash frozen in liquid nitrogen and stored at −80 °C.

C. Manipulation of the Extract

Membrane sheets provide direct access to the cytosolic leaflet of the plasma membrane. Thus cytosolic composition and functionality can be easily manipulated without potential complications of compensatory mechanisms, an inevitable problem during the chronic perturbations of endocytosis. Many methods can be chosen to perturb individual factors or interactions in the cell-free system, including addition of proteins or of peptides that can achieve dominant negative effects and of chemical inhibitors. Antibodies can be added to cytosolic extracts block the function of endogenous proteins or can be used to immunodeplete them from the cytosolic extracts, thus producing a loss-of-function condition.

IV. Cell-Free Reaction

A. Materials

- Cytosolic extracts; stored at −80 °C

  The following stocks are stored at −20 °C:
- ATP (Sigma A9187-1G, 150 mM, i.e., 100x stock)
- Phosphocreatine (Sigma P-7936, 0.67 M, i.e., 40x stock)
- Creatine phosphokinase (Sigma C-3755, 668 U/mL, i.e., 40x stock)
- GTPγS (Roche 10220647001, 15 mM, i.e., 100x stock)
- GTP (Sigma G8877-250 mg, 32 mM, i.e., 20x stock)
- Phosphopyruvate (Sigma P0564, 80 mM, i.e., 80x stock)
- Pyruvate kinase (Sigma P9136, 1600 unit/mL, i.e., 80x stock)

  The following stocks are purchased and stored at 4 °C:
- Paraformaldehyde (Electron Microscopy Sciences 15710, 16%)
- Glutaraldehyde (Electron Microscopy Sciences 16220, 25%)

B. Procedure

1. Use rinsed membrane sheets for the *in vitro* assay within 20 min from their preparation. Prolonged delay decreases activity.

2. Prepare the final cytosolic mixture by supplementing the cytosolic extracts with an ATP regenerating system (ATP, phosphocreatine, and creatine phosphokinase) and GTPγS on ice. For a typical reaction in MatTek 160 μg of cytosolic proteins are used. Adjust the total volume of the mixture (about 40 μL) by adding cytosolic buffer.
3. Apply cytosolic mixture to the membranes; incubate at 37 ° for the desired duration. Maximum activity of tubulation is observed around 15–20 min. Temperature is important for tubulation. Tubulation is inhibited at room temperature. A temperature-controlled incubator kept at 37 ° should be used for this reaction.
4. Prepare the second cytosolic mixture by supplementing cytosolic extracts with ATP regenerating system and GTP regenerating system (GTP, phosphopyruvate, and pyruvate kinase).
5. Replace the first cytosolic mixture (the one that contains GTPγS) with plenty of cytosolic buffer. Repeat once. Add the second cytosolic mixture to allow fission to occur. This reaction is also sensitive to temperature. 37 ° is required.
6. The budding and fission reactions can be stopped at any time by a wash in cytosolic buffer followed by fixation for the analysis of structural intermediates at specific time points. For immunofluorescence, when samples are imaged on the same day, we fix them with 4% paraformaldehyde at room temperature for 10–15 min. For microscopy analysis after longer storage, we fix them with 4% paraformaldehyde plus 0.1% glutaraldehyde at room temperature for 20 min, followed by standard immunofluorescence (usually 30 min incubation with primary and secondary antibody) and a post-fixation (again in with 4% paraformaldehyde plus 0.1% glutaraldehyde for 20 min at room temperature) in order to fix the antibodies. For EM, we fix them with 2.5% glutaraldehyde prior to postfixation in OsO4 and embedding in epon.

C. Conditions for Membrane Budding

Presence of nucleotides is critical for the endocytic reaction in this cell-free system. The endocytic reaction requires energy and thus ATP. One of the ATP requirements is for the synthesis of PI(4,5)P_2 at the plasma membrane (Di Paolo and De Camilli, 2006), primarily via the action of PI(4)P 5-kinases (type I PIP kinases) and their regulatory proteins (Krauss *et al.*, 2003, 2006; Thieman *et al.*, 2009). Plasma membrane sheets incubated with cytosol, similar to plasma membranes in living cells, contain free PI(4,5)P_2, as indicated by the recruitment of exogenously added recombinant PH domain (Fig. 3). Guanyl nucleotides are needed to activate many regulatory GTPases as well as dynamin, which mediates the fission reaction of endocytosis. GTPγS, the non-hydrolysable analogs of GTP, blocks the fission reaction by locking dynamin in its GTP-bound state, but also drastically enhances the formation of endocytic intermediates, most likely by acting on multiple targets, including small GTPases that regulate PI(4,5)P_2 production (Krauss *et al.*, 2003) and Rho family GTPases that regulate actin dynamics.

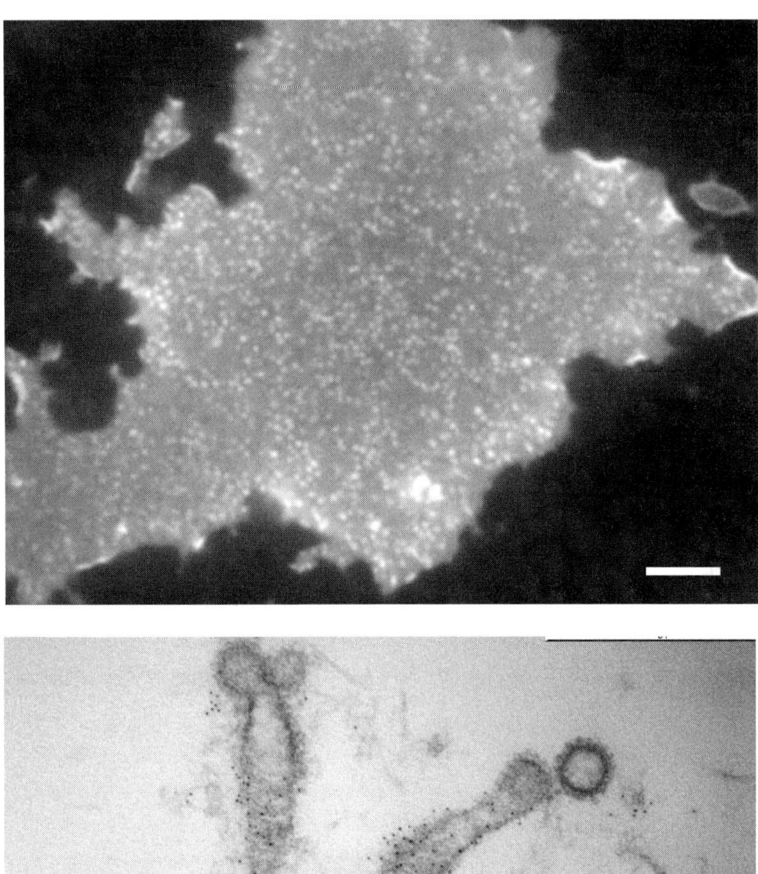

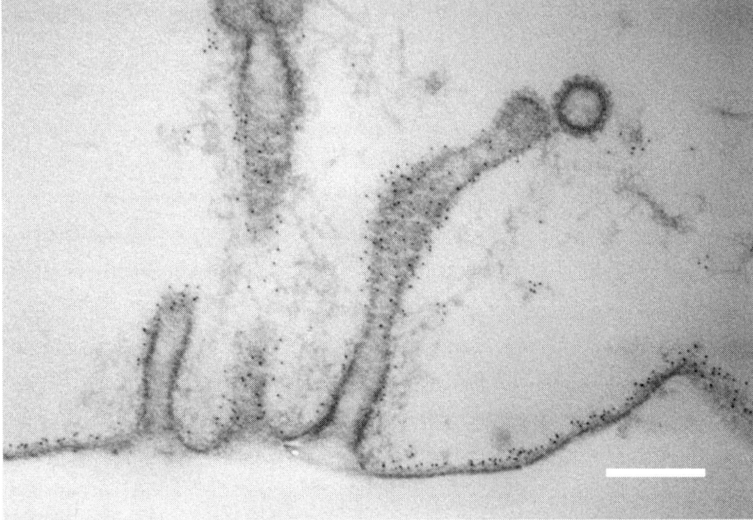

Fig. 3 Localization of PI(4,5)P2 by fluorescence microscopy (top) and immuno EM (bottom) on plasma membrane sheets. In both cases, membrane sheets from PTK2 cells were incubated with cytosol supplemented with ATP and GTPγS, and labeled using recombinant PH-GFP from PLCδ as a marker for PI(4,5)P2. PH-GFP was visualized by fluorescence microscopy (top) or by electron microscopy following fixation and immunogold labeling for GFP (bottom). The EM micrograph shows representative examples of the tubular invaginations that form under these conditions. Note that gold particles are missing from the clathrin-coated tips of the tubular invaginations, possibly because of the lack of accessibility due to the densely packed coat proteins. Scale bar: (top) 5 μm; (bottom) 200 nm. (For color version of this figure, the reader is referred to the web version of this book.)

D. Methods to Observe Membrane Budding

The planar organization of the membrane sheets makes them ideally suited for fluorescence imaging. To characterize membrane deformations, we routinely use fluorescent membrane probes as markers. As invaginations occur (typically tubular invaginations under our assay conditions), they grow perpendicularly to the substrate and appear as bright puncta on a homogeneous background. This increase in fluorescence intensity is largely due to membrane folding, although one cannot rule out the contribution of an enrichment of the probes in the invaginated membrane patch. Epifluorescence microscopy is well suited to document this process. Visualization of the fluorescent puncta by confocal microscopy is dependent on the precise z-position (Fig. 4): the punctate appearance of the membrane, which reflects the occurrence of the invaginations, is barely visible when the focal plane is at the plane of the sheets, but become pronounced above the membrane. It is worth noting that TIRF (total internal reflection fluorescence microscopy), the method of choice to visualize plasma membrane localized events, is not suited for documenting membrane invaginations. This is because the power of the evanescent illumination decays as the distance from the surface increases, practically reducing the contrast of membrane puncta compared to the rest of the membrane. Polarized TIRF may represent an attractive modification to increase the sensitivity of TIRF for the detection of membrane deformation using fluorescent probes (Anantharam *et al.*, 2010). However, the use of this method is complicated by the need to separate contributions from local probe concentration, lipid orientation within the membrane and membrane topology.

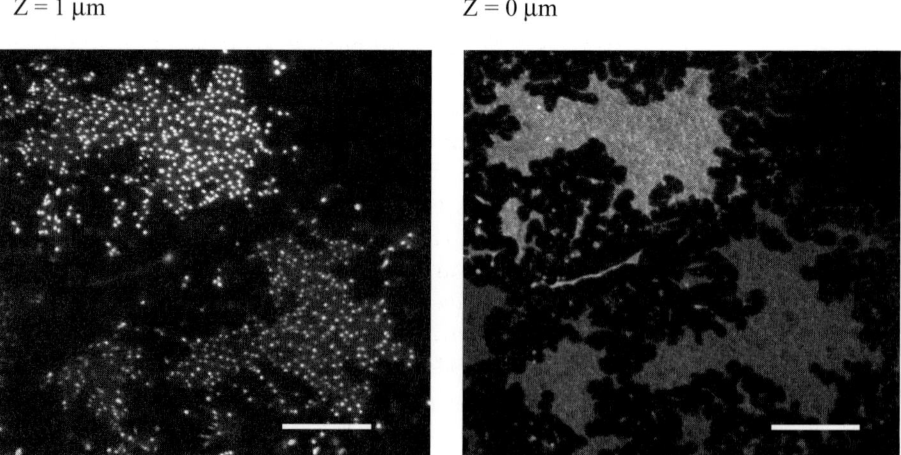

Fig. 4 Visualization of the endocytic tubulated buds by spinning disk confocal fluorescence microscopy depends on the precise z position of the focal plane. Membrane buds on membrane sheets derived from PTK2 cells stably transfected with PM-GFP. Visualization of the buds is optimal when the focal plane is above (Z = 1 μm) the plane of the sheets (Z = 0 μm). Scale bar: 5 μm.

To monitor the fission reaction, it is best to image at a plane that is higher than the plane of the membrane for better signal-to-noise ratio. Both spinning disk confocal fluorescence microscopy and epifluorescence microscopy would work, but confocal microscopy is more sensitive. We usually place the focal plane 20 μm above the plane of the membrane for confocal microscopy and 80 μm for epifluorescence. We find it convenient, when using plasma membrane sheets derived from cells expressing PM-GFP, to add some beads pre-coated with anti-GFP antibodies to the incubation used for the fission reaction. Accumulation of vesicles on the beads can help differentiating free vesicles, from vesicles that are still tethered to the membrane.

It is relatively straightforward to couple our membrane budding assay with super-resolution fluorescent imaging methods such as STORM (stochastic optical reconstruction microscopy) (Wu *et al.*, 2010), PALM (Photo-activated localization microscopy) (Lillemeier *et al.*, 2010), STED (stimulated emission depletion) (Sieber *et al.*, 2007), and near-field scanning optical microscopy (Hoppener and Novotny, 2008). The assay can also be coupled to AFM (Avery *et al.*, 2000), or to EM including immuno EM (Wu *et al.*, 2010). While some of these high-resolution techniques can be performed in intact cells, membrane budding intermediates *in vivo* are rare, heterogeneous, and labile. Thus, one additional advantage of our method is the better statistical confidence in the interpretation of microscopic signals with nano-scale details due to the accumulation of large numbers of synchronized events.

V. Conclusions

We devised a simple cell-free system that supports membrane budding and fission events from isolated plasma membranes and is optimally suited to fluorescence imaging studies. The value of this system extends beyond what is described here and will depend on the ability to combine well-controlled perturbations with high precision readouts. For instance, when combined with methods aimed at manipulating the lipid composition of the plasma membranes, the system can be used to elucidate the role of specific lipids in endocytosis. One can anticipate that these "top-down" cell-free reconstitution methods will complement and possibly guide "bottom-up" reconstitution strategies based on purified component and synthetic lipid templates. Results obtained by these approaches will be essential to elucidate the fundamental organizational principles of the endocytic machineries by bridging information at different length scales.

Acknowledgments

We thank Ruben Fernandez-Busnadiego for careful reading of the manuscript, Morven Graham and Louise Lucast for critical technical assistance. This work was supported by the G. Harold and Leila Y. Mathers Charitable Foundation, the W.M. Keck Foundation, the National Institutes of Health grants (NS36251 and DK45735) and Howard Hughes Medical Institute.

References

Anantharam, A., Onoa, B., Edwards, R. H., Holz, R. W., and Axelrod, D. (2010). Localized topological changes of the plasma membrane upon exocytosis visualized by polarized TIRFM. *J. Cell Biol.* **188**, 415–428.

Avery, J., Ellis, D. J., Lang, T., Holroyd, P., Riedel, D., Henderson, R. M., Edwardson, J. M., and Jahn, R. (2000). A Cell-Free System for Regulated Exocytosis in Pc12 Cells. *J. Cell Biol.* **148**, 317–324.

Bezrukov, L., Blank, P. S., Polozov, I. V., and Zimmerberg, J. (2009). An adhesion-based method for plasma membrane isolation: evaluating cholesterol extraction from cells and their membranes. *Anal. Biochem.* **394**, 171–176.

Chen, H., and De Camilli, P. (2005). The association of epsin with ubiquitinated cargo along the endocytic pathway is negatively regulated by its interaction with clathrin. *Proc. Natl. Acad. Sci.* **102**, 2766–2771.

Collins, A., Warrington, A., Taylor, K. A., and Svitkina, T. (2011). Structural organization of the actin cytoskeleton at sites of clathrin-mediated endocytosis. *Curr. Biol.* **21**, 1167–1175.

Crabb, J. H., and Jackson, R. C. (1985). In vitro reconstitution of exocytosis from plasma membrane and isolated secretory vesicles. *J. Cell Biol.* **101**, 2263–2273.

Di Paolo, G., and De Camilli, P. (2006). Phosphoinositides in cell regulation and membrane dynamics. *Nature* **443**, 651–657.

Drees, F., Reilein, A., and Nelson, W. J. (2005). Cell-adhesion assays: fabrication of an E-cadherin substratum and isolation of lateral and Basal membrane patches. *Methods Mol. Biol.* **294**, 303–320.

Farsad, K., Ringstad, N., Takei, K., Floyd, S. R., Rose, K., and De Camilli, P. (2001). Generation of high curvature membranes mediated by direct endophilin bilayer interactions. *J. Cell Biol.* **155**, 193–200.

Ford, M. G. J., Mills, I. G., Peter, B. J., Vallis, Y., Praefcke, G. J. K., Evans, P. R., and McMahon, H. T. (2002). Curvature of clathrin-coated pits driven by epsin. *Nature* **419**, 361–366.

Ford, M. G. J., Pearse, B. M. F., Higgins, M. K., Vallis, Y., Owen, D. J., Gibson, A., Hopkins, C. R., Evans, P. R., and McMahon, H. T. (2001). Simultaneous binding of PtdIns(4,5)P-2 and clathrin by AP180 in the nucleation of clathrin lattices on membranes. *Science* **291**, 1051–1055.

Gilbert, A., Paccaud, J. P., and Carpentier, J. L. (1997). Direct measurement of clathrin-coated vesicle formation using a cell-free assay. *J. Cell Sci.* **110**(Pt 24), 3105–3115.

Heuser, J. (2000). The production of 'cell cortices' for light and electron microscopy. *Traffic* **1**, 545–552.

Hoppener, C., and Novotny, L. (2008). Antenna-based optical imaging of single Ca2+ transmembrane proteins in liquids. *Nano Lett.* **8**, 642–646.

Itoh, T., Erdmann, K. S., Roux, A., Habermann, B., Werner, H., and De Camilli, P. (2005). Dynamin and the Actin Cytoskeleton Cooperatively Regulate Plasma Membrane Invagination by BAR and F-BAR Proteins. *Dev. Cell* **9**, 791–804.

Izumi, G., Sakisaka, T., Baba, T., Tanaka, S., Morimoto, K., and Takai, Y. (2004). Endocytosis of E-cadherin regulated by Rac and Cdc42 small G proteins through IQGAP1 and actin filaments. *J. Cell Biol.* **166**, 237–248.

Jacobson, B. S. (1977). Isolation of plasma membrane from eukaryotic cells on polylysine-coated polyacrylamide beads. *Biochim. Biophys. Acta* **471**, 331–335.

Jacobson, B. S., and Branton, D. (1977). Plasma membrane: rapid isolation and exposure of the cytoplasmic surface by use of positively charged beads. *Science* **195**, 302–304.

Jahraus, A., Egeberg, M., Hinner, B., Habermann, A., Sackman, E., Pralle, A., Faulstich, H., Rybin, V., Defacque, H., and Griffiths, G. (2001). ATP-dependent membrane assembly of F-actin facilitates membrane fusion. *Mol. Biol. Cell* **12**, 155–170.

Krauss, M., Kinuta, M., Wenk, M. R., De Camilli, P., Takei, K., and Haucke, V. (2003). ARF6 stimulates clathrin/AP-2 recruitment to synaptic membranes by activating phosphatidylinositol phosphate kinase type Igamma. *J. Cell Biol.* **162**, 113–124.

Krauss, M., Kukhtina, V., Pechstein, A., and Haucke, V. (2006). Stimulation of phosphatidylinositol kinase type I-mediated phosphatidylinositol (4,5)-bisphosphate synthesis by AP-2mu-cargo complexes. *Proc. Natl. Acad. Sci. U. S. A.* **103**, 11934–11939.

Kurzchalia, T. V., Gorvel, J. P., Dupree, P., Parton, R., Kellner, R., Houthaeve, T., Gruenberg, J., and Simons, K. (1992). Interactions of rab5 with cytosolic proteins. *J. Biol. Chem.* **267**, 18419–18423.

Lang, T., Bruns, D., Wenzel, D., Riedel, D., Holroyd, P., Thiele, C., and Jahn, R. (2001). SNAREs are concentrated in cholesterol-dependent clusters that define docking and fusion sites for exocytosis. *EMBO J.* **20**, 2202–2213.

Lang, T., Margittai, M., Holzler, H., and Jahn, R. (2002). SNAREs in native plasma membranes are active and readily form core complexes with endogenous and exogenous SNAREs. *J. Cell Biol.* **158**, 751–760.

Lillemeier, B. F., Mortelmaier, M. A., Forstner, M. B., Huppa, J. B., Groves, J. T., and Davis, M. M. (2010). TCR and Lat are expressed on separate protein islands on T cell membranes and concatenate during activation. *Nat. Immunol.* **11**, 90–96.

Lillemeier, B. F., Pfeiffer, J. R., Surviladze, Z., Wilson, B. S., and Davis, M. M. (2006). Plasma membrane-associated proteins are clustered into islands attached to the cytoskeleton. *Proc. Natl. Acad. Sci. U. S. A.* **103**, 18992–18997.

Lin, H. C., Moore, M. S., Sanan, D. A., and Anderson, R. G. (1991). Reconstitution of clathrin-coated pit budding from plasma membranes. *J. Cell Biol.* **114**, 881–891.

Malhotra, V., Serafini, T., Orci, L., Shepherd, J. C., and Rothman, J. E. (1989). Purification of a novel class of coated vesicles mediating biosynthetic protein transport through the Golgi stack. *Cell* **58**, 329–336.

Moore, M. S., Mahaffey, D. T., Brodsky, F. M., and Anderson, R. G. W. (1987). Assembly of clathrin-coated pits onto purified plasma membranes. *Science* **236**, 558–563.

Morone, N., Fujiwara, T., Murase, K., Kasai, R. S., Ike, H., Yuasa, S., Usukura, J., and Kusumi, A. (2006). Three-dimensional reconstruction of the membrane skeleton at the plasma membrane interface by electron tomography. *J. Cell Biol.* **174**, 851–862.

Prior, I. A., Muncke, C., Parton, R. G., and Hancock, J. F. (2003). Direct visualization of Ras proteins in spatially distinct cell surface microdomains. *J. Cell Biol.* **160**, 165–170.

Pucadyil, T. J., and Schmid, S. L. (2008). Real-time visualization of dynamin-catalyzed membrane fission and vesicle release. *Cell* **135**, 1263–1275.

Reilein, A., Yamada, S., and Nelson, W. J. (2005). Self-organization of an acentrosomal microtubule network at the basal cortex of polarized epithelial cells. *J. Cell Biol.* **171**, 845–855.

Rodal, A. A., Kozubowski, L., Goode, B. L., Drubin, D. G., and Hartwig, J. H. (2005). Actin and septin ultrastructures at the budding yeast cell cortex. *Mol. Biol. Cell* **16**, 372–384.

Roux, A., Uyhazi, K., Frost, A., and De Camilli, P. (2006). GTP-dependent twisting of dynamin implicates constriction and tension in membrane fission. *Nature* **441**, 528–531.

Sanan, D. A., and Anderson, R. G. (1991). Simultaneous visualization of LDL receptor distribution and clathrin lattices on membranes torn from the upper surface of cultured cells. *J. Histochem. Cytochem.* **39**, 1017–1024.

Schmid, S., and Smythe, E. (1991). Stage-specific assays for coated pit formation and coated vesicle budding in vitro. *J. Cell Biol.* **114**, 869–880.

Seaman, M. N., Ball, C. L., and Robinson, M. S. (1993). Targeting and mistargeting of plasma membrane adaptors in vitro. *J. Cell Biol.* **123**, 1093–1105.

Shi, G., Faundez, V., Roos, J., Dell'Angelica, E. C., and Kelly, R. B. (1998). Neuroendocrine synaptic vesicles are formed in vitro by both clathrin-dependent and clathrin-independent pathways. *J. Cell Biol.* **143**, 947–955.

Sieber, J. J., Willig, K. I., Kutzner, C., Gerding-Reimers, C., Harke, B., Donnert, G., Rammner, B., Eggeling, C., Hell, S. W., and Grubmuller, H., *et al.* (2007). Anatomy and dynamics of a supramolecular membrane protein cluster. *Science* **317**, 1072–1076.

Smythe, E., Pypaert, M., Lucocq, J., and Warren, G. (1989). Formation of coated vesicles from coated pits in broken A431 cells. *J. Cell Biol.* **108**, 843–853.

Takei, K., McPherson[ast], P. S., Schmid, S. L., and Camilli, P. D. (1995). Tubular membrane invaginations coated by dynamin rings are induced by GTP-[gamma]S in nerve terminals. *Nature* **374**, 186–190.

Takei, K., Mundigl, O., Daniell, L., and De Camilli, P. (1996). The synaptic vesicle cycle: a single vesicle budding step involving clathrin and dynamin. *J. Cell Biol.* **133**, 1237–1250.

Takei, K., Slepnev, V. I., Haucke, V., and De Camilli, P. (1999). Functional partnership between amphiphysin and dynamin in clathrin-mediated endocytosis. *Nat. Cell Biol.* **1**, 33–39.

Thieman, J. R., Mishra, S. K., Ling, K., Doray, B., Anderson, R. A., and Traub, L. M. (2009). Clathrin regulates the association of PIPKIgamma661 with the AP-2 adaptor beta2 appendage. *J. Biol. Chem.* **284**, 13924–13939.

Traub, L. M., Ostrom, J. A., and Kornfeld, S. (1993). Biochemical dissection of AP-1 recruitment onto Golgi membranes. *J. Cell Biol.* **123**, 561–573.

Wilson, B. S., Pfeiffer, J. R., and Oliver, J. M. (2000). Observing FcepsilonRI signaling from the inside of the mast cell membrane. *J. Cell Biol.* **149**, 1131–1142.

Wu, M., Huang, B., Graham, M., Raimondi, A., Heuser, J. E., Zhuang, X., and De Camilli, P. (2010). Coupling between clathrin-dependent endocytic budding and F-BAR-dependent tubulation in a cell-free system. *Nat. Cell Biol.* **12**, 902–908.

Yamada, S., Pokutta, S., Drees, F., Weis, W. I., and Nelson, W. J. (2005). Deconstructing the cadherin-catenin-actin complex. *Cell* **123**, 889–901.

CHAPTER 2

Studying Lipids Involved in the Endosomal Pathway

Christin Bissig, Shem Johnson and Jean Gruenberg

Biochemistry Department, University of Geneva, 30 quai Ernest Ansermet, 1211 Geneva 4, Switzerland

Abstract

Endosomes along the degradation pathway exhibit a multivesicular appearance and differ in their lipid compositions. Association of proteins to specific membrane lipids and presumably also lipid–lipid interactions contribute to the formation of functional membrane platforms that regulate endosome biogenesis and function. This chapter provides a brief review of the functions of endosomal lipids in the degradation pathway, a discussion of techniques that allow studying lipid-based mechanisms and a selection of step-by-step protocols for *in vivo* and *in vitro* methods commonly used to study lipid roles in endocytosis. The techniques described here have been used to elucidate the function of the late endosomal lipid lysobisphosphatidic acid and allow the monitoring of lipid distribution, levels and dynamics, as well as the characterization of lipid-binding partners.

Copyright 2012, Elsevier Inc. All rights reserved.

0091-679X/10 $35.00
DOI 10.1016/B978-0-12-386487-1.00002-X

I. Introduction

A. The Endocytic Pathway

Endocytosis controls many cellular processes, such as uptake of nutrients, turnover of membrane lipids, intracellular signaling, development, maintenance of cell polarity, and antigen presentation in animal cells (Gruenberg and Stenmark, 2004). At the plasma membrane (PM) signaling receptors, components of the PM and nutrients are internalized into clathrin-coated vesicles and other non-coated vesicles that are transported to and fuse with early endosomes (Hayer *et al.*, 2010; Mayor and Pagano, 2007). There cargo that has to be reutilized accumulates in tubular structures and is recycled back to the PM or transported to the *trans*-Golgi network (TGN). Alternatively, most proteins to be downregulated, in particular activated signaling receptors, are sorted into intralumenal vesicles by endosomal sorting complexes required for transport (ESCRT) and are transported to late endosomes by multivesicular endosomes (MVEs) (Hurley *et al.*, 2010; Katzmann *et al.*, 2002). Late endosomes serve as a second sorting station from where cargo can be recycled back to the TGN, sent to lysosomes for degradation or transported to other cellular destinations (Falguieres *et al.*, 2008b; Gibbings *et al.*, 2009; Trajkovic *et al.*, 2008) (Fig. 1A). This direct route

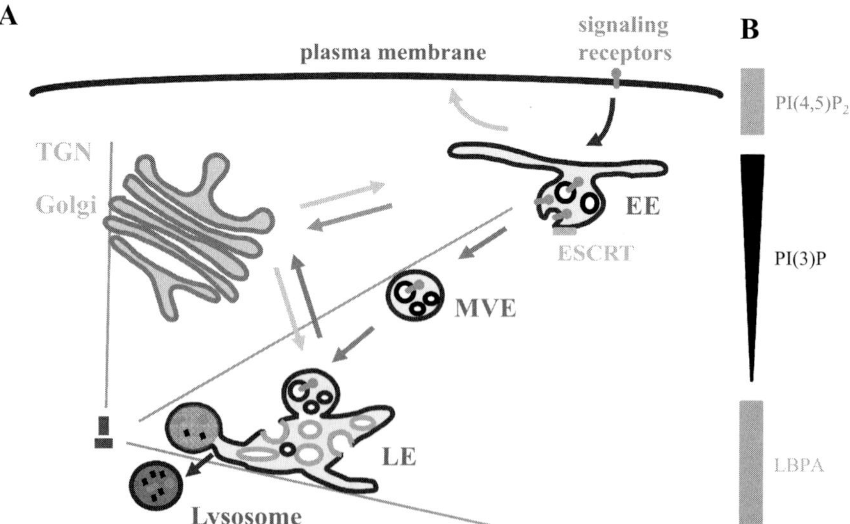

Fig. 1 Lipid populations along the endocytic pathway. (A) Cargo internalized from the plasma membrane is transported to early endosomes (EE), from there it can be recycled to the plasma membrane, sent to the trans-Golgi network (TGN) or routed towards lysosomes for degradation. Signaling receptors that have to be downregulated are sorted into intralumenal vesicles by endosomal sorting complexes required for transport (ESCRT) and are transported to late endosomes (LE) by multivesicular endosomes (MVEs). From LE cargo can be recycled to the TGN, transported to other cellular destinations or sent to lysosomes, where it is finally degraded. (B) PI(4,5)P_2 is predominantly found at the plasma membrane, PI(3)P is abundant in EE and MVEs, whereas LBPA is enriched in LE. The localization of PI(3,5)P_2 is not yet clear. (See color plate.)

from the PM to late endosomes is also hijacked by viruses and toxins, such as vesicular stomatitis virus (VSV) and anthrax lethal factor, respectively. Viral capsids and toxins are believed to be transported to perinuclear late endosomes in intralumenal vesicles of MVEs and released into the cytosol by backfusion of intralumenal vesicles with the limiting membrane (Le Blanc *et al.*, 2005; van der Goot and Gruenberg, 2006).

Functional and morphological differences of endosomes along the endocytic pathway are also reflected by their particular lipid compositions. For instance, phosphatidylinositol 4,5-bisphosphate (PI(4,5)P_2) plays a role in the formation of clathrin-coated pits at the PM (Jost *et al.*, 1998). Whereas, phosphatidylinositol 3-phosphate (PI(3)P) and lysobisphosphatidic acid (LBPA; also called bis(monoacylglyceryl)phosphate (BMP)) are involved in the biogenesis of early endosomes and late endosomes, respectively (Gruenberg, 2003) (Fig. 1B). In addition, endosomes contain also other functional important lipids like phosphatidylserine (PS), which is enriched in early endosomes and the cytoplasmic leaflet of the PM, and phosphatidylinositol 3,5-bisphosphate (PI(3,5)P_2), whose precise distribution is unclear (Gagescu *et al.*, 2000; Leventis and Grinstein, 2010; Shisheva, 2008).

Specialized lipid compositions together with lipid–lipid, protein–lipid, and protein–protein interactions presumably create microenvironments and functional platforms that are essential for organelle architecture and function (Gruenberg, 2001). Here we first discuss the role of endocytic lipids in the degradative pathway and then focus on approaches to study lipid functions in endocytosis.

B. Roles of Lipids in Early Endosomes

PI(3)P is highly enriched in early endosomes, intralumenal vesicles of MVEs and yeast vacuoles (Gillooly *et al.*, 2000) and has been shown to play a role in autophagy (Yang and Klionsky, 2010). The function of PI(3)P is tightly linked to its effector proteins that bind the lipid via specific binding domains. Two such PI(3)P-binding motifs have been described previously: the FYVE domain and the PX- or PHOX-homology domain. The FYVE domain is named after the first letter of the first four proteins in which it was originally found: Fab1p, YOTB, Vac1p, EEA1 (FYVE) (Stenmark *et al.*, 1996). The PX domain was first described in two subunits of the neutrophil oxidase: p40^{phox} and p47^{phox} (Ellson *et al.*, 2001; Kanai *et al.*, 2001) and in proteins of the SNX family (van Weering *et al.*, 2010; Worby and Dixon, 2002).

PI(3)P is generated on early endosomes mostly by the action of the class III PI 3-kinase Vps34 (Schu *et al.*, 1993; Shin *et al.*, 2005) and its formation is directly controlled by the small GTPase Rab5, since Vps34 is a Rab5 effector. Moreover, several other Rab5 effectors including early endosome antigen 1 (EEA1), are themselves FYVE-containing PI(3)P-binding proteins, leading to the notion that a positive feedback loop controls PI(3)P function in early endosomal dynamics (Zerial and McBride, 2001). Indeed, EEA1 and rabenosin-5, another FYVE-containing Rab5 effector, are recruited to early endosomes by PI(3)P and are involved in endosome fusion (Nielsen *et al.*, 2000; Simonsen *et al.*, 1998). In addition, PI(3)P also controls the degradation pathway leading to lysosomes directly. Interfering with PI(3)P by wortmannin, a drug that inhibits PI 3-kinase, impairs intralumenal vesicle formation (Fernandez-Borja *et al.*, 1999) and signaling receptor degradation, but does not affect bulk transport from early to late endosomes (Petiot *et al.*, 2003). This function of PI(3)P depends on Hrs (Hepatocyte growth factor Regulated tyrosine kinase Substrate) another FYVE domain-containing protein, which recruits ESCRTs onto endosomes and binds ubiquitinated receptors (Bache *et al.*, 2003; Lloyd *et al.*, 2002; Razi and Futter, 2006). In turn, ESCRTs sort these activated receptors into intralumenal vesicles, and may contribute to the biogenesis of the intralumenal vesicles themselves (Hurley and Hanson, 2010).

Sorting nexin 3 (SNX3) interacts with PI(3)P via its PX domain (Xu *et al.*, 2001) and is involved in early endosomal dynamics. It is required for the formation of intralumenal vesicles within endosomes, but is dispensable for signaling receptor downregulation. In contrast, Hrs is necessary for lysosomal targeting, but not for MVE formation (Pons *et al.*, 2008). Furthermore, SNX3 was recently shown to mediate Wntless sorting by the retromer (Harterink *et al.*, 2011). Besides SNX3 also other SNX family members, like SNX1 and SNX2 have been shown to bind PI(3)P by their PX domain (Carlton *et al.*, 2005; Cozier *et al.*, 2002). These proteins are part of the retromer complex that mediates retrograde endosome-to-TGN transport (Attar and Cullen, 2010). Thus, PI(3)P and its effector proteins control basic endosome functions, including intralumenal vesicle formation, receptor downregulation, docking, fusion, and motility.

In addition, PI(3)P is metabolized to PI(3,5)P_2 by PI(3)P 5-kinases Fab1p in yeast (Odorizzi *et al.*, 1998) and mammalian PIKfyve (Sbrissa *et al.*, 2002). PIKfyve itself contains a PI(3)P-binding FYVE domain and, like other FYVE-containing proteins, is presumably recruited to early endosomes where its substrate is found (Cabezas *et al.*, 2006; Ikonomov *et al.*, 2006; Rutherford *et al.*, 2006). However, the kinase may well be further transported elsewhere (Shisheva, 2008). Although the precise distribution of PI(3,5)P_2 is not yet known, PI(3,5)P_2 synthesis may begin in early endosomes and continue at later steps of the pathway. Neither is the exact role of PI(3,5)P_2 known. The lipid was reported to play a role in trafficking via activation of a mucolipin transient receptor potential ion channel (TRPML) (Dong *et al.*, 2010), in autophagy in the nervous system (Ferguson *et al.*, 2010), and in cardiac contractility via activation of the ryanodine receptor (Touchberry *et al.*, 2010).

In addition, cholesterol and cholesterol-binding proteins are also involved in endosome dynamics (Pichler and Riezman, 2004). Cholesterol will be discussed separately in another chapter (17). Annexin A2 for example binds early endosomes in a cholesterol-dependent manner (Harder *et al.*, 1997) and was shown to play an important role in the formation and detachment of MVEs without affecting the formation of intralumenal vesicles (Mayran *et al.*, 2003). Furthermore, annexin A2 binds (Gerke and Weber, 1984; Hayes *et al.*, 2004) and nucleates actin on endosomes and these actin patches drive the endosome formation process (Morel *et al.*, 2009).

C. Roles of Lipids in Late Endosomes

Late endosomes contain little PI(3)P (Gillooly *et al.*, 2000), but their intralumenal membranes are highly enriched in the poorly degradable phospholipid LBPA (15 mole per cent of phospholipids) (Kobayashi *et al.*, 1998). The biosynthetic pathway of LBPA is unknown, but since it is only detected in late endosomes one may speculate that it might be synthesized there from a phospholipid precursor. Interfering with LBPA functions by endocytosis of anti-LBPA antibodies inhibits cholesterol export from endosomes, as well as transport of proteins in transit destined to compartments other than the lysosomes (Kobayashi *et al.*, 1999) without interfering with epidermal growth factor receptor (EGFR) transport to lysosomes and degradation (Luyet *et al.*, 2008). Likewise, knockdown of the LBPA partner protein Alix (ALG-2-interacting protein X), an ESCRT-associated protein, does not affect EGFR degradation, but interferes with transport from endosomes to other cellular destinations (Bowers *et al.*, 2006; Gibbings *et al.*, 2009; Luyet *et al.*, 2008; Schmidt *et al.*, 2004). *In vitro* studies have shown that 2,2′ LBPA, but not other isoforms and isomers, has the capacity to trigger the formation of multivesicular liposomes in presence of a pH gradient, similar to the one found in late endosomes. Furthermore, this process is inhibited by addition of recombinant Alix that binds LBPA-containing membranes (Matsuo *et al.*, 2004). Similarly, LBPA and Alix control the formation of intralumenal vesicles within endosomes (Falguieres *et al.*, 2008a).

Knockdown of Alix in mammalian cells results in a decrease in the number of late endosomes containing intralumenal membranes and causes a reduction in LBPA and cholesterol levels (Chevallier *et al.*, 2008; Matsuo *et al.*, 2004). Interestingly, LBPA and cholesterol levels, as well as intralumenal membranes can be restored by feeding cells with liposomes containing 2,2′ LBPA (Chevallier *et al.*, 2008). Hence, LBPA controls directly or indirectly endosomal cholesterol levels. Besides controlling the formation of intralumenal vesicles, LBPA and Alix also regulate their backfusion with the limiting membrane, since Alix knockdown and endocytosis of anti-LBPA antibody inhibit VSV capsid and anthrax lethal factor release into the cytoplasm (Le Blanc *et al.*, 2005; Luyet *et al.*, 2008; van der Goot and Gruenberg, 2006). One may speculate that intralumenal vesicle backfusion with the limiting membrane also controls cholesterol levels in endosomes. These observations suggest that LBPA and Alix play a role in the recycling of components that do not have to be degraded in the lysosome.

Thus, endosomes contain different populations of intralumenal vesicles, which can be discriminated by their lipid and protein content. PI(3)P, which is clearly enriched on the early endosome limiting membrane, is also abundant in intralumenal vesicles of early endosomes, while, in the late endosomes, it is less abundant and present on fewer intralumenal vesicles (Gillooly *et al.*, 2000). Cholesterol seems to follow a similar distribution within endosomes (Mobius *et al.*, 2003). LBPA on the other hand is only detected in late endosomes and is absent in early endosomes (Kobayashi *et al.*, 1998). Interestingly, electron microscopic analysis of late endosomal intralumenal vesicles revealed that vesicles labeled for PI(3)P or cholesterol are distinct from LBPA containing vesicles, arguing for different populations of intralumenal vesicles (Gillooly *et al.*, 2000; Mobius *et al.*, 2003). Furthermore, evidence indicates that intralumenal vesicles can be formed by different mechanisms. PI(3)P clearly initiates the formation of intralumenal vesicles early in the pathway, before LBPA can be detected, by recruiting PI(3)P-binding proteins, for example, ESCRTs. LBPA on the other hand is involved in the formation of intralumenal vesicles in late endosomes, where it is localized. Future work will be required to determine whether LBPA contributes to the regulation of an ESCRT-dependent pathway within late endosomes, or whether LBPA- and PI3P-dependent pathways are separate, but share some common constituents (e.g., Alix) (Falguieres *et al.*, 2008a; Luyet *et al.*, 2008). Moreover, further evidence for the existence of different mechanisms controlling intralumenal vesicle formation is coming from the studies of exosomes, which are believed to represent intralumenal vesicles released in the extracellular space by fusion of MVEs with the PM. This process was found to be ESCRT- and Alix-independent, but ceramide-dependent in oligodendrocytes (Trajkovic *et al.*, 2008), whereas an Alix-dependent, ESCRT-independent process was found in monocytes (Gibbings *et al.*, 2009).

D. Studying Lipids Involved in Endocytosis

During previous years much progress has been made in developing tools to study lipid functions. Nevertheless, compared to protein and DNA, investigation of lipids still remains challenging, because of technical limitations. However, new methods are emerging quickly and the next years may well see revolutionary developments in the field of lipid biology. In this section, we briefly summarize tools to investigate lipid functions *in vivo* and *in vitro*.

1. Synthetic Lipids

In vitro studies, where lipid and protein functions are studied under precisely controlled conditions, became possible due to a large selection of synthetic lipids. In recent years elaborate *in vitro* methods have been designed to reveal many lipid-based mechanisms. Here we briefly summarize some of the latest techniques and their important contributions towards understanding the role of lipids in endocytosis.

The rapid analysis of protein–lipid binding can be conveniently achieved using strips with immobilized lipids (Dowler *et al.*, 2002). This technique facilitated the analysis of lipid-binding specificity of many proteins, like the specificity of the PX domains of Vam7t-SNARE (Cheever *et al.*, 2001) and p40^{phox} (Zhan *et al.*, 2002) towards PI(3)P. This method is often effective as a first step, but is not very precise and certainly not quantitative and is best combined with other techniques.

Lipid microarrays are based on the principle of lipid strips, but contain many different lipid species (Feng, 2005). They therefore allow the identification of novel lipid-binding partners (Kanter *et al.*, 2006). Similarly, proteome microarrays can be used to screen for novel lipid-binding motifs (Zhu *et al.*, 2001). These methods, which are more demanding, provide a comprehensive and global view of lipid–protein interactions. Also, one should not forget that immobilized lipids and proteins may not mimic their natural conformation and therefore, it is best to confirm the existence of a specific interaction with other methods.

A more precise method to characterize lipid-binding proteins is a classical protein–liposome binding experiment that separates liposome-bound from free protein by either floatation or sedimentation (Fig. 3). Using this technique Alix was identified as an effector of LBPA and its specificity for 2,2′ LBPA-containing membranes, over other LBPA isoforms and isomers was demonstrated. In further experiments, Alix was shown to control LBPA-mediated membrane invagination within acidic liposomes (Matsuo *et al.*, 2004).

Due to the large diameter (10–30 μm) of giant unilamellar vesicles (GUVs), microscopic observations of lipid-based mechanisms, such as protein–lipid binding, lipid-domain formation (Kaiser *et al.*, 2009) and lipid- or protein-induced membrane deformation (Ewers *et al.*, 2010; Romer *et al.*, 2007) became possible. Using this approach Wollert and Hurley recently proposed a molecular mechanism of MVE biogenesis by ESCRT complexes, where ESCRT-0 forms domains of clustered cargo, ESCRT-I and –II induce membrane budding and ESCRT-III catalyzes the fission of membrane buds (Wollert and Hurley, 2010; Wollert *et al.*, 2009). (See chapter 4 by Wollert in this issue.)

Structural information of lipid–protein interactions at atomic resolution has been gained from co-crystallization of lipid-protein complexes (Bravo *et al.*, 2001) and from solution NMR studies that monitor protein docking on micelles (Dancea *et al.*, 2008).

2. Fluorescent Lipid Analogs

Fluorescently labeled lipid analogs are widely used to study cellular lipid transport and metabolism (Pagano and Chen, 1998). Today a large variety of lipids bearing fluorescent groups linked to diverse lipid backbone positions are commercially available. Fluorescent tags as well as the position of lipid-linkage (head group or fatty acid tail) may strongly affect biophysical and biochemical characteristics of the lipid analog. Therefore properties of fluorescent lipid analogs have to be studied carefully and compared to endogenous lipid properties, in order to use them as a reliable live-cell imaging tool. With this in mind Pagano and co-workers showed that fluorescent lipid analogs can be powerful tools to study membrane microdomains (Marks *et al.*, 2008) and intracellular lipid trafficking (Singh *et al.*, 2007). Using Bodipy-labeled sphingolipid analogs they depicted the intracellular trafficking route of these lipids from the PM to the Golgi complex via endosomes. Interestingly, this route is impaired in some sphingolipid storage diseases, where the lipid analogs accumulate in endosomes and lysosomes (Pagano *et al.*, 2000). Bodipy-lipid analogs are well suited for trafficking studies, because in addition to lipid localization, information about lipid concentration is obtained, due to a concentration-dependent spectral shift in emission wavelengths (Pagano *et al.*, 1991). The disadvantages of Bodipy-fluorophores are its bulky size that may influence lipid behaviour and its spectral overlap with green fluorescent proteins.

Recently, polyene lipids have been introduced as promising alternatives to established fluorescent labels. Polyene lipids, containing linear hydrocarbons with five conjugated double bonds, closely mimic natural lipids and allowed for the first time microscopic visualization of ether lipids (Kuerschner *et al.*, 2005). Unfortunately, polyene lipids have small quantum yields and are sensitive to photobleaching, which makes imaging demanding.

In addition, chemical biology strategies are being developed that allow the incorporation of lipids and lipid analogs into cells in a controlled fashion, including particular phosphatidylinositols (Subramanian *et al.*, 2010). These strategies have made it possible to modulate the cell content of these lipids over a short time-course *in vivo*, and thus to study their precise functions.

3. Lipid-Binding Proteins

Detection of lipids in living cells is still challenging and few techniques are available. The most widely used tool takes advantage of phospholipid-binding domains fused to fluorescent proteins to visualize cellular lipid dynamics and localization (Fig. 4). Using this approach, phosphatidylinositols have been widely studied over the past 10 years (Halet, 2005) and nowadays an almost complete toolbox for their detection is available (Overduin *et al.*, 2001). Also other domains that bind phospholipids have been identified and used for lipid visualization. One example is the discoidin C2 domain of lactadherin that recognizes PS with high affinity and specificity (Leventis and Grinstein, 2010). One of the best characterized phosphatidylinositol-binding domains is the tandem FYVE domain of Hrs. Due to its high affinity and specificity that made fluorescence and electron microscopy possible, this domain has been proven an extremely valuable tool for the precise localization of PI(3)P in yeast and mammalian cells (Gillooly *et al.*, 2000). As protein–lipid interactions are in general weak, one key component for the high affinity of the tandem FYVE domain is the use of two consecutive FYVE domains, that allows simultaneous interaction with two PI(3)P molecules. However, data have to be interpreted with caution and under consideration of the following issues: (i) the specificity of phospholipid-binding domains might be different *in vivo* and *in vitro*, because of additional protein–protein and lipid–protein interactions that will determine the cellular localization; (ii) the overexpressed reporter domain might not have access to all cellular phospholipid pools, because they are already occupied by endogenous effector proteins; (iii) overexpression of phospholipid-binding domains may interfere with lipid-mediated processes, because of their competition with endogenous effectors (Varnai and Balla, 2006). Nevertheless, the use of phospholipid-binding domains helped to unravel cellular localizations and functions of many phospholipid species (Lemmon, 2008).

Alternatively, phospholipids can be visualized in living cells by endocytosis or microinjection of anti-phospholipid antibodies (Kobayashi *et al.*, 1998; Matuoka *et al.*, 1988). The drawback of this method is the requirement of anti-phospholipid antibody with high affinity and specificity. The monoclonal mouse anti-LBPA antibody (6C4) fulfils this requirement and has been used to describe the function of LBPA in membrane trafficking and endosome biogenesis (Kobayashi *et al.*, 1998; Le Blanc *et al.*, 2005; Luyet *et al.*, 2008) (Figs. 4 and 5C, D). Endocytosed anti-LBPA antibodies interfere with cellular LBPA functions in a dose-dependent manner, providing an efficient tool to study cellular LBPA localization and function *in vivo*.

In addition, toxins that bind lipids have been successfully used to analyze the function and distribution of sphingolipids (Shogomori and Kobayashi, 2008) and cholesterol (Mobius *et al.*, 2003; Ohno-Iwashita *et al.*, 2004). However, the most widely used dye for the visualization of cholesterol remains filipin III, which is a polyene macrolide isolated from cultures of *S. filipinensis*. Filipin III binds to cholesterol, presumably clustered cholesterol, in membranes and can be conveniently visualized by UV light, providing a highly valuable tool to visualize cholesterol by fluorescence microscopy (Kobayashi *et al.*, 1999) (Figs. 2B and 5).

Importantly, working with lipid-binding probes one should always keep in mind that the probes themselves may interfere with protein sorting and membrane dynamics. Nevertheless, reagents, whether antibodies, recombinant proteins, fusion proteins or toxins that bind lipids *in vitro* or *in vivo* are important components of an appropriate lipid tool box.

4. Mass Spectrometry Combined with Genetic Approaches

One of the most promising techniques in the lipid research field is probably mass spectrometry (MS). Enormous progress has been made in developing MS that is suitable to successfully analyze complex lipid mixtures (Wenk, 2010). MS makes it possible to obtain precise notions about the lipid composition of a sample, to detect minor lipid classes, and possibly to identify and characterize yet unknown lipid species. High-throughput MS combined with genetic approaches will lead towards comprehensive understanding of lipid functions and regulations. Guan and co-workers have taken this approach one step further. They analyzed yeast sterol biogenesis mutants by MS and showed that cells have a mechanism to react to changes in their membrane sterol composition by modifying their sphingolipid composition (Guan *et al.*, 2009).

II. Methods

In this section, we describe different *in vitro* and *in vivo* techniques to monitor endosomal lipids with a focus on the late endosomal lipid LBPA. The methods summarized here provide a mean to identify LBPA binding partners and monitor LBPA levels, localization, and dynamics.

A. Cell Culture

1. BHK Cells

Monolayers of baby hamster kidney (BHK-21) cells were cultured in Glasgow Minimum Essential Medium supplemented with 5% fetal calf serum and penicillin–streptomycin. Cells were maintained at 37 °C in a 5% CO_2 incubator.

2. HeLa Cells

HeLa cells were cultured in Dulbeco's Modified Eagle Medium supplemented with 5% fetal calf serum and penicillin–streptomycin. Cells were maintained at 37 °C in a 5% CO_2 incubator.

B. *In vitro* Assays

1. Subcellular Fractionation

Subcellular fractionation remains essential in order to characterize an organelle's lipid composition. However, not all organelles are easily isolated in sufficient purity and quantity to allow lipid analysis. Fortunately, fractions containing late and early endosomes can be purified from BHK cells, making lipid analysis possible by MS, ELISA, TLC, and other methods. Here we describe a well established fractionation protocol for preparation of endosomal fractions from BHK cells.

Materials

- 10 cm Petri dishes (Nunc, Thermo Fisher Scientific, Rosklide, Denmark)
- 15 mL tubes (Techno Plastic Products, Trasadingen, Switzerland)
- BHK cell culture medium: Glasgow Minimum Essential Medium (Sigma-Aldrich, St. Louis, MO)
- PBS^{--}: 137 mM NaCl, 2.7 mM KCl, 1.5 mM KH_2PO_4, 6.5 mM Na_2HPO_4 pH 7.4
- Cell scraper (homemade: silicone rubber piece of about 2 cm, cut at a sharp angle and attached to a metal bar or held with forceps)
- Centrifuges: low speed centrifuge and ultracentrifuge with SW60 rotor
- SW60 centrifuge tubes (Beckman Coulter, Brea, CA)
- 22G needles and 1 mL syringes (BD, San Jose, CA)
- Plastic Pasteur pipettes (Assistent, Sondheim, Germany)
- Homogenization buffer (HB): 8.5% sucrose in 3 mM imidazole
- 62% sucrose: 62% sucrose in 3 mM imidazole

- 35% sucrose: 35% sucrose in 3 mM imidazole
- 25% sucrose: 25% sucrose in 3 mM imidazole
- HB^{+++}: HB supplemented with protease inhibitors: 10 μg/mL aprotinin, 5 μg/mL leupeptin and 1 μM pepstatin (Roche, Basel, Switzerland)
- Phase contrast microscope with 40x objective (Carl Zeiss Inc., Jena, Germany)
- Glass slides and coverslips (18 × 18 mm) (Assistent, Sondheim, Germany)
- Refractometer (Carl Zeiss Inc., Jena, Germany)
- 96 well-plate (Nunc, Thermo Fisher Scientific, Rosklide, Denmark)
- Bradford solution (BioRad Labs, Hercules, CA)
- Mouse IgG antibody (1 mg/mL) (Sigma-Aldrich, St. Louis, MO)
- Microplate spectrophotometer (Bucher Biotec AG, Basel, Switzerland)

Procedure

Note: All steps have to be carried out on ice.

1. Prepare per condition three Petri dishes (10 cm diameter) of BHK cells at 1/4 dilution (surface:surface) 16 h before the experiment (dishes are seeded with approx. 4×10^5 cells).
2. Wash cells 2x with 5 mL ice-cold PBS^{--}.
3. Remove all PBS^{--} from the last wash and add 2.5 mL/dish PBS^{--}. Scrape cells by using a cell scraper. First scrape the outside of the Petri dish by doing one circular movement and then scrape the remaining middle part from top to down. The cells should detach in "sheets" from the Petri dish. Gently transfer cells from three dishes into a 15 mL tube using a plastic Pasteur pipette.
4. Centrifuge at 900 rpm for 5 min at 4 °C and remove the supernatant.
5. Resuspend the cell pellet by adding 2.5 mL HB and gently pipetting up and down two times using a plastic Pasteur pipette.
6. Centrifuge at 2000 rpm for 10 min at 4 °C and remove the supernatant.
7. Add 400 μL HB^{+++} and pipette five times gently up and down using a 1 mL pipette.
8. Homogenize the cells using a 22G needle and a 1 mL syringe, that have been washed with HB^{+++}, by slowly pulling up the cell suspension and expelling it while holding the tip of the needle against the wall of the tube. Repeat this procedure 2–5 times. In order to monitor the homogenization, put 3 μL of homogenate on a glass slide, cover it with a glass coverslip and observe by a phase-contrast microscope using a 40x objective. For a good homogenization the PM should be broken, but the nuclear membranes should stay intact.
9. Centrifuge at 2000 rpm for 10 min at 4 °C and carefully collect the post-nuclear supernatant (PNS).
10. Adjust sucrose concentration of the PNS to 40.6% by adding about 1.1 volume of 62% sucrose solution per volume of PNS. Check sucrose concentration using a refractometer.

11. Place adjusted PNS in the bottom of SW60 centrifuge tubes. Using a plastic Pasteur pipette overlay the PNS gently with first 1.5 mL 35% sucrose solution, then 1 mL 25% sucrose solution and fill up with HB. Note that the interphases should not be disturbed and be clearly visible.
12. Centrifuge in a SW60 rotor at 35000 rpm for 1 h at 4 °C.
13. After centrifugation the interphases should appear white, because they contain membranes. Carefully remove the layer of white lipids on the top of the gradient. Collect 300 μL of HB/25% interphase (≈1.06 g/cm^3), which is highly enriched in late endosomes and MVEs, using a 1 mL pipette. Gently remove the solution until the 25% sucrose layer starts. Repeat the same for the 25%/35% interphase containing light membranes (≈1.104 g/cm^3) and for the 35%/40.6% interphase containing heavy membranes (≈1.15 g/cm^3). The light membrane fraction is enriched in early endosomes, but also contains other membranes (Rojo *et al.*, 2000). The heavy membrane fraction contains rough endoplasmic reticulum, lysosomes, peroxisomes, and mitochondria.
14. Measure protein concentration in a 96 well-plate by Bradford using the following dilutions:

    Prepare duplicates of:

Standards:	20 μL H_2O
	19 μL H_2O + 1 μL 1 mg/mL IgG
	18 μL H_2O + 2 μL 1 mg/mL IgG
	16 μL H_2O + 4 μL 1 mg/mL IgG
Samples:	PNS: 20 μL of a 1/100 dilution
	Heavy membranes: 18 μL H_2O + 2 μL heavy membrane fraction
	Light membranes: 10 μL H_2O + 10 μL light membrane fraction
	Late endosomes : 20 μL late endosome fraction

    Add to each well 180 μL of the following Bradford dilution:

    40 μL Bradford + 140 μL H_2O.
    Read the absorbance at 595 nm after 10 min.

15. Fractions can be frozen in liquid nitrogen and stored at −80 °C.

2. ELISA

The LBPA levels in light membrane and late endosomal fractions can be analyzed by an ELISA assay using anti-LBPA antibody. As LBPA is localized to late endosomes, ~ 10x more LBPA is found in late endosomal fractions compared to light membrane fractions (Fig. 2A). This high LBPA ratio between light membrane and late endosomal fractions also indicates a good subcellular fractionation.

Materials

- 96 well plate (Nunc, Thermo Fisher Scientific, Rosklide, Denmark)
- 20 mM Hepes-NaOH pH 7.4, 150 mM NaCl
- 3% BSA in 20 mM Hepes-NaOH pH 7.4, 150 mM NaCl
- 1% BSA in 20 mM Hepes-NaOH pH 7.4, 150 mM NaCl
- 10 mM Tris-HCl pH 7.4, 150 mM NaCl
- Mouse anti-LBPA antibody (6C4) (Echelon Bioscience Inc., Salt Lake City, UT) (Kobayashi *et al.*, 1998)
- Biotin-conjugated anti-mouse antibody (Chemicon (Millipore) Billerica, MA)
- Streptavidin-conjugated alkaline phosphatase (Roche, Basel, Switzerland)
- UMP: 4-methylumbellyphenylphosphate (Biosynth, Itasca, IL)
- Diethanolamine (Sigma-Aldrich, St. Louis, MO)
- 1 M $MgCl_2$
- Microplate spectrofluorometer (Bucher Biotec AG, Basel, Switzerland)

Procedure

Use a 96 well plate and make duplicates of each sample:

1. Give 1 μg protein of the late endosomal and light membrane fractions into wells and fill up to 200 μL with 20 mM Hepes-NaOH pH 7.4, 150 mM NaCl.
2. Incubate over night at 4 °C to allow the membranes to bind to the wells.
3. Discard the solution and add 200 μL 3% BSA in 20 mM Hepes-NaOH pH 7.4, 150 mM NaCl to each well.
4. Incubate for 2 h at RT.
5. Wash 3x with 10 mM Tris-HCl pH 7.4, 150 mM NaCl.
6. Dilute anti-LBPA antibody (6C4) in 1% BSA in 20 mM Hepes-NaOH pH 7.4, 150 mM NaCl and add 50 μL to each well.
7. Incubate for 2 h at RT.
8. Wash 6x with 10 mM Tris-HCl pH 7.4, 150 mM NaCl.
9. Dilute biotin-conjugated anti-mouse antibody in 1% BSA in 20 mM Hepes-NaOH pH 7.4, 150 mM NaCl and add 50 μL to each well.
10. Incubate for 90 min at RT.
11. Wash 6x with 10 mM Tris-HCl pH 7.4, 150 mM NaCl.
12. Dilute streptavidin-conjugated alkaline phosphatase in 1% BSA in 20 mM Hepes-NaOH pH 7.4, 150 mM NaCl and add 50 μL to each well.
13. Incubate for 1 h at RT.
14. Wash 6x with 10 mM Tris-HCl pH 7.4, 150 mM NaCl.
15. Prepare a solution of 0.25 mM 4-methylumbellyphenylphosphate (UMP) in 10% diethanolamine, 2 mM $MgCl_2$.
16. Add 100 μL of the above-mentioned solution to each well.
17. Incubate for 20 min at RT in the dark.
18. Read the fluorescence using a microplate spectrofluorometer:

Excitation wavelength = 365 nm
Emission wavelength = 450 nm

Results

There is ~ 10x more LBPA in the late endosomal fraction compared to the light membrane fraction (Fig. 2A). This indicates that the light endosomal fraction is not contaminated with late endosomes and the subcellular fractionation was efficient.

3. Liposome Floatation Assay

Liposome floatation assays are used to characterize protein–lipid interactions in terms of lipid specificity, liposome size preferences etc. Proteins bound to liposomes float in a step sucrose gradient due to the low buoyant density of liposomes,

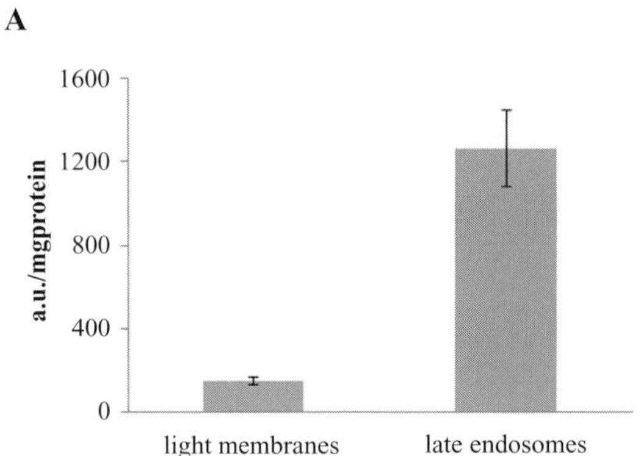

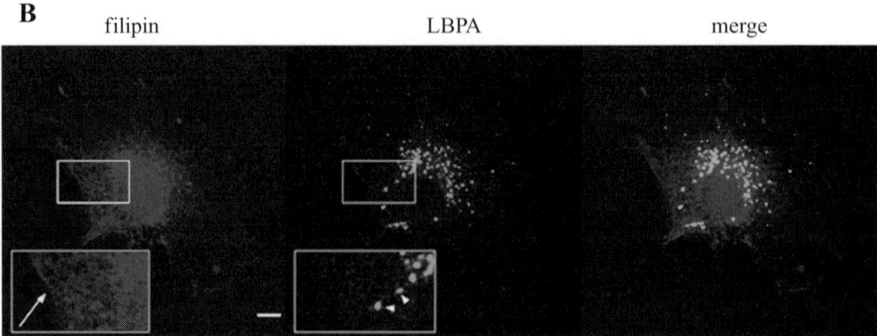

Fig. 2 Localization of LBPA and cholesterol. (A) LBPA levels in light and late endosomal fractions were analyzed by ELISA using anti-LBPA antibody. Means (± SEM) of a typical triplicate. (B) BHK cells were fixed and stained with anti-LBPA antibody and filipin. Bar: 10 μm. Symbols: arrow and arrowheads point at plasma membrane and late endosomes, respectively. (See color plate.)

whereas free protein stays at the bottom. An alternative protocol is sedimentation of protein–liposome complexes by high-speed centrifugation, but as protein aggregates also sediment, protein–liposome binding might be overestimated. To overcome this risk, we describe a liposome floatation assay. Here we bind anti-LBPA antibody and control IgG antibody to liposomes containing LBPA, but the assay can also be performed with any lipid mixture and recombinant protein or cytosol preparation.

Materials

- Mouse anti-LBPA antibody (6C4) (Echelon Bioscience Inc., Salt Lake City, UT) (Kobayashi *et al.*, 1998)
- Mouse IgG antibody (Sigma-Aldrich, St. Louis, MO)
- Lipids (e.g., DOPC, DOPE, PI, 2,2′ LBPA in chloroform) (Avanti Polar Lipids Inc., Alabaster, AL and Echelon Bioscience Inc., Salt Lake City, UT)
- Pyrex test tubes (VWR international, West Chester, PA)
- 1.7 mL test tubes (Axygen Inc., Union city, CA)
- Argon
- Bath sonicator
- Rotating wheel
- Vortex
- Ultracentrifuge with TL55 rotor
- TL55 centrifuge tubes (Beckman Coulter, Brea, CA)
- 25 mM Hepes-NaOH pH 7.4
- Homogenization buffer (HB): 8.5% sucrose in 3 mM imidazole
- 62% sucrose: 62% sucrose in 3 mM imidazole
- 35% sucrose: 35% sucrose in 3 mM imidazole
- Refractometer (Carl Zeiss Inc., Jena, Germany)
- SDS–PAGE and western blot equipment (BioRad Labs, Hercules, CA)

Procedure

1. If necessary change buffer of antibody and concentrate it.

   → Buffer of anti-LBPA antibody was changed by dialysis to 25 mM Hepes-NaOH pH 7.4 and then concentrated using centrifugal concentrator (e.g., Centriprep-100 concentrator; Amicon).

2. Mix per condition $\geq$50 μg of lipids in a pyrex test tube, vortex and dry under Argon.

   → 50 μg DOPC, 19 μg DOPE, 11 μg PI, and 26 μg 2,2′ LBPA (total: 106 μg).

3. Add 25 mM Hepes-NaOH pH 7.4 to a final concentration of 0.5 mg/mL to dried lipids and do not pipette up and down.
4. Let lipids rehydrated by incubation on ice in the dark for 1 h.

5. Increase the formation of liposomes by 2 × 5 min sonication in a bath sonicator.
6. Vortex 30 s.
7. Give 100 μL liposome solution (= 50 μg lipids) into a 1.7 mL test tube.
8. Add 1.8 μg antibodies and fill up to 200 μL with 25 mM Hepes-NaOH pH 7.4.
9. Allow protein–liposome binding by incubation for 2 h at 4 °C on a rotating wheel in the dark.
10. Adjust sucrose concentration of the protein–liposome mix to 40.6% by adding about 1.35 volumes of 62% sucrose solution per volume of protein–liposome mix. Check sucrose concentration using a refractometer.
11. Place the adjusted protein–liposome mix in the bottom of TL55 centrifuge tubes. Using a plastic Pasteur pipette overlay the protein–liposome mix gently with 1 mL 35% sucrose solution and fill up with HB. Note that the interphases should not be disturbed and be clearly visible.
12. Mark the sucrose interphases.
13. Centrifuge in a TL55 rotor at 55000 rpm for 1 h at 4 °C.
14. Collect 110 μL of the HB/35% interphase that contains liposome-bound protein using a cut tip of a 200 μL pipette. Gently remove the solution until the 35% sucrose layer starts. Collect in the same manner 110 μL of the 35% sucrose layer. Gently remove the solution until the 40.6% sucrose layer starts. And collect 110 μL of the 40.6% sucrose layer containing free protein.
15. Use 55 μL of the collected fractions for western blot.
16. Prepare 0.25 μg antibody as a loading control for western blot.
17. Run samples in non-reducing sample buffer on an 8% SDS-polyacrylamide gel and immunoblot against anti-mouse antibody.
18. Quantification of western blot:
    - Use software (e.g., ImageJ) to quantify bands.
    - To obtain the total amount of protein per fraction, multiply each signal with (total volume of fraction)/(volume loaded).

→ Our case:	Liposome fraction: 110/55
	35% sucrose: 1000/55
	40.6% sucrose: 470/55

    - Calculate the sum of all three fractions (total protein).
    - Calculate the percentage of protein in each fraction.

Results

The anti-LBPA antibody shows a strong signal in the liposome fraction meaning that the antibody binds to liposomes containing 2,2′ LBPA. In contrast, control IgG antibody does not bind liposomes since the antibody is only found in the 40.6%

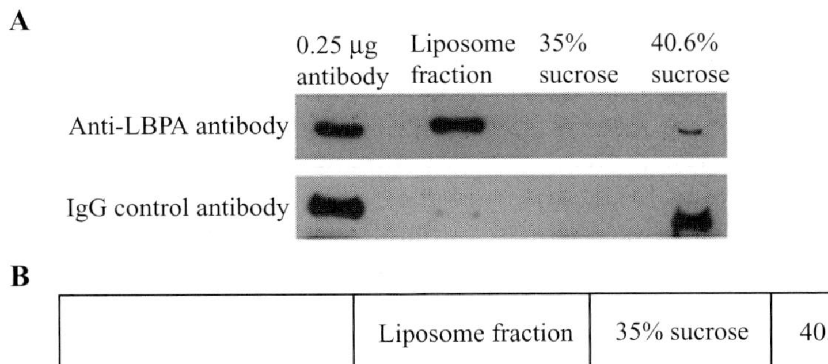

	Liposome fraction	35% sucrose	40.6% sucrose
Anti-LBPA antibody	39%	0%	61%
IgG control antibody	0%	0%	100%

Fig. 3 Liposome floatation assay. (A) Liposome, 35% sucrose and 40.6% sucrose fractions were analyzed by SDS–PAGE and immunoblotting against anti-mouse. **(B)** Quantification of **A** showing the percentage of antibody in each fraction.

sucrose that contains free protein (Fig. 3A). The quantification of the western blot signals shows that 39% of the total anti-LBPA antibody is bound to liposomes, whereas no binding occurred with control IgG antibody (Fig. 3B).

C. *In vivo* Assays

1. Immunofluorescence

Cellular LBPA can easily be visualized *in vivo* by immunofluorescence using anti-LBPA antibody. Here we describe a protocol for immunostaining of fixed BHK cells, where filipin III is used to visualize cholesterol. Filipin is easily visualized, but requires UV light, which is not always available.

Materials

- Glass coverslips with 12 mm diameter (Assistent, Sondheim, Germany)
- BHK cell culture medium: Glasgow Minimum Essential Medium (Sigma-Aldrich, St. Louis, MO)
- PBS^{--}: 137 mM NaCl, 2.7 mM KCl, 1.5 mM KH_2PO_4, 6.5 mM Na_2HPO_4 pH 7.4
- PFA: 3% (w/v) paraformaldehyde in PBS^{--}
- 50 mM NH_4Cl in PBS^{--}
- 3% (w/v) BSA in PBS^{--}
- 1% (w/v) BSA in PBS^{--}
- Humid chamber with glass plate
- Parafilm

- Mouse anti-LBPA antibody (6C4) (Echelon Bioscience Inc., Salt Lake City, UT) (Kobayashi *et al.*, 1998)
- Fluorophore-conjugated secondary antibody (Jackson ImmunoResearch Laboratories Inc., West Grove, PA)
- Filipin III (2.5 mg/mL in DMSO) (Sigma-Aldrich, St. Louis, MO)
- Glass slides (Assistent, Sondheim, Germany)
- Mowiol (CalBiochem (EMD Biosciences Inc.), San Diego, CA)
- Confocal micorsocope LSM510 (Carl Zeiss Inc., Jena, Germany)

Procedure

1. Plate BHK cells at the desired density directly onto 12 mm diameter coverslips 16 h before the experiment.
2. Wash cells 2x with PBS^{--}.
3. Fix cells by covering them with 3% PFA and incubate for 20 min at RT.
4. Wash cells 3x with PBS^{--}.
5. Quench for 10 min with 50 mM NH_4Cl in PBS^{--}.
6. Wash cells 3x with PBS^{--}.
7. Block to avoid non-specific antibody binding, by incubation in 3% BSA in PBS^{--} for 20 min at RT.
8. Wash cells 1x with 1% BSA in PBS^{--}.
9. Warp a glass plate with parafilm and place it in a humid chamber.
10. Prepare primary antibody (anti-LBPA (6C4)) at the desired dilution in 1% BSA, 60 μg/mL filipin in PBS^{--} and give drops of 40 μL onto the previously prepared glass plate. Put the coverslips down on the drops of antibody (cells are facing the glass plate) and incubate for 30 min at RT in the dark.
11. Wash cells 3x with 1% BSA in PBS^{--} by transferring the coverslips to three drops of 1% BSA in PBS^{--}.
12. Prepare fluorophore-conjugated secondary antibody at the desired concentration in 1% BSA in PBS^{--} and give drops of 40 μL onto the glass plate. Transfer the coverslips onto the drops and incubate for 30 min at RT in the dark.
13. Wash cells 3x with PBS^{--} by transferring the coverslips to three drops of PBS^{--}.
14. Rinse the coverslips in H_2O and dry them on a tissue.
15. Mount in 6 μL of Mowiol on a glass slide.
16. Take pictures at a confocal microscope (LSM510).

Results

Filipin shows high PM (arrow in Fig. 2B) and some intracellular staining – consistent with the high cholesterol content of the PM. The anti-LBPA antibody stains late endosomes localized in the perinuclear area (arrowhead in Fig. 2B). While these markers clearly do not co-localize under normal, physiological conditions, U18666A (Fig. 5A, B) or endocytosed anti-LBPA antibody (Fig. 5C, D) cause the accumulation of cholesterol in late endosomes containing LBPA.

2. Visualization of PI(3)P and LBPA in Living Cells

In this part we describe double labeling of early and late endosomal lipids PI(3)P and LBPA for live-cell time-lapse microscopy. Early endosomes containing PI(3)P can be stained by the highly selective PI(3)P-binding domain tandem FYVE-GFP (Gillooly *et al.*, 2000) and LBPA in late endosomes is visualized by endocytosis of anti-LBPA antibody that binds LBPA facing the endosomal lumen. At low over-expression levels of FYVE-GFP and low anti-LBPA antibody concentration endosome functions and dynamics are not affected.

Materials

- 3.5 mm glass bottom dishes (MatTek Corporation, Ashland, MA)
- HeLa cell culture medium: Dulbeco's Modified Eagle Medium (Sigma-Aldrich, St. Louis, MO)
- Mouse anti-LBPA antibody (6C4) (Echelon Bioscience Inc., Salt Lake City, UT) (Kobayashi *et al.*, 1998)
- Fluorophore-conjugated anti-mouse antibody (1 mg/mL) (Jackson ImmunoResearch Laboratories Inc., West Grove, PA)
- OptiMEM (Gibco (Invitrogen Corporation), Carlsbad, CA)
- FuGENE HD (Roche, Basel, Switzerland)
- Tandem FYVE-GFP plasmid (Gillooly *et al.*, 2000)
- Spinning disc microscope (3i Marinanas) (Leica Microsystems, Wetzlar, Germany)

Procedure

1. Plate HeLa cells at the desired density into 3.5 mm glass bottom dishes 24 h before the experiment.
2. Incubate 0.5 mL mouse anti-LBPA antibody (5 μg/mL) with 4 μL fluorophore-conjugated anti-mouse antibody (1 mg/mL) for 30 min at RT.
3. Dilute the previously prepared antibody mix 1/2 in HeLa medium without antibiotics and add it to the cells.
4. Transfect the cells with tandem FYVE-GFP plasmid (Gillooly *et al.*, 2000) using FuGENE HD and OptiMEM 16 h before the experiment. (Per 3.5 mm dish: 29 μL OptiMEM + 1.5 μg DNA + 3 μL FuGENE HD in medium without antibiotics.)
5. Incubate overnight at 37 °C.
6. Live-cell time-lapse confocal microscopy:

   Image cells at a spinning disc confocal microsope (3i Marianas) with a frame rate of 0.6 s for 4–5 min.

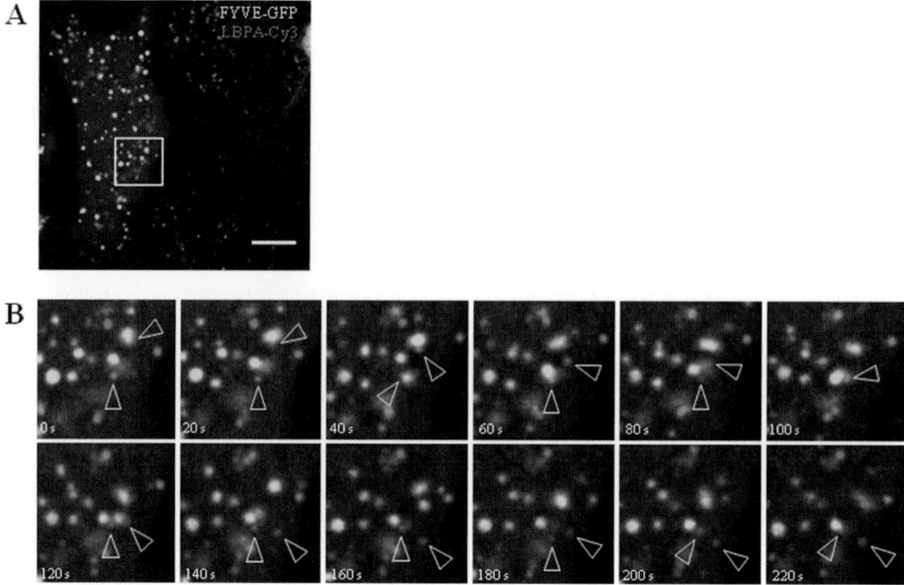

Fig. 4 Visualization of PI(3)P and LBPA in live cells. (A) HeLa cells were transfected with tandem FYVE-GFP and incubated overnight with mouse anti-LBPA and anti-mouse Cy3 antibodies. The cells were imaged *in vivo* and movies were taken at a frame rate of 0.6 s by time-lapse confocal microscopy. Panel A shows the first frame of supplementary movie 1. Both cells in this frame contain endocytosed anti-LBPA antibodies but only the left cell expresses tandem FYVE-GFP. (B) Selected frames of magnified region. Arrowheads are showing movements of LBPA-positive endosomes. Bar: 10 μm. (See color plate.)

Results

Dynamics of PI(3)P- and LBPA-containing endosomes are visualized in a live-cell microscopy assay using the PI(3)P-binding domain tandem FYVE-GFP and endocytosed anti-LBPA antibody (Fig. 4 and supplementary movie 1). In agreement with the notion of PI(3)P and LBPA being in different endosome populations, there is no co-localization between the two lipids.

3. *In vivo* Modification of Endosomal Lipid Levels

Endosomal lipid levels (in particular LBPA and cholesterol) can be modified in various ways. Here we describe three approaches: Drug treatment (Kobayashi *et al.*, 1999), internalization of anti-LBPA antibody (Kobayashi *et al.*, 1998), and RNAi of endosomal proteins (Matsuo *et al.*, 2004).

U18666A Treatment

Cells treated with the hydrophobic amines U18666A accumulate LDL (low-density-lipoprotein)-derived cholesterol and LBPA in late endosomes (Kobayashi *et al.*, 1999; Liscum and Faust, 1989). This accumulation can easily be monitored by immunofluorescence using anti-LBPA antibody and filipin III to stain cholesterol (Fig. 5A, B).

Materials

- Glass coverslips with 12 mm diameter (Assistent, Sondheim, Germany)
- BHK cell culture medium: Glasgow Minimum Essential Medium (Sigma-Aldrich, St. Louis, MO)
- U18666A (10 mg/mL) (Sigma-Aldrich, St. Louis, MO)

Procedure

1. Plate BHK cells at the desired density directly onto 12 mm diameter coverslips 36 h before the experiment.
2. Incubate cells with 3 μg/mL U18666A for 0, 2, 4, and 8 h at 37 °C.
3. Perform immunofluorescence as described.

Internalization of Anti-LBPA Antibody

Endocytosis of low doses of anti-LBPA antibody can be used to visualize LBPA *in vivo* as described previously (Fig. 4). In contrast, the internalization of high doses of anti-LBPA antibody leads to a concomitant accumulation of cholesterol and LBPA in late endosomes (Kobayashi *et al.*, 1999), which leads to impaired endosomal dynamics (Fig. 5C, D).

Materials

- Glass coverslips with 12 mm diameter (Assistent, Sondheim, Germany)
- BHK cell culture medium: Glasgow Minimum Essential Medium (Sigma-Aldrich, St. Louis, MO)
- Mouse anti-LBPA antibody (6C4) (Echelon Bioscience Inc., Salt Lake City, UT) (Kobayashi *et al.*, 1998)

Procedure

1. Plate BHK cells at the desired density directly onto 12 mm diameter coverslips 36 h before the experiment.
2. Prepare 50 μg/mL mouse anti-LBPA antibody in BHK cell culture medium.

3. Give anti-LBPA antibody onto cells and incubate for 0, 4, 8, and 24 h at 37 °C.
4. Perform immunofluorescence as described. But use only secondary antibody for LBPA staining.

Alix Knockdown by RNAi

Knockdown of the LBPA binding protein Alix, results in a reduction of the endosomal lipid LBPA (Matsuo *et al.*, 2004), which is accompanied by a decrease in cholesterol (Chevallier *et al.*, 2008) (Fig. 5E, F).

Materials

- Glass coverslips with 12 mm diameter (Assistent, Sondheim, Germany)
- BHK cell culture medium: Glasgow Minimum Essential Medium (Sigma-Aldrich, St. Louis, MO)
- Oligofectamine (Invitrogen Corporation, Carlsbad, CA)
- OptiMEM (Gibco (Invitrogen Corporation), Carlsbad, CA)
- Alix siRNA (20 μM) (Matsuo *et al.*, 2004) (Qiagen, Germantown, MD)

Procedure

1. Plate BHK cells at the desired density directly onto 12 mm diameter coverslips (immunofluorescence) or in 10 cm dishes (ELISA) 96 h before the experiment.
2. Transfect the cells 24 h after plating with siRNA against Alix using Oligofectamine, OptiMEM, and BHK medium without antibiotics.

Per 10 cm dish:
1) 80 μL siRNA (stock: 20 μM) + 1330 μL OptiMEM
2) 80 μL Oligofectamine + 320 μL OptiMEM
→ Incubate for 7 min, mix tube 1 in tube 2, incubate for 20 min, add 850 μL OptiMEM per tube and add onto cells.

3. Incubate 72 h at 37 °C.
4. Perform immunofluorescence and fractionation followed by ELISA as described.

Results

Treatment of BHK cells with U18666A (Fig. 5A, B) or endocytosis of high doses of anti-LBPA antibody (Fig. 5C, D) leads to cholesterol and LBPA accumulation in endosomes. In contrast, knockdown of Alix, an effector of LBPA, results in decreased LBPA and cellular cholesterol levels (Fig. 5E, F).

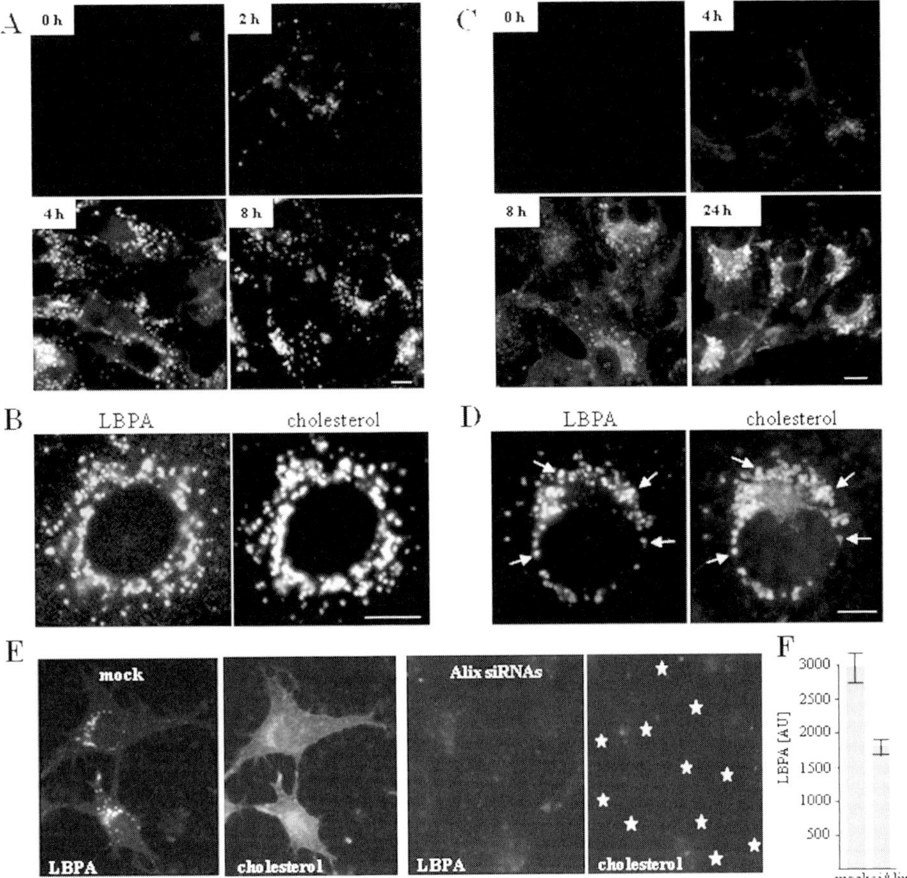

Fig. 5 *In vivo* modification of endosomal lipid levels. (A) BHK cells were incubated with U18666A for indicated times and then fixed and stained for filipin. Fig. 3 A–D in (Kobayashi *et al.*, 1999); reproduced with permission from Nature publishing group. (B) As (A) but fixation after 8 h of U18666A treatment and double labeling with filipin and anti-LBPA antibody. Fig. 3 G, H in (Kobayashi *et al.*, 1999); reproduced with permission from Nature publishing group. (C) BHK cells were incubated with anti-LBPA antibody for indicated times and then fixed and stained for filipin. Fig. 6A in (Kobayashi *et al.*, 1999); reproduced with permission from Nature publishing group. (D) As (C) but double labeled with filipin and anti-LBPA antibody. Fig. 6A in (Kobayashi *et al.*, 1999); reproduced with permission from Nature publishing group. (E) BHK cells were treated for 72 h with siRNA against Alix or mock siRNA, fixed and stained with filipin and anti-LBPA antibody. Stars indicate the position of cell nuclei. Fig. 1A in (Chevallier *et al.*, 2008); reproduced with permission from the American Society for Biochemistry and Molecular Biology. (F) LBPA levels were analyzed by ELISA using anti-LBPA antibody. Fig. 1B in (Chevallier *et al.*, 2008); reproduced with permission from the American Society for Biochemistry and Molecular Biology. Bar: 10 μm. (See color plate.)

III. Conclusion

In past years techniques, like *in vitro* experiments using synthetic lipids, fluorescent lipid analogs, lipid-binding domains, genetic approaches and MS elucidated the importance of lipids for intracellular trafficking. Although the involvement of endosomal lipids, like phosphatidylinositols, cholesterol, and LBPA in receptor downregulation and intralumenal vesicles formation is evident (Gruenberg and Stenmark, 2004), underlying molecular mechanisms remain poorly understood, but with the rapid development of sophisticated tools, mechanisms and functions will hopefully be unraveled within in the next few years. Here we described some techniques that allow studying the role of lipids in endocytosis and can be performed in any cell biology laboratory.

Acknowledgment

Support was from the Swiss National Science Foundation, PRISM from the EU Sixth Framework Program, NCCR in Chemical Biology and LipidX from Swiss SystemsX.ch initiative, evaluated by the Swiss National Science Foundation (J.G).

Appendix A. Supplementary Movies

Supplementary data associated with this chapter can be found, in the online version, at doi:10.1016/B978-0-12-386487-1.00002-X.

References

Attar, N., and Cullen, P. J. (2010). The retromer complex. *Adv. Enzyme Regul.* **50**, 216–236.

Bache, K. G., Brech, A., Mehlum, A., and Stenmark, H. (2003). Hrs regulates multivesicular body formation via ESCRT recruitment to endosomes. *J. Cell Biol.* **162**, 435–442.

Bowers, K., Piper, S. C., Edeling, M. A., Gray, S. R., Owen, D. J., Lehner, P. J., and Luzio, J. P. (2006). Degradation of endocytosed epidermal growth factor and virally ubiquitinated major histocompatibility complex class I is independent of mammalian ESCRTII. *J. Biol. Chem.* **281**, 5094–5105.

Bravo, J., Karathanassis, D., Pacold, C. M., Pacold, M. E., Ellson, C. D., Anderson, K. E., Butler, P. J., Lavenir, I., Perisic, O., and Hawkins, P. T., *et al.* (2001). The crystal structure of the PX domain from p40 (phox) bound to phosphatidylinositol 3-phosphate. *Mol. Cell* **8**, 829–839.

Cabezas, A., Pattni, K., and Stenmark, H. (2006). Cloning and subcellular localization of a human phosphatidylinositol 3-phosphate 5-kinase, PIKfyve/Fab1. *Gene* **371**, 34–41.

Carlton, J. G., Bujny, M. V., Peter, B. J., Oorschot, V. M., Rutherford, A., Arkell, R. S., Klumperman, J., McMahon, H. T., and Cullen, P. J. (2005). Sorting nexin-2 is associated with tubular elements of the early endosome, but is not essential for retromer-mediated endosome-to-TGN transport. *J. Cell Sci.* **118**, 4527–4539.

Cheever, M. L., Sato, T. K., de Beer, T., Kutateladze, T. G., Emr, S. D., and Overduin, M. (2001). Phox domain interaction with PtdIns(3)P targets the Vam7 t-SNARE to vacuole membranes. *Nat. Cell Biol.* **3**, 613–618.

Chevallier, J., Chamoun, Z., Jiang, G., Prestwich, G., Sakai, N., Matile, S., Parton, R. G., and Gruenberg, J. (2008). Lysobisphosphatidic acid controls endosomal cholesterol levels. *J. Biol. Chem.* **283**, 27871–27880.

Cozier, G. E., Carlton, J., McGregor, A. H., Gleeson, P. A., Teasdale, R. D., Mellor, H., and Cullen, P. J. (2002). The phox homology (PX) domain-dependent, 3-phosphoinositide-mediated association of sorting nexin-1 with an early sorting endosomal compartment is required for its ability to regulate epidermal growth factor receptor degradation. *J. Biol. Chem.* **277**, 48730–48736.

Dancea, F., Kami, K., and Overduin, M. (2008). Lipid interaction networks of peripheral membrane proteins revealed by data-driven micelle docking. *Biophys. J.* **94**, 515–524.

Dong, X. P., Shen, D., Wang, X., Dawson, T., Li, X., Zhang, Q., Cheng, X., Zhang, Y., Weisman, L. S., and Delling, M., *et al.* (2010). PI(3,5)P(2) controls membrane trafficking by direct activation of mucolipin Ca(2+) release channels in the endolysosome. *Nat. Commun.* **1**, 38.

Dowler, S., Kular, G., and Alessi, D. R. (2002). Protein lipid overlay assay. *Sci. STKE* **2002**, pl6.

Ellson, C. D., Gobert-Gosse, S., Anderson, K. E., Davidson, K., Erdjument-Bromage, H., Tempst, P., Thuring, J. W., Cooper, M. A., Lim, Z. Y., and Holmes, A. B., *et al.* (2001). PtdIns(3)P regulates the neutrophil oxidase complex by binding to the PX domain of p40(phox). *Nat. Cell Biol.* **3**, 679–682.

Ewers, H., Romer, W., Smith, A. E., Bacia, K., Dmitrieff, S., Chai, W., Mancini, R., Kartenbeck, J., Chambon, V., and Berland, L., *et al.* (2010). GM1 structure determines SV40-induced membrane invagination and infection. *Nat. Cell Biol.* **12**, 11–18; sup pp 11–12.

Falguieres, T., Luyet, P. P., Bissig, C., Scott, C. C., Velluz, M. C., and Gruenberg, J. (2008a). In vitro budding of intralumenal vesicles into late endosomes is regulated by Alix and Tsg101. *Mol. Biol. Cell* **19**, 4942–4955.

Falguieres, T., Luyet, P. P., and Gruenberg, J. (2008b). Molecular assemblies and membrane domains in multivesicular endosome dynamics. *Exp. Cell Res.* **24**, 24.

Feng, L. (2005). Probing lipid-protein interactions using lipid microarrays. *Prostaglandins Other Lipid Mediat.* **77**, 158–167.

Ferguson, C. J., Lenk, G. M., and Meisler, M. H. (2010). PtdIns(3,5)P2 and autophagy in mouse models of neurodegeneration. *Autophagy* **6**, 170–171.

Fernandez-Borja, M., Wubbolts, R., Calafat, J., Janssen, H., Divecha, N., Dusseljee, S., and Neefjes, J. (1999). Multivesicular body morphogenesis requires phosphatidyl-inositol 3-kinase activity. *Curr. Biol.* **9**, 55–58.

Gagescu, R., Demaurex, N., Parton, R. G., Hunziker, W., Huber, L. A., and Gruenberg, J. (2000). The recycling endosome of Madin-Darby canine kidney cells is a mildly acidic compartment rich in raft components. *Mol. Biol. Cell* **11**, 2775–2791.

Gerke, V., and Weber, K. (1984). Identity of p36K phosphorylated upon Rous sarcoma virus transformation with a protein purified from brush borders; calcium-dependent binding to non-erythroid spectrin and F-actin. *EMBO J.* **3**, 227–233.

Gibbings, D. J., Ciaudo, C., Erhardt, M., and Voinnet, O. (2009). Multivesicular bodies associate with components of miRNA effector complexes and modulate miRNA activity. *Nat. Cell Biol.* **11**, 1143–1149.

Gillooly, D. J., Morrow, I. C., Lindsay, M., Gould, R., Bryant, N. J., Gaullier, J. M., Parton, R. G., and Stenmark, H. (2000). Localization of phosphatidylinositol 3-phosphate in yeast and mammalian cells. *EMBO J.* **19**, 4577–4588.

Gruenberg, J. (2001). The endocytic pathway: a mosaic of domains. *Nat. Rev. Mol. Cell Biol.* **2**, 721–730.

Gruenberg, J. (2003). Lipids in endocytic membrane transport and sorting. *Curr. Opin. Cell Biol.* **15**, 382–388.

Gruenberg, J., and Stenmark, H. (2004). The biogenesis of multivesicular endosomes. *Nat. Rev. Mol. Cell Biol.* **5**, 317–323.

Guan, X. L., Souza, C. M., Pichler, H., Dewhurst, G., Schaad, O., Kajiwara, K., Wakabayashi, H., Ivanova, T., Castillon, G. A., and Piccolis, M., *et al.* (2009). Functional interactions between sphingolipids and sterols in biological membranes regulating cell physiology. *Mol. Biol. Cell* **20**, 2083–2095.

Halet, G. (2005). Imaging phosphoinositide dynamics using GFP-tagged protein domains. *Biol. Cell* **97**, 501–518.

Harder, T., Kellner, R., Parton, R. G., and Gruenberg, J. (1997). Specific release of membrane-bound annexin II and cortical cytoskeletal elements by sequestration of membrane cholesterol. *Mol. Biol. Cell* **8**, 533–545.

Harterink, M., Port, F., Lorenowicz, M. J., McGough, I. J., Silhankova, M., Betist, M. C., van Weering, J. R., van Heesbeen, R. G., Middelkoop, T. C., and Basler, K., *et al.* (2011). A SNX3-dependent retromer pathway mediates retrograde transport of the Wnt sorting receptor Wntless and is required for Wnt secretion. *Nat. Cell Biol.* **13**, 914–923.

Hayer, A., Stoeber, M., Ritz, D., Engel, S., Meyer, H. H., and Helenius, A. (2010). Caveolin-1 is ubiquitinated and targeted to intralumenal vesicles in endolysosomes for degradation. *J. Cell Biol.* **191**, 615–629.

Hayes, M. J., Rescher, U., Gerke, V., and Moss, S. E. (2004). Annexin-actin interactions. *Traffic* **5**, 571–576.

Hurley, J. H., Boura, E., Carlson, L. A., and Rozycki, B. (2010). Membrane budding. *Cell* **143**, 875–887.

Hurley, J. H., and Hanson, P. I. (2010). Membrane budding and scission by the ESCRT machinery: it's all in the neck. *Nat. Rev. Mol. Cell Biol.* **11**, 556–566.

Ikonomov, O. C., Sbrissa, D., and Shisheva, A. (2006). Localized PtdIns 3,5-P2 synthesis to regulate early endosome dynamics and fusion. *Am. J. Physiol. Cell Physiol.* **291**, C393–C404.

Jost, M., Simpson, F., Kavran, J. M., Lemmon, M. A., and Schmid, S. L. (1998). Phosphatidylinositol-4,5-bisphosphate is required for endocytic coated vesicle formation. *Curr. Biol.* **8**, 1399–1402.

Kaiser, H. J., Lingwood, D., Levental, I., Sampaio, J. L., Kalvodova, L., Rajendran, L., and Simons, K. (2009). Order of lipid phases in model and plasma membranes. *Proc. Natl. Acad. Sci. U. S. A.* **106**, 16645–16650.

Kanai, F., Liu, H., Field, S. J., Akbary, H., Matsuo, T., Brown, G. E., Cantley, L. C., and Yaffe, M. B. (2001). The PX domains of p47phox and p40phox bind to lipid products of PI(3)K. *Nat. Cell Biol.* **3**, 675–678.

Kanter, J. L., Narayana, S., Ho, P. P., Catz, I., Warren, K. G., Sobel, R. A., Steinman, L., and Robinson, W. H. (2006). Lipid microarrays identify key mediators of autoimmune brain inflammation. *Nat. Med.* **12**, 138–143.

Katzmann, D. J., Odorizzi, G., and Emr, S. D. (2002). Receptor downregulation and multivesicular-body sorting. *Nat. Rev. Mol. Cell Biol.* **3**, 893–905.

Kobayashi, T., Beuchat, M. H., Lindsay, M., Frias, S., Palmiter, R. D., Sakuraba, H., Parton, R. G., and Gruenberg, J. (1999). Late endosomal membranes rich in lysobisphosphatidic acid regulate cholesterol transport. *Nat. Cell Biol.* **1**, 113–118.

Kobayashi, T., Stang, E., Fang, K. S., de Moerloose, P., Parton, R. G., and Gruenberg, J. (1998). A lipid associated with the antiphospholipid syndrome regulates endosome structure and function. *Nature* **392**, 193–197.

Kuerschner, L., Ejsing, C. S., Ekroos, K., Shevchenko, A., Anderson, K. I., and Thiele, C. (2005). Polyene-lipids: a new tool to image lipids. *Nat. Methods* **2**, 39–45.

Le Blanc, I., Luyet, P. P., Pons, V., Ferguson, C., Emans, N., Petiot, A., Mayran, N., Demaurex, N., Faure, J., and Sadoul, R., *et al.* (2005). Endosome-to-cytosol transport of viral nucleocapsids. *Nat. Cell Biol.* **7**, 653–664.

Lemmon, M. A. (2008). Membrane recognition by phospholipid-binding domains. *Nat. Rev. Mol. Cell Biol.* **9**, 99–111.

Leventis, P. A., and Grinstein, S. (2010). The distribution and function of phosphatidylserine in cellular membranes. *Annu. Rev. Biophys.* **39**, 407–427.

Liscum, L., and Faust, J. R. (1989). The intracellular transport of low density lipoprotein-derived cholesterol is inhibited in Chinese hamster ovary cells cultured with 3-beta-[2-(diethylamino)ethoxy]androst-5-en-17-one. *J. Biol. Chem.* **264**, 11796–11806.

Lloyd, T. E., Atkinson, R., Wu, M. N., Zhou, Y., Pennetta, G., and Bellen, H. J. (2002). Hrs regulates endosome membrane invagination and tyrosine kinase receptor signaling in Drosophila. *Cell* **108**, 261–269.

Luyet, P. P., Falguieres, T., Pons, V., Pattnaik, A. K., and Gruenberg, J. (2008). The ESCRT-I subunit TSG101 controls endosome-to-cytosol release of viral RNA. *Traffic* **9**, 2279–2290.

Marks, D. L., Bittman, R., and Pagano, R. E. (2008). Use of Bodipy-labeled sphingolipid and cholesterol analogs to examine membrane microdomains in cells. *Histochem. Cell Biol.* **130**, 819–832.

Matsuo, H., Chevallier, J., Mayran, N., Le Blanc, I., Ferguson, C., Faure, J., Blanc, N. S., Matile, S., Dubochet, J., and Sadoul, R., *et al.* (2004). Role of LBPA and Alix in multivesicular liposome formation and endosome organization. *Science* **303**, 531–534.

Matuoka, K., Fukami, K., Nakanishi, O., Kawai, S., and Takenawa, T. (1988). Mitogenesis in response to PDGF and bombesin abolished by microinjection of antibody to PIP2. *Science* **239**, 640–643.

Mayor, S., and Pagano, R. E. (2007). Pathways of clathrin-independent endocytosis. *Nat. Rev. Mol. Cell Biol.* **8**, 603–612.

Mayran, N., Parton, R. G., and Gruenberg, J. (2003). Annexin II regulates multivesicular endosome biogenesis in the degradation pathway of animal cells. *EMBO J.* **22**, 3242–3253.

Mobius, W., van Donselaar, E., Ohno-Iwashita, Y., Shimada, Y., Heijnen, H. F., Slot, J. W., and Geuze, H. J. (2003). Recycling compartments and the internal vesicles of multivesicular bodies harbor most of the cholesterol found in the endocytic pathway. *Traffic* **4**, 222–231.

Morel, E., Parton, R. G., and Gruenberg, J. (2009). Annexin A2-dependent polymerization of actin mediates endosome biogenesis. *Dev. Cell* **16**, 445–457.

Nielsen, E., Christoforidis, S., Uttenweiler-Joseph, S., Miaczynska, M., Dewitte, F., Wilm, M., Hoflack, B., and Zerial, M. (2000). Rabenosyn-5, a novel Rab5 effector, is complexed with hVPS45 and recruited to endosomes through a FYVE finger domain. *J. Cell Biol.* **151**, 601–612.

Odorizzi, G., Babst, M., and Emr, S. D. (1998). Fab1p PtdIns(3)P 5-kinase function essential for protein sorting in the multivesicular body. *Cell* **95**, 847–858.

Ohno-Iwashita, Y., Shimada, Y., Waheed, A. A., Hayashi, M., Inomata, M., Nakamura, M., Maruya, M., and Iwashita, S. (2004). Perfringolysin O, a cholesterol-binding cytolysin, as a probe for lipid rafts. *Anaerobe* **10**, 125–134.

Overduin, M., Cheever, M. L., and Kutateladze, T. G. (2001). Signaling with phosphoinositides: better than binary. *Mol. Interv.* **1**, 150–159.

Pagano, R. E., and Chen, C. S. (1998). Use of BODIPY-labeled sphingolipids to study membrane traffic along the endocytic pathway. *Ann. N. Y. Acad. Sci.* **845**, 152–160.

Pagano, R. E., Martin, O. C., Kang, H. C., and Haugland, R. P. (1991). A novel fluorescent ceramide analogue for studying membrane traffic in animal cells: accumulation at the Golgi apparatus results in altered spectral properties of the sphingolipid precursor. *J. Cell Biol.* **113**, 1267–1279.

Pagano, R. E., Puri, V., Dominguez, M., and Marks, D. L. (2000). Membrane traffic in sphingolipid storage diseases. *Traffic* **1**, 807–815.

Petiot, A., Faure, J., Stenmark, H., and Gruenberg, J. (2003). PI3P signaling regulates receptor sorting but not transport in the endosomal pathway. *J. Cell Biol.* **162**, 971–979.

Pichler, H., and Riezman, H. (2004). Where sterols are required for endocytosis. *Biochim. Biophys. Acta* **1666**, 51–61.

Pons, V., Luyet, P. P., Morel, E., Abrami, L., van der Goot, F. G., Parton, R. G., and Gruenberg, J. (2008). Hrs and SNX3 functions in sorting and membrane invagination within multivesicular bodies. *PLoS Biol.* **6**, e214.

Razi, M., and Futter, C. E. (2006). Distinct roles for Tsg101 and Hrs in multivesicular body formation and inward vesiculation. *Mol. Biol. Cell* **17**, 3469–3483.

Rojo, M., Emery, G., Marjomaki, V., McDowall, A. W., Parton, R. G., and Gruenberg, J. (2000). The transmembrane protein p23 contributes to the organization of the Golgi apparatus. *J. Cell Sci.* **113**(Pt 6), 1043–1057.

Romer, W., Berland, L., Chambon, V., Gaus, K., Windschiegl, B., Tenza, D., Aly, M. R., Fraisier, V., Florent, J. C., and Perrais, D., *et al.* (2007). Shiga toxin induces tubular membrane invaginations for its uptake into cells. *Nature* **450**, 670–675.

Rutherford, A. C., Traer, C., Wassmer, T., Pattni, K., Bujny, M. V., Carlton, J. G., Stenmark, H., and Cullen, P. J. (2006). The mammalian phosphatidylinositol 3-phosphate 5-kinase (PIKfyve) regulates endosome-to-TGN retrograde transport. *J. Cell Sci.* **119**, 3944–3957.

Sbrissa, D., Ikonomov, O. C., and Shisheva, A. (2002). Phosphatidylinositol 3-phosphate-interacting domains in PIKfyve. Binding specificity and role in PIKfyve. Endomenbrane localization. *J. Biol. Chem.* **277**, 6073–6079.

Schmidt, M. H., Hoeller, D., Yu, J., Furnari, F. B., Cavenee, W. K., Dikic, I., and Bogler, O. (2004). Alix/AIP1 antagonizes epidermal growth factor receptor downregulation by the Cbl-SETA/CIN85 complex. *Mol. Cell Biol.* **24**, 8981–8993.

Schu, P. V., Takegawa, K., Fry, M. J., Stack, J. H., Waterfield, M. D., and Emr, S. D. (1993). Phosphatidylinositol 3-kinase encoded by yeast VPS34 gene essential for protein sorting. *Science* **260**, 88–91.

Shin, H. W., Hayashi, M., Christoforidis, S., Lacas-Gervais, S., Hoepfner, S., Wenk, M. R., Modregger, J., Uttenweiler-Joseph, S., Wilm, M., and Nystuen, A., *et al.* (2005). An enzymatic cascade of Rab5 effectors regulates phosphoinositide turnover in the endocytic pathway. *J. Cell Biol.* **170**, 607–618.

Shisheva, A. (2008). PIKfyve: Partners, significance, debates and paradoxes. *Cell Biol. Int.* **32**, 591–604.

Shogomori, H., and Kobayashi, T. (2008). Lysenin: a sphingomyelin specific pore-forming toxin. *Biochim. Biophys. Acta* **1780**, 612–618.

Simonsen, A., Lippe, R., Christoforidis, S., Gaullier, J. M., Brech, A., Callaghan, J., Toh, B. H., Murphy, C., Zerial, M., and Stenmark, H. (1998). EEA1 links PI(3)K function to Rab5 regulation of endosome fusion. *Nature* **394**, 494–498.

Singh, R. D., Marks, D. L., and Pagano, R. E. (2007). Using fluorescent sphingolipid analogs to study intracellular lipid trafficking. *Curr. Protoc. Cell Biol.* **Chapter 24,** Unit 24 21.

Stenmark, H., Aasland, R., Toh, B. H., and D'Arrigo, A. (1996). Endosomal localization of the autoantigen EEA1 is mediated by a zinc-binding FYVE finger. *J. Biol. Chem.* **271**, 24048–24054.

Subramanian, D., Laketa, V., Muller, R., Tischer, C., Zarbakhsh, S., Pepperkok, R., and Schultz, C. (2010). Activation of membrane-permeant caged PtdIns(3)P induces endosomal fusion in cells. *Nat. Chem. Biol.* **6**, 324–326.

Touchberry, C. D., Bales, I. K., Stone, J. K., Rohrberg, T. J., Parelkar, N. K., Nguyen, T., Fuentes, O., Liu, X., Qu, C. K., and Andresen, J. J., *et al.* (2010). Phosphatidylinositol 3,5-bisphosphate (PI(3,5)P2) potentiates cardiac contractility via activation of the ryanodine receptor. *J. Biol. Chem.* **285**, 40312–40321.

Trajkovic, K., Hsu, C., Chiantia, S., Rajendran, L., Wenzel, D., Wieland, F., Schwille, P., Brugger, B., and Simons, M. (2008). Ceramide triggers budding of exosome vesicles into multivesicular endosomes. *Science* **319**, 1244–1247.

van der Goot, F. G., and Gruenberg, J. (2006). Intra-endosomal membrane traffic. *Trends Cell Biol.* **16**, 514–521.

van Weering, J. R., Verkade, P., and Cullen, P. J. (2010). SNX-BAR proteins in phosphoinositide-mediated, tubular-based endosomal sorting. *Semin. Cell Dev. Biol.* **21**, 371–380.

Varnai, P., and Balla, T. (2006). Live cell imaging of phosphoinositide dynamics with fluorescent protein domains. *Biochim. Biophys. Acta* **1761**, 957–967.

Wenk, M. R. (2010). Lipidomics: new tools and applications. *Cell* **143**, 888–895.

Wollert, T., and Hurley, J. H. (2010). Molecular mechanism of multivesicular body biogenesis by ESCRT complexes. *Nature* **464**, 864–869.

Wollert, T., Wunder, C., Lippincott-Schwartz, J., and Hurley, J. H. (2009). Membrane scission by the ESCRT-III complex. *Nature* **458**, 172–177.

Worby, C. A., and Dixon, J. E. (2002). Sorting out the cellular functions of sorting nexins. *Nat. Rev. Mol. Cell Biol.* **3**, 919–931.

Xu, Y., Hortsman, H., Seet, L., Wong, S. H., and Hong, W. (2001). SNX3 regulates endosomal function through its PX-domain-mediated interaction with PtdIns(3)P. *Nat. Cell Biol.* **3**, 658–666.

Yang, Z., and Klionsky, D. J. (2010). Eaten alive: a history of macroautophagy. *Nat. Cell Biol.* **12**, 814–822.

Zerial, M., and McBride, H. (2001). Rab proteins as membrane organizers. *Nat. Rev. Mol. Cell Biol.* **2**, 107–117.

Zhan, Y., Virbasius, J. V., Song, X., Pomerleau, D. P., and Zhou, G. W. (2002). The p40phox and p47phox PX domains of NADPH oxidase target cell membranes via direct and indirect recruitment by phosphoinositides. *J. Biol. Chem.* **277**, 4512–4518.

Zhu, H., Bilgin, M., Bangham, R., Hall, D., Casamayor, A., Bertone, P., Lan, N., Jansen, R., Bidlingmaier, S., and Houfek, T., *et al.* (2001). Global analysis of protein activities using proteome chips. *Science* **293**, 2101–2105.

CHAPTER 3

Studying *In Vitro* Membrane Curvature Recognition by Proteins and its Role in Vesicular Trafficking

Jean-Baptiste Manneville[*], **Cécile Leduc**[†], **Benoit Sorre**[‡] **and Guillaume Drin**[§]

[*]Unité Mixte de Recherche 144, CNRS and Institut Curie, 26 rue d'Ulm, 75248 Paris Cedex 05, France

[†]Laboratoire Photonique, Numérique et Nanosciences (LP2N), Institut d'Optique Graduate School, Université de Bordeaux and CNRS, 33400 Talence Cedex, France

[‡]Laboratory of Theoretical Condensed Matter Physics, The Rockefeller University, New York, NY 10065

[§]Institut de Pharmacologie Moléculaire et Cellulaire, Université de Nice Sophia-Antipolis and CNRS, 660 route des lucioles, 06560 Valbonne, France

---

Copyright 2012, Elsevier Inc. All rights reserved.

0091-679X/10 $35.00
DOI 10.1016/B978-0-12-386487-1.00003-1

Abstract

In recent years, the interest for proteins that exert key functions in vesicular trafficking through their ability to sense or induce positive membrane curvature has expanded. In this chapter, we first present simple protocols to determine whether a protein targets positively curved membranes with liposomes of well-defined size. Next we describe more sophisticated approaches based on the controlled deformation of giant liposomes. These approaches allow visualization and quantification of protein binding to membrane regions of high curvature by real-time fluorescence microscopy. Last we describe several functional assays to measure how membrane curvature controls the activation state of Arf1 *via* ArfGAP1 or the asymmetric tethering between flat and curved membranes *via* the golgin GMAP-210.

I. Introduction

In the cell, molecular coats shape flat lipid membrane patches into vesicles filled with cargo proteins. For example the COPI coat generates vesicles that convey proteins from the Golgi apparatus to the endoplasmic reticulum. First, GTP substitutes for GDP in the small cytosolic G protein Arfl. Arfl-GTP binds tightly to the Golgi membrane and recruits coatomers that collect cargo and self-assemble into a spherical coat. This coat forces the membrane to bud into a vesicle of ~40 nm in radius. Once the COPI vesicle is pinched off the Golgi membrane, the coat must be disassembled. *In vitro,* ArfGAPl, a GTPase Activating Protein for Arfl, triggers coat disassembly by hydrolyzing GTP in Arfl. This disassembly occurs slowly on large liposomes (radius $R = 150$ nm) but 100-times faster on smaller liposomes close in size to COPI vesicles. The positive curvature acquired by a vesicle was proposed to program coat disassembly through ArfGAPl (Bigay *et al.*, 2003). Next we described how ArfGAPl recognizes curvature *via* two ALPS motifs (Amphipathic Lipid Packing Sensor) of 35 amino acids ALPS1 and ALPS2 (Ambroggio *et al.*, 2010; Bigay *et al.*, 2005). The ALPS1 motif is individually a powerful curvature sensor. More recently we used lipid nanotubes pulled from giant unilamellar vesicles (GUVs) by kinesin motors or with optical tweezers at controlled membrane tension. Interestingly, a tube (of adjustable curvature) connected to a GUV (a flat membrane) has a geometry comparable to that of a budding vesicle (Pinot *et al.*, 2010). Arfl-GTP binds independently of curvature. In contrast, ArfGAPl or its ALPS1-ALPS2 motif binds only to tubes above a threshold curvature of $1/35\ \text{nm}^{-1}$. We showed next that the flat GUV membrane acts as a reservoir of Arfl-GTP (protected from ArfGAPl) that diffuses in the tube from where it is detached by ArfGAPl. The competition between these two processes creates a concentration gradient of membrane-bound Arfl-GTP along the tube (high on the base, lower at the tip). This suggests that coat disassembly by ArfGAPl can only be completed after vesicle release when Arfl-GTP cannot populate the vesicle by diffusion (Ambroggio *et al.,* 2010). In distinct studies, we identified other curvature-sensing proteins (Drin *et al.,* 2007). One of them, GMAP-210 is a molecular rope with a N-terminal ALPS-like motif and a C-terminal GRAB domain that recognizes Arfl-GTP. With these two ends GMAP-210 likely bridges curved transport vesicles to flat Golgi-cisternae (Drin *et al.,* 2008). Here we describe our *in vitro* assays to study the curvature sensitivity of proteins. We also present functional assays to measure how this sensitivity governs ArfGAPl or GMAP-210 activity.

A. Rationale

Intracellular trafficking events imply rapid and transient membrane deformations governed by complex machineries. A membrane remodeling activity is often assigned to a protein if its overexpression in cells induces a clear phenotype such as membrane tubulation. In contrast, no straightforward cell biology approach exists to identify a curvature-sensing protein. *In vitro* assays with purified proteins and artificial membranes whose curvature can be defined and controlled are thus instrumental to study this particular category of proteins.

II. Preparation of proteins and liposomes

A. Purification and Labeling of Proteins

1. Purification of ArfGAPl

The full-length ArfGAPl (from rat, residue 1–415) or the shorter construct ArfGAP[l-257] are our reference curvature sensors. They integrate a N-terminal catalytic GAP domain coupled to two ALPS motifs or only the ALPS1 motif, respectively. N-terminal His-tagged constructs are cloned in the pET16b vector (Novagen), expressed in *E.coli,* purified from inclusion bodies on Ni^{2+}-NTA column (Qiagen) under denaturing conditions, refolded and purified on an ion exchange column (Bigay *et al.*, 2005).The nucleotide sequence of full-length ArfGAPl integrates silent mutations to be expressed in bacteria.

2. Purification and Fluorescent Labeling of ALPS Motifs

We purify ArfGAPl fragments encompassing only the ALPS1 motif [192–257] or the ALPS1-ALPS2 motif [192–304]. These fragments in fusion with GST are cloned in pGEX-2T vectors (Pharmacia) and include a thrombin cleavage site to remove

GST. For NBD-based binding assays, they are labeled with an environment-sensitive probe (NBD) on a cysteine residue introduced in the face of the ALPS1 motif that inserts into membrane (A236C). For microscopy assays, we purified a mutant (K297C) of the ALPS1-ALPS2 construct labeled with a bright, photostable fluorophore (Alexa488). It should be noted that the ALPS motif adopts a helical structure only upon binding to membranes. This absence of secondary structure in solution allows purification methods usually dedicated to synthetic peptides. Briefly, the constructs are expressed for 3–4 h at 37 °C in *E.coli* that are next resuspended in TN buffer (50 mM Tris pH 7.4 and 150 mM NaCl), supplemented with protease inhibitors (1 mM PMSF, 1 mM pepstatin, 10 mM bestatin, 10 mM phosphoramidon) and 2 mM DTT. Bacteria are lysed with a French press. The lysate is ultracentrifuged at 160,000 g for 60 min. The supernatant is incubated for 3 h with glutathione-sepharose 4B beads (Amersham) that are then washed three times with TN buffer containing 2 mM DTT. The constructs are then separated from GST by thrombin (25 U, GE Healthcare) overnight at 4 °C. It is crucial to use for all these steps fresh, degassed buffers to protect cysteine from oxydation. The eluate is then purified by HPLC on a Chromolith Performance RP-18e 100–4.6 mm column (Merck) with a 0–80% linear acetonitrile (ACN) gradient (buffer A: ddH_20 with 0.1% trifluoroacetic acid (TFA)/buffer B: ACN/ddH_20 (80/20) with 0.08% TFA). The collected fractions identified by SDS–PAGE as corresponding to the peptide are pooled, lyophilized and stored at −20 °C. Peptide identity is checked by MALDITOF mass spectrometry. The peptide is resuspended in HK (50 mM Hepes pH 7.2, 120 mM K-Acetate) buffer (without DTT) at 40–200 μM and mixed with a 10-fold molar excess of IANBD amide (Invitrogen) or of Alexa488-C5-maleimide (Invitrogen) and incubated at room temperature for 1–2 h in the dark. The stock concentration of the thiol-reactive probe must be 5–6 mM in DMF to limit the final amount of DMF in the reaction mix (no more than 5% v/v). The reaction is stopped by adding an excess of L-cysteine (10-fold more than the probe). The free probe reacts with L-cysteine and becomes highly soluble, poorly retained on a C18 column and easily separated from peptide by HPLC. The labeled peptide is lyophilized, stored at −20 °C in the dark and its identity is confirmed by mass spectrometry.

For NBD-based binding assays, one can also express a His-tagged ArfGAP1 [137–237] (A236C) fragment (but other constructs can be envisaged) in *E.coli*. Bacteria are next lyzed in 6 M guanidine and the protein is purified on a Ni^{2+}-NTA column. The eluted fractions are dialyzed three times against 25 mM Tris pH 7.5, 400 mM NaCl containing 10% glycerol and 2 mM DTT. The peptide is then purified by HPLC, labeled and repurified as described above.

3. Purification and Labeling of Arf1

Myristoylated Arf1-GDP is purified from *E. coli* coexpressing N-Myristoyl Transferase and bovine Arf1 as described previously (Franco *et al.*, 1995). A construct carrying an extra cysteine (C182) is labeled with OregonGreen (Manneville *et al.*, 2008).

4. Purification of mGMAP

GMAP-210 is difficult to express in *E. coli* and poorly soluble. We use a shorter construct, mGMAP, that consists of amino acids 1–375 and 1597–1843 of GMAP-210 separated by a short linker (PGSTRAAAS), cloned in pGEX-2T vector in fusion with GST. mGMAP has the two functional ends of GMAP-210 (ALPS motif and GRAB domain). It is expressed overnight at 17 °C in *E.coli.* Purification on GST beads is done as described above. After cleavage, the eluates are pooled, concentrated and purified by gel filtration chromatography (Sephacryl S-300) to eliminate aggregates.

5. Kinesin Molecular Motors

Biotinylated and truncated kinesin-1 (KinBio401 from *D. melanogaster*) motors are purified as described previously (Surrey *et al.,* 1998) or from *E. coli* expressing the kinesin-BCCP-H6 plasmid (Addgene, Plasmid 15960: pWC2) using a standard protocol for His-tagged proteins. Motors remain active for a few months when stored in liquid nitrogen.

B. Liposome Preparation

1. Preparation of Unilamellar Liposomes of Defined Diameter

Natural and biotinylated lipids are from Avanti Polar Lipids (http://avantilipids.com). Fluorescent lipids (NBD-PE, BODIPY® TR-ceramide) are from Invitrogen. Lipid stock solutions (usually in chloroform) are aliquoted in 2 mL glass vials filled with argon and tightly sealed with a Teflon cap. Vials are stored at −20 °C.

To make liposomes of defined composition, lipids taken from stock solutions with glass syringes (Hamilton) are mixed at the desired molar ratio in a pear-shaped glass flask (25 mL, 14/23 Duran). The mixture adjusted to 1 mL with chloroform contains 1–10 μmoles of lipid. Next the solvent is removed in a rotary evaporator at 20–35 °C at 500 rpm for at least 30 min. A lipid film appears on the glass surface. The evaporator and the flask are filled with argon and the flask is removed and placed in a vacuum chamber (for 45 min) to remove solvent traces. Then the film is hydrated with 1–4 mL HK buffer and vortexed to obtain a suspension of multilamellar lipid vesicles (MLVs, lipid concentration 1–5 mM). The addition of 4 mm diameter glass beads (Sigma) in the flask optimizes the resuspension. The suspension is frozen and thawed five times (using liquid nitrogen and a water bath at 37 °C), extruded or stored at −20 °C. To work with ArfGAPl and mGMAP, we prepare "Golgi-mix" liposomes whose composition (phosphatidylcholine, phosphatidylethanolamine, phosphatidylinositol, phosphatidylserine, cholesterol (49/19/10/5/16 mol/mol %)) mimics the composition of Golgi membranes.

MLVs are extruded with a mini-extruder (Avanti Polar Lipids). A 19 mm polycarbonate filter (Millipore) with pores of defined size is sandwiched between two pre-filters maintained by two Teflon rings. The suspension is passed 21 times

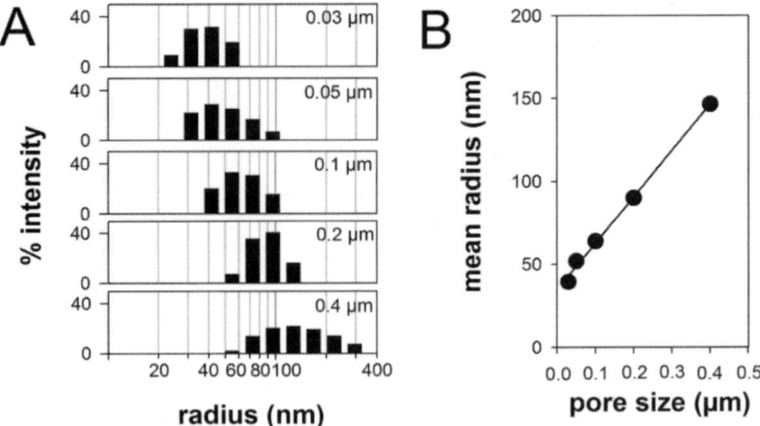

Fig. 1 Determination by dynamic light scattering (DLS) of the size of liposomes produced by sequential extrusion. (A) Size distribution of liposomes extruded with filters of decreasing pore size. (B) Mean hydrodynamic radius (R_H) as a function of the pore size.

through the filter by using 0.25 or 1 mL Hamilton syringes to fragment the MLVs into smaller and unilamellar liposomes of defined diameter (MacDonald *et al.*, 1991). To make liposomes of different diameters, the extrusion is performed sequentially, first with a 0.4 μm filter (pore size) and then with filters of decreasing pore size (0.2, 0.1, 0.05, and 0.03 μm). After each extrusion a liposome aliquot is stored for the experiment and has to be used within 2 days. The use of a 0.25 mL Hamilton syringe is recommended with filters of 0.05 or 0.03 μm pore size to limit the manual force required for extrusion.

Dynamic light scattering (DLS) is used to determine the radius of liposome. The fluctuations in intensity of light scattered by particles along time is informative on their movement. The rate of liposome diffusion and thus the hydrodynamic radius (i.e., the radius of a hard sphere that could diffuse at the same speed) is obtained by autocorrelating the fluctuating intensity trace. Liposomes are diluted at 0.1 mM (total lipids) in 20 μL of HK in a small cuvette (Hellma 105.252-QS) and the measurement is performed with a Dynapro apparatus (http://www.wyatt.com). Twelve measures are performed in 2 min and processed by the Dynamics v5 or v6 software (Dynapro) to determine the mean radius and the polydispersity of a sample (see Fig. 1).

2. Preparation of Giant Unilamellar Vesicles (GUVs)

Several techniques exist to prepare GUVs of 5-50 μm in diameter (Walde *et al.*, 2010). We use the electroformation technique (Angelova *et al.*, 1992) based on the swelling of dried lipid films rehydrated in a sucrose solution under an alternating electric field. This protocol has two main limitations. First, its yield decreases sharply

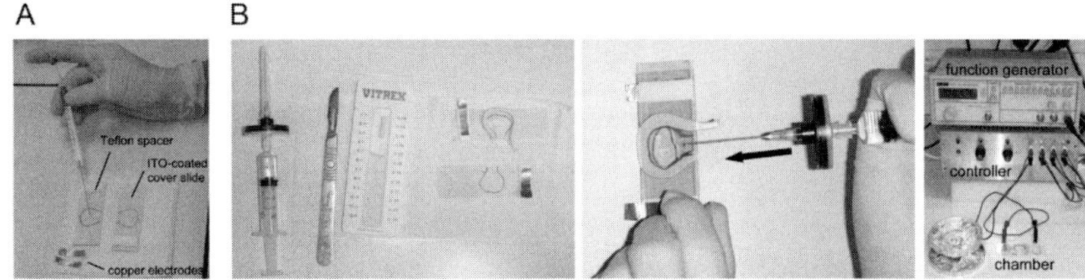

Fig. 2 Preparation of giant unilamellar liposomes. (A) Spreading the lipids on ITO-coated cover slides. The picture shows the two ITO-coated cover slides. Lipids in chloroform are deposited with a 10 μL Hamilton syringe. The two copper electrodes are shown on the bottom left. (B) Set-up for electroformation. Left panel: Materials needed for assembling and filling the electroformation chamber. From left to right: sucrose solution in a syringe equipped with a 0.2 μm filter and needle, scalpel, Vitrex Sigilum wax, dried lipid films on ITO-coated cover slides. Middle panel: Injection of the sucrose solution (about 200 μL) inside the electroformation chamber. Right panel: After closing the chamber with Sigilum wax, the copper electrodes are connected to a homemade controller that delivers the ramp voltage described in the text from the function generator situated on top of the controller. (For color version of this figure, the reader is referred to the web version of this book.)

if the membrane contains more than 30% mol/mol charged lipid. Second, it has to be modified to prepare GUVs in physiological buffers (Meleard *et al.,* 2009; Montes *et al.,* 2007, 2010).

Step 1 – *Spreading and drying lipid mixtures on ITO-coated cover slides*: Cover slides coated with Indium Tin Oxide (ITO) are purchased from Prazisions Glas & Optik GmbH (Germany). Two slides are used to build an electroformation chamber and lipids can be deposited on the conductive ITO-coated side of one or both slides (Fig. 2A).

a. Clean thoroughly an ITO-coated cover slide with ethanol and water several times then once with chloroform.
b. Mark the spot where lipids will be deposited on the non-conductive side of the cover glass.
c. Take 10 μL of lipid mix (0.5 mg/mL in chloroform) using a glass syringe.
d. Under a chemical hood, slowly spread the lipids on ~1 cm^2 of the ITO-coating.
e. Dry the lipid film for at least 4 h at room temperature in a vacuum oven.

Step 2 – *Preparing electroformation*: Because GUVs are very sensitive to osmotic pressure shocks, the osmotic pressure of the sucrose solution should match that of the experimental buffer. Transferring GUVs into a hypoosmotic or hyperosmotic medium respectively increases or decreases the membrane tension. The osmotic pressure of the HKM buffer (50 mM Hepes pH 7.2, 120 mM KAcetate, 1 mM $MgCl_2$) used in our experiments, measured using a micro-osmometer (Roebling, Germany), is around 280 mOsm. For experiments with Arfl alone or the ALPS constructs, the sucrose is adjusted to 260 mOsm to slightly deflate the GUVs. For experiments with coatomer or full-length ArfGAPl, electroformation had to be performed with 480 mOsm sucrose to match the high osmotic pressure (500 mOsm) of the protein buffer due to the presence of glycerol and salts (Manneville *et al.,* 2008).

a. Warm the sucrose solution (stored at 4 °C) at room temperature and take 1–2 mL of this solution in a 5 mL syringe.
b. Mount a 0.2 μm filter and a needle on the syringe.
c. Check that the voltage ramp used to generate the electric field is functioning properly.

Step 3 – *Assembling the chamber for GUV electroformation*: Exposure to air of the dried lipid films should be minimized. The electroformation chamber is built with two ITO cover slides with their conductive sides facing each other, separated by a 1 mm Teflon spacer, connected to a low frequency generator (TG315 function generator, TTi Thurlby Thandar Instruments, United Kingdom) *via* adhesive copper electrodes and sealed with Sigilum wax (Vitrex Medical A/S, Denmark, see Fig. 2B).

a. Prepare a ~5 cm long and ~2 mm thick roll of Sigilum wax and two ~3 cm long adhesive copper electrodes.
b. Remove the ITO slides from the vacuum oven and stick the copper electrodes to the Teflon spacer.
c. Place the roll of Sigilum wax around the lipid film (but not too close) on one of the ITO cover slide, leaving an opening to introduce the sucrose solution.
d. Close the chamber by placing the second ITO cover slide on top of the first one and pressing on the top cover slide to seal with the Sigilum wax.

Step 4 – *Growing GUVs*: The lipid film is rehydrated by injecting the sucrose solution in the chamber. An alternative electric field is then applied quickly to avoid spontaneous multilamellar vesicle formation (Fig. 2B). The voltage ramp classically used for electroformation generates eight 5 min voltage steps from 20 mV to 1.1 Vat 10–20 Hz (sine wave). The voltage can be set manually or *via* a homemade controller (see Fig. 2B). The frequency (10–20 Hz) is adjusted according to the osmolarity of the sucrose solution (20 Hz below 100 mOsm, 10 Hz at 200–300 mOsm and 8 Hz at 500 mOsm). Above 500 mOsm, the yield of GUV electroformation drops. The voltage is then kept constant at 1.1 for 2–3 h. Finally, to help separate the GUVs from each other and from the lipid film, a 4 Hz 1.4 V^2 voltage is applied for 30 min. This last step, as well as the initial increase of the voltage, optimize GUV production but are not compulsory. To incorporate charged lipids, applying a 700 mV voltage for 30 min gives the best results.

a. Inject about 200–500 μl of sucrose to fill the chamber.
b. Close the chamber with Sigilum wax.
c. Connect the chamber to the voltage ramp generator *via* the copper electrodes.
d. Start the voltage ramp. Check the connections and the voltage between the two ITO conducting sides. Check that the chamber is not leaking.
e. When using fluorescent lipids, cover the chamber to keep it in the dark.
f. Leave for 2–3 h.

g. Check that GUVs have grown using a microscope equipped with phase contrast and a 40X or 60X long working distance (> 1 mm) air objective.

GUVs are very sensitive to mechanical or osmotic shocks and must be carefully handled. After electroformation, GUVs are either stored in their electroformation chamber at 4 °C for a few days or transferred in an Eppendorf tube and stored at 4 °C or further concentrated by gentle centrifugation or sedimentation (see VI-B). For microscopy, GUVs are transferred to an observation chamber (see below and Fig. 4A).

III. Binding Assays for Testing Curvature Recognition by a Protein

A. Flotation Assays

Flotation assays, contrary to sedimentation assays, allow the use of liposome of various radii. We improved the original protocol (Matsuoka and Schekman, 2000) to visualize the liposomes during the assay. Liposomes doped with 0.2% fluorescent NBD-PE are extruded sequentially (see II-B). The protein (0.5 to 1 μM) is incubated at room temperature with 0.5 to 1 mM liposomes in HKM buffer (volume 150 μL) in a polycarbonate tube adapted to a TLS 55 Beckman swing-rotor (http://beckmancoulter.com/). After 5 min, the mix is adjusted to 30% sucrose (w/v) by adding 100 μL of 2.2 M sucrose in HKM buffer (vortex gently). Two cushions are delicately overlaid on the mix: 200 μL of HK buffer containing 0.75 M sucrose and 50 μL of sucrose-free HK (Fig. 3A). Tubes are photographed with a fluorescence imaging system (FUJIFILM LAS-3000, http://home.fujifilm.com/); the liposomes must be in the 30% sucrose cushion and not in the upper ones. The tubes are centrifuged at 240,000 g (55,000 rpm) for 1 h. The presence of liposomes at the top of the sucrose gradient is then assessed. Then, three fractions are collected from bottom to top using a Hamilton syringe (bottom fraction: 250 μL, middle one: 150 μL, top one: 100 μL). Five microliters of each fraction are spotted on a black plastic plate to measure liposome fluorescence. Only spots corresponding to the top fraction must give a signal. Next 30 μL of top fractions are analyzed by SDS–PAGE to determine the amount of membrane-bound protein, stained by SYPRO-orange. This is determined for each lane by comparing the density of protein bands to that of a reference lane containing 100% of the initial amount of protein (Fig. 3B). A curvature-sensing protein binds preferentially to the smallest liposomes.

B. Fluorescence Assays

NBD fluorescent measurements are performed in a standard fluorimeter (90° format; e.g., Shimadzu RF5301PC) with a temperature-controlled cell holder. The emission spectrum of NBD-labeled proteins (0.75–1 μM in 100 μL HKM buffer) is recorded at 20–37 °C in a small quartz cell (Hellma 105.251-QS). In Fig. 3C, spectra are measured from 520 to 650 nm (bandwidth 5 nm) upon excitation at 495 nm

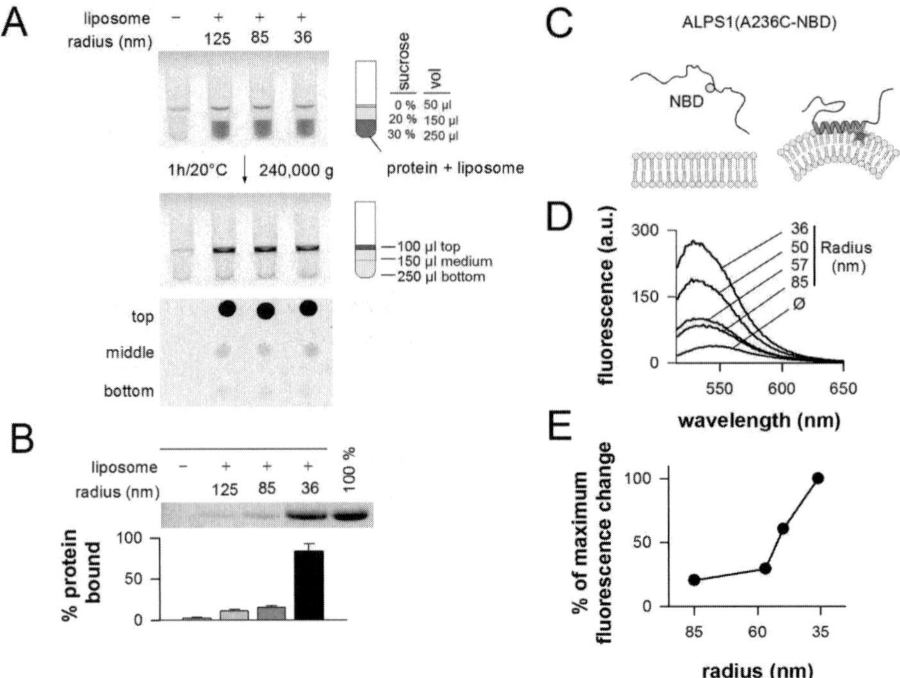

Fig. 3 Detecting a curvature-sensing protein by binding assays. (A) ArfGAPl[1-257] (0.75 μM) is incubated with or without fluorescent liposomes (0.75 mM total lipids) of different radius. The sucrose gradient is generated as described in the text. The tubes are illuminated with lateral blue diodes and the fluorescence of the liposome suspension is imaged by a CCD camera. After centrifugation the liposomes are at the top of the gradient. Fractions are collected from bottom to top and lipid fluorescence in each fraction is quantified from 5 μL spots. (B) Lipid-bound proteins recovered in the top fraction are analyzed by SDS–PAGE and SYPRO-orange staining. (C) The fluorescence of a NBD probe located in the hydrophobic side of the ALPS motif increases in contact with a curved membrane. (D) Emission spectra of 0.75 μM ArfGAP [137-257](A236C-NBD) without or with liposomes (2.5 mM) of various radius. (E) Normalized intensity at 529 nm of the protein as a function of liposome radius. Adapted from (Drin *et al.*, 2007). (For interpretation of the references to color in this figure legend, the reader is referred to the web version of this book.)

(bandwidth 5 nm). The spectral features of NBD could slightly differ depending on the protein and the labeling site. It is important first to define the liposome concentration for which 100% of protein is membrane-bound. This is done by incubating the construct with increasing concentrations of small liposome (extruded through a 0.03 μm filter). Alternatively smaller liposomes prepared by sonication can be used (Mesmin *et al.,* 2007). Usually the emission spectrum increases strongly in intensity and shifts toward shorter wavelengths if the construct binds to membrane. The reference concentration is found when adding extra liposomes does not induce an additional increase in intensity and blue shift of the maximal wavelength of emission. Next the experiment is repeated at the selected lipid concentration with liposomes of different radii (Fig. 3D). For each condition, a spectrum of liposomes alone is recorded and subtracted from that of the NBD-labeled construct. For a curvature-sensing protein, a sharp increase in fluorescence is observed as the radius decreases (Fig. 3E).

IV. Distribution of a Curvature-Sensing Protein on Tube Networks Pulled by Kinesin Motors

Roux *et al.* first reported the *in vitro* formation of membrane tube networks with a machinery consisting of (i) stabilized microtubules (MTs), (ii) kinesin motors, and (iii) GUVs which provide a membrane reservoir (Roux *et al.,* 2002). GUVs covered with kinesin motors are sedimented onto a fixed MT network. After ATP is added, kinesins move along MTs and extract tubes from GUVs (Fig. 4B). Initially biotinylated-kinesin motors were coupled to biotinylated lipids of the GUV *via* 100 nm streptavidin-coated latex beads. We describe below a more recent protocol in which beads are replaced by streptavidin in excess (Koster *et al.,* 2003; Leduc *et al.,* 2004). Two parameters are critical for a good yield of tube extraction. First, MTs have to be strongly attached to the coverslip. Second, the force exerted by the motors has to overcome the elastic resistance of the membrane imposed by its bending rigidity and its tension. To obtain a large number of tubes, GUVs are slightly deflated to decrease their tension. To get individual tubes, a higher membrane tension is necessary. Tension can be adjusted by tuning the osmotic pressure inside and outside the GUVs (see II-B).

A. Elongation of Tubes by Kinesin

1. Preparation of Biotinylated GUVs and Microtubules

GUVs are composed of 98 mol% DOPC, 1 mol% Bodipy-TR-Ceramide (red fluorescent lipid), and 1 mol% biotinylated lipid Biot-Cap-DOPE. The number of biotinylated sites in the lipid membrane must be much lower than the number of kinesin-streptavidin complexes and therefore saturated with kinesin motors (Leduc *et al.,* 2004). Adjusting the quantity of biotinylated lipids offers a way to control the surface density of motors on the GUVs.

Tubulin dimers self-assemble into MTs in the presence of GTP and $MgCl_2$ at 37 °C. Taxol is used to stabilize the filaments by stopping their depolymerization. Tubulin is purified from animal brains (Hyman, 1991) or purchased from Cytoskeleton (www.cytoskeleton.com).

a. Mix 5 μL of tubulin (4 mg/ml tubulin units) in BRB80 buffer (PIPES pH 6.9, 1 mM EGTA, 1 mM $MgCl_2$) with 4 mM $MgCl_2$, 1 mM Mg-GTP, and 5% DMSO on ice.
b. The mix is incubated for 30 min at 37 °C, diluted 100-fold in BRB80 containing 30 μM taxol and quickly vortexed at room-temperature.

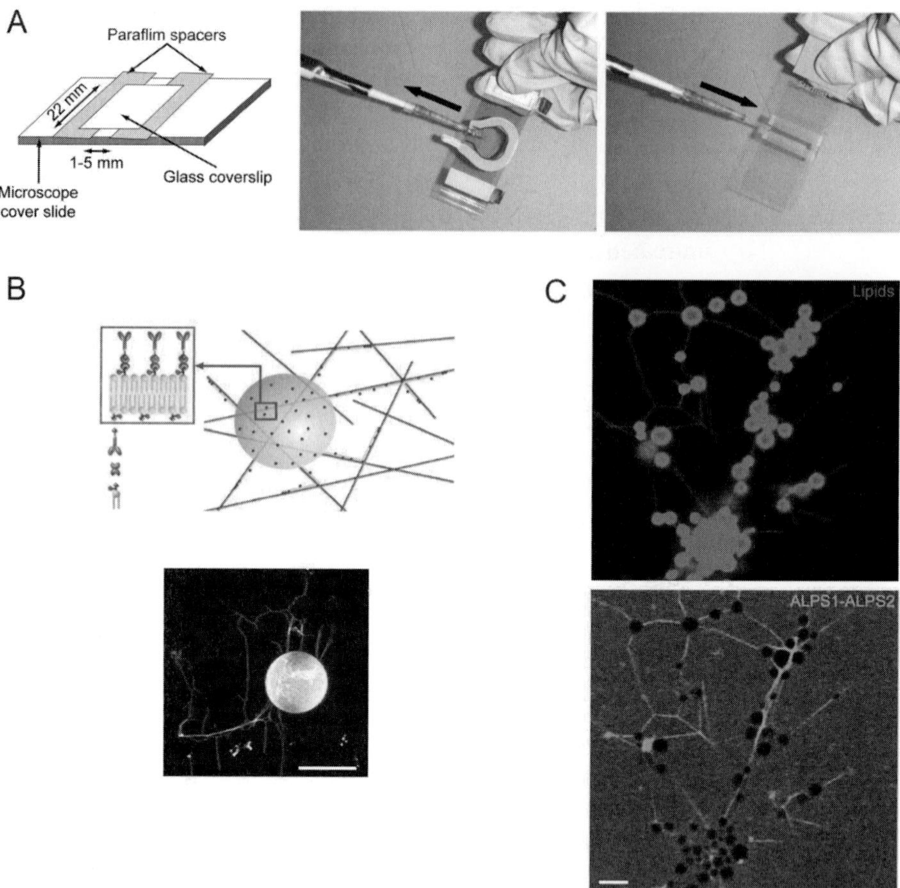

Fig. 4 Membrane tube formation by kinesin motors moving along microtubules. (A) Transfer of GUVs for observation under an optical microscope and for tube pulling by kinesin motors. The observation chamber is built by melting two Parafilm spacers between a microscope slide and a glass cover slip (left panel). GUVs are first aspirated from the electroformation chamber in a capillary mounted on a pipette tip (middle panel) then pushed into the observation chamber (right panel). (B) Sketch of the experimental set up: biotinylated kinesin motors (purple) are bound to GUVs (yellow) containing DHPE-Biot-Rhod lipids *via* streptavidin molecules (red). Membrane tubes grow along the stabilized microtubule network (green), pulled by kinesin motors (pink). Confocal image (2D projection of a z-stack) of a membrane tube network. The membrane of the GUV is uniformly labeled with fluorescent lipids. The GUV membrane is slightly under tension (20 mOsm excess osmotic pressure inside compared to outside the GUVs). Adapted from (Leduc *et al.*, 2010). (C) Visualization of ALPS1-ALPS2-Alexa^{488} on tube networks pulled by kinesin motors. The ALPS1-ALPS2 construct (green, lower panel) binds only to lipid nanotubes and not to the GUVs. Lipids are shown in red (upper panel). Bars in B and C, 10 μm. (See color plate.)

c. Remove non-polymerized tubulin by centrifuging the MTs at 37 °C (70,000 rpm for 15 min at 37 °C). MTs can be stored for one week at room temperature.

2. Tube Networks Pulled by Kinesin Motors in a Flow Chamber

Build a chamber with two parafilm spacers sandwiched between a cover slide and a 22 × 22 mm glass coverslip and melted on a hot plate (Fig. 4A). MTs bind to untreated coverslips. If a stronger and more reproducible binding is required, the coverslip can be plasma cleaned for 30 s, incubated for 1–2 min in a poly(L-lysine) solution (0.01% w/v) and dried under a nitrogen flux. Several buffers (200 μL) are prepared:

HKM: 50 mM Hepes pH 7.2, 120 mM K-Acetate, 1 mM $MgCl_2$ with 2 mM EGTA.
HKM-Tx: 30 μM taxol in HKM.
HKM-casein-Tx: 5 mg/mL casein, 30 μM taxol in HKM.
HKM-DTT-Tx: 1 mM DTT (or TCEP-HC1), 30 μM taxol in HKM.
Motility buffer: 1 mM ATP, 30 μM taxol, 5 mM DTT with 0.18 mg/mL catalase, 0.37 mg/mL glucose oxydase and 25 mM glucose (as an oxygen scavenger to limit photobleaching) in HKM.

All injections have to be performed very slowly to avoid detaching the microtubules by shear flow.

a. Incubate biotinylated kinesins ($\sim 2 \times 10^{-12}$ mol) with streptavidin ($\sim 2 \times 10^{-11}$ mol) for 10 min on ice. Because streptavidin is in large excess compared to kinesin, there is at most one kinesin per streptavidin molecule. In the case of biotinylated kinesins purified from the kinesin - BCCP-H6 plasmid, incubate 2.5 μL kinesin (at ~2 mg/mL) with 5 μL streptavidin (Pierce) at 1 mg/mL. After incubation, dilute the kinesin-streptavidin mix to 30 μL with HKM-DTT-Tx buffer. Unused diluted kinesins bound to streptavidin can be flash-frozen in liquid nitrogen and reused once. The aliquot of concentrated kinesins can be flash-frozen in liquid nitrogen and reused at least 10 times.
b. Fill the chamber with 5 μL of MT solution for 10 min to bind MTs to the coverslip.
c. Incubate the chamber with 10 μL HKM-casein-Tx buffer for 5 min (casein covers the exposed glass surface and prevents non specific binding of kinesin).
d. Rinse the chamber with 10 μL HKM –Tx buffer.
e. Inject 5 μL of diluted kinesin-streptavidin solution (kinesins bind to microtubules).
f. Optional: rinse with 10 μL HKM-Tx buffer (to remove unbound kinesins and streptavidin molecules).
g. Rinse with 10 μL motility buffer.
h. GUVs are slowly pipetted from their storing location with a 1 mm outer diameter thin-walled glass capillary tube coupled to a pipet tip using parafilm to minimize shear stress (Fig. 4A).
i. Inject 1 μL of GUVs at matching osmotic pressure (slightly hypoosmotic sucrose compared to the motility buffer to yield a large number of tubes, or slightly

hyperosmotic to obtain few individual tubes). The volume of GUVs should be less than 10–20% of the total volume of the observation chamber to avoid dilution of proteins. Keep the flow chamber vertical so that GUVs fall into the chamber by gravity.

j. Close the chamber with Sigilum wax and store horizontally before observation.
k. Image tube networks as soon as possible after GUVs injection to visualize the dynamics of tube extraction and/or wait 10–15 min to observe tube networks at steady state.

B. Visualization of a Curvature-Sensing Protein on Tube Networks

The curvature-sensing protein ArfGAPl or the ALPS1-ALPS2-Alexa^{488} construct is added in "motility buffer" at 0.5–1 μM before pulling tubes. ArfGAPl is labeled with a rabbit polyclonal antibody targeting the 1–257 region of ArfGAPl and a secondary fluorescent (Alexa^{488}) anti-rabbit antibody. Images of tube networks are acquired by confocal microscopy (Fig. 4C). The distribution ratio of the protein in the tube compared to the vesicle is equal to $(I_{protein}/I_{lipid})_{tube}/(I_{protein}/I_{lipid})_{GUV}$ where $I_{protein}$ and I_{lipid} are the fluorescence intensities of the protein (green channel) and of the membrane (red channel) measured in a region of interest along the tube or on the GUV using the \Analyze\Measure function in ImageJ. Alternatively intensities in the GUV can be determined with the OvalProfile ImageJ plug-in (downloadable at http://rsbweb.nih.gov/ij/plugins/download/Oval_Profile.java). Intensities are corrected by subtracting the background intensity measured in an area close to the tube or the GUV. With highly curvature-sensitive proteins, such as ArfGAPl, the fluorescence on the GUV is comparable to background levels. In that case, the distribution ratio appears infinite.

V. Distribution of a Curvature-Sensing Protein on a Tube Elongated by Optical Tweezers

Lipid nanotubes are generated by using an optically trapped bead to apply a point force necessary to extract the tube while holding the vesicle with a micropipette to set membrane tension (Cuvelier *et al.*, 2005; Heinrich and Waugh, 1996; Roux *et al.*, 2005; Waugh, 1982). This technique allows controlling the tube radius in a range between 10 and 200 nm and is a method of choice to address questions related to membrane curvature. The typical experimental configuration is sketched in Fig. 5A.

A. Measurement of the Tube Radius

It is well established (see (Derényi *et al.*, 2007) for a review) that under conditions in which membrane tension is fixed by pipette aspiration (Fig. 5A) and for a single component GUV, the force necessary to hold the tube and the tube radius R_t depend

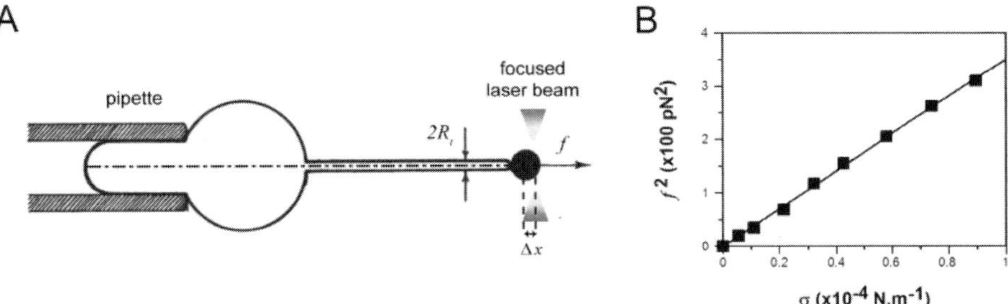

Fig. 5 Membrane tube formation at controlled membrane tension using optical tweezers and micropipette aspiration. (A) Experimental configuration of nanotube pulling by optical tweezers from a vesicle aspirated in a micropipette. The GUV is held on the left side by a micropipette connected to a water tank. On the right side, a membrane nanotube is pulled using a bead (black circle) trapped in optical tweezers. R_t: tube radius; Δx: displacement of the bead relative to the center of the optical trap; f: force necessary to hold the tube. Adapted from (Svetina *et al.*, 1998). (B) Evolution of the tube force f as a function of membrane tension σ. Typical variation f^2 versus σ for a single component membrane (EggPC). A linear fit according to Eq. (1) yields the membrane bending rigidity κ (here: $\kappa = 10 \pm 1\ k_B$T for EggPC). (For interpretation of the references to color in this figure legend, the reader is referred to the web version of this book.)

only on the bending rigidity (κ) and the membrane tension (σ) but not on the tube length:

$$f = 2\pi\sqrt{2\kappa\sigma} \tag{1}$$

(Fig. 5B) and

$$R_t = \sqrt{\frac{\kappa}{2\sigma}} \tag{2}$$

Combining Eqs. (1) and (2) gives the tube radius as a function of the force f and the membrane tension σ.

$$R_t = \frac{f}{4\pi\sigma}. \tag{3}$$

These two quantities, f and σ, are experimentally measurable so Eq. (3) yields an indirect measure of the tube radius. Note that this expression is in principle valid only for an uncoated tube.

B. Experimental Set-up Combining Optical Tweezers and Micropipette Aspiration on a Confocal Microscope

The set-up is built on a commercial inverted microscope (Nikon TE2000 or Ti-E) modified with the optional stage riser in order to create an extra port (Fig. 6). The confocal head is the eCl confocal system with two laser lines (488 nm and 543 nm). A complete description can be found in (Sorre *et al.*, 2009).

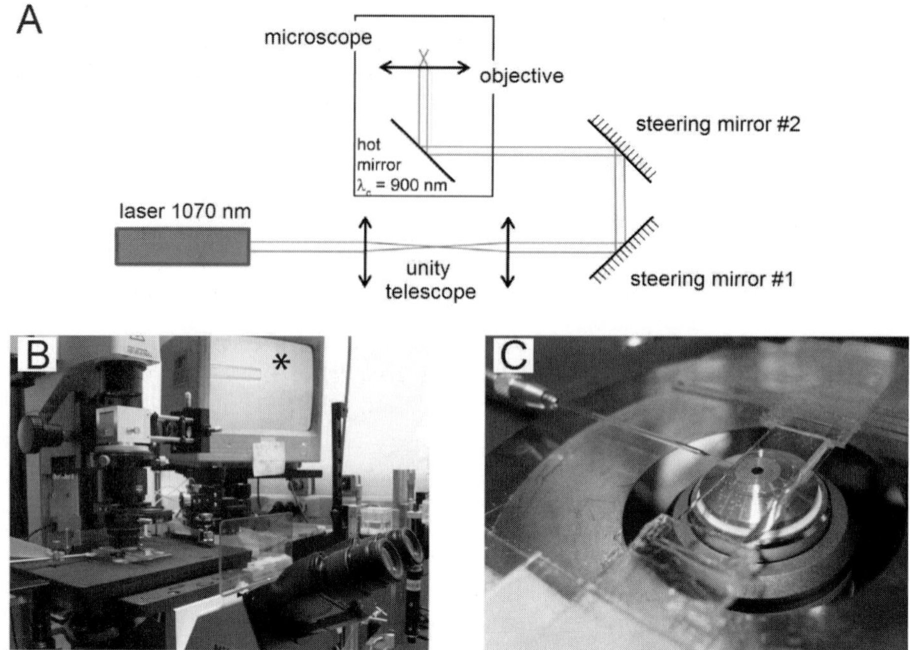

Fig. 6 Set-up for tube pulling with optical tweezers. (A) Optical path of a single, non-moving optical trap. (B) General view of the set-up. The pipette used to hold the GUV is on the left and marked on the screen with a star. (C) Zoom on the pipette inside the manipulation chamber. (For color version of this figure, the reader is referred to the web version of this book.)

The micropipette aspiration system is based on the set-up developed by Evans *et al.* (Kwok and Evans, 1981; Waugh and Evans, 1979). A suction pressure is applied between the inside and the outside of the pipette by lowering a water reservoir compared to a reference level. Varying the pipette aspiration from 0 to 200 Pa (corresponding to a vertical displacement of the water reservoir between 0 and 2 cm) allowed varying the membrane tension of the GUV from 5×10^{-6} to 2×10^{-4} N.m^{-1}, corresponding to tube radii from 200 to 10 nm.

Pipettes are prepared using borosilicate glass capillaries (internal radius 0.7 mm, external radius 1 mm, Kimble Glass Inc. Vineland, New Jersey, ref 46485-1). Capillaries are pulled using a pipette puller (Sutter instrument P-2000) and the pipette radius is set to the appropriate size (about 4 μm) using a microforge (MF-800 Narishige, Japan). Before starting experiments, the pipette is backfilled with buffer and incubated in the manipulation chamber in a buffer containing 5–10 mg/mL casein in order to passivate glass surfaces and prevent membrane adhesion.

The optical tweezers described in Fig. 6A are created by a focused infrared laser beam (Ytterbium fibre laser 1070 nm, 5W, IPG GmBH Germany) according to Lee and co-workers (Lee *et al.*, 2007). The force applied on the trapped bead is deduced from the displacement of the bead Δx compared to its equilibrium position at the centre of the trap: $f = k\ \Delta x$ where k is the trap stiffness (450 ± 30 pN/nm/W after calibration of our tweezers using the Stokes hydrodynamic drag force method (Neuman and Block, 2004)). Video tracking of the bead position with a homemade Matlab routine is used to obtain the displacement Δx with a 35 nm and 40 ms spatio-temporal resolution.

C. Distribution of ALPS1–ALPS2 on a Tube of Controlled Curvature

GUVs are made of 99 mol% DOPC, 1 mol% Bodipy-TR-Ceramide, and 0.03 mol% DSPE-PEG(2000)-Biotin. Experiments are performed at room temperature (21 ± 1 °C).

a. Build a 10 mm wide, 20 mm long, and 1 mm thick open micromanipulation chamber with two clean glass coverslips (see Fig. 6B and C) spaced by a glass slide.
b. Coat the chamber with β-casein (5 mg/mL, 15 min) to prevent adhesion of the GUVs to the surface of the chamber.
c. Rinse the chamber with HKM buffer.
d. Fill the chamber with 200 μL HKM buffer containing 1 μM ALPS1-ALPS2-Alexa^{488}.
e. Inject 5 μL of 3.2 μm diameter streptavidin-coated polystyrene beads (Spherotech Inc., commercial solution diluted 100 times).
f. Insert the micropipette in the chamber.
g. Inject 5 μL of GUV solution at matching osmotic pressure. Wait a few minutes to slightly evaporate the solution. This creates a slight osmotic pressure difference (higher in the buffer) that deflates the GUVs and reduces their tension.
h. Seal the manipulation chamber with mineral oil to prevent further water evaporation.
i. Select a GUV with a low membrane tension (optically fluctuating $\sigma \sim 5\times10^{-6}$ N.m^{-1}).
j. Set the zero reference pressure in the pipette by adjusting the water level such as no movement of a bead in the pipette is detected.
k. Aspirate the GUV with a low suction pressure.
l. Trap a bead with the optical tweezers and contact the GUV with the bead to adhere the streptavidin-coated bead to the biotinylated GUV.
m. After nucleation of the tube, extend the tube to 10 μm by moving the pipette away from the trap. Increase membrane tension by steps (typically by 500 μm vertical displacements of the water reservoir corresponding to 5 Pa steps).
n. For each tension, record the position of the bead relative to the trap centre by video microscopy and acquire the fluorescent signals of the tube and the GUV in both green (proteins) and red (lipids) channels by confocal microscopy. Wait at least 1 min between each tension step to reach equilibrium.

The tube radius R is calculated knowing the force f and the membrane tension σ (see V-A above). The fluorescence intensities of the protein I_{protein} and of lipids

I_{lipid} in the tube are measured from a rectangular region of interest that includes the horizontal tube. To increase the signal-to-noise ratio, the fluorescence signal can be averaged along the tube length (Roux *et al.*, 2010; Sorre *et al.*, 2009). The background noise is subtracted from the fluorescence peak for each dye. The ratio $(I_{protein}/I_{lipid})_{tube}$ is plotted as a function of the tube curvature ($1/R_t$).

VI. Assays to Measure the Curvature-Dependant Activity of ArfGAP1 and GMAP-210

A. Measuring the Arf1 Inactivation by ArfGAP1 in Response to Curvature

The fluorescence of a tryptophan (Trp78) in the switch II region of Arf1 doubles when the protein undergoes a conformational change upon GDP-to-GTP exchange (Vetter and Wittinghofer, 2001). This signal change allows to follow in real time the activation/inactivation of Arf1. Measurements are performed in a standard fluorimeter (as for III-B). The signal is recorded at 340 nm with a large bandwidth (10–30 nm) to maximize the signal-to-noise ratio. To prevent light absorption by nucleotide the sample is excited at $\lambda > 290$ nm (e.g., 297.5 nm; bandwidth ≤ 5 nm). To measure kinetics accurately, it is key to continuously stir the sample at a constant temperature (37 °C). Thus the fluorimeter must be equipped with a cell holder connected to a thermostat and a magnetic stirrer. The cuvette is equipped with a small magnetic bar (2 × 7 mm, Hellma). We use classical quartz cuvettes (10 × 10 mm, Hellma) or more often custom-made cylindrical quartz cuvettes (internal diameter 8 mm) inserted in a 3-window metal holder to minimize the sample volume (600 μL). It is important to use filtered and degassed buffers to minimize light-scattering artifacts caused by dust particles or bubbles. Figure 7 shows a typical measurement. Arf1-GDP (0.5 μM) is injected from a stock solution to liposomes in HKM buffer (0.2 mM total lipids) and a fluorescence signal is observed. Next Arf1 is activated by adding sequentially GTP (40 μM final concentration, Roche) and lowering the concentration of free Mg^{2+} with 2 mM EDTA. The switch of Arf1 from a GDP to a GTP-bound state induces an increase in fluorescence. This is coupled to the binding of Arf1 on liposomes. Arf1 at 0.2–1 μM gives a fluorescence emission with a good signal-to-noise ratio. It is important to choose a lipid concentration accordingly (between 0.2 and 1 mM) to work at a reasonable lipid-to-protein ratio (typically 400). At lower ratios, the lack of membrane surface can limit the amount of Arf1 attached to liposomes. Once Arf1 is activated, the concentration of free Mg^{2+} is restored by adding 2 mM $MgCl_2$. Finally ArfGAP1 (50 nM) is added to hydrolyze GTP in Arf1. Arf1, GTP, EDTA, $MgCl_2$, and ArfGAP1 are injected from concentrated stock solutions with Hamilton syringes (10–50 μL) through a guide in the cover of the fluorimeter. The guide is set up to position the tip of the needle in contact with the meniscus of the sample. As the needle does not cross the light beam, the measurement is not interrupted by injections: kinetics can be correctly recorded with a ~1 s time resolution.

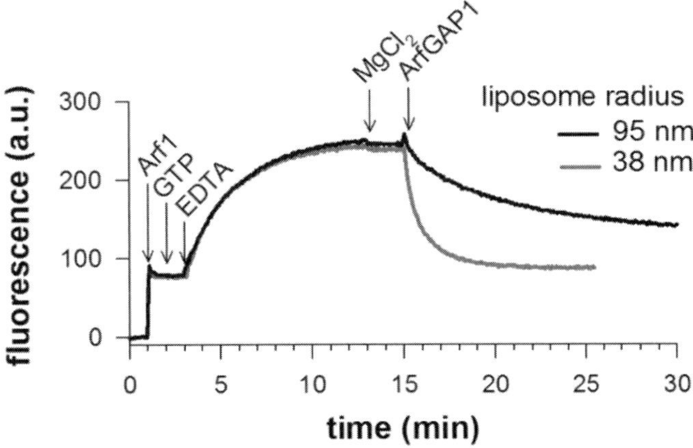

Fig. 7 Curvature-dependant inactivation of Arfl by ArfGAPl. Myristoylated Arfl-GDP (0.5 μM) is added to small or large Golgi-mix liposomes (0.2 mM total lipids). Arfl is activated by adding 40 μM GTP and by lowering the concentration of free Mg^{2+} with 2 mM EDTA. After 15 min, the Mg^{2+} concentration is raised back to millimolar levels. GTP hydrolysis in Arfl is triggered by adding 50 nM ArfGAPl [l–257]. Adapted from (Drin *et al.*, 2007). (For color version of this figure, the reader is referred to the web version of this book.)

To see how curvature influences Arfl inactivation by ArfGAPl, the assay is repeated with liposomes of different radii (Fig. 7). The activation of Arfl is not sensitive to the liposome radius. In contrast the kinetics of GTP hydrolysis is dramatically affected. With large liposomes ($R = 95$ nm), the $t_{1/2}$ of Arfl inactivation by ArfGAPl is of several minutes whereas with small liposomes ($R = 38$ nm), the $t_{1/2}$ is of a few seconds. This constitutes a typical example of a curvature-dependant activity that directly mirrors the higher avidity of ArfGAPl for positively curved membranes.

B. Generation of Arfl Gradient on Curved Membranes by ArfGAPl Activity

Adding ArfGAPl to a membrane tube pulled from a GUV covered by Arfl-GTP generates an Arfl-GTP concentration gradient along the tube (Fig. 8). This is caused by the competition between diffusion of Arfl-GTP from the flat GUV and its dissociation in the curved tube by ArfGAPl (Ambroggio *et al.,* 2010). Two conditions are required to observe the gradient. First, ArfGAPl has to be used in catalytic amounts (10 nM). Second, the tubes have to be long enough, since the concentration gradient of Arf 1 has a characteristic length of ~13 μm in our assays. To prepare the experiment, Arfl labeled with OregonGreen is attached to GUVs in a GTP-bound state and unbound Arfl-OregonGreen is washed by gentle centrifugation.

a. Prepare GUVs made of DOPC/Bodipy-TR-Ceramide/DSPE-PEG(2000)-Biotin (99/1/0.03% mol/mol as in V-C).
b. Incubate a 0.5 mL Eppendorf tube with 10 mg/mL β-casein in HKM buffer for 10 min at room temperature. Rinse with HKM.

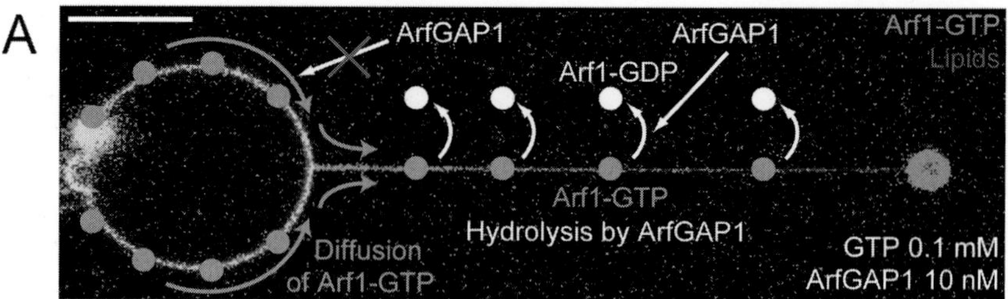

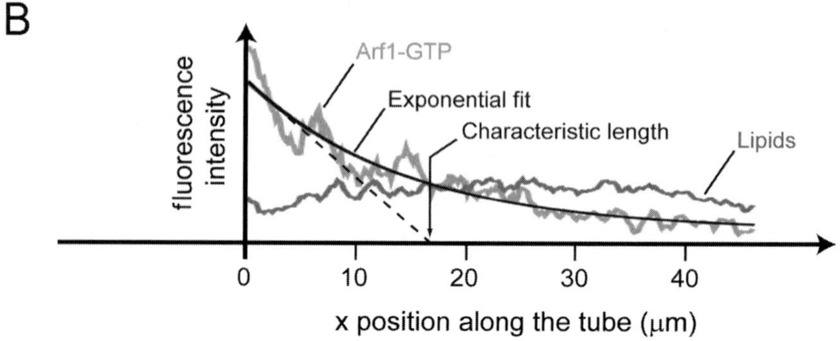

Fig. 8 Generation of concentration gradients of Arf1 along tubes pulled from giant vesicles. (A) A lipid nanotube was pulled using a 3 μm streptavidin-coated bead (green sphere on the right of the image) from a GUV containing biotinylated lipids held at a constant membrane tension by micropipette aspiration (on the left). GUVs (labeled with red fluorescent lipids) were pre-incubated with 2 μM Arf1-Oregon Green (in green). When a tube is pulled in the presence of 10 nM ArfGAP1 and 0.1 mM GTP, a concentration gradient of Arf1-GTP appears along the tube, with Arf1-GTP concentration decreasing from the base of the tube (close to the GUV) to the tip of the tube (close to the bead). The gradient is due to the competition between diffusion of Arf1-GTP (green circles) from the GUV into the tube and the dissociation of Arf1-GTP induced by ArfGAP1 (white circles) specifically in the tube (region of high curvature). The low curvature of the GUV membrane prevents ArfGAP1 binding and activity on the GUV and protects Arf1-GTP from hydrolysis in this region. (B) Exponential fitting of the data yields the characteristic length of the gradient. The plot shows Arf1-Oregon Green fluorescence (green trace), lipid fluorescence (red trace), and the exponential fit (black line) as a function of the distance x from the base of the tube to its tip. Adapted from (Ambroggio *et al.*, 2010). (See color plate.)

c. Transfer 5 μL GUVs in 95 μL HKM supplemented with 0.5 μM Arf1-OregonGreen, 0.1 mM GTP and 2 mM EDTA in the tube and incubate for 20 min at room temperature.
d. Centrifuge at 600 rpm for 10 min at room temperature.
e. Discard 90 μL of the supernatant and resuspend the pellet in 90 μl of HKM, EDTA 2 mM, GTP 0.1 mM.
f. Repeat steps d and e three times.
g. During the last wash, do not resuspend the pellet and keep 5–10 μL GUV.

h. Complete the 5–10 μL GUVs to 100 μL with 90–95 μL HKM, EDTA 2 mM, GTP 0.1 mM and supplement this suspension with 10 nM ArfGAPl.
i. Inject the 100 μL sample in the observation chamber (see V-C).
j. Inject 3.2 μm streptavidin-coated beads and insert the micropipette in the chamber (see V-C).
k. Using the micropipette and optical tweezers, extract a ~40 μl long tube from a low tension GUV held into the micropipette (see V-C).
l. Increase the tension above 10^{-4} N.m^{-1} to obtain a tube with a radius below 15 nm (below the threshold radius for ArfGAPl binding of 35 nm).
m. Acquire the fluorescent signals of the tube and the GUV in both green (Arfl-OregonGreen) and red (lipids) channels by confocal microscopy.
n. Measure the fluorescence intensity profiles of Arf1-OregonGreen and fluorescent lipids along the tube using ImageJ\Analyze\Plot Profile function. Fit with an exponential to deduce the characteristic length of the Arfl concentration gradient.

C. Measuring Tethering Between Flat and Curved Membranes by mGMAP

We present here an assay to follow in real-time the ability of mGMAP to aggregate small (R ~ 35 nm) and large liposomes (R ~ 145 nm) and thereby show its tethering ability. The aggregation process generates objects whose size is easily monitored by DLS. For this assay, it is key to improve the size homogeneity of large liposomes obtained by extrusion, often contaminated by smaller liposomes. Golgi-mix liposomes (1 mL, 3 mM total lipids) including 0.2% NBD-PE are extruded through a 0.4 μl filter and centrifuged at 50,000 g for 15 min with a TLS 55 rotor at 20 °C to pellet the largest liposomes. Once the supernatant is discarded, the pellet is resuspended in 1 mL HK buffer. The fluorescence of the liposome suspension before and after centrifugation is measured to estimate the liposome fraction recovered in the pellet (usually 55–65%). It is useful to compare the mean radius and polydispersity of liposomes before and after centrifugation. These large liposomes are not strictly unilamellar. To determine the lamellarity of liposomes, we measure the extinction of NBD-PE fluorescence by 10 mM dithionite in a fluorimeter at 533 nm (l_{ex} = 470 nm) (Mclntyre and Sleight, 1991). Typically 30% of the fluorescence of large liposomes (0.1 mM) is quenched within a minute. This corresponds to the fraction of exposed lipids. Knowing the amount of total lipids after centrifugation, one can calculate the liposome concentration in terms of accessible lipid. One can then estimate the occupancy of liposome surface by proteins and relate this to aggregation levels (Drin *et al.*, 2008). Arfl (0.5 μM*)* is mixed with liposomes (50 μM accessible lipids), activated with GTP (see VI-A) for 15 min, supplemented with $MgCl_2$ and the liposomes are collected. Small liposomes are prepared by sequential extrusion through 0.03 μm filter and are perfectly unilamellar (the accessible lipids correspond to one-half of total lipids).

DLS is suitable to detect if a protein tethers liposomes by measuring if it aggregates liposomes. However, this approach cannot tell directly which are the connected

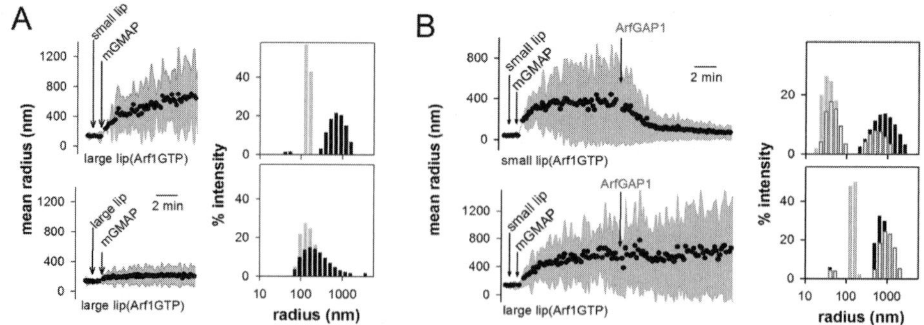

Fig. 9 Tethering of flat and curved membranes by mGMAP. (A) mGMAP (125 nM) is added to a suspension of large liposomes (25 μM accessible lipids) covered with Arfl-GTP (0.25 μM) quickly mixed with small or large liposomes (25 μM accessible lipids). (B) Small or large liposomes covered by Arfl-GTP (0.25 μM) are mixed with naked small liposomes. mGMAP (125 nM) is added to trigger the aggregation. After 10 min, ArfGAPl full-length (250 nM) is added. Gray bars, initial size distribution; black bars, size distribution after aggregation; white bars with red edge, final size distribution after adding ArfGAPl. Small liposomes, $R_H = 36 \pm 7$ nm; large liposomes, $R_H = 143 \pm 45$ nm. Adapted from (Drin *et al.*, 2008). (For interpretation of the references to color in this figure legend, the reader is referred to the web version of this book.)

liposomes. Several experiments are necessary to show if a certain type of connection takes place between two distinct populations of liposome. First, large liposomes covered by Arfl-GTP are mixed with a second population of liposomes, either small or large. One observes respectively either a massive aggregation or almost nothing (Fig. 9A). mGMAP recruited on large liposomes by Arfl through its C-terminal GRAB can connect only small liposomes with its other extremity (Drin *et al.*, 2008). Second, small liposomes are mixed with small or large liposomes covered with Arfl-GTP. In both cases aggregation occurs. After 10 min, ArfGAPl (250 nM) is added. If Arfl-GTP is anchored on small liposomes, aggregates disassemble within minutes because ArfGAPl removes Arfl from these highly-curved liposomes by GTP hydrolysis. In contrast, if Arfl-GTP is on large liposomes, aggregates are protected from ArfGAPl (Fig. 9B). Thus a stable tethering in the presence of ArfGAPl only occurs between flat membranes covered with Arfl-GTP and small vesicles.

Aggregation kinetics are measured at 25 °C in a Dynapro apparatus (as in II-B) by mixing liposomes (25 μM accessible lipids) covered with Arfl (0.25 μM) with naked liposomes (25 μM accessible lipids) and mGMAP (0.25 μM). Modulating lipid and protein concentrations affects the aggregation level. mGMAP must be added quickly once the two populations of liposome are mixed because Arfl-GTP can dissociate slowly from membranes ($k_{off} \sim 200$–300 s) and thus populate the second population of liposome initially devoid of Arfl-GTP.

a. Mix 10 μL of liposomes covered with Arfl-GTP with 6.5 μL HKM buffer (with 1 mM DTT) in the cuvette.
b. Measure liposome size every 10 s.

c. After 2 min, quickly add 1 μL of naked liposome (stock 500 μM accessible lipids). The suspension is mixed 2–3 times with a Gilson pipette. Replace quickly the cell in its holder to resume the measurement.
d. After 1 min, mix 2 μL of mGMAP (2.5 μM) thoroughly with liposomes.
e. If necessary, after 10 min, mix ArfGAP1 (full-length, 250 nM) with the suspension.

Autocorrelation curves calculated every 10 s are fitted with a cumulant algorithm that gives the mean radius (*R*) and the polydispersity (shaded area, Fig. 9) of the sample. Only one liposome population with a Gaussian size distribution is assumed to exist. If free liposomes co-exist with large aggregates, the polydispersity reaches very large values. To better analyze the aggregation process one can re-analyze autocorrelation functions (12 measures) corresponding to parts of the kinetics by a regularization algorithm (Dynamics v6.1) that resolves the size distribution (Fig. 9, histogram) of different populations (free liposomes, aggregates). Because proteins are stored in glycerol-containing buffers, all calculations must take into account the final percentage of glycerol in the sample as it influences the viscosity of the medium.

VII. Summary and Conclusion

The methods presented here can be applied to any protein to give clear-cut and quantitative results on its ability to sense curvature. It is important however to note that the influence of curvature on a protein could be modulated or even abolished by the degree of lipid packing and the presence of negatively charged lipids. The liposome composition must be cautiously defined to best mimic the features of cellular membranes targeted by the protein of interest and to avoid irrelevant results. It should be also noted that numerous cellular processes resulting in a change of membrane structure are known to be often induced by proteins whose local density on membrane surface is high (i.e., at low lipid-to-protein ratio). For this reason and because most of the assays presented here are dedicated to measure a simple membrane-binding step for membrane of defined curvature, we recommend working preferentially at high lipid-to-protein molar ratio (500–1000).

Acknowledgments

We thank Bruno Antonny, Bruno Goud, Patricia Bassereau, Joëlle Bigay, Bruno Mesmin, Ernesto Ambroggio, and David Guet for participating in the experiments and methods presented in this paper. We thank Mathieu Pinot for assistance with the photographic panels of figures.

References

Ambroggio, E., *et al.* (2010). ArfGAP1 generates an Arf1 gradient on continuous lipid membranes displaying flat and curved regions. *EMBO J.* **29**, 292–303.

Angelova, M. I., *et al.* (1992). Preparation of giant vesicles by external AC electric fields. Kinetics and applications. *Progr. Colloid. Polym. Sci.* **89**, 127–131.

Bigay, J., *et al.* (2003). Lipid packing sensed by ArfGAP1 couples COPI coat disassembly to membrane bilayer curvature. *Nature* **426**, 563–566.

Bigay, J., *et al.* (2005). ArfGAP1 responds to membrane curvature through the folding of a lipid packing sensor motif. *EMBO J.* **24**, 2244–2253.

Cuvelier, D., *et al.* (2005). Coalescence of membrane tethers: experiments, theory, and applications. *Biophys. J.* **88**, 2714–2726.

Derenyi, I., *et al.* (2007). Membrane nanotube. *In* "Controlled Nanoscale Motion. Lecture notes in Physics," (H. Linke, and A. Mansson, eds.), Vol. 711, pp. 141–159. Springer-Verlag, Berlin.

Drin, G., *et al.* (2007). A general amphipathic alpha-helical motif for sensing membrane curvature. *Nat. Struct. Mol. Biol.* **14**, 138–146.

Drin, G., *et al.* (2008). Asymmetric tethering of flat and curved lipid membranes by a golgin. *Science* **320**, 670–673.

Franco, M., *et al.* (1995). Myristoylation of ADP-ribosylation factor 1 facilitates nucleotide exchange at physiological Mg2+ levels. *J. Biol. Chem.* **270**, 1337–1341.

Heinrich, V., and Waugh, R. E. (1996). A piconewton force transducer and its application to measurement of the bending stiffness of phospholipid membranes. *Ann. Biomed. Eng.* **24**, 595–605.

Hyman, A. A. (1991). Preparation of marked microtubules for the assay of the polarity of microtubule-based motors by fluorescence. *J. Cell Sci. Suppl.* **14**, 125–127.

Koster, G., *et al.* (2003). Membrane tube formation from giant vesicles by dynamic association of motor proteins. *Proc. Natl. Acad. Sci. U. S. A.* **100**, 15583–15588.

Kwok, R., and Evans, E. (1981). Thermoelasticity of large lecithin bilayer vesicles. *Biophys. J.* **35**, 637–652.

Leduc, C., *et al.* (2004). Cooperative extraction of membrane nanotubes by molecular motors. *Proc. Natl. Acad. Sci. U. S. A.* **101**, 17096–17101.

Leduc, C., *et al.* (2010). Mechanism of membrane nanotube formation by molecular motors. *Biochim. Biophys. Acta* **1798**, 1418–1426.

Lee, W. M., *et al.* (2007). Construction and calibration of an optical trap on a fluorescence optical microscope. *Nat. Protoc.* **2**, 3226–3238.

MacDonald, RC., *et al.* (1991). Small-volume extrusion apparatus for preparation of large, unilamellar vesicles. *Biochim. Biophys. Acta* **1061**, 297–303.

Manneville, J. B., *et al.* (2008). COPI coat assembly occurs on liquid-disordered domains and the associated membrane deformations are limited by membrane tension. *Proc. Natl. Acad. Sci. U. S. A.* **105**, 16946–16951.

Matsuoka, K., and Schekman, R. (2000). The use of liposomes to study COPII- and COPI-coated vesicle formation and membrane protein sorting. *Methods* **20**, 417–428.

Mclntyre, J. C., and Sleight, R. G. (1991). Fluorescence assay for phospholipid membrane asymmetry. *Biochemistry* **30**, 11819–11827.

Meleard, P., *et al.* (2009). Giant unilamellar vesicle electroformation from lipid mixtures to native membranes under physiological conditions. *Methods Enzymol.* **465**, 161–176.

Mesmin, B., *et al.* (2007). Two lipid-packing sensor motifs contribute to the sensitivity of ArfGAP1 to membrane curvature. *Biochemistry* **46**, 1779–1790.

Montes, L. R., *et al.* (2007). Giant unilamellar vesicles electroformed from native membranes and organic lipid mixtures under physiological conditions. *Biophys. J.* **93**, 3548–3554.

Montes, L. R., *et al.* (2010). Electroformation of giant unilamellar vesicles from native membranes and organic lipid mixtures for the study of lipid domains under physiological ionic-strength conditions. *Methods Mol. Biol.* **606**, 105–114.

Neuman, K. C., and Block, S. M. (2004). Optical trapping. *Rev. Sci. Instrum.* **75**, 2787–2809.

Pinot, M., *et al.* (2010). Physical aspects of COPI vesicle formation. *Mol. Membr. Biol.* **27**, 428–442.

Roux, A., *et al.* (2002). A minimal system allowing tabulation with molecular motors pulling on giant liposomes. *Proc. Natl. Acad. Sci. U. S. A.* **99**, 5394–5399.

Roux, A., *et al.* (2005). Role of curvature and phase transition in lipid sorting and fission of membrane tubules. *EMBO J.* **24**, 1537–1545.

Roux, A., *et al.* (2010). Membrane curvature controls dynamin polymerization. *Proc. Natl. Acad. Sci. U. S. A.* **54**(107), 4141–4146.

Sorre, B., *et al.* (2009). Curvature-driven lipid sorting needs proximity to a demixing point and is aided by proteins. *Proc. Natl. Acad. Sci. U. S. A.* **106**, 5622–5626.

Surrey, T., *et al.* (1998). Chromophore-assisted light inactivation and self-organization of microtubules and motors. *Proc. Natl. Acad. Sci. U. S. A.* **95**, 4293–4298.

Svetina, S., *et al.* (1998). Theoretical analysis of the effect of the transbilayer movement of phospholipid molecules on the dynamic behavior of a microtube pulled out of an aspirated vesicle. *Eur. Biophys. J.* **21**, 197–209.

Vetter, I. R., and Wittinghofer, A. (2001). The guanine nucleotide-binding switch in three dimensions. *Science* **294**, 1299–1304.

Walde, P., *et al.* (2010). Giant vesicles: preparations and applications. *Chembiochem* **11**, 848–865.

Waugh, R., and Evans, E. A. (1979). Thermoelasticity of red blood cell membrane. *Biophys. J.* **26**, 115–131.

CHAPTER 4

Reconstituting Multivesicular Body Biogenesis with Purified Components

Thomas Wollert

Molecular Membrane and Organelle Biology, Max Planck Institute of Biochemistry, Martinsried, Germany

Abstract

Activated cell surface receptors are rapidly removed from the plasma membrane through clathrin mediated endocytosis and transported to the endosome where they are either recycled or sorted to the lysosomal pathway to be degraded. Receptors, destined for degradation in the lysosome, are packaged into intraluminal vesicles (ILVs) of endosomes by a reaction that is topologically unrelated to other budding reactions in cells. First, receptors are clustered at the endosomal membrane and receptor-rich membrane patches then bud towards the lumen of the endosome. The nascent membrane buds are finally cleaved from the limiting membrane to release cargo-bearing vesicles into the endosomal interior.

Copyright 2012, Elsevier Inc. All rights reserved.

0091-679X/10 $35.00
DOI 10.1016/B978-0-12-386487-1.00004-3

The molecular machinery that drives multivesicular body biogenesis, the endosomal sorting complex required for transport (ESCRT) machinery, has been identified through genetic screens. It consists of the cytoplasmic, hetero-multimeric complexes ESCRT-0, -I, -II, and -III, and of the Vps4/VtaI complex. Although the ESCRT machinery has been characterized extensively using cell-biological and biochemical approaches, the molecular mechanism of multivesicular body biogenesis remained unclear. In this chapter, I will present *in vitro* reconstitution systems that we used to study ESCRT-driven membrane remodeling reactions with purified components on artificial membranes. This includes generation of large and giant unilamellar liposomes, as well as *in vitro* reconstitution reactions of fluorescently labeled proteins on such membranes. I will discuss both, the potential of *in vitro* systems to analyze membrane-remodeling events and also their limitations.

I. Introduction

Plasmamembrane proteins, including transmembrane receptors or channels, destined for lysosomal degradation, are delivered to endosomes where they are sorted into intraluminal vesicles (ILVs), giving rise to multivesicular bodies (MVBs). Packaging of those transmembrane proteins from the limiting membrane of endosomes into ILVs requires a topologically unusual budding reaction. The primary sorting signal for proteins to be delivered to ILVs is ubiquitination (Katzmann *et al.*, 2001). Mono- or poly-ubiquitin moieties are conjugated to the cytoplasmic tail of transmembrane proteins to initiate their uptake from the plasma membrane through clathrin-mediated endocytosis (Hicke and Dunn, 2003; Huang *et al.*, 2006) giving rise to clathrin-coated vesicles, which transport cargo proteins to the endosome. At the endosomal membrane ubiquitin is again the primary sorting signal required to deliver cargo into ILVs. The unusual budding reaction involved, requires clustering of cargo molecules at the endosomal membrane and budding of cargo-rich membrane patches towards the lumen. Nascent membrane buds are cleaved from the cytoplasmic leaflet of the lipid bilayer to release ILVs into the endosomes (Hurley and Hanson, 2010).

The endosomal sorting complex required for transport (ESCRT) machinery has been identified to catalyze this reaction. It comprises the complexes ESCRT-0, -I, -II, -III, and the Vps4/VtaI complex which are highly conserved from yeast to humans (Leung *et al.*, 2008). Experimental evidence suggests that flat clathrin coats at the early endosomal membrane are involved in receptor clustering and in recruiting the first complex of the ESCRT-machinery, ESCRT-0, to the endosomal membrane by binding its subunit Hrs (Vps27 in yeast, Raiborg *et al.*, 2006). ESCRT-0 is a heterodimeric complex of Vps27 and Hse1 in yeast and Hrs and STAM1/2 in humans, which possesses five Ub-binding domains. Recent studies demonstrated that ESCRT-0 binds poly-ubiquitinated cargo with high avidity (Ren and Hurley, 2010) and clusters a mono-ubiquitin model-cargo on artificial membranes *in vitro* (Wollert and Hurley, 2010). To package cargo into ILVs, a budding reaction needs to be induced which is topologically opposite from other budding reaction in cells in a sense that budding takes place away from the cytosol (Hurley and Hanson, 2010). Interestingly, budding of virions from the plasma membrane, such as HIV-I virions, is topologically related to MVB biogenesis (Bieniasz, 2009). Whereas the viral Gag protein is responsible for the formation of the membrane bud in viral budding (Briggs *et al.*, 2009), a super-complex consisting of ESCRT-I and -II has been shown to induce membrane budding on artificial membrane substrates *in vitro* (Wollert and Hurley, 2010). ESCRT-I is a hetero-tetrameric complex consisting of the four subunits Vps23, Vps28, Vps37, and Mvb12. Both, Mvb12 and Vps23 possess an ubiquitin interacting motif (Shields *et al.*, 2009), and the ubiquitin binding domain of Vps23 (UEV-domain) also recognizes the PTAP-like motif in Vps27 of ESCRT-0, providing a physical link between both complexes (Bilodeau *et al.*, 2003). The recruitment of ESCRT-I by ESCRT-0 appears to be important to coordinate ESCRT-0 mediated cargo clustering and membrane-bud formation by ESCRT-I and -II in order to sort cargo into ILVs during MVB-biogenesis (Wollert and Hurley, 2010). ESCRT-II consists of two copies of Vps25 and one copy of each of Vps22 and Vps36. Ubiquitin recognition is mediated by the GLUE-domain in human and the NZF2-domain in yeast Vps36, respectively (Wollert *et al.*, 2009b). The exact mechanism of cargo delivery from ESCRT-0 domains into ESCRT-I/-II stabilized membrane buds remains to be elucidated, but probably involves transient interactions of the ubiquitin-moieties on cargo-molecules with ESCRT-I and -II, as well as deubiquitination by deubiquitinating enzymes such as yeast Doa4 (Luhtala and Odorizzi, 2004). Vps25 of ESCRT-II recruits the first subunit of the ESCRT-III complex, Vps20, to the nascent membrane bud (Wollert and Hurley, 2010; Yorikawa *et al.*, 2005). In contrast to the upstream ESCRT-complexes, ESCRT-III subunits are monomers in the cytosol adopting an autoinhibited conformation. Their sequentially recruitment and assembly at the membrane (Teis *et al.*, 2008) is initiated by ESCRT-II in MVB-biogenesis and by the ESCRT-associated protein Alix in cytokinesis and in retroviral budding (McCullough *et al.*, 2008; Teo *et al.*, 2004). Evolutionarily, the ESCRT-III complex is the most conserved part of the ESCRT-machinery as it is also present in a subset of Archaea where its only function is to cut the membrane bridge between dividing cells (Samson *et al.*, 2008). During MVB biogenesis Vps20 recruits and activates Snf7, which polymerizes into spiral like homo-oligomers (Hanson *et al.*, 2008) that are capped by Vps24. The spirals are thought to align with the bud-neck in a way that each turn of the inward directed spiral constricts the membrane neck until spontaneous fusion of the membrane occurs and the nascent bud is released into the lumen of MVBs (Wollert and Hurley, 2010; Wollert *et al.*, 2009a). The last core-subunit of ESCRT-III, Vps2, is recruited after scission of the membrane neck has occurred and recruits the Vps4/VtaI complex to the membrane (Yeo *et al.*, 2003). The AAA-ATPase Vps4 recycles the ESCRT-III complex from the limiting membrane and releases ESCRT-III subunits in their autoinhibited states into the cytosol (Lata *et al.*, 2008). Apart from the core ESCRT-machinery, a number of regulatory and associated proteins are required *in vivo* to modulate ESCRT-function in MVB-biogenesis and cytokinesis in order to provide temporal and spatial control (Roxrud *et al.*, 2010).

II. Rationale

The membrane remodeling activity of proteins can be analyzed by reconstituting them on artificial lipid bilayers *in vitro*. The proteins of interest are expressed recombinantly, purified to homogeneity and, depending on the experimental system, labeled with fluorescent dyes. They are assembled into functional complexes on model membranes including supported lipid bilayers (membranes that are immobilized on a surfaces, such as cover slips) and vesicles. The latter have the advantage that the lipid bilayer can be bent or remodeled without interference of a support, allowing positive and negative membrane curvature to be induced, as well as membrane fusion and scission reactions to be studied.

Membrane vesicles with defined lipid compositions can be produced by hydrating dried lipid films with aqueous solutions. This “swelling” approach results in multilamellar vesicles (MLVs) of various sizes, which can be extruded through membranes with defined pore-sizes to generate large unilamellar vesicles (LUVs) with narrow size distribution. These LUVs are hundreds of nanometers in diameter and are characterized by both, high membrane-curvature and -tension. Some membrane binding proteins are able to sense the curvature of membranes, which is reflected in their preferential association with membranes of certain curvature, and LUVs have been used to study this relationship (Peter *et al.*, 2004). However, the high membrane tension of LUVs might interfere with membrane remodeling reactions to be reconstituted with purified proteins. Because of their size, LUVs, require electron microscopy to be visualized. Giant unilamellar vesicles (GUVs) are in the range of tens of micrometers in diameter allowing microscopic techniques to be used to visualize them. They are generated by hydrating lipid films in the presence of alternating electric fields. GUVs are characterized by their low membrane tension, which facilitates the reconstitution of membrane remodeling reactions. Fluorescent-labeled lipids can be incorporated into the lipid bilayer of GUVs in order to visualize them under the fluorescence microscope. The combination of fluorescent GUVs with fluorescent-labeled proteins is particularly suited to study membrane remodeling reaction in real time.

In this chapter, I will describe different approaches to reconstitute molecular machines on both, LUVs and GUVs in order to reveal their mechanisms of action. The LUV-system can be used to quantify binding affinities of proteins to membranes using Fluorescence Resonance Energy Transfer (FRET), which provides valuable information on spatiotemporal organization of proteins and complexes on membranes. The membrane is labeled with a fluorescent dye (the donor) that absorbs light of shorter wavelength than the fluorescent-label (acceptor) that is attached to the protein of interest. If the absorption spectrum of the acceptor overlaps with the emission spectrum of the donor, energy can be transferred from the donor- to the acceptor-dye non-radiatively. This resonant energy transfer only occurs if donor and acceptor are in close proximity, which usually requires protein and membrane to be in direct contact. Under these circumstances, the excited donor is not emitting energy in form of fluorescence, but transferring it to the acceptor dye, which emits light with its characteristic emission-wavelength that is detected as FRET-signal.

Another insightful property of proteins at membranes can be analyzed by reconstituting the fluorescently labeled protein at GUVs. Direct observation under the fluorescent microscope allows Fluorescent Recovery After Photobleaching (FRAP) experiments to be performed in order to reveal the mobility of proteins at the membrane and to estimate their lateral diffusion coefficients (Guo *et al.*, 2008). Since membrane bending and remodeling usually requires a concerted action of multiple protein-subunits or copies, which frequently self-organize in order to induced and/or maintain membrane-shapes such as tubules or buds, their mobility is strongly reduced compared to free lateral diffusion. The extent of immobilization of those proteins at membranes can be measured by FRAP. The fluorescence of the (fluorescently labeled) protein of interest at the GUV-membrane is photobleached with high intensity laser light in a defined area and the recovery of fluorescence is detected and quantified as a function of time. Photobleached molecules irreversibly loose their fluorescence. Recovery of the fluorescence occurs by exchange with the bulk solution due to constant association and dissociation of proteins, and by lateral diffusion of proteins from non-bleached membrane areas.

Reconstituting biological processes with purified components *in vitro* provides highly dissectible systems, which can be combined with a number of microscopic and biochemical approaches to holistically characterize the functions of the proteins involved. The major limitation of the system lies in the fact that the components that are required, including lipids, proteins, and cofactors, need to be known. Another requirement pertains protein expression and purification, which might be limited as well. Genetic and proteomic screens have, on the other hand, identified many components of these pathways and technical advance has been made in protein production and lipid-synthesis or -isolation. For these reasons *in vitro* reconstitutions just begin to be explored in order to reveal molecular mechanisms involved in membrane trafficking and remodeling.

III. Methods

A. Lipid Handling and Lipid Mixtures

Liposomes are generated from dried lipid-films with a defined lipid composition that usually mirrors the composition of the organelle of interest. To produce such lipid films, mixtures of lipids, dissolved in suitable solvents, are spread on glass surfaces and dried under a nitrogen-stream or under vacuum. Phospholipids and sterols, such as ergosterol in yeast, cholesterol in animal membranes and sterylglucosides in plants, constitute the bulk of the lipids in biological membranes. Due to their hydrophobic nature they need to be dissolved in chloroform. To avoid evaporation of the organic solvent, lipids should be stored at $< -20\ ^{\circ}C$ in glass vials with Teflon sealing. Exposure to oxygen and light leads to oxidation of unsaturated lipids. For this reason lipids need to be kept under nitrogen-atmospheres, protected from light, for long-term storage. All lipids that are dissolved in chloroform should also be protected from any contact with plastic devices or parafilm. Pipetting of lipids should be carried out with microdispensers equipped with glass-capillaries and all glassware should be rinsed with chloroform before usage. Glycolipids and highly charged lipids like phosphatidylinositol phosphates are, depending on their hydrophobicity, soluble in methanol, choloroform:methanol or chloroform:methanol:water mixtures. If more hydrophilic lipids need to be incorporated into lipid mixtures containing hydrophobic phospholipids, care must be taken regarding the composition of the solvent to avoid phase separation of the solvent and partitioning of lipids.

For reconstitution experiments of the ESCRT-machinery, a lipid mixture has been used that was, on the one hand, similar to the composition of endosomes, but produced, on the other hand, high-quality liposomes. This becomes particularly challenging in case of GUVs. The preparation of GUVs involves electroformation that limits the composition of the lipid films from which vesicles are being generated. The amount of charged lipids like phosphatidylserine or phosphatidylinositol phosphates should not exceed 20 mol%, otherwise quality and yield of GUVs are dramatically reduced. If sphingolipids (e.g., sphingomyelin) need to be incorporated into the membrane to produce Golgi-, ER-, or plasma-membrane-like compositions, phase separation into liquid ordered and disordered phases may occur. Hydrogen bond formation between sphingomyelin and cholesterol appears to drive phase formation and stabilizes the liquid ordered phase. Lipid-microdomains (or lipid rafts) also exist in biological membranes where they are required to segregate processes like vesicle budding to defined membrane areas of organelles (Bonnon *et al.*, 2010). Such processes, including COPI mediated budding of transport vesicles from the Golgi, have been reconstituted on GUVs with phase separated membrane domains, highlighting the importance of lipid-domains in organizing and spatially localizing membrane remodeling activities of proteins (Manneville *et al.*, 2008).

The physical property of the membrane can be influenced by other factors than phase separations as well, for example by the amount of cholesterol and the ratio of saturated and unsaturated lipids. Increasing amounts of cholesterol decrease membrane fluidity but prevent membranes with high content of saturated lipids to crystallize under physiological conditions. Apart from these considerations, one has to take into account that biological membranes are much more complex then membranes *in vitro*. The biggest concern to translate *in vitro* findings into the *in vivo* situation relates to transmembrane proteins, which are usually absent in model membranes but highly abundant in biological membranes, where they have great impact on the physical properties of membranes.

The lipid mixture, that have been used to reconstitute MVB-biogenesis consisted of 62% 1-palmitoyl-2-oleoyl-sn-glycero-3-phosphocholine (POPC), 25% cholesterol, 10% 1-palmitoyl-2-oleoyl-sn-glycero-3-phospho-L-serine (POPS), and 3% phosphatidylinositol-3-phosphate (PI(3)P). Due to the absence of sphingomyeline, phase separation did not occur and the membrane appeared to be in the liquid disordered phase.

B. Preparation of LUVs

LUVs are a widely used system to study protein–lipid interactions in solution. Different techniques have been developed in order to generate LUVs, but the most frequently used one involves extrusion of MLVs, which consist of concentric lipid bilayers separated by solvent. MLVs can be produced by hydration of thin lipid films that have been generated in a borosilicate glass tube by evaporating the solvent of a lipid mixture under a nitrogen-stream. The total lipid amount used to generate the lipid film varies usually between 1 and 10 mg. Lipid mixtures which contain apart from chloroform and methanol also water need to be dried more extensively, for example, under vacuum for several hours or overnight. After the solvents have been removed, the lipid film is swollen by incubating 1 mL buffer for 30 min with frequent vortexing. The incubation temperature needs to exceed the phase transition temperature of the used lipids. Phase transition of lipids describes the change from the ordered gel-phase, where the hydrocarbon chains are highly ordered, to the disordered liquid crystalline state, in which hydrocarbon chains display higher conformational disorder. The transition temperature of most lipids is below room temperature; however, phosphatidylinositol phosphates have transition temperatures of 60 °C and above, requiring vesicle swelling and subsequent extrusion to be carried out at elevated temperatures to avoid lipids with higher transition temperature to be excluded from the forming lipid bilayer. Swelling of lipid films produces usually large MLVs with diameters of >200 nm.

After swelling, five freezing (dry ice plus ethanol) thawing (60 °C warm water bath) cycles should be performed with vortexing the sample in between. Through different freezing characteristics of intraluminal and interlamellar water, osmotic stress between concentric lipid bilayers causes them to fuse, reducing the number of bilayers in MLV (Kaasgaard *et al.*, 2003). To produce LUVs with defined diameters, the suspension needs to be extruded 10–20 times through a membrane with a defined pore size (mini extruder set, Avanti polar lipids). During extrusion MLVs are squeezed through the pore of the membrane, which leads to fusion of the bilayers, making them unilamellar, and reduction of their size, leading to a homogeneous population of LUVs with defined diameters. LUVs can be stored in microcentrifuge tubes for up to 5 days at 4 °C.

C. Preparation of GUVs

Two different concepts have been developed to generate GUVs, gentle swelling and electroformation (Angelova and Dimitrov, 1986). The first approach has the advantage that fewer restrictions in lipid composition apply; however, the method suffers from low GUV yield and quality. For this reason, the most commonly used approach is electroformation and will therefore be explained in detail. A variety of different devices, including commercially available once, have been developed to generate GUVs by electroformation. All of them have in common that a dried lipid film on a conductive carrier material, such as a platinum wire or conductive glass slides, is exposed to an electric AC-field in the presence of aqueous solution. The electric field stimulates hydration of the lipid film and subsequent formation of unilamellar vesicle by an unknown mechanism. Depending on the experimental setup, homogeneous GUV preparations with diameters between 10 and 100 μm can be generated, which are easily observable by light microscopy.

The highest yield of GUVs can be obtained by using conductive glass slides (coated with, e.g., indium tin oxide) with a resistance of 0.4 ± 0.08 Ω/cm^2 as carrier of the dried lipid film (manufactured by Delta Technologies, Loveland, CO). A lipid mixture with a total lipid concentration of 10 mg/mL should be used to generate a thin, homogenous lipid film. The quality of GUVs depends critically on the quality of the lipid film and best results are obtained if thin glass capillaries are used to spread 5–10 μL of the lipid mix on the slides. After drying under vacuum (<10 mbar, in a vacuum oven), the chamber is being assembled and slides are arranged in parallel, separated by a 1 mm gap, which is filled with a non-conductive aqueous solution. The configuration is similar to that of a plate-capacitor (Fig. 1). The glass slides are connected to a function generator and an alternating current of 1.2 V and 10 Hz is applied. If the lipid mixture contains higher amounts of charged lipids or buffer with salt is being used for swelling, the voltage can be stepwise ramped up from 0.25 V and 5 Hz to 1.2 V and 10 Hz with 10 min incubation intervals. This procedure increases both, quality and quantity (fewer multivesicular vesicles and reduced number of GUVs containing intraluminal structures) of

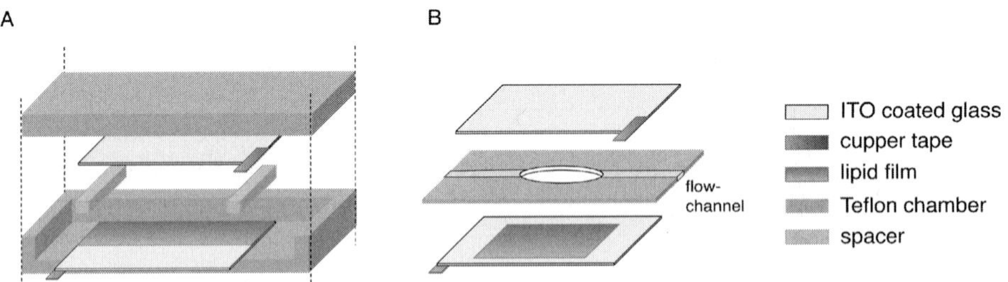

Fig. 1 Chamber design for the electroformation of GUVs. Schematic drawing of the two Teflon chamber designs used for the electroformation of GUVs from lipid films. Lipid mixtures are streaked out onto ITO-coated glass slides and dried under vacuum. The homogeneity of the dried lipid film determines both, homogeneity in terms of lipid and size distribution of GUVs and their quality. Multilamellar GUVs and intra- or extra-luminal tube/vesicles may be observed if the amount of charged lipids in the lipid film is too high. The slides are assembled with the lipid film carrying conductive surfaces facing each other, separated by a 1-mm wide gap, which is filled with an aqueous solution (e.g., 600 mM sucrose solution). High ion strength (>50 mM) of the hydration solution severely impacts on the quality of GUVs. (A) Closed chamber. The Teflon chamber consists of two pieces. Spacers maintain a distance of 1 mm between the two glass slides. Copper foil is attached to the glass slides, providing a flexible link to connect them with the function generator using alligator-clip wires. Contact faces of the Teflon chamber might be sealed with grease to avoid leakages. (B) Flow chamber: The two glass slides are separated by a Teflon slide. Its central hole generates a chamber that is connected to flow channels that are embedded in the Teflon slide. (See color plate.)

GUVs (Mitov *et al.*, 1993), which would otherwise suffer from the high ion strength of the lipid film or the aqueous medium. The maximal ion concentration of the buffer that can be tolerated is, however, 50 mM. In addition, an ambient temperature that exceeds the phase transition temperature of the used lipids (Goni *et al.*, 2008) should be maintained throughout electroformation and may require the procedure to be performed in an oven or incubator. For most phospholipids, including POPC, POPS, or POPE, phase transition temperatures are below room temperature. However, phophatidylinositol-phosphates have phase transition temperatures that exceed 50 °C. Incubation times depend on the lipid film and its composition but GUV formation is usually completed within 3–4 h. If the chamber was heated during electroformation it needs to be cooled down slowly to room temperature, otherwise artifacts in lipid distribution may be induced (Morales-Penningston *et al.*, 2010). Two basic types of chambers have been constructed to generate GUVs, a flow chamber that allows GUVs to be examined under the microscope *in situ* without transferring them into an observation chamber, and a closed system, which require GUVs to be harvested and transferred to an observation chamber after electroformation (Fig. 1). The advantage of the flow cell system is that GUVs stay connected with the glass slide and buffer or protein solutions can be added and washed out easily (Fig. 1B). However, remnants of the lipid film, which have not been transformed into GUVs, causes high background fluorescence and unproductively bind significant amount of proteins. In addition a high percentage of GUVs show due to their attachment to the glass substrate defects in their lipid bilayers, making them leaky so that buffer and added proteins can diffuse into the lumen of GUVs. This can cause membrane-remodeling proteins to act on both, the lumen and the extraluminal leaflet of the GUV-membrane at the same time. Another source of membrane damage comes from shearing forces that are produced by the flow itself.

The major disadvantage of the closed electroformation system (Fig. 1A) is caused by its design, which requires GUVs to be transferred to a chambered microscopy slides in order to be imaged. Pipetting of GUVs can cause membrane-rupture or -alterations, including the generation of intra- or extraluminal membrane-structures. This can be avoided by directly transferring GUVs from the Teflon chamber into the observation chamber using pipette-tips with a capacity >200 μL. The GUVs also need to be mixed with buffer or protein solution (e.g., 100 μL GUV suspension + 100 μL buffer), which introduces a second source of membrane disturbance. In addition, GUVs need to settle down to the bottom of the microscopy chamber to be observed under an inverted, fluorescent microscope, which requires differences in the density of GUVs and buffer. For this reason GUVs are usually generated in the presence of 600 mM sucrose and diluted with a buffer that is isoosmotic to the sucrose solution in a one-to-one ratio. The bottom of the chamber may be coated with BSA (0.5 mg/mL for 5 min followed by washing with buffer) to avoid damage of the GUVs upon contact with the bottom of the observation-chamber. The biggest advantage of the system is that different reactions can be set up in parallel and analyzed in short periods of time with identical conditions. This becomes particularly important if multi-component systems are being reconstituted. In addition, control reactions can be set up by incubating buffer blanks instead of proteins with GUVs. Almost all GUVs can be transferred to the observation chamber without being disrupted so that no buffer or protein solution can diffuse into their lumen. Taken together, both systems have advantages and disadvantages and the choice depend on experimental requirements.

D. Liposome Sedimentation Assay

Protein–lipid interactions can be analyzed in a qualitative or semi-quantitative way by liposome sedimentation assays (Peter *et al.*, 2004). LUVs with defined diameters, for example, 100 μm, are incubated with the protein or complex of interest for 30 min at room temperature. LUVs are sedimented by ultracentrifugation for 30 min at 300,000 *g* and the supernatant containing the unbound protein fraction is removed. The pellet is washed with buffer and both, supernatant and pellet are subjected to SDS-polyacrylamide-gel-electrophoresis (PAGE). The pellet contains the bound protein fraction and quantification of the intensities of the corresponding protein bands can be used for semi-quantitative analysis and to estimate binding affinities of proteins to membranes. The physical separation of bound and unbound protein through ultracentrifugation has the disadvantage, that weak or transient interactions cannot be analyzed. Due to a constant re-adjustment of the chemical equilibrium, bound protein dissociates from the LUV-membrane during centrifugation and the actual amount of protein being recovered from the pelleted fraction underestimates binding affinity or makes it impossible to detect any interaction. For this reason, quantitative measurements of protein-membrane interactions, which do not rely on separation of bound and unbound protein, need to be performed if reliable and precise data are needed. Care must also been taken if proteins tend to aggregate since these protein aggregates are co-sedimented with LUVs without a physical interaction between protein and membrane. The presence of protein in the pelleted fraction, therefore, not necessarily indicates binding of the protein to LUVs.

E. Quantifying Affinities of Proteins to Membranes

The system of choice to reliably determine binding affinities of proteins to membranes uses FRET to quantify the amount of membrane-bound protein at a given total protein concentration (Coutinho *et al.*, 2011). FRET has, compared to other techniques that have been used to study protein-membrane interactions (e.g., fluorescence correlation spectroscopy or surface plasmon resonance) the advantage that liposomes can be used as membrane substrates. The membrane of liposomes can be deformed, allowing affinities of proteins to be determined, which not only bind, but also deform membranes. In order to perform FRET experiments, fluorescent labels, which are acting as donor and acceptor, need to be incorporated into the membrane and covalently attached to the protein, respectively. The FRET-donor is excited with light of the corresponding excitation wavelength. If the emission spectrum of the donor and excitation spectrum of the acceptor overlap and both are in close proximity, the excitation energy is transmitted from donor to acceptor over space without light being emitted. The excited acceptor emits light of its corresponding emission wavelength, which is detected as FRET-signal. The transfer-efficiency depends also on the orientation of the dipole moments of donor and acceptor. Maximal efficiency occurs if they are arranged in parallel, no transfer takes place if they are perpendicular to each other. This introduces a geometric relationship between FRET efficiency and donor–acceptor orientation, which, under experimental conditions, often averages out due to rotational freedom of the attached fluorescent dyes. The covalent modification of proteins with fluorescent dyes bears, on the other hand, the risk that membrane binding sites might be masked. In order to overcome this possible limitation, the fluorophore can be attached to different regions of the protein. Alternatively, tryptophan fluorescence of unlabeled proteins might be used. The latter approach requires, however, tryptophan residues to be in close vicinity to the membrane binding site for FRET to occur.

Liposomes are being produced as described, but 1 mol% fluorescein-PE (or another suitable fluorescently labeled lipids) is added to the lipid mixture. FRET titrations are carried out using standard fluorescence spectrophotometer. The cuvette is filled with LUVs (total lipid concentration ~145 μM) and the protein of interest is being added (labeled with, e.g., tetramethylrhodamine-iodoacetamide) by titrating it into the LUV suspension. At least 20 titration steps should be performed to obtain enough data points to cover a broad range of the binding isotherm. The concentration of the protein solution that is titrated into the LUV solution depends on the affinity of that protein to membranes. Useful titration curves are being recorded if the final protein-concentration in the cuvette is 10-fold higher than its affinity (K_D) to the membrane. Initially, affinities are unknown, requiring a series of experiments with different protein concentrations to be performed. To evaluate binding affinities, the detected FRET signal in the absence of protein needs to be determined (I_0) and used for correction of FRET signals after each titration step ($\Delta I = I - I_0$). If the amount of added protein exceeds available binding places by a factor of at least two and the used protein concentrations are higher than the K_D of the protein to the membrane, the titration curve has an asymptotic shape and saturation of the FRET signal I_∞ occurs, which is also corrected for I_0 ($\Delta I_{max} = I_\infty - I_0$). The binding isotherm is obtained by plotting the ratio $\Delta I/\Delta I_{max}$ as a function of the protein concentration and can be evaluated using the Langmuir equation (Eq. (1)) where [P] is the concentration of the titrated protein, Θ is the occupancy of binding places, and K_A is the association constant, the reciprocal of the dissociation constant K_D. This mathematical model, however, only explains experimental data if all binding places are equal and no cooperative or anti-cooperative effects are present.

$$\Theta = \frac{I_0}{I_{max}} = \frac{K_A \cdot [P]}{1 + K_A \cdot [P]} \tag{1}$$

Interaction analysis with FRET also requires control measurements to be performed to avoid misinterpretations of the FRET signal. It is recommended to titrate the protein into a solution of buffer (corresponding to, e.g., acceptor without donor) and buffer into the solution of liposomes (corresponding to, e.g., donor without acceptor) as controls for dilution effects. Another consideration that needs to be taken into account concerns the membrane curvature of LUVs. It has been shown for a number of proteins that they sense the curvature of membranes and bind those with high curvatures more tightly than flat membranes and *vice versa*. Examples include ArfGAP1 which initiates the disassembly of the COPI coat from vesicles, and the membrane scission machine dynamin which cuts nascent clathrin-coated vesicles from the source membrane (Bigay *et al.*, 2003; Roux *et al.*, 2010). For these proteins, the determined binding affinity is a function of the LUV-diameter because small diameters correspond to high membrane curvature and *vice versa*. Valuable insights into the mechanism of action of membrane binding proteins can therefore be revealed if they are reconstituted on LUVs with different diameters and corresponding binding affinities are being determined and compared. Due to technical limitations in producing LUVs with average diameters larger than 200 nm, LUV-membranes are generally highly curved compared to GUVs. The latter can be considered as flat membranes compared to the size-scale of proteins, which also needs to be taken into account if proteins that prefer bent membranes are reconstituted on GUVs.

F. Reconstituting Proteins on GUVs

After electroformation of GUVs is completed, the chamber is cooled down to room temperature. If the closed system is used (Fig. 1A) most GUVs are floating in the sucrose solution, and those that are still attached to the glass support can be harvested by carefully sliding a Pasteur pipette-tip along the surface of the glass slide. However, GUVs cannot be stored and need to be used immediately. Due to their low membrane tension, GUVs are rather fragile and the quality of the preparation rapidly decreases over time. For each experiment 100 μL of the GUV-suspension are mixed with 100 μL of buffer (50 mM TrisHCl pH 7,4; 300 mM NaCl) in a Lab-Tek 8-well chambered microscopy slide (Borosilicate 1.0). Due to the higher density of the luminal sucrose solution of GUVs (600 mM sucrose) compared to the density of the bulk solution (300 mM sucrose, 25 mM TrisHCl, and 125 mM NaCl), GUVs are settling down to the bottom of the chamber and can be imaged with an inverted fluorescence microscope. Proteins or protein complexes of interest can be added to the GUV-suspension by pipetting and gentle stirring to facilitate mixing. If possible, proteins should be labeled with fluorescent dyes, which allow their localization to be followed. This is advantageous because the lipid composition of GUVs has frequently been observed to be inhomogeneous, particularly if phosphatidylinositol-phosphates or other hydrophilic lipids are incorporated into the membrane. The inhomogeneity directly impacts on protein recruitment to GUVs if those lipids are specifically recognized, causing an apparently diverse protein density on GUVs, which cannot be detected if proteins are not fluorescently labeled. On the other hand, membrane bending or remodeling phenomena can be correlated with protein function if proteins are labeled and their localization at the membrane can be detected precisely. Evaluating the spatial localization of proteins with respect to the deformed membrane may reveal valuable insights into the mechanism of membrane remodeling by those proteins.

G. Mobility of Proteins at Membranes

Peripheral and integral membrane proteins are not statically localizing to a certain membrane area, they are mobile due to lateral diffusion. In addition, peripheral membrane proteins are in a chemical equilibrium with non-bound proteins in the bulk solution. The kinetics of their association with and dissociation from the membrane in turn, determines their affinity to membranes. However, this equilibrium can be influenced by factors such as binding partners that are immobilizing peripheral membrane proteins at the membrane. Determining the mobility of proteins and macromolecular assemblies at membranes could therefore reveal valuable information on their function. The method of choice to do so is FRAP of fluorescent-labeled proteins that have been reconstituted on GUVs. Recovery of fluorescence after photobleaching of these proteins occurs due to lateral diffusion and exchange with the bulk solution. Since both factors cannot be determined independently, the exchange rate with the bulk solution needs to be analyzed first. After adding the protein of interest to GUVs and allowing the binding-equilibrium to be reached, the fluorescence of the dye that is attached to the protein is bleached by high intensity laser light covering the entire area of the GUV (Fig. 2A). Since fluorescence of the bleached molecules cannot recover, which is not true for all fluorescent dyes, but represents a crucial requirement, all molecules associated with the membrane of that particular GUV are inactivated. Due to the fact that photobleached proteins constantly exchange with the bulk solution, fluorescence recovers over time. By determining the fluorescence intensity of the bleached area as a function of time, the exchange rate and the amount of exchangeable proteins can be determined. The influence of other factors, for example, interaction partners, on recovery characteristics can be assessed by comparing recovery profiles in the presence and absence of those factors. If the exchange rate is slow, FRAP can be used to determine two-dimensional diffusion rates of the fluorescently labeled proteins at the membrane. The protein of interest is again reconstituted on GUVs but only a part, for example a quarter of the area covered by the GUV, is photobleached (Fig. 2B). The recovery of that area is observed over time and because exchange with the bulk solution is slow, recovery is dominated by lateral diffusion of proteins from the non-bleached into the bleached area. Again, the influence of other factors such as interaction partners can be assessed by comparing recovery profiles in the presence and absence of that factor. To reveal quantitative data such as recovery half-times form FRAP-

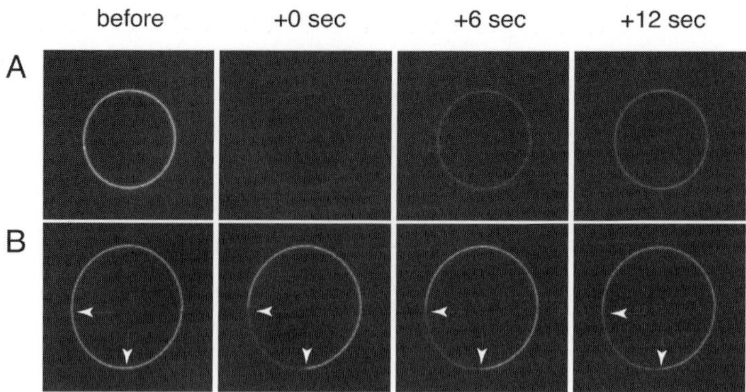

Fig. 2 FRAP experiment of fluorescent-labeled proteins on GUV-membranes. ESCRT-0 labeled with Alexa-488 using native Cys-residues and maleimide coupling chemistry binds GUV-membranes in the absence of cargo efficiently. The fluorescence of ESCRT-0–Alexa-488 was bleached by 20 iterations of high-intensity laser light pulses (100% laser power of laser lines 458 nm, 488 nm, and 514 nm). Imaging before and after bleaching was performed using a LSM510 confocal microscope (Zeiss) with 63× Plan Apochromat 1.4 NA objective and a 488/543 dichroic mirror. Images were captured at three-fold zoom, a resolution of 124 × 124 pixels, an open pinhole, and 14 μW laser power of the 488 nm laser line every 149 msec for 60 sec. The shown images have been taken before (first column) and at indicated time-points after photobleaching (columns 2, 3, and 4). (A) An area covering the entire GUV was photobleached. Recovery occurs only through exchange of membrane-bound proteins with the bulk solution and is incomplete and slow. (B) A quarter of the GUV was photobleached (indicated by arrows) to analyze recovery by both, exchange with the bulk solution and diffusion of proteins from non-bleached areas of the GUV-membrane into the bleached area.

profiles, determined fluorescence intensities need to be corrected for background fluorescence. This is done by integrating the fluorescence of a non-bleached area of similar size but without membrane in each recorded frame. This background fluorescence is subtracted from the integrated intensity of the bleached area of the same frame. A second correction that needs to be done eliminates the influence of photobleaching while recovery-images are being acquired. For this purpose the integrated intensity of a non-bleached GUV area of similar size is determined per frame and also corrected for background fluorescence. Dividing the corrected fluorescence of the photobleached area by that of the non-photobleached area yield the intensity that is also corrected for photobleaching during imaging. The intensities are finally normalized by setting the integrated and corrected fluorescence before the area of interest was photobleached to one and by converting corrected intensities after photobleaching to respective fractions of one. To determine model-cargo mobility at GUV membranes in the absence and presence of ESCRT-complexes, ten and one thousand images were obtained before and after the area of interest was photobleached, respectively. The duration of image acquisition was 0.15 s.

FRAP experiments on GUVs have the disadvantage, that high laser intensities might cause local heating of the photobleached membrane area. This can result in membrane deformations because of different membrane tensions between photobleached and non-bleached areas.

H. Reconstituting the ESCRT-Machinery on Artificial Membranes

From all membrane remodeling reactions that have been reconstituted so far, the generation of MVBs through the action of ESCRT-proteins is the most complex one. It involved clustering of cargo protein on model-membranes, budding of cargo containing vesicles, and scission of the nascent membrane buds from the limiting membrane to release them into the lumen of GUVs. The substrates of ESCRT proteins are, apart from membranes, ubiquitinated transmembrane proteins. However, incorporating transmembrane proteins into the membrane of GUVs is difficult and high densities can usually not be achieved. To simplify the reconstitution reaction we tethered His-tagged ubiquitin to GUV-membranes using DOGS-(Ni)NTA (1,2-dioleoyl-sn-glycero-3-[(N-(5-amino-1-carboxypentyl) iminodiacetic acid) succinyl] nickel salt; Avanti Polar Lipids). The (Ni)NTA head group forms a strong complex with His-tags, allowing their efficient recruitment to the membrane, whereas the phospholipid DOGS functions as membrane anchor. Five mol% DOGS-(Ni)NTA have been added to the standard lipid mixture, described above, by corresponding reduction in POPC concentrations. To track the localization of the model-cargo, we fused cyan-fluorescent protein (CFP) to the C-terminus of ubiquitin. The model cargo His-Ub-CFP was efficiently recruited to the membrane of GUVs at concentrations as low as 130 nM, as judged by the distribution of CFP-fluorescence on membranes and in the bulk-solution (Fig. 3A). Although tight, the interaction of His-Ub-CFP and DOGS-(Ni)NTA was highly dynamic. Photobleaching of the entire CFP-fluorescence of His-Ub-CFP on one vesicle recovered within seconds, indicating that a fast exchange of the inactivated cargo with non-bleached cargo molecules from the bulk solution occurs. To visualize ESCRTs at membranes, subunits in the corresponding complexes have been labeled with Alexa-488 dyes using native (ESCRT-0, -I, and -II) or engineered Cys-residues (ESCRT-III subunits) within the proteins or complexes after purification. This allowed us to observe binding and co-localization of ESCRTs (green) and cargo (cyan) at membranes (red, Fig. 3). The ESCRT-complexes were incubated with GUVs by adding them stepwise to the observation chamber with 5-minute incubation intervals in the order ESCRT-0, -I, -II, Vps20, Snf7, Vps24, Vps2, and Vps4. Concentrations were in the nanomolar range. The fluorescence of membrane-associated Alexa488 labeled ESCRT-complexes recovered, in contrast to that of the model cargo His-Ub-CFP, very slowly after photobleaching (Fig. 2). Once associated with the membrane, ESCRTs appear to exchange very slowly with the bulk solution. The inefficient recovery of ESCRT-Alexa488 fluorescence also suggested slow lateral diffusion of these complexes on membranes.

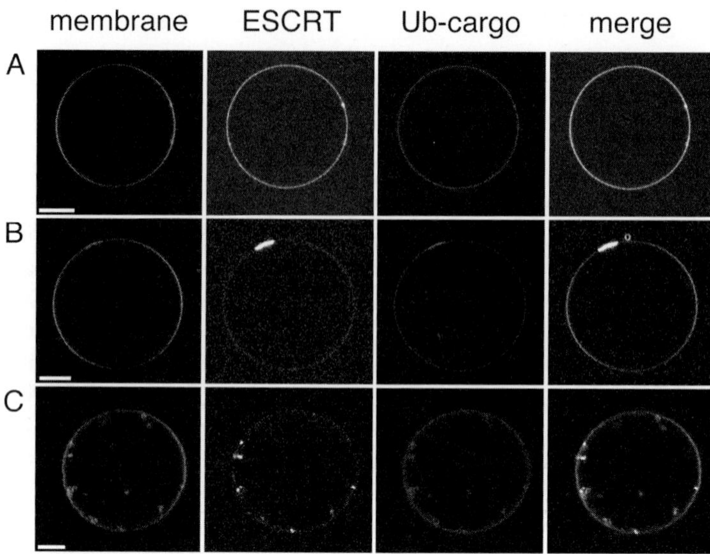

Fig. 3 Co-localization with and deformation of GUV-membranes by fluorescent-labeled ESCRT complexes. After decorating GUVs (membrane stained with rhodamine-PE, red) with the model cargo His-Ub-CFP (cyan) by tethering it to the lipid DOGS-(Ni)NTA, Alexa488-labeled ESCRT-complexes (green) have been added. (A) ESCRT-I is efficiently recruited to the membrane but does not induce membrane deformations. (B) ESCRT-0 clusters cargo at the membrane and initiates the formation of large cargo- and ESCRT-0-rich domains. C) The ESCRT-I/-II supercomplex deforms the GUV-membrane into membrane buds. Its localization, however, is restricted to the bud neck without entering the lumen of the nascent membrane buds. Scale bar = 10 μm. (See color plate.)

IV. Materials

Phospholipids including POPC, POPS, Cholesterol, DOGS-(Ni)NTA, and fluorescein-PE have been purchased from Avanti Polar Lipid, PI(3)P from Echelon. For lipid-pipetting VWR single-channel digital microdispensers have been used. Dyes for fluorescent labeling of proteins (Alexa Fluor 488 C_5 maleimide, tetrametylrhodamine-iodoacetamide) have been purchased from Molecular Probes. The Teflon chamber for GUV electroformation was manufactured according to Fig. 1A, and indium tin oxide coated glass slides (50 mm $\times$ 50 mm $\times$ 1.1 mm, 10 $\pm$ 2 Ω) have been obtained from Delta Technologies. The alternating current for electroformation has been generated with a BK Precision 3 MHz function generator and monitored with a Tenma 20 MHz oscilloscope. Imaging has been performed using Lab-Tek chambered coverglasses (1.0 Borosilicate) and an LSM 510 confocal fluorescence microscope (Zeiss). Liposomes have been generated using the Avanti Mini-Extruder, for liposome sedimentation a Beckman Coulter Optima TLX-120

Ultracentrifuge has been used. FRET data were obtained using a Fluorolog-3 fluorescence spectrometer (HoribaJobinYvon).

V. Discussion

Membrane remodeling is not only essential for MVB biogenesis and receptor-downregulation, but for almost all aspects of cellular homeostasis. The integrity of both, organelles and the plasma membrane, depend on a vesicular transport system to supply them with building blocks like proteins and lipids. Organelles themselves are highly dynamic structures and their function critically depend on tightly regulated dynamic rearrangements as well as organelle fusion and fission (Stenmark, 2009). Over the last decades it became clear that identifying the molecular mechanisms of membrane trafficking and remodeling in cells is required to understand fundamental cellular functions. Genetic and proteomics screens have been used to identify the proteins that are involved in membrane remodeling and organelle homeostasis (Andersen and Mann, 2006; Bankaitis *et al.*, 1986; Novick *et al.*, 1980). These proteins have further been characterized biochemically, cell-biologically and, in part, structurally. However, it has often been difficult to reveal the underlying molecular mechanisms by assigning protein functions to distinct membrane remodeling events. In the case of MVB-biogenesis, the ESCRT-complexes have been characterized extensively using cell-biological and biochemical assays as well as X-ray crystallography and electron microscopy (Williams and Urbe, 2007), but specific contributions of each complex to cargo clustering, membrane budding, and membrane scission remained elusive. We reconstituted the ESCRT-machinery on artificial membranes *in vitro* using purified proteins and GUVs, decorated with an ubiquitin-model cargo, as membrane substrate. These experiments allowed us to establish the molecular mechanism of MVB-biogenesis involving cargo clustering by ESCRT-0 (Fig. 3B), membrane budding by the ESCRT-I/-II supercomplex (Fig. 3C), scission of nascent membrane buds by ESCRT-III, and recycling of ESCRT-III by Vps4 (Wollert and Hurley, 2010; Wollert *et al.*, 2009a). However, reconstituting membrane-remodeling pathways on GUVs has its limitations. The spatial dimensions of the induced effects need to exceed the resolution limit of confocal microscopy and distinct changes in membrane morphology smaller than 1 μm are barely visible due to dynamic fluctuations of the membranes. The membrane buds that have been induced by ESCRT-I and -II were bigger than those generated *in vivo*, which enabled us to visualize them. *In vivo*, ILVs of yeast MVBs are ~25 nm in diameter (Nickerson *et al.*, 2006), which is well below the limit of resolution of confocal fluorescence microscopy. The same holds true for many other biological membrane remodeling reactions making it difficult or impossible to visualize them by confocal microscopy. In these cases reconstitution on LUVs with subsequent analysis by cryo-EM will be more appropriate. Limitations of the LUV-system, however, include high membrane tension and high membrane curvature, both of which could prevent remodeling reactions from being induced.

VI. Summary and Outlook

In vitro reconstitutions of membrane remodeling machines on artificial membranes have proven successful in deciphering molecular mechanisms not only in MVB biogenesis but also in vesicular trafficking. The COPI induced budding reaction, which is involved in vesicular transport from the Golgi complex to the endoplasmic reticulum, has been reconstituted on GUVs (Manneville *et al.*, 2008) to study its assembly at membranes and to reveal how this molecular machine remodels membranes. In addition insights into the pathogenesis of *Shigella dysenteria* has been revealed by reconstituting the uptake of Shiga-toxin into GUVs (Römer *et al.*, 2007). Many other membrane remodeling reactions need to be reconstituted and further characterized until the complex membrane trafficking network of cells is being understood comprehensively. This will be one of the important tasks in the membrane trafficking field to be undertaken during the next decade.

Acknowledgment

ESCRT-reconstitution experiments have been carried out in the laboratories of James H. Hurley (NIDDK, NIH) and Jennifer Lippincott-Schwartz (NICHD, NIH) in collaboration with Christian Wunder (NICHD, NIH). The work was supported by the Intramural Program of the NIH to James H. Hurley and by an EMBO long-term fellowship to T.W. I thank the Max-Planck Society for financial support.

References

Andersen, J. S., and Mann, M. (2006). Organellar proteomics: turning inventories into insights. *EMBO Rep.* **7**, 874–879.

Angelova, M. I., and Dimitrov, D. S. (1986). Liposome Electroformation. *Faraday Discuss. Chem. Soc* **81**, 303–311.

Bankaitis, V. A., Johnson, L. M., and Emr, S. D. (1986). Isolation of yeast mutants defective in protein targeting to the vacuole. *Proc. Natl. Acad. Sci. U. S. A.* **83**, 9075–9079.

Bieniasz, P. D. (2009). The cell biology of HIV-1 virion genesis. *Cell Host Microbe* **5**, 550–558.

Bigay, J., Gounon, P., Robineau, S., and Antonny, B. (2003). Lipid packing sensed by ArfGAP1 couples COPI coat disassembly to membrane bilayer curvature. *Nature* **426**, 563–566.

Bilodeau, P. S., Winistorfer, S. C., Kearney, W. R., Robertson, A. D., and Piper, R. C. (2003). Vps27-Hse1 and ESCRT-I complexes cooperate to increase efficiency of sorting ubiquitinated proteins at the endosome. *J. Cell Biol.* **163**, 237–243.

Bonnon, C., Wendeler, M. W., Paccaud, J. P., and Hauri, H. P. (2010). Selective export of human GPI-anchored proteins from the endoplasmic reticulum. *J. Cell Sci.* **123**, 1705–1715.

Briggs, J. A., Riches, J. D., Glass, B., Bartonova, V., Zanetti, G., and Krausslich, H. G. (2009). Structure and assembly of immature HIV. *Proc. Natl. Acad. Sci. U. S. A.* **106**, 11090–11095.

Coutinho, A., Loura, L. M., and Prieto, M. (2011). FRET studies of lipid-protein aggregates related to amyloid-like fibers. *J. Neurochem.* **116**, 696–701.

Goni, F. M., Alonso, A., Bagatolli, L. A., Brown, R. E., Marsh, D., Prieto, M., and Thewalt, J. L. (2008). Phase diagrams of lipid mixtures relevant to the study of membrane rafts. *Biochim. Biophys. Acta* **1781**, 665–684.

Guo, L., Har, J. Y., Sankaran, J., Hong, Y., Kannan, B., and Wohland, T. (2008). Molecular diffusion measurement in lipid bilayers over wide concentration ranges: a comparative study. *Chemphyschem* **9**, 721–728.

Hanson, P. I., Roth, R., Lin, Y., and Heuser, J. E. (2008). Plasma membrane deformation by circular arrays of ESCRT-III protein filaments. *J. Cell Biol.* **180**, 389–402.

Hicke, L., and Dunn, R. (2003). Regulation of membrane protein transport by ubiquitin and ubiquitin-binding proteins. *Annu. Rev. Cell Dev. Biol.* **19**, 141–172.

Huang, F., Kirkpatrick, D., Jiang, X., Gygi, S., and Sorkin, A. (2006). Differential regulation of EGF receptor internalization and degradation by multiubiquitination within the kinase domain. *Mol. Cell* **21**, 737–748.

Hurley, J. H., and Hanson, P. I. (2010). Membrane budding and scission by the ESCRT machinery: it's all in the neck. *Nat. Rev. Mol. Cell Biol.* **11**, 556–566.

Kaasgaard, T., Mouritsen, O. G., and Jorgensen, K. (2003). Freeze/thaw effects on lipid-bilayer vesicles investigated by differential scanning calorimetry. *Biochim. Biophys. Acta* **1615**, 77–83.

Katzmann, D. J., Babst, M., and Emr, S. D. (2001). Ubiquitin-dependent sorting into the multivesicular body pathway requires the function of a conserved endosomal protein sorting complex, ESCRT-I. *Cell* **106**, 145–155.

Lata, S., Schoehn, G., Jain, A., Pires, R., Piehler, J., Gottlinger, H. G., and Weissenhorn, W. (2008). Helical structures of ESCRT-III are disassembled by VPS4. *Science* **321**, 1354–1357.

Leung, K. F., Dacks, J. B., and Field, M. C. (2008). Evolution of the multivesicular body ESCRT machinery; retention across the eukaryotic lineage. *Traffic* **9**, 1698–1716.

Luhtala, N., and Odorizzi, G. (2004). Bro1 coordinates deubiquitination in the multivesicular body pathway by recruiting Doa4 to endosomes. *J. Cell Biol.* **166**, 717–729.

Manneville, J. B., Casella, J. F., Ambroggio, E., Gounon, P., Bertherat, J., Bassereau, P., Cartaud, J., Antonny, B., and Goud, B. (2008). COPI coat assembly occurs on liquid-disordered domains and the associated membrane deformations are limited by membrane tension. *Proc. Natl. Acad. Sci. U. S. A.* **105**, 16946–16951.

McCullough, J., Fisher, R. D., Whitby, F. G., Sundquist, W. I., and Hill, C. P. (2008). ALIX-CHMP4 interactions in the human ESCRT pathway. *Proc. Natl. Acad. Sci. U. S. A.* **105**, 7687–7691.

Mitov, M. D., Meleard, P., Winterhalter, M., Angelova, M. I., and Bothorel, P. (1993). Electric-field-dependent thermal fluctuations of giant vesicles. *Phys. Rev. E Stat. Phys. Plasmas Fluids Relat. Interdiscip. Topics* **48**, 628–631.

Morales-Penningston, N. F., Wu, J., Farkas, E. R., Goh, S. L., Konyakhina, T. M., Zheng, J. Y., Webb, W. W., and Feigenson, G. W. (2010). GUV preparation and imaging: minimizing artifacts. *Biochim. Biophys. Acta* **1798**, 1324–1332.

Nickerson, D. P., West, M., and Odorizzi, G. (2006). Did2 coordinates Vps4-mediated dissociation of ESCRT-III from endosomes. *J.Cell Biol.* 175.

Novick, P., Field, C., and Schekman, R. (1980). Identification of 23 complementation groups required for post-translational events in the yeast secretory pathway. *Cell* **21**, 205–215.

Peter, B. J., Kent, H. M., Mills, I. G., Vallis, Y., Butler, P. J., Evans, P. R., and McMahon, H. T. (2004). BAR domains as sensors of membrane curvature: the amphiphysin BAR structure. *Science* **303**, 459–495.

Raiborg, C., Wesche, J., Malerod, L., and Stenmark, H. (2006). Flat clathrin coats on endosomes mediate degradative protein sorting by scaffolding Hrs in dynamic microdomains. *J. Cell Sci.* **119**, 2414–2424.

Ren, X., and Hurley, J. H. (2010). VHS domains of ESCRT-0 cooperate in high-avidity binding to polyubiquitinated cargo. *EMBO J.* **29**, 1045–1054.

Römer, W., Berland, L., Chambon, V., Gaus, K., Windschiegl, B., Tenza, D., Aly, M. R. E., Fraisier, V., Florent, J. C., Perrais, D., Lamaze, C., Raposo, G., Steinem, C., Sens, P., Bassereau, P., and Johannes, L. (2007). Shiga toxin induces tubular membrane invaginations for its uptake into cells. *Nature* **450**, 670–675.

Roux, A., Koster, G., Lenz, M., Sorre, B., Manneville, J. B., Nassoy, P., and Bassereau, P. (2010). Membrane curvature controls dynamin polymerization. *Proc. Natl. Acad. Sci. U. S. A.* **107**, 4141–4146.

Roxrud, I., Stenmark, H., and Malerod, L. (2010). ESCRT & Co. *Biol. Cell* **102**, 293–318.

Samson, R. Y., Obita, T., Freund, S. M., Williams, R. L., and Bell, S. D. (2008). A role for the ESCRT system in cell division in archaea. *Science* **322**, 1710–1713.

Shields, S. B., Oestreich, A. J., Winistorfer, S., Nguyen, D., Payne, J. A., Katzmann, D. J., and Piper, R. (2009). ESCRT ubiquitin-binding domains function cooperatively during MVB cargo sorting. *J. Cell Biol.* **185**, 213–224.

Stenmark, H. (2009). Rab GTPases as coordinators of vesicle traffic. *Nat. Rev. Mol. Cell Biol.* **10**, 513–525.

Teis, D., Saksena, S., and Emr, S. D. (2008). Ordered Assembly of the ESCRT-III Complex on Endosomes Is Required to Sequester Cargo during MVB Formation. *Dev. Cell* **15**, 578–589.

Teo, H., Perisic, O., Gonzalez, B., and Williams, R. L. (2004). ESCRT-II, an endosome-associated complex required for protein sorting: Crystal structure and interactions with ESCRT-III and membranes. *Dev. Cell* **7**, 559–569.

Williams, R. L., and Urbe, S. (2007). The emerging shape of the ESCRT machinery. *Nat. Rev. Mol. Cell Biol.* **8**, 355–368.

Wollert, T., and Hurley, J. H. (2010). Molecular mechanism of multivesicular body biogenesis by ESCRT complexes. *Nature* **464**, 864–869.

Wollert, T., Wunder, C., Lippincott-Schwartz, J., and Hurley, J. H. (2009a). Membrane scission by the ESCRT-III complex. *Nature* **458**, 172–177.

Wollert, T., Yang, D., Ren, X., Lee, H. H., Im, Y. J., and Hurley, J. H. (2009b). The ESCRT machinery at a glance. *J. Cell Sci.* **122**, 2163–2166.

Yeo, S. C., Xu, L., Ren, J., Boulton, V. J., Wagle, M. D., Liu, C., Ren, G., Wong, P., Zahn, R., Sasajala, P., Yang, H., Piper, R. C., and Munn, A. L. (2003). Vps20p and Vta1p interact with Vps4p and function in multivesicular body sorting and endosomal transport in Saccharomyces cerevisiae. *J. Cell Sci.* **116**, 3957–3970.

Yorikawa, C., Shibata, H., Waguri, S., Hatta, K., Horii, M., Katoh, K., Kobayashi, T., Uchiyama, Y., and Maki, M. (2005). Human CHMP6, a myristoylated ESCRT-III protein, interacts directly with an ESCRT-II component EAP20 and regulates endosomal cargo sorting. *Biochem. J.* **387**, 17–26.

CHAPTER 5

Approaches to the Study of Atg8-Mediated Membrane Dynamics *In Vitro*

Anjali Jotwani*, Diana N. Richerson*, Isabelle Motta†, Omar Julca-Zevallos* and Thomas J. Melia*

*Department of Cell Biology, Yale School of Medicine, Connecticut, USA

†Laboratoire de Physique Statistique, Ecole Normale Supérieure, Paris, France

Copyright 2012, Elsevier Inc. All rights reserved.

0091-679X/10 $35.00
DOI 10.1016/B978-0-12-386487-1.00005-5

Abstract

Macro-autophagy is the intracellular stress-response pathway by which the cell packages portions of the cytosol for delivery into the lysosome. This "packaging" is carried out by the *de novo* formation of a new organelle called the autophagosome that grows and encapsulates cytosolic material for eventual lysosomal degradation. How autophagosomes form, including especially how the membrane expands and eventually closes upon itself is an area of intense study. One factor implicated in both membrane expansion and membrane fusion is the ubiquitin-like protein, Atg8. During autophagy, Atg8 becomes covalently bound to phosphatidylethanolamine (PE) on the pre-autophagosomal membrane and remains bound through the maturation process of the autophagosome. In this chapter, we discuss two approaches to the *in vitro* reconstitution of this lipidation reaction. We then describe methods to study Atg8-PE mediated membrane tethering and fusion, two functions implicated in Atg8's role in autophagosome maturation.

I. Introduction

Autophagy is an evolutionarily conserved process that enables the sequestration of cytosol and cellular organelles within a double-membrane vesicle or autophagosome and facilitates its delivery to the lytic compartment (Klionsky, 2005; Mizushima *et al.*, 2002; Nakatogawa *et al.*, 2009). This action is exacerbated in response to intracellular stressors including nutrient deprivation and has been implicated in a variety of cell functions including organelle clearance, cell death, and tumor suppression.

How the autophagosome forms is only poorly understood. Much of the basic machinery was first characterized through yeast genetic approaches (Harding *et al.*, 1995; Thumm *et al.*, 1994; Tsukada and Ohsumi, 1993) and there are more than 30 autophagy-related (ATG) genes identified in *S. cerevisiae*. Biochemical approaches have segregated these gene products into different functional groups, but how these proteins act collaboratively to drive the membrane dynamics of autophagosome biogenesis has not been established (Xie *et al.*, 2008b).

Of the ATG proteins identified, Atg8 is one of two ubiquitin-like proteins implicated in the formation of the autophagosome ((Geng and Klionsky, 2008) and Fig. 1). In yeast, Atg8 is synthesized with an arginine residue at its C terminus, which is subsequently removed by Atg4 to expose a glycine residue ($Atg8^{G116}$) (Kirisako *et al.*, 2000). In a ubiquitin-like conjugation reaction, E1-like and E2-like enzymes Atg7 and Atg3 couple this processed form of Atg8 to the lipid phosphatidylethanolamine (PE) (Ichimura *et al.*, 2000) and the resulting proteo-lipid conjugate, Atg8-PE, is tightly associated with the autophagosomal membrane. Surprisingly, Atg8 is the only known stable, membrane-bound protein associated with the autophagosome, providing a reliable marker for the study of autophagy (Kirisako *et al.*, 1999). Yeast lacking Atg8 form abnormally small autophagosomes (Abeliovich *et al.*, 2000;

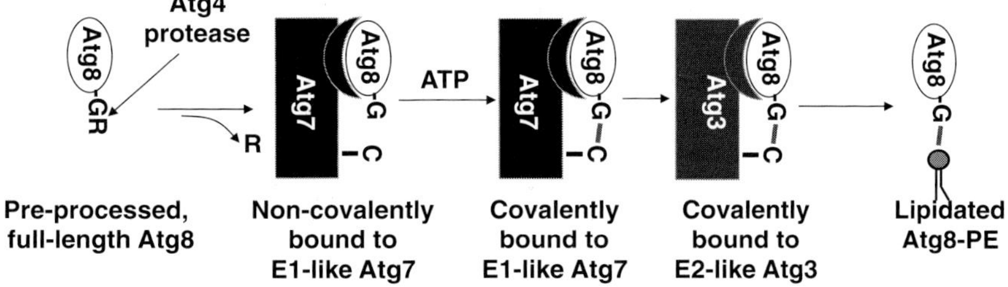

Fig. 1 Atg8-PE conjugation pathway in yeast – Cartoon model of the ubiquitin-like lipidation of Atg8 with phosphatidylethanolamine (PE). (For color version of this figure, the reader is referred to the web version of this book.)

Kirisako *et al.*, 2000) as do starving yeast expressing attenuated amounts of Atg8 (Xie *et al.*, 2008a), suggesting that a major function of this protein is to regulate membrane expansion. Mammalian cells depleted for a subset of Atg8 homologues by RNAi produce autophagosomes of different sizes (depending upon the subset present) (Weidberg *et al.*, 2010), also consistent with a role in expansion. In addition, the loss of a subset of Atg8 also results in more omegasome-like structures, thought to be "open" autophagosomes (Weidberg *et al.*, 2010). The same phenotype is observed if lipidation is blocked for all mammalian homologues (Fujita *et al.*, 2008). These results indicate that the Atg8 family may also control a late step in autophagosome maturation that may include closure/fusion of the mature autophagosome. Using *in vitro* reconstitution of Atg8 lipidation, Ohsumi and colleagues recently established that Atg8 can tether liposomes, bringing the membranes into close apposition (Nakatogawa *et al.*, 2007). They also concluded that Atg8 is itself a membrane fusogen, potentially establishing a new protein paradigm for membrane fusion machineries and providing the first mechanistic explanation for the various roles of the Atg8 family in cell biology (Nakatogawa *et al.*, 2007). Our recent studies suggest that these *in vitro* functions of Atg8 are strongly dependent upon the design of the experiment (Nair *et al.*, 2011), and additional studies may be required to establish whether Atg8 plays a direct fundamental role in membrane dynamics. Thus herein we describe these *in vitro* assays of Atg8 lipidation and function and discuss the strengths and limitation of each assay when exploring the potential direct role of the Atg8 family in autophagosome membrane dynamics.

II. Using Liposomes as *In Vitro* Mimics of Autophagosome Membrane

A. Rationale

Our current understanding of the membranes (the Pre-Autophagosomal Structure (PAS) also termed phagophore or isolation membrane) to which Atg8 naturally couples is limited. Already in the 1960 s, electron microscopy revealed cup-like intermediate membrane structures in the autophagosome biogenesis events occurring in the liver of starved rats. A variety of experiments designed to block autophagosome maturation have yielded similar structures, sometimes termed omegasomes. Several models suggest these structures arise by budding off an existing organelle like the endoplasmic reticulum (Axe *et al.*, 2008) or possibly the mitochondria (Hailey *et al.*, 2010). By cryo-electron tomography, these structures appear to have almost no luminal volume, forming a cisternae with the two opposite membranes very closely apposed (Hayashi-Nishino *et al.*, 2009; Yla-Anttila *et al.*, 2009). In contrast, in yeast, the phagophore assembly site appears instead to be a cluster of vesicles and tubules that will eventually fuse into the cup-like phagophore (Mari *et al.*, 2010). Furthermore, autophagosomes can engulf cargoes of various shapes and sizes and thus its growth and form are not likely to be restricted by a coat-complex assembly (like COPs or clathrin) but instead will rely upon a flexible assembly strategy.

Thus, to replicate features of this biogenesis *in vitro*, one would like to employ potentially flexible membrane substrates. To this end, Atg8 or its mammalian homologues have been coupled to PE on unilamellar liposomes of varying size and lipid composition (Fujioka *et al.*, 2008; Hanada *et al.*, 2007; Ichimura *et al.*, 2004; Nair *et al.*, 2011; Nakatogawa *et al.*, 2007; Shao *et al.*, 2007).

Liposomes have the advantages that they can (1) range more than three orders of magnitude in diameter, from about 25 nm to more than 50 μm (so called Giant unilamellar vesicles or GUVs), (2) comprise a very homogeneous population, especially for small (SUVs) and large (LUVs) unilamellar vesicles, (3) be prepared from synthetic lipids with an absolutely controlled lipid composition, and (4) are relatively stable – storage at 4 °C over a period of weeks is not uncommon. Liposomes are thus desirable as they are both architecturally flexible and experimentally tractable and well suited to the purpose of developing autophagosome mimics.

B. Method – Liposome Preparation

A number of review articles are available on the preparation and storage of liposomes including for example their use in SNARE fusion experiments (Scott *et al.*, 2003). Our preparation of liposomes for Atg8-related experiments is summarized below.

1. Lipids

Except for unusual chemical modifications, all lipids are purchased from Avanti Polar Lipids already dissolved in chloroform and delivered in sealed ampules. After opening, lipids are stored in Pierce Reactivials with Teflon-covered caps, each sample is maintained under an argon blanket to minimize lipid oxidation and the vials are stored at −20 °C.

2. Liposomes

To prepare liposomes with various phospholipid compositions, lipids are mixed in pre-cleaned borosilicate glass tubes and dried to even film under nitrogen gas. To remove residual chloroform, samples are further dried under vacuum at room temperature for 1 h. The lipid films are resuspended in "Liposome resuspension buffer" (50 mM Tris-HCl, pH 8.0, 100 mM NaCl, 1 mM $MgCl_2$) to a final concentration of 5–30 mM at which point the suspension will consist of mixtures of large and small liposomes that are predominantly multi-lamellar. To generate mostly unilamellar liposomes, the sample is next subjected to seven cycles of freezing (in liquid nitrogen) and thawing (in warm water bath).

If one expects to use the same lipid composition over a period of weeks or months, this is also a convenient place to aliquot the mixture (already having experienced several rounds of freeze/thaw) and freeze. To prepare liposomes with a relatively homogeneous size, a single aliquot is thawed and the mixture is extruded 19 times through two polycarbonate membranes of equal pore size using the LiposoFast-Basic extruder (Avestin). We typically use 100 nm because at this size the liposome population will have a mean diameter of about 95 nm and a small variance distribution (Hope *et al.*, 1985). Larger pore cutoffs will give more heterogeneous distributions (Mayer *et al.*, 1986).

3. Liposome Storage

Large liposomes can be kept at 0 °C (on ice in a 4 °C refrigerator) for several weeks, however small liposomes, especially those approaching the curvature limit of ~25 nm diameter, will fuse spontaneously and must be prepared fresh for each use. If the liposome size is not a concern, liposomes and even proteo-liposomes can be frozen in liquid nitrogen (in that case we include 10% w/v glycerol) and stored at −80 °C for months (note however, that freeze/thaw cycles promote membrane fusion and thus small liposomes are likely to fuse together in this instance).

III. Proteins

A. Rationale

Only three proteins are absolutely required for the lipidation reaction; co-expression in *E. coli* of a processed form of Atg8, $Atg8^{G116}$ (Atg8 with a COOH-terminal deletion to expose the reacting glycine at position 116 – simply referred to as Atg8 in the rest of this manuscript) along with the enzymes Atg7 and Atg3 will result in lipidated Atg8 (Ichimura *et al.*, 2004). To reconstitute completely *in vitro*, each protein can be individually expressed in and purified from bacteria (Ichimura *et al.*, 2004); however, we have noted significantly higher specific activities of both the yeast and mammalian Atg7 proteins when they are expressed in *Drosophila* SF9 cells via the baculoviral expression system. We generally express each protein with an amino-terminal affinity tag. Although these tags are often removed by specific proteolysis, we have not observed significant reductions in lipidation when either Atg7 or Atg8 is amino-terminally modified, consistent with the *in vivo* expression of functionally relevant GFP-tagged Atg8 (Geng *et al.*, 2008).

B. Method – Protein Expression/Purification

Molecular biology and protein production are performed via standard protocols and only summarized below. We cloned the gene encoding each protein directly out of purified *S cerevisiae* genomic DNA. The expression of Atg3, Atg8^{G116}, and Atg8^{G116C} is fairly robust and relatively insensitive to the choice of tag. For simplicity, we typically express each with an N-terminal GST-tag followed by a protease cleavage site (either thrombin or PreScission Protease). To induce expression, Codon-plus RIL BL21-DE3 *E. coli* cells are transformed with the appropriate plasmid. Cells are used to serially inoculate increasing volumes of Luria Broth (LB) until we have 4 L of LB with an OD between 0.6 and 0.8. Protein expression is induced with 0.5 mM isopropyl β-D(-)-thiogalactopyranoside (IPTG) for 3 h and cells are collected by centrifugation and frozen at −80 °C until use.

To purify protein, cells are resuspended in 50 mL Breaking Buffer (20 mM Tris, pH 7.6, 100 mM NaCl, 5 mM $MgCl_2$, and 2 mM $CaCl_2$, 1 mM DTT) containing one complete EDTA-free Protease Inhibitor Cocktail Tablet (Roche) and broken via five passes through the Avestin cell disrupter operating at >10000 PSI. Cell lysate is cleared of debris by centrifugation at 40,000 rpm and 4 °C for 35 min, and the supernatant incubated with 1 mL Glutathione Sepharose 4B (GE Healthcare) for 1 h at 4 °C. Beads are then washed with a 20x bed volume of appropriate elution buffer: for PreScission Protease (20 mM Tris, pH 7.6, 100 mM NaCl, 5 mM $MgCl_2$, 2 mM EDTA, 1 mM DTT), for thrombin (20 mM Tris, pH 7.6, 100 mM NaCl, 5 mM $MgCl_2$, 2 mM $CaCl_2$, 1 mM DTT). Protein is eluted by cleavage with PreScission Protease or thrombin at 4 °C. Eluted protein is mixed with glycerol for a final glycerol concentration of 20% (and with 0.2 mM AEBSF to quench residual thrombin). Protein is stored at −80 °C.

Atg7 is expressed by baculoviral infection of SF9 cells. Typically, we infect SF9 cells with viral stock having titre ranges from 1 x 10^6 to 1 x 10^7 pfu/mL using an MOI = 0.1. After 72–96 h of infection, cells are collected and resuspended in 20 mL Lysis Buffer (20 mM Tris, pH 7.6, 0.5 M NaCl, 10% glycerol, 20 mM Imidizole, 1 mM β-mercaptoethanol (β-Me) (Sigma)) per liter of cells with one complete EDTA-free Protease Inhibitor Cocktail Tablet (Roche) per 50 mL buffer. Cells are lysed by sonication and the lysate is cleared by centrifugation at 18,000 rpm and 4 °C for 1 h. The resulting supernatant is incubated with 1 mL 50% Ni-NTA Agarose (Qiagen) per 1 L of original cells for 2 h at 4 °C. Beads are then collected and washed in batch, three times each, against a 10 bead-volume of Wash Buffer (20 mM Tris, pH 7.6, 10 mM NaCl, 20 mM Imidizole, 1 mM β-Me). To elute protein, beads are incubated with 1 mL Atg7 Elution Buffer (20 mM Tris, pH 7.6, 10 mM NaCl, 500 mM Imidizole, 1 mM β-Me) at room temperature for 5–10 min and centrifuged at 900 rpm and 4 °C for 5 min. Purified protein is stored at −80 °C in 20% glycerol.

IV. The Lipidation Reaction

A. Rationale

Although Atg8 shares no sequence homology with ubiquitin, its lipidation reaction is perfectly analogous to the well-described coupling of ubiquitin to its substrate protein. The COOH-terminal glycine of ubiquitin reacts in an ATP-dependent manner with the reactive cysteine of its E1 enzyme to form a covalent sulfhydryl bond. The energy of this bond is used to transfer ubiquitin to the reactive cysteine of its E2 enzyme. The E2 enzyme brings ubiquitin and its substrate into close apposition and facilitates the covalent association of the COOH-terminal glycine of ubiquitin with a free amine (a lysine) on the substrate protein. The autophagic lipidation reaction is structurally similar, including an E1-like (Atg7) and E2-like (Atg3) enzyme (Fig. 1). Unlike all other known ubiquitin-related proteins however, its final substrate is not a free amine in a protein but rather an amine on a lipid. Physiologically, this appears to be the lipid headgroup of phosphatidylethanolamine (but see below).

The efficiency and the substrate specificity of the E2 enzymatic step of ubiquitination can often be enhanced by the inclusion of an E3 adaptor, a protein complex designed to bring the E2-ubiquitin conjugate into close proximity with the final substrate protein. For Atg8, the membrane itself already plays much of this role; the membrane contains the substrate lipid, and exhibits a natural, albeit low, affinity for the E2 enzyme, Atg3 (Hanada *et al.*, 2009). Interestingly, this affinity can be enhanced by an additional protein complex (the Atg5-12 conjugate, often called E3-like), which increases the overall reaction efficiency *in vitro* (Hanada *et al.*, 2007), and appears to be essential *in vivo* (Suzuki *et al.*, 2001). Nonetheless, lipidation reactions without Atg5-12 can still proceed to completion even at high densities of Atg8-PE (see below) and thus we do not further consider the *in vitro* utility of Atg5-12 in this chapter.

B. Method – Reaction Conditions

In vitro, the lipidation reaction is extremely robust, working over a wide-range of protein and lipid concentrations. Both Atg7 and Atg3 work catalytically, such that under conditions where lipid is not limiting, these enzymes can drive a 10-fold or greater excess of Atg8 completely into the PE-bound form in a matter of tens of minutes. A typical reaction protocol is as follows:

1. Mix liposomes (1 mM total lipid), plus Atg7 (1 μM), Atg3 (1 μM), and Atg8^{G116} (2–20 μM) in a reaction microcentrifuge tube. Our standard reactions are run in

SN buffer (100 mM NaCl, 20 mM HEPES, pH7.4, 1 mM DTT), but we have used HEPES, Tris, and phosphate-based buffers, and varied salt from 10 to 400 mM, all with reasonable efficiency.
2. Initiate the lipidation reaction with freshly prepared ATP-Mg^{2+} (1 mM in H_2O).
3. Run at 30 °C in water bath for 15–90 min.
4. Reactions are stopped by moving to ice.

The extent of lipidation can be monitored by gel electrophoresis, because the lipidated form of Atg8 runs faster on a urea–gel (Nakatogawa *et al.*, 2007) (but exhibits an essentially unchanged migration on standard SDS–PAGE; Fig. 2A). An 8 M urea–gel protocol can be found here (Hafiz, 2005), which we have adapted only with the minor modification of also including Urea in the sample buffer.

More conclusive evidence that the protein has become lipid-associated can be gleaned from density float-up experiments (Fig. 2B). Liposomes are loaded into the bottom of a nycodenz density step gradient and then spin at high speed in an ultracentrifuge using a protocol we modified from work with other proteo-liposomes (Melia *et al.*, 2002). At equilibrium, liposomes will collect at the interface between 30% nycodenz and nycodenz-free buffer, near the top of the gradient. If Atg8 is covalently associated with the liposomes (through coupling to PE) it will also collect in this band. On the other hand, if Atg8 is free in solution, it will remain at the bottom of the gradient. Importantly, this same strategy can be employed when tagged variants of Atg8 (or its homologues) are used. For example, the addition of a GST to Atg8 results in a protein so large that the coupling to a small molecular weight PE is less easily resolvable by electrophoresis *in vitro*. In this instance, flotation experiments can confirm that the protein has become membrane-associated.

C. Considerations on Lipid Compositions

In autophagy, phosphatidylethanolamine (PE) plays an essential role as the substrate for the ubiquitin-like lipidation reaction of Atg8. PE also has been implicated as a possible Atg5-Atg12 binding site (Hanada *et al.*, 2007) and as a limiting substrate for autophagosome formation in yeast (Nebauer *et al.*, 2007), and not surprisingly, *in vitro* reactions are highly dependent upon the PE surface density of the liposomes. PE is a cone-shaped lipid, meaning its head group occupies a volume significantly smaller than its hydrophobic acyl chains. High molar concentrations of cone-shaped lipids are not conducive to the formation of stable, lamellar membrane surfaces because the cone-shape either forces the membrane into new architectures or else leaves the hydrophobic core of the membrane exposed (Chernomordik and Kozlov, 2003). Cellular membranes are instead complex mixtures that also include inverted cone-shaped lipids (like lyso-lipids), cholesterol, and proteins. Thus when designing reductionist *in vitro* systems, one must be mindful of the unique physicality of high mole percent PE membranes.

When only considering the simplest phospholipid mixtures, combinations of the neutral lipids PE and phosphatidylcholine (PC), the *in vitro* lipidation reaction

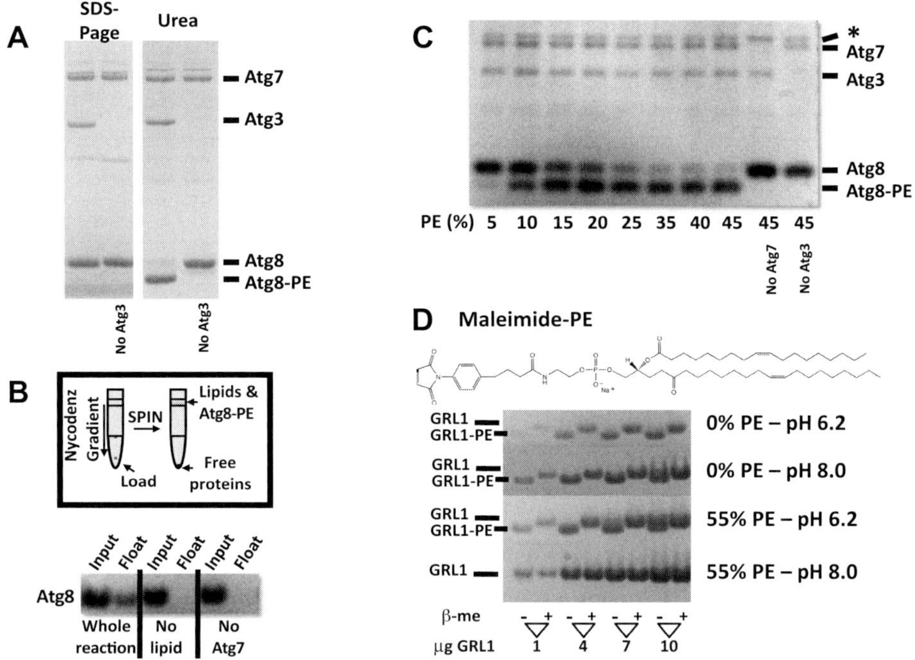

Fig. 2 *In vitro* reconstitution of Atg8-membrane conjugation. (A) Visualization of Atg8 lipidation by urea–SDS–PAGE versus SDS–PAGE and coomassie stain. The *in vitro* reaction mixture (1 mM ATP, 1.0 mM DTT, 1 μM Atg7, 4 μM Atg8, and liposomes containing 30 mol% PE) was incubated at 30 °C for 90 min in the presence or absence of Atg3 (2.5 μM). An additional, faster-traveling Atg8 band, which designates the lipidated product, is visible via urea–SDS–PAGE; however, this band cannot be resolved using conventional SDS–PAGE technique. (B) Density float-up assay permits visualization of Atg8-lipid conjugate by SDS–PAGE. The *in vitro* reaction mixtures (1 mM ATP, 0.2 mM DTT, 8 μM Atg7, 6 μM Atg3, 5 μM Atg8, and liposomes containing 55 mol% PE) were incubated at 30 °C for 90 min. Samples were mixed with 80% nycodenz to prepare a 40% nycodenz solution, then included in a nycodenz gradient (0–30%–40% nycodenz). Each reaction mixture was centrifuged at 48,000 RPM at 4 °C for 4 h; this separates lipidated protein (e.g., Atg8-PE) from free protein (e.g., uncoupled Atg8). Lipidated protein was visualized using SDS–PAGE and Coomassie stain. A band associated with float-up sample, running parallel to Atg8, designates Atg8-PE. (C) Effect of negatively charged lipids on *in vitro* Atg8 lipidation. 1 mM ATP, 0.2 mM DTT, 0.5 μM Atg7, 0.5 μM Atg 3, and 7.5 μM Atg8 were incubated at 30 °C for 90 min in the presence of liposomes containing 10% PI (from bovine liver), 0–55% PE, and the remaining percentage of PC. The reaction mixture was incubated in the presence (+) or absence (−) of proteins Atg3 and Atg7 and visualized using urea–SDS–PAGE and coomassie stain. (D) An alternate strategy for lipidation of the mammalian Atg8 homologue GABARAP L1 (GRL1): direct chemical linkage to a maleimide-coupled lipid. (Top): Structure of commercially available maleimide-PE (Avanti). (Bottom): SDS–Page of lipidation reactions using liposomes with 5% maleimide lipid, 10% PI, 0 or 55% PE, and the remaining percentage of PC. Liposomes (30 mM lipid) are prepared in SN buffer at a pH of either 8.0 or 6.2. Reactions are run at 37 °C for 90 min on 1 mM lipid in SN buffer at pH 8.0 with or without quenching concentrations of β-me, and are initiated by the addition of $GRL1^{G116C}$ at the indicated amounts in a total volume of 12 μL. Qualitatively similar results are observed with $Atg8^{G116C}$ (Nair *et al.*, 2011). (For color version of this figure, the reader is referred to the web version of this book.)

requires very high molar concentrations of PE; however, inclusion of a modest amount of charged lipids will facilitate Atg8 lipidation at much lower PE surface densities (Ichimura *et al.*, 2004). For example, Fig. 2C illustrates lipidation reactions on liposomes comprised of 10% acidic phospholipids and various amounts of PE. When ATP is included to drive the reaction, Atg8 is efficiently coupled to the lipid PE in a PE concentration-dependent manner. Here, even with as little as 20 mol% PE, we can conjugate approximately 60% of the available Atg8 pool. At 30% mol PE, we can conjugate 84% $\pm$ 8% of the Atg8 to lipid, and this efficiency does not change with higher surface densities of PE. It is important to note that under these conditions, the amount of PE as a substrate is never limiting. Half of the total lipid of our unilamellar liposomes (the lipid comprising the outer leaflet) is accessible amounting to 500 μM total accessible lipid. Even at only 5 mol% PE, there is an eightfold excess of accessible PE over Atg8.

The choice of acidic lipid is complicated by the fact that the acidic lipid present at the highest molar concentrations in *S cerevisiae*, phosphatidylserine (PS) is also a substrate lipid for the *in vitro* lipidation reaction (Oh-oka *et al.*, 2008; Sou *et al.*, 2006). For this reason, we and others typically use phosphatidylinositol (PI) as an alternative. Interestingly however, the reactivity with PS can be diminished by altering the pH of the reaction (Oh-oka *et al.*, 2008), such that at physiologic relevant pH of 7.4, relatively little Atg8-PS product is formed. We have tested the reaction with a wide variety of different lipids, including phosphatidic acid and phosphatidylglycerol and see very little preference for the structure of the lipid, only a requirement for a negative charge.

D. Considerations on Protein and Lipid Densities

The cellular concentration of Atg8 appears to influence the ultimate size of the resulting autophagosome, suggesting that the surface density of Atg8 may be integral to its role in expansion. The number of Atg8 proteins residing at the PAS structure is approximately 300 proteins (Geng *et al.*, 2008). Given this information, we can set lower limits for the amount of PE necessary to maintain an autophagic response, and an upper limit for the density of protein on our liposomes that would best mimic an autophagosome. We do not know the precise dimensions of the PAS, but if we estimate the size of the membrane based on the size of the final autophagosome, we can imagine that this is a membrane roughly comparable to a 200 nm liposome. Such a liposome would have around 320,000 lipids, half of which would reside on the inner leaflet of the liposome and be unavailable for reaction. Further, a natural membrane could include significant amounts of trans-membrane protein and non-phospholipids such as ergosterol, possibly as much as half of the total mass. Thus, there might be only about 80,000 lipids available for reaction. Of these, 300 are needed to couple Atg8, putting the upper limit of PE covalently bound to Atg8 at the developing PAS at about 0.4 mol% of all exposed lipids. Even at this low density, Atg8 coating of the PAS resembles the distribution of coat proteins in COP complex assemblies on budding vesicles (Xie *et al.*, 2009). For our purposes, it means that the apparent physiologic upper-limit of Atg8 lipidation would be equivalent to running the reaction in Fig. 2B with 10-fold less protein. To put it another way, *in vitro* we can quantitatively lipidate Atg8 under conditions where our final protein:lipid density is more than an order-of-magnitude beyond what is apparently realized *in vivo*.

V. Alternative Lipidation Approach for PE and Enzyme Independence

A. Rationale

Although the enzymatic addition of PE to Atg8 is the most physiologically relevant, Atg8-driven membrane activities may also be revealed by artificial lipidation of Atg8 (Nair *et al.*, 2011), or of its mammalian homologues (Ma *et al.*, 2010; Weidberg *et al.*, 2011). Such an approach may even present certain advantages when considering functions like membrane fusion (below) where either high PE concentrations or high enzyme concentrations can confound interpretation. This alternative lipid-anchoring strategy is similar to experiments we performed over several years with the SNARE fusogens (i.e., Li *et al.*, 2007; McNew *et al.*, 2000; Melia *et al.*, 2002). The idea is to replace the COOH-terminal glycine of Atg8 with a cysteine residue and then use sulfhydryl chemistry to link this cysteine to a phospholipid carrying a reactive maleimide on its headgroup (Fig. 2D). In this way the concentration of membrane-associated Atg8 can be varied independently of PE composition and without the aid of any enzymes. The reaction is robust, allowing quantitative coupling of $Atg8^{G116C}$ to maleimide liposomes in under 10 min. Importantly, this maleimide-anchored protein retains the capacity to tether membranes (i.e., to form trans-complexes; see below) making it a useful mimetic of the naturally coupled protein.

B. Method

In order to facilitate this approach, protocols for both protein and liposome preparation require some modest alterations:

1. Site directed mutagenesis is used to mutate the one natural cysteine in $Atg8^{G116C}$ to a serine residue (C33S mutation), thus ensuring that the maleimide-lipid reaction is specific for the COOH-terminal cysteine.
2. Proteins are purified in the presence of the reducing agent tris (2-carboxyethyl) phosphine (TCEP) rather than β-mercaptoethanol or DTT. Reducing agents will quench the reactivity of the components; however, at low concentrations TCEP has been shown to prevent disulfide formation in purified proteins but not interfere significantly with the maleimide reaction (Getz *et al.*, 1999). TCEP also has the additional advantage of being stable in solution for days to weeks at 4 °C. We maintain all proteins and buffers in 0.2 mM TCEP.

3. Liposomes are prepared at a lower pH. In order to minimize the quenching of the maleimide by the relatively high local concentration of amine-containing lipids (i.e., PE), we prepare liposomes in a buffer with a pH of 6.2 (See Fig. 2D and (Nair *et al.*, 2011). In the absence of PE, this is not necessary.

C. Considerations When Using Maleimide Linkages

Structurally, this linkage differs from the native coupling by the addition of the maleimide ring. However, several experiments (described below) suggest that this linker does not impede the function of the protein, at least at the level of activities currently described *in vitro*.

For the experiments in Figs. 2E, 3D, E and (Nair *et al.*, 2011), we set the maleimide mole percentage to 5%. At this density, complete coupling will approximate the highest density of Atg8 observed in the original fusion experiments described by Nakatogawa *et al.* (2007), where maximum coupling peaked at about 1 Atg8 per 17 accessible (outward-facing) lipids, and essentially be at the sterically limited maximum. Thus, there will be very little free outward-facing reactive lipid when coupling is complete. If these proteo-liposomes are going to be used for subsequent biochemistry (e.g., pull downs of Atg8-interacting factors), the trace remaining reactive lipid can be quenched with β-me.

There are a variety of reasons to avoid experimental conditions with a large excess of maleimide lipid. In our hands, the lipid is not entirely benign; most notably, there is likely some interaction between the maleimide lipid and PE (or other amine-containing lipids). We have observed that when the surface density of maleimide lipid is significantly lower than the surface density of PE, the coupling efficiency of Atg8 goes down. Indeed, on liposomes having 55% PE and 5% maleimide-lipid (Fig. 2D), reactivity to the maleimide approaches zero within a few hours of preparation. We interpret this quenching of maleimide reactivity by PE as deriving from a low-efficiency side reaction; Maleimides are generally used for site-specific labeling of cysteines because although these chemicals will react with either free amines or free sulfhydryls, the reaction with the sulfhydryl is about 1000 times faster at neutral pH and thus in a typical biochemical experiment labeling of amine groups will be negligible. However, in our liposome set-up, the local concentration of free amines (in the form of PE) is very high relative to the lipid-bound maleimide, and worse, at the moment of liposome formation, there is no competing sulfhydryl whatsoever. We have been able to overcome this quenching by shifting the pH at which we prepare our liposomes to 6.2; the selectivity of a sulfhydryl versus amino reaction increases with lower pH. These acidic liposomes, prepared at high lipid concentration, can then be added back into a standard coupling reaction in neutral pH buffer and now effective coupling of Atg8 is maintained even to liposomes with 55 mole% PE (Fig. 2D). Note, we can also overcome this PE-dependent quenching by simply increasing the proportion of maleimide-lipid relative to PE, however in this instance the resulting liposomes are likely to be some combination of free PE and PE-maleimide adducts. How and whether these adducts contribute to membrane dynamic events like fusion still needs to be determined.

VI. Membrane Tethering

A. Rationale

Atg8 and its mammalian homologues are able to homo-multimerize *in vitro* (Nair *et al.*, 2011; Nakatogawa *et al.*, 2007; Nymann-Andersen *et al.*, 2002; Pacheco *et al.*, 2010; Weidberg *et al.*, 2011). By gel electrophoresis, it is apparent that multimers of Atg8 subunits form and further, mutants that abrogate this multimerization also appear to abrogate autophagy (Nakatogawa *et al.*, 2007). Multimers that form in the plane of the membrane via interactions that we would describe as "*cis*" have not yet been experimentally described. In contrast, interactions that occur in "*trans,*" between membranes, give rise to the lipidation-dependent aggregation of liposomes *in vitro* (Nakatogawa *et al.*, 2007), a process that is phenotypically analogous to vesicle or membrane tethering. In the course of autophagosome maturation, one can imagine a critical role for tethering either in the recruitment of vesicles targeting the expanding PAS or in the closure of the autophagosome itself. To characterize tethering *in vitro*, one can consider three independent assays, each with varying advantages and disadvantages.

B. Method 1 – Macroscopic Precipitation

When liposomes are decorated with membrane tethers, they form aggregates whose size and shape reflect the biochemistry of the tethers {Chiruvolu, 1994 #1552}. For example, the SNARE family of membrane fusing proteins must first tether liposomes before fusion can occur. These tethered aggregates can only be visualized by electron microscopy (e.g., Shen *et al.*, 2007), because the aggregates on average have <10 liposomes. The size of the aggregate likely reflects the relatively low copy number of SNARE proteins that can be reconstituted into liposomes and the fact that multiple SNAREs are needed and/or recruited to each liposome-liposome interface. Atg8, on the other hand, can be reconstituted to much higher surface densities as described above. Thus Atg8-decorated liposomes are highly multivalent, each capable of interacting with many other liposomes. The resulting aggregates, given sufficient time and concentrations, are visible by simple phase contrast light microscopy (Nakatogawa *et al.*, 2007). For example, we incubated approximately 10^{14} 100 nm liposomes with increasing concentrations of Atg8 and fixed concentrations of Atg7, Atg3, and ATP (Fig. 3B). In the absence of Atg8, these liposomes are below the resolution limit of conventional microscopy and are invisible. However, in the presence of Atg8, liposome tethering occurs in an Atg8 concentration-dependent manner such that at the highest concentrations essentially all of the 10^{14} liposomes are driven into a few 10's of readily visible liposome

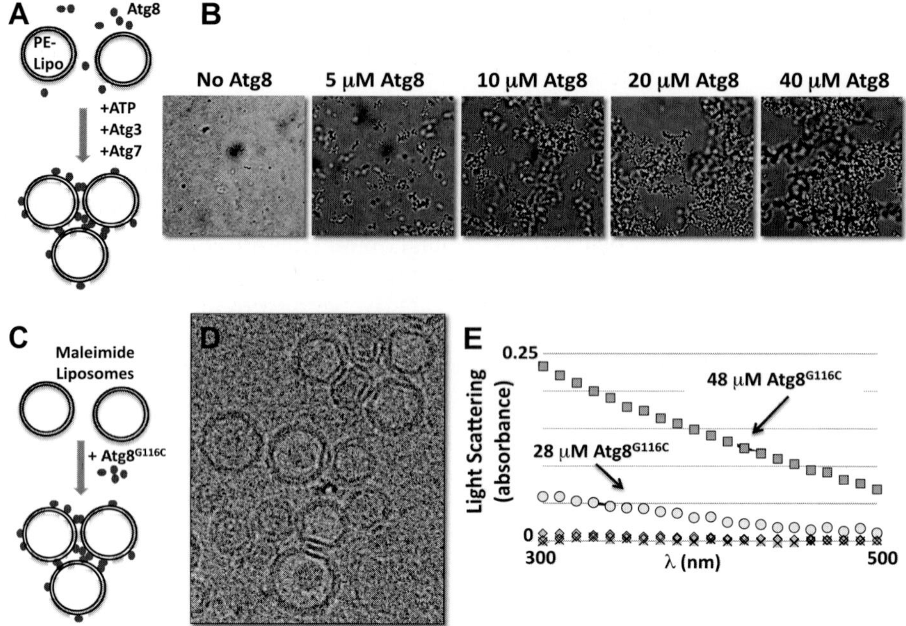

Fig. 3 Assays to follow Atg8-mediated membrane tethering – Cartoon models of tethering mediated by enzymatic lipidation of Atg8 (A) and maleimide-lipidation of Atg8^{G116C} (C). Atg8 is an equally effective tether with either strategy (Nair *et al.*, 2011). (B) Light microscopy of macroscopic aggregates – purified proteins Atg7 (2 μM) and Atg3 (2 μM) were incubated at 30 °C for 90 min with 1 mM DTT, liposomes containing 30 mol% PE, and purified protein Atg8 (at indicated concentrations) in the presence or absence of ATP (1 mM). Images are of phase contrast microscopy with a 40x objective. (D) Cryo-electron microscopy of a maleimide-lipid reaction with 0% PE liposomes confirms that proteo-liposomes containing no PE are still capable of vesicle tethering, including the formation of tight, extensive contact areas between adherent liposomes. Liposomes were prepared as in Fig. 1D (0% PE, pH 6.2), and incubated with 3 μM Atg8^{G116C} for 5 min at 30 °C before the sample was diluted 1:10 in SN buffer. Five microliters of the dilution was transferred onto a carbon-coated perforated electron microscope grid and immediately plunged into liquid ethane. Images were collected on a Tecnai 20 FEG operating in Low Dose mode at 50,000x magnification. (E). Light scattering turbidity assays from liposome reactions run as in Fig. 3D with 0% PE liposomes demonstrate that the concentration of protein coupling to maleimide lipid is closely associated with the extent of liposome aggregation. Spectrophotometer readings are of a sample allowed to react for 1.5 h at 30 °C. The spectrophotometer was blanked by an equivalent sample pre-treated with β-me in order to quench reactivity. (For color version of this figure, the reader is referred to the web version of this book.)

aggregates. In order to assay several conditions at once, we run these reactions in individual microcentrifuge tubes incubated at 30 °C. Aggregation is apparent as a visible precipitate within as little as 30 min. For microscopy, the reaction is transferred to a glass-bottom petri dish, using wide-bore pipette tips and diluted 20–40 fold in buffer, such that individual aggregates can be resolved. The petri dishes are pre-treated with 10% w/v BSA (in water) for 10 min to block exposed surfaces and then rinsed 5x in water and 1x in buffer before applying sample. This approach is a convenient readout of active lipidation and of the general propensity of these proteins to tether membranes, but does not lend itself to further quantitative analysis.

C. Method 2 – Cryo-Electron Microscopy of Tethered Liposomes

Electron microscopy allows the nature of the contact region between liposomes to be interrogated, such that very close apposition and possibly even hemi-fusion might be visible. It has the additional advantage of directly establishing the size and morphology of the liposomes in the experiment. While many groups have used negative stain electron microscopy to study liposome dynamics, it is well-known that the fixative properties of the stains destabilize and dehydrate the liposome samples such that the bulk of the sample will appear as "deflated beachballs," a kind of folded up liposome. Thus, we prefer to use cryo-electron microscopy, where the spherical nature of the liposome is maintained and where each leaflet of the bilayer can be clearly visualized all while maintaining the liposomes in the same buffer conditions as the experiment (e.g., Fig. 3D). In addition to standard cryo-electron microscopy techniques, imaging Atg8-mediated tethering requires the following:

1. Reactions are run in individual microcentrifuge tubes (as in Fig. 2), diluted at least 10-fold in SN buffer to a final lipid concentration of 50–100 μM and then transferred onto carbon-coated perforated grids (Quantifoil) that have been glow-discharge in the presence of amyl-acetate, which allows moderately thicker ice to accumulate. Under these conditions, ice with a thickness of >100 nm is full of liposomes. We limit these studies to extruded liposomes of 50 or 100 nm.
2. Because Atg8 will eventually drive the formation of macroscopic aggregates, structures that are far too large to capture or image in the vitreous ice, we plunge samples within 10 min of beginning the coupling reaction. Indeed, even at 5 min, much of the sample is already in aggregates beyond our imaging capacity, and appears as impenetrable thick ice. However, in areas where the ice remains thin enough, the majority of the liposomes will have already entered into small aggregates, distinct from structures observed in the absence of Atg8-lipidation.

D. Method 3 – Turbidity Assay to Follow the Light-Scattering of Aggregating Liposomes

To follow vesicle aggregation over time, light-scattering is the method of choice and if one is not interested in the final size of the aggregates (they are likely to be heterogeneous), a simple turbidity measurement is sufficient (Nakatogawa *et al.*, 2007). Using a spectrophotometer with a 1 mL cuvette and a conical stir bar, we can follow turbidity in real-time as the reaction proceeds. However, in most cases we are interested in running several reactions in parallel, each at an elevated temperature, and thus our typical data are of light-scattering endpoints collected after incubation in microcentrifuge tubes at 30 °C (for Atg8) or 37 °C (for mammalian homologues of Atg8). In each case, we blank the instrument with a non-aggregating control. Our preference is to use samples lacking Atg3 because a relatively small fraction of the total protein (and thus a small fraction of the absorbance at 280 nm) is contributed by Atg3 and because in this way we can blank out the large 260 nm absorption caused by the presence of ATP. When following aggregation by maleimide-dependent coupling (Fig. 3E), we blank with a β-me containing sample. The scattering signal from these experiments manifests as an increase in absorption from 300 nm out to beyond 600 nm. In order to quantify aggregation and compare samples, we follow the total absorption at a fixed wavelength – 450 nm is convenient in that it is far from protein absorption signatures and still retains significant signal in our assays. This assay is sufficiently quantitative to reveal a tight correlation between the extent of tethering and the extent of Atg8 lipidation, each as a function of the total surface density of PE (Nair *et al.*, 2011).

E. Membrane Fusion

"Membrane fusion" encompasses several steps – the close apposition and exchange of lipids from the outer leaflets of each of the interacting liposomes (hemi-fusion), followed by the merger and exchange of lipids between the inner leaflets and finally the opening of an aqueous pore between the two membranes and the free exchange of content. During the maturation of the autophagosome, there are several places where membrane fusion may be essential and where Atg8 is likely to already be present on the growing membrane. First, membrane expansion depends in part on the vesicle mediated trafficking of Atg9 to and from the pre-autophagosomal structure (Legakis *et al.*, 2007). Second, the closing of the autophagosome likely involves a unique fission/fusion event wherein the edges of the cup-like cisternae are brought into close apposition and fused. Finally, the autophagosome must fuse into the endolysosomal system to degrade its contents. Which proteins are involved in each of these steps are currently unknown, although SNARE proteins are known to be essential at both early and late stages of autophagosome maturation (e.g., Darsow *et al.*, 1997; Fader *et al.*, 2009; Furuta *et al.*, 2010; Ishihara *et al.*, 2001; Nair *et al.*, 2011; Moreau, 2011 #1562}). *In vitro*, lipid-mixing is a good proxy for total membrane fusion as it can reveal each of the first two steps (hemi-fusion and full mixing of lipids), and indeed it is this system that has provided the first evidence that Atg8 may have membrane-fusing capabilities (Nakatogawa *et al.*, 2007).

1. Method

The best characterized lipid-mixing assay is a dequenching assay developed 30 years ago (Struck *et al.*, 1981 and reviewed in Scott *et al.*, 2003). In this assay, two fluorescently tagged lipids are included in one population of liposomes such that the local concentration of the rhodamine-tagged lipid is high enough to allow rhodamine-dependent quenching of the NBD-tagged lipid (Fig. 4A). Fusion of these "donor" liposomes with non-fluorescent "acceptor" liposomes leads to a dilution

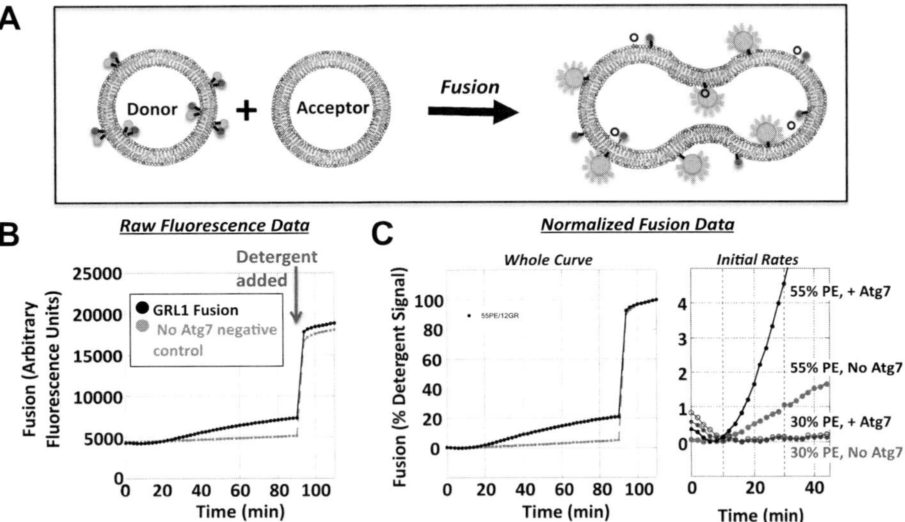

Fig. 4 Lipid-mixing between liposomes tethered by lipidated Atg8 or Atg8 homologues. (A) Fluorescence-dequenching based lipid-mixing assay: In close proximity, rhodamine-tagged lipid quenches the fluorescent signal of NBD within the donor liposome population. When fluorescent liposomes fuse with non-fluorescent acceptor liposomes, the average distance between the two fluorophores is increased and NBD's fluorescence signal is dequenched. NBD fluorescence becomes a read-out for lipid-mixing and accordingly, membrane fusion. (B) GRL1 significantly augments the intrinsic lipid-mixing of 55 mol% PE liposomes. The *in vitro* reaction mixture (1 mM ATP, 1 mM DTT, 2 μM Atg3, 2.5 μM Atg8 homologue GABARAP-L1 (GRL1), and donor and acceptor liposomes containing 55 mol% PE) was incubated at 37 °C in the presence or absence of Atg7 (1 μM). Fluorescence was measured every 2 min with excitation and emission wavelengths of 460 and 538 nm, respectively. After 90 min, 0.25% w/v dodecylmaltoside detergent was added to solubilize liposomes and eliminate rhodamine quenching of NBD. (C – left) Normalization of the signal to this detergent-dependent fluorescent maximum provides a convenient readout to compare experiments, even across laboratories or fluorimeters. (C – right) Extent of fusion is best characterized as the rate of lipid-mixing at early time points, before reactants are consumed. In this instance, it is clear that the negative control (pink) already exhibits significant lipid-mixing as has been observed in other publications with liposomes of this composition (Nair *et al.*, 2011; Nakatogawa *et al.*, 2007). GRL1 augments this basal fusogenicity by a factor of about 7 in rate (black). For comparison, identical reactions are shown using liposomes with only 30 mol% PE. In this case, reactions with Atg7 (filled blue circles) and without Atg7 (empty blue circles) are equivalent, indicating that GRL1 cannot fuse these membranes. Furthermore, both curves are below the negative control condition of the 55% PE liposomes, again indicating that membranes with that lipid composition are intrinsically unstable. (See color plate.)

of each of the fluorophores in the new fused membrane and a dequenching of NBD. The amount of NBD fluorescence then becomes a direct readout of the extent of lipid-mixing in the system. Variations on this assay have been used to follow Atg8-mediated membrane fusion via either enzymatic (Nakatogawa *et al.*, 2007) or maleimide lipidation strategies (Nair *et al.*, 2011). Figure 4B includes an example of a comparable experiment in which the mammalian homologue for Atg8, GABARAP L1 (GRL1), is the putative fusogen. Below we will summarize our protocol for Atg8 or its homologues and then discuss the elements that are unique or particularly advantageous to following Atg8-mediated membrane fusion.

When using Atg8 or its homologues, we use the following protocol:

1. All reaction components except for ATP are mixed in a total volume of 60–70 μL in a white-bottom 96-well dish (Nunc #437591) and the dish is pre-warmed to our reaction temperature (30 °C for Atg8 or 37 °C for mammalian homologues). An easy method for pre-warming is to introduce the 96-well plate into the plate reader for 7–10 min.
2. Reactions are initiated by the addition of ATP to each well (or by the addition of $Atg8^{G116C}$ for the maleimide lipidation strategy) and the plates are returned to the plate-reader. Wells are illuminated with 460 nm excitation light and fluorescence emission is followed at 538 nm.
3. Typical fusion kinetics of the Atg8 proteins are on the order of minutes, thus it is sufficient to measure fluorescence every 1–2 min following ATP addition. Note that with either the enzymatic coupling reaction (Fig. 4B) or with maleimide chemistries (Nair *et al.*, 2011), the reactions first exhibit a lag phase of several minutes before fusion begins.
4. *Normalization*: The fluorescence increase does not scale linearly with the extent of fusion. This is because the fluorescence signal is a measure of dequenching, which depends upon the extent of dilution of the lipids (following fusion) and scales with the 6th power of distance. Furthermore, actual fluorescence values are arbitrary and machine dependent. Thus to compare data across systems or across laboratories several parameters must be normalized. Experimentally, the liposome composition (especially the concentration of fluorophores) must be identical across experiments, to account for the non-linear increase in fluorescence following dequenching. In Fig. 4 and in previously published Atg8 fusion studies (Nakatogawa *et al.*, 2007), donor liposomes contain 1% NBD and 1.5% rhodamine-tagged lipids. With lower concentrations of rhodamine, the assay can be tweaked to be more sensitive to a first round of fusion, but each subsequent fusion event will have a markedly reduced further increment in fluorescence. In our experiments, we use a fourfold excess of acceptor over donor that facilitates the rapid consumption of all fluorescent liposomes but also likely results in some donor liposomes fusing multiple times to distinct acceptor liposomes. The donor and acceptor liposomes should also be the same size, so that each fusion event affords a single round of dilution. Finally, to compare results across different fluorimeters, all experiments can be normalized to a common maximum, the maximum fluorescence observed following complete dequenching due to the addition of detergent at the end of the assay. Here, we use 0.5% w/v dodecylmaltoside (arrow) and re-plot the data in the form of "% Detergent maximum" (Fig. 4B).
5. As can be seen in Fig. 4 for GRL1 or in (Nakatogawa *et al.*, 2007) and (Nair *et al.*, 2011) for Atg8, the fluorescence profile eventually plateaus. Thus, it is most useful to compare rates of fusion rather than endpoints. As illustrated in Fig. 4C,

the full reaction achieves a maximal rate about seven times faster than the "No Atg7" control reaction. Notably however, even the negative control has appreciable fusion, owing to the use of 55% PE liposomes (see below).

2. The Effect of Lipid Composition on the Intrinsic Fusogenicity of Liposomes

Liposomes of high PE and high charge density do not comprise a useful system for evaluating membrane fusion. Others have demonstrated that liposomes of very similar composition can be made to fuse with a wide variety of proteins and small molecules, each of which has no known role in membrane trafficking, much less fusion (Brugger *et al.*, 2000). In fact among the first papers to study membrane fusion with *in vitro* systems demonstrated that membranes exceptionally rich in PE can be made to fuse spontaneously with the addition of divalent cations or modest osmotic pressure, that is, without the intervention of any protein whatsoever (Cohen *et al.*, 1984). Thus, the base-line state of a system with these types of lipids is already fusogenic (as observed in the Fig. 4 no Atg7 control as well as no ATP controls in references (Nair *et al.*, 2011; Nakatogawa *et al.*, 2007)), and in that context, Atg8 is simply manipulating this baseline, accelerating the spontaneous fusion that is already underway (presumably through liposome aggregation). Furthermore, the bulk of the Atg8-dependent signal is hemifusion (Nakatogawa *et al.*, 2007), a common end state for systems driven primarily by PE-instabilities. To better understand whether Atg8 or GRL1 might be a physiologically relevant fusogen, we tested liposomes with 30 mol% PE, the highest concentration observed in yeast or mammalian organelles (see (Nair *et al.*, 2011) and references within). Both Atg8 (Nair *et al.*, 2011) and GRL1 (Fig. 4C, blue circles) fail to fuse membranes with this composition. Indeed, the baseline drift in these samples with or without active lipidation machinery is always lower than in the negative control reactions on 55% PE liposomes. Contrast this behavior with other established fusogens: Ari Hellenius demonstrated in the early 1980 s that membrane-enveloped influenza viral particles could fuse to liposomes with and without hexagonal-phase inducing lipids like PE (Stegmann *et al.*, 1990; White *et al.*, 1982). Isolated purified fusion protein (HA2) was eventually shown to have the same ability (Epand *et al.*, 1999). By the late 1990 s, the Rothman lab had added SNARE proteins to the list of membrane fusogens, which could overcome physiologically relevant energies to drive the fusion of liposomes with liquid-crystalline phase favoring lipids like pure PC (Weber *et al.*, 1998). Importantly, both the viral fusogens and the SNAREs drive full fusion of membranes, thus compared to Atg8 they are functional in a more physiologically relevant lipid setting and they drive the more energetically challenging event of complete lipid-mixing of both leaflets.

There is also an interesting historical parallel to putative fusogenicty of Atg8 in the study of intracellular vesicular membrane fusion. The AAA ATPases, p97 and NSF were each found to have intrinsic membrane fusion activity when reconstituted with liposomes of high charge (to facilitate binding of the proteins) and high PE (50%)

(Otter-Nilsson *et al.*, 1999). Like Atg8, these proteins were already known to play integral roles in membrane traffic and thus a role in fusion could have profound implications. However, these activities were subsequently shown to be lost upon modest reduction of the PE content of the membranes (Brugger *et al.*, 2000). Indeed, the apparent fusogenicity of NSF and p97 could be replicated with a variety of other compounds including the proteins GAPDH and lactate dehydrogenase, as well as divalent cations (Brugger *et al.*, 2000). Together, these experiments suggest that liposomes with a very high molar PE composition are not meaningful substrates for determining whether a protein or compound is likely to exhibit physiologically relevant membrane fusion activities.

Intriguingly, using the maleimide approach the mammalian Atg8s have recently been shown to fuse liposomes with moderate levels of PE (Weidberg *et al.*, 2011). This result appears inconsistent with our GRL1 experiments presented here, as well as the previously published Atg8 and LC3 maleimide experiments (Nair *et al.*, 2011). It now becomes important to work out what additional elements of the experimental design may be contributing to these conflicting results as we strive to understand how Atg8 influences membrane dynamics *in vivo*.

VII. Conclusion

What then is the role of the Atg8 family in autophagosome maturation? Atg8 is properly situated spatially and temporally to impact the dramatic membrane rearrangements occurring in the later stages of autophagosome maturation. Precisely how Atg8 affects these maturation events remains contentious, but a variety of labs using complementary approaches are converging on some common themes including a clear capacity to tether membranes. This tethering suggests a trans-interaction state for the Atg8 family, where it is able to homooligomerize with proteins located on distinct, or distal membrane surfaces perhaps analogous to dimeric structures captured by crystallography of a mammalian homologue, GABARAP (Coyle *et al.*, 2002). Thus one possibility is that the Atg8 family still plays a critical role in closing the autophagosome, capturing or gluing the various surfaces together (as a tether) ahead of whatever machinery will finally drive membrane fission. Indeed, if we consider the fusion Nakatogowa *et al.* observe with unstable 55% PE liposomes as a proxy readout of this tethering activity, then the correlation between Atg8 mutants that fail to fuse *in vitro* and that fail to develop autophagosomes *in vivo* can be reconsidered as a failure to tether. Alternatively, the phagophore may comprise either a unique lipid composition or an energetically unstable architecture such that the modest energies associated with tethering could suffice to initiate lipid-mixing. Such a low-energy fusogen would be quite distinct from the essentially universal SNARE and viral fusogens already described. An intriguing possibility along these lines could be that other cone-shaped lipids are produced locally in a dynamic fashion to facilitate fusion at the proper time and place (e.g., Dall'armi *et al.*, 2010). Thus determining the lipidomics {Wenk, 2005 #1553} of autophagosome formation and maturation will be an important direction in the future regardless of which proteins are ultimately found to drive membrane fusion in this system.

Finally, Atg8 also plays a role as a cargo adaptor or cargo-capturing platform via its interactions with Atg19 (e.g., Chang and Huang, 2007). Thus the same interfaces that are essential for trans-interaction may play a role in capturing proteins essential for autophagosome maturation including closure. More biochemistry is needed to establish whether the tethering we observe *in vitro* is compatible with binding and capture of putative cargo adaptors.

Acknowledgments

Funding for this work was provided by a Kingsley Medical Fellowship (TJM), Yale Summer Fellowships including the STARS program, and the George J. Schulz Summer Fellowship in the Physical Sciences (AJ) and the REPU program – Research Experience for Peruvian Undergrads (OJZ). Cryo-electron microscopy was performed at the New York Structural Biology Center. We would also like to thank Shanta Nag and Dr. Sangeeta Nath for help in experiments.

References

Abeliovich, H., Dunn Jr., W. A., Kim, J., and Klionsky, D. J. (2000). Dissection of autophagosome biogenesis into distinct nucleation and expansion steps. *J. Cell Biol.* **151**, 1025–1034.

Axe, E. L., Walker, S. A., Manifava, M., Chandra, P., Roderick, H. L., Habermann, A., Griffiths, G., and Ktistakis, N. T. (2008). Autophagosome formation from membrane compartments enriched in phosphatidylinositol 3-phosphate and dynamically connected to the endoplasmic reticulum. *J. Cell Biol.* **182**, 685–701.

Brugger, B., Nickel, W., Weber, T., Parlati, F., McNew, J. A., Rothman, J. E., and Sollner, T. (2000). Putative fusogenic activity of NSF is restricted to a lipid mixture whose coalescence is also triggered by other factors. *EMBO J.* **19**, 1272–1278.

Chang, C. Y., and Huang, W. P. (2007). Atg19 mediates a dual interaction cargo sorting mechanism in selective autophagy. *Mol. Biol. Cell* **18**, 919–929.

Chernomordik, L. V., and Kozlov, M. M. (2003). Protein-lipid interplay in fusion and fission of biological membranes. *Annu. Rev. Biochem.* **72**, 175–207.

Cohen, F. S., Akabas, M. H., Zimmerberg, J., and Finkelstein, A. (1984). Parameters affecting the fusion of unilamellar phospholipid vesicles with planar bilayer membranes. *J. Cell Biol.* **98**, 1054–1062.

Coyle, J. E., Qamar, S., Rajashankar, K. R., and Nikolov, D. B. (2002). Structure of GABARAP in two conformations: implications for GABA(A) receptor localization and tubulin binding. *Neuron* **33**, 63–74.

Dall'armi, C., Hurtado-Lorenzo, A., Tian, H., Morel, E., Nezu, A., Chan, R. B., Yu, W. H., Robinson, K. S., Yeku, O., Small, S. A., Duff, K., Frohman, M. A., Wenk, M. R., Yamamoto, A., and Di Paolo, G. (2010). The phospholipase D1 pathway modulates macroautophagy. *Nat. Commun.* **1**, 142.

Darsow, T., Rieder, S. E., and Emr, S. D. (1997). A multispecificity syntaxin homologue, Vam3p, essential for autophagic and biosynthetic protein transport to the vacuole. *J. Cell Biol.* **138**, 517–529.

Epand, R. F., Macosko, J. C., Russell, C. J., Shin, Y. K., and Epand, R. M. (1999). The ectodomain of HA2 of influenza virus promotes rapid pH dependent membrane fusion. *J. Mol. Biol.* **286**, 489–503.

Fader, C. M., Sanchez, D. G., Mestre, M. B., and Colombo, M. I. (2009). TI-VAMP/VAMP7 and VAMP3/cellubrevin: two v-SNARE proteins involved in specific steps of the autophagy/multivesicular body pathways. *Biochim. Biophys. Acta* **1793**, 1901–1916.

Fujioka, Y., Noda, N. N., Fujii, K., Yoshimoto, K., Ohsumi, Y., and Inagaki, F. (2008). In vitro reconstitution of plant Atg8 and Atg12 conjugation systems essential for autophagy. *J. Biol. Chem.* **283**, 1921–1928.

Fujita, N., Hayashi-Nishino, M., Fukumoto, H., Omori, H., Yamamoto, A., Noda, T., and Yoshimori, T. (2008). An Atg4B mutant hampers the lipidation of LC3 paralogues and causes defects in autophagosome closure. *Mol. Biol. Cell* **19**, 4651–4659.

Furuta, N., Fujita, N., Noda, T., Yoshimori, T., and Amano, A. (2010). Combinational soluble N-ethylmaleimide-sensitive factor attachment protein receptor proteins VAMP8 and Vti1b mediate fusion of antimicrobial and canonical autophagosomes with lysosomes. *Mol. Biol. Cell* **21**, 1001–1010.

Geng, J., Baba, M., Nair, U., and Klionsky, D. J. (2008). Quantitative analysis of autophagy-related protein stoichiometry by fluorescence microscopy. *J. Cell Biol.* **182**, 129–140.

Geng, J., and Klionsky, D. J. (2008). The Atg8 and Atg12 ubiquitin-like conjugation systems in macroautophagy. 'Protein modifications: beyond the usual suspects' review series. *EMBO Rep.* **9**, 859–864.

Getz, E. B., Xiao, M., Chakrabarty, T., Cooke, R., and Selvin, P. R. (1999). A comparison between the sulfhydryl reductants tris(2-carboxyethyl)phosphine and dithiothreitol for use in protein biochemistry. *Anal. Biochem.* **273**, 73–80.

Hafiz, A. (2005). Principles and Reactions of Protein Extraction, Purification, and Characterization. CRC Press, Boca raton, FL, pp. 71–131.

Hailey, D. W., Rambold, A. S., Satpute-Krishnan, P., Mitra, K., Sougrat, R., Kim, P. K., and Lippincott-Schwartz, J. (2010). Mitochondria supply membranes for autophagosome biogenesis during starvation. *Cell* **141**, 656–667.

Hanada, T., Noda, N. N., Satomi, Y., Ichimura, Y., Fujioka, Y., Takao, T., Inagaki, F., and Ohsumi, Y. (2007). The Atg12-Atg5 conjugate has a novel E3-like activity for protein lipidation in autophagy. *J. Biol. Chem.* **282**, 37298–37302.

Hanada, T., Satomi, Y., Takao, T., and Ohsumi, Y. (2009). The amino-terminal region of Atg3 is essential for association with phosphatidylethanolamine in Atg8 lipidation. *FEBS Lett.* **583**, 1078–1083.

Harding, T. M., Morano, K. A., Scott, S. V., and Klionsky, D. J. (1995). Isolation and characterization of yeast mutants in the cytoplasm to vacuole protein targeting pathway. *J. Cell Biol.* **131**, 591–602.

Hayashi-Nishino, M., Fujita, N., Noda, T., Yamaguchi, A., Yoshimori, T., and Yamamoto, A. (2009). A subdomain of the endoplasmic reticulum forms a cradle for autophagosome formation. *Nat. Cell Biol.* **11**, 1433–1437.

Hope, M. J., Bally, M. B., Webb, G., and Cullis, P. R. (1985). Production of large unilamellar vesicles by a rapid extrusion procedure. Characterization of size distribution, trapped volume and ability to maintain a membrane potential. *Biochim. Biophys. Acta* **812**, 55–65.

Ichimura, Y., Imamura, Y., Emoto, K., Umeda, M., Noda, T., and Ohsumi, Y. (2004). In vivo and in vitro reconstitution of Atg8 conjugation essential for autophagy. *J. Biol. Chem.* **279**, 40584–40592.

Ichimura, Y., Kirisako, T., Takao, T., Satomi, Y., Shimonishi, Y., Ishihara, N., Mizushima, N., Tanida, I., Kominami, E., Ohsumi, M., Noda, T., and Ohsumi, Y. (2000). A ubiquitin-like system mediates protein lipidation. *Nature* **408**, 488–492.

Ishihara, N., Hamasaki, M., Yokota, S., Suzuki, K., Kamada, Y., Kihara, A., Yoshimori, T., Noda, T., and Ohsumi, Y. (2001). Autophagosome requires specific early Sec proteins for its formation and NSF/SNARE for vacuolar fusion. *Mol. Biol. Cell* **12**, 3690–3702.

Kirisako, T., Baba, M., Ishihara, N., Miyazawa, K., Ohsumi, M., Yoshimori, T., Noda, T., and Ohsumi, Y. (1999). Formation process of autophagosome is traced with Apg8/Aut7p in yeast. *J. Cell Biol.* **147**, 435–446.

Kirisako, T., Ichimura, Y., Okada, H., Kabeya, Y., Mizushima, N., Yoshimori, T., Ohsumi, M., Takao, T., Noda, T., and Ohsumi, Y. (2000). The reversible modification regulates the membrane-binding state of Apg8/Aut7 essential for autophagy and the cytoplasm to vacuole targeting pathway. *J. Cell Biol.* **151**, 263–276.

Klionsky, D. J. (2005). The molecular machinery of autophagy: unanswered questions. *J. Cell Sci.* **118**, 7–18.

Legakis, J. E., Yen, W. L., and Klionsky, D. J. (2007). A cycling protein complex required for selective autophagy. *Autophagy* **3**, 422–432.

Li, F., Pincet, F., Perez, E., Eng, W. S., Melia, T. J., Rothman, J. E., and Tareste, D. (2007). Energetics and dynamics of SNAREpin folding across lipid bilayers. *Nat. Struct. Mol. Biol.* **14**, 890–896.

Ma, P., Mohrluder, J., Schwarten, M., Stoldt, M., Singh, S. K., Hartmann, R., Pacheco, V., and Willbold, D. (2010). Preparation of a functional GABARAP-lipid conjugate in nanodiscs and its investigation by solution NMR spectroscopy. *Chembiochem* **11**, 1967–1970.

Mari, M., Griffith, J., Rieter, E., Krishnappa, L., Klionsky, D. J., and Reggiori, F. (2010). An Atg9-containing compartment that functions in the early steps of autophagosome biogenesis. *J. Cell Biol.* **190**, 1005–1022.

Mayer, L. D., Hope, M. J., and Cullis, P. R. (1986). Vesicles of variable sizes produced by a rapid extrusion procedure. *Biochim. Biophys. Acta* **858**, 161–168.

McNew, J. A., Weber, T., Parlati, F., Johnston, R. J., Melia, T. J., Sollner, T. H., and Rothman, J. E. (2000). Close is not enough: SNARE-dependent membrane fusion requires an active mechanism that transduces force to membrane anchors. *J. Cell Biol.* **150**, 105–117.

Melia, T. J., Weber, T., McNew, J. A., Fisher, L. E., Johnston, R. J., Parlati, F., Mahal, L. K., Sollner, T. H., and Rothman, J. E. (2002). Regulation of membrane fusion by the membrane-proximal coil of the t-SNARE during zippering of SNAREpins. *J. Cell Biol.* **158**, 929–940.

Mizushima, N., Ohsumi, Y., and Yoshimori, T. (2002). Autophagosome formation in mammalian cells. *Cell Struct. Funct.* **27**, 421–429.

Nair, U., Jotwani, A., Geng, J., Gammoh, N., Richerson, D., Yen, W. L., Griffith, J., Nag, S., Wang, K., Moss, T., Baba, M., McNew, J. A., Jiang, X., Reggiori, F., Melia, T. J., and Klionsky, D. J. (2011). SNARE Proteins Are Required for Macroautophagy. *Cell* **146**, 290–302.

Nakatogawa, H., Ichimura, Y., and Ohsumi, Y. (2007). Atg8, a ubiquitin-like protein required for autophagosome formation, mediates membrane tethering and hemifusion. *Cell* **130**, 165–178.

Nakatogawa, H., Suzuki, K., Kamada, Y., and Ohsumi, Y. (2009). Dynamics and diversity in autophagy mechanisms: lessons from yeast. *Nat. Rev. Mol. Cell Biol.* **10**, 458–467.

Nebauer, R., Rosenberger, S., and Daum, G. (2007). Phosphatidylethanolamine, a limiting factor of autophagy in yeast strains bearing a defect in the carboxypeptidase Y pathway of vacuolar targeting. *J. Biol. Chem.* **282**, 16736–16743.

Nymann-Andersen, J., Wang, H., and Olsen, R. W. (2002). Biochemical identification of the binding domain in the GABA(A) receptor-associated protein (GABARAP) mediating dimer formation. *Neuropharmacology* **43**, 476–481.

Oh-oka, K., Nakatogawa, H., and Ohsumi, Y. (2008). Physiological pH and acidic phospholipids contribute to substrate specificity in lipidation of Atg8. *J. Biol. Chem.* **283**, 21847–21852.

Otter-Nilsson, M., Hendriks, R., Pecheur-Huet, E. I., Hoekstra, D., and Nilsson, T. (1999). Cytosolic ATPases, p97 and NSF, are sufficient to mediate rapid membrane fusion. *EMBO J.* **18**, 2074–2083.

Pacheco, V., Ma, P., Thielmann, Y., Hartmann, R., Weiergraber, O. H., Mohrluder, J., and Willbold, D. (2010). Assessment of GABARAP self-association by its diffusion properties. *J. Biomol. NMR* **48**, 49–58.

Scott, B. L., Van Komen, J. S., Liu, S., Weber, T., Melia, T. J., and McNew, J. A. (2003). Liposome fusion assay to monitor intracellular membrane fusion machines. *Methods Enzymol.* **372**, 274–300.

Shao, Y., Gao, Z., Feldman, T., and Jiang, X. (2007). Stimulation of ATG12-ATG5 conjugation by ribonucleic acid. *Autophagy* **3**, 10–16.

Shen, J., Tareste, D. C., Paumet, F., Rothman, J. E., and Melia, T. J. (2007). Selective activation of cognate SNAREpins by Sec1/Munc18 proteins. *Cell* **128**, 183–195.

Sou, Y. S., Tanida, I., Komatsu, M., Ueno, T., and Kominami, E. (2006). Phosphatidylserine in addition to phosphatidylethanolamine is an in vitro target of the mammalian Atg8 modifiers, LC3, GABARAP, and GATE-16. *J. Biol. Chem.* **281**, 3017–3024.

Stegmann, T., White, J. M., and Helenius, A. (1990). Intermediates in influenza induced membrane fusion. *EMBO J.* **9**, 4231–4241.

Struck, D. K., Hoekstra, D., and Pagano, R. E. (1981). Use of resonance energy transfer to monitor membrane fusion. *Biochemistry* **20**, 4093–4099.

Suzuki, K., Kirisako, T., Kamada, Y., Mizushima, N., Noda, T., and Ohsumi, Y. (2001). The pre-autophagosomal structure organized by concerted functions of APG genes is essential for autophagosome formation. *EMBO J.* **20**, 5971–5981.

Thumm, M., Egner, R., Koch, B., Schlumpberger, M., Straub, M., Veenhuis, M., and Wolf, D. H. (1994). Isolation of autophagocytosis mutants of Saccharomyces cerevisiae. *FEBS Lett.* **349**, 275–280.

Tsukada, M., and Ohsumi, Y. (1993). Isolation and characterization of autophagy-defective mutants of Saccharomyces cerevisiae. *FEBS Lett.* **333**, 169–174.

Weber, T., Zemelman, B. V., McNew, J. A., Westermann, B., Gmachl, M., Parlati, F., Sollner, T. H., and Rothman, J. E. (1998). SNAREpins: minimal machinery for membrane fusion. *Cell* **92**, 759–772.

Weidberg, H., Shpilka, T., Shvets, E., Abada, A., Shimron, F., and Elazar, Z. (2011). LC3 and GATE-16 N termini mediate membrane fusion processes required for autophagosome biogenesis. *Dev. Cell* **20**, 444–454.

Weidberg, H., Shvets, E., Shpilka, T., Shimron, F., Shinder, V., and Elazar, Z. (2010). LC3 and GATE-16/GABARAP subfamilies are both essential yet act differently in autophagosome biogenesis. *EMBO J.* **29**, 1792–1802.

White, J., Kartenbeck, J., and Helenius, A. (1982). Membrane fusion activity of influenza virus. *EMBO J.* **1**, 217–222.

Xie, Z., Nair, U., Geng, J., Szefler, M. B., Rothman, E. D., and Klionsky, D. J. (2009). Indirect estimation of the area density of Atg8 on the phagophore. *Autophagy* **5**, 217–220.

Xie, Z., Nair, U., and Klionsky, D. J. (2008a). Atg8 controls phagophore expansion during autophagosome formation. *Mol. Biol. Cell* **19**, 3290–3298.

Xie, Z., Nair, U., and Klionsky, D. J. (2008b). Dissecting autophagosome formation: the missing pieces. *Autophagy* **4**, 920–922.

Yla-Anttila, P., Vihinen, H., Jokitalo, E., and Eskelinen, E. L. (2009). 3D tomography reveals connections between the phagophore and endoplasmic reticulum. *Autophagy* **5**, 1180–1185.

CHAPTER 6

Reconstitution Assay System for Ceramide Transport With Semi-Intact Cells

Keigo Kumagai[*], Masahiro Nishijima[†] and Kentaro Hanada[*]

[*]Department of Biochemistry and Cell Biology, National Institute of Infectious Diseases, Shinjuku-ku, Tokyo, Japan

[†]National Institute of Health Sciences, Setagaya-ku, Tokyo, Japan

Abstract

The intracellular transport of lipids from the sites of their synthesis to their appropriate destination is a critical step for lipid metabolism. One well-defined inter-organelle lipid movement is the transport of ceramide by ceramide transport protein (CERT). Ceramide, a key intermediate for both sphingomyelin and glycosphingolipids, is synthesized at the endoplasmic reticulum and delivered to the Golgi apparatus to be converted to sphingomyelin. CERT delivers ceramide from the ER to the Golgi apparatus in a non-vesicular and ATP-dependent manner. This chapter describes a reconstitution assay system for ceramide transport with semi-intact cells, which is useful for the study of the CERT-mediated inter-organelle transport of ceramide.

Copyright 2012, Elsevier Inc. All rights reserved.

0091-679X/10 $35.00
DOI 10.1016/B978-0-12-386487-1.00006-7

I. Introduction

The intracellular transport of lipids from the sites of their synthesis to their appropriate destination is a critical step for lipid metabolism, because various steps of lipid biosynthesis occur in different intracellular compartments (van Meer *et al.*, 2008). The cellular lipids are transported by vesicular and non-vesicular mechanisms, however, little is known about the molecular machinery for the inter-organelle movement of lipids. One well-defined inter-organelle lipid movement is the transport of ceramide in sphingomyelin (SM) biosynthesis. Ceramide, a key intermediate for both SM and glycosphingolipids, is synthesized at the cytosolic face of the endoplasmic reticulum (ER) (Futerman and Riezman, 2005; Mandon *et al.*, 1992). It is then delivered to the luminal side of the Golgi apparatus to be converted to SM by SM synthase 1, which catalyzes the transfer of phosphocholine from phosphatidylcholine to ceramide (Huitema *et al.*, 2004; Yamaoka *et al.*, 2004). Ceramide transport protein (CERT) delivers ceramide from the ER to the Golgi apparatus in a non-vesicular and ATP-dependent manner (Hanada *et al.*, 2009). In this chapter, we describe a reconstitution assay system for ceramide transport with semi-intact cells, which is useful for the study of the CERT-mediated inter-organelle transport of ceramide (Funakoshi *et al.*, 2000; Hanada *et al.*, 2009).

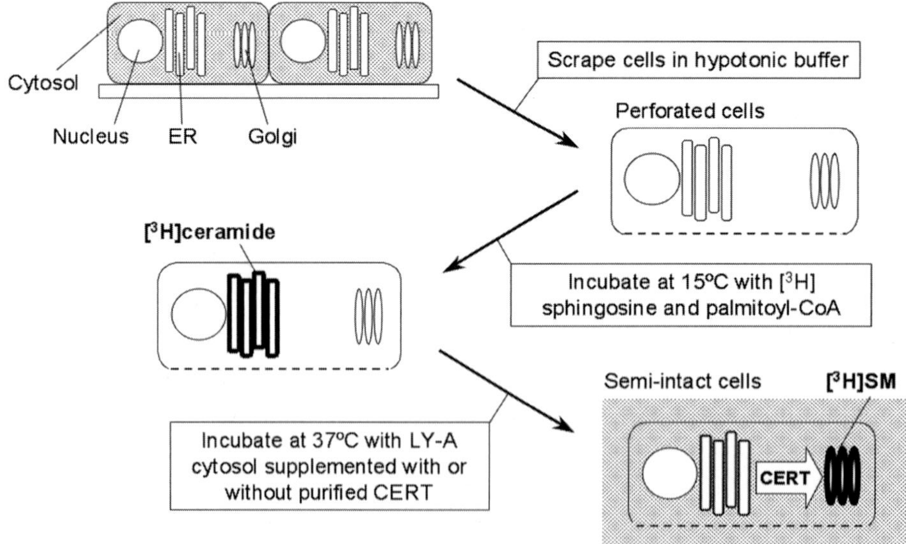

Fig. 1 An outline of the reconstitution assay system for ceramide transport with semi-intact cells. The ER and Golgi apparatus are surrounded by the cytosol in the intact cells. LY-A cells are scraped under hypotonic conditions to become perforated and the cytosolic factors leak out. The perforated cells are incubated with [^{3}H]sphingosine and palmitoyl-CoA at 15 °C. [^{3}H]Ceramide is synthesized and accumulated mainly in the ER membrane. Then, the pre-labeled perforated cells are added into the LY-A cytosol with exogenous factors such as purified His$_6$-tagged CERT to start the ceramide transport reaction. [^{3}H]Ceramide transported to the Golgi apparatus is converted to [^{3}H]SM by SM synthase in the semi-intact cell system. Cellular lipids are extracted, and analyzed. In this figure, the semi-intact cell system is defined as the system composed of the perforated cells, cytosol, and various supplementary factors (see **Note 16**).

The reconstitution assay system is outlined in Fig. 1. LY-A, a CHO-K1-derived cell line, is deficient in ceramide transport activity due to a mutation converting glycine 67 to glutamic acid in CERT cDNA (Fukasawa *et al.*, 1999; Hanada *et al.*, 1998; Hanada *et al.*, 2003). Perforated cells are prepared by scraping adhered LY-A cells under hypotonic conditions. Cytosolic factors leak from the small pores formed at the plasma membrane while other cellular structures and the localization of various membrane-bound enzymes are maintained (Beckers *et al.*, 1987). When [^{3}H]sphingosine is added to the perforated cells, [^{3}H]ceramide is synthesized and accumulated in the ER membrane where most of the ceramide synthases exist (Mizutani *et al.*, 2005; Riebeling *et al.*, 2003; Venkataraman *et al.*, 2002). It is important to keep the temperature below 15 °C at this step to suppress ceramide transport from the ER. Then, exogenous CERT diluted in the cytosol fraction of LY-A is added to the perforated cells. Because some yet to be identified cytosolic components seem necessary for efficient ATP-dependent ceramide transport, the cytosol fraction of LY-A, of which endogenous CERT is deficient, is used for the reconstitution system when the activity of exogenous CERT is examined (Fig. 2). When the perforated cells

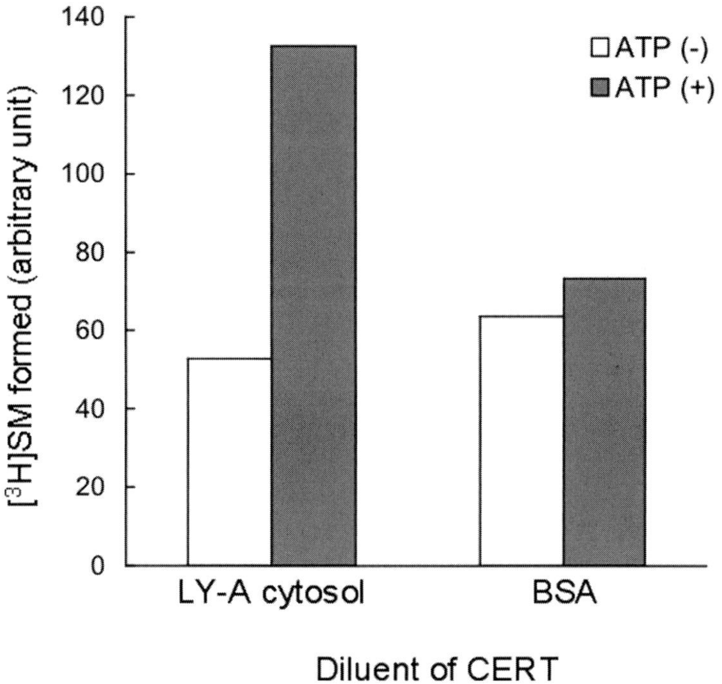

Fig. 2 Cytosolic factors are necessary for ATP-dependent efficient ceramide transport. Reconstitution assay of ceramide transport with perforated cells is carried out with bovine serum albumin (BSA) (50 μg) or the LY-A cytosol (100 μg) as a diluent of purified His_6-tagged CERT (3.5 ng). In the presence of the LY-A cytosol, the conversion of [^{3}H]ceramide to [^{3}H]SM is enhanced by ATP (left columns). Replacement of the LY-A cytosol with BSA abrogates the ATP-dependent enhancement (right columns). ATP (-), addition of apyrase (1 U/assay) without addition of the ATP-regenerating system; ATP (+), addition of the ATP-regenerating system.

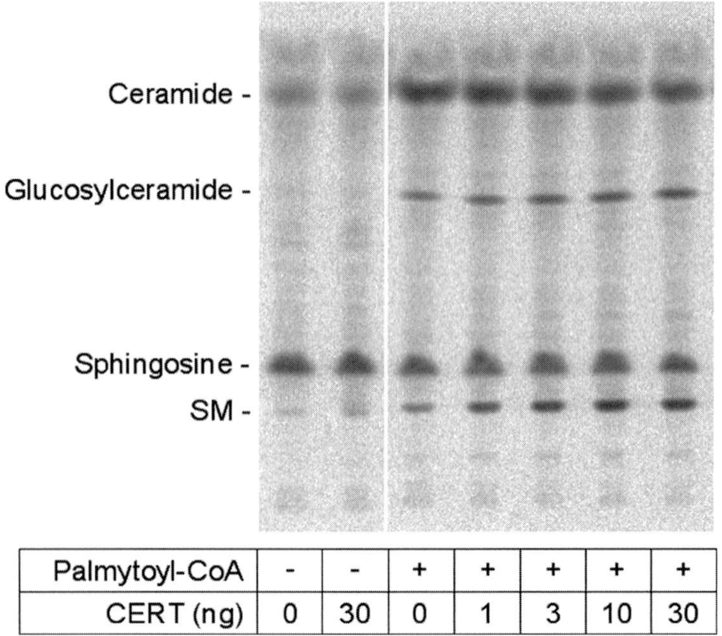

Fig. 3 Radioactive lipids analyzed by TLC. The amount of [^{3}H]SM synthesized increases in a CERT-dependent manner, while [^{3}H]ceramide is consumed. [^{3}H]Glucosylceramide also increases moderately related to the dose of CERT. In the control experiment without addition of palmitoyl-CoA, small amounts of [^{3}H]ceramide and [^{3}H]SM are detected independently of the added CERT, which means the existence of a small number of unperforated intact cells in the preparation.

are added to the cytosol fraction including functional CERT, [^{3}H]ceramide is transported from the ER to the Golgi apparatus to synthesize [^{3}H]SM in an ATP-dependent manner (Fukasawa *et al.*, 1999; Funakoshi *et al.*, 2000). The cellular lipids are extracted and analyzed by thin layer chromatography (TLC) (Fig. 3). Ceramide transport activity is estimated by quantifying [^{3}H]SM with an image analyzer.

The materials and methods for the assay of the activity of ER-to-Golgi transport of ceramide of purified recombinant CERT are described below.

II. Materials

All chemicals of reagent grade or better are used. Water purified with Milli-Q (Millipore) is used.

1. ES medium (Nissui Pharmaceutical Co., Tokyo, Japan), prepared just before use (see **Subheading III B 2 and Note 1**).
2. Ham's F-12 medium (Gibco), stored at 4 °C

3. HEPES-NaOH (pH 7.4), sterilized by filtration (0.22 μm filter), aseptically stored at room temperature at a concentration of 1 M.
4. Ten percent $NaHCO_3$, sterilized by filtration (0.22 μm filter), aseptically stored at room temperature.
5. 10 × Phosphate-buffered saline (pH 7.4) (PBS): 80 g NaCl, 29 g $Na_2HPO_4 \cdot 12H_2O$, 2 g KCl and 2 g KH_2PO_4 are dissolved in water at the final volume of 1 L, autoclaved, stored at room temperature.
6. PBS: diluted 10 × PBS with water, autoclaved, stored at room temperature.
7. Suspension buffer: 10 mM Tris-HCl (pH 7.4) containing 250 mM sucrose, stored at 4 °C.
8. Luria-Bertani broth containing kanamycin (25 μg/mL), stored at room temperature.
9. Isopropyl β-D-thiogalactoside sterilized by filtration (0.22 μm filter, Millipore), stored at −20 °C at a concentration of 1 M.
10. Lysis buffer: 25 mM Tris-HCl (pH 7.4) containing 1% Triton X-100, 1 mM sodium orthovanadate, 50 mM NaF, 5 mM sodium pyrophosphate, 270 mM sucrose, 2.5 mM 2-mercaptoethanol, and protease inhibitors (one tablet of EDTA-free Complete™ protease inhibitor cocktail (Roche Diagnostics) per 50 ml), stored at 4 °C.
11. TALON® metal affinity resin (Takara Bio Inc., Ohtsu, Japan, or Clontech).
12. Buffer A: 50 mM Na_2HPO_4-NaH_2PO_4 buffer (pH 7.0), 300 mM NaCl, stored at 4 °C.
13. Buffer A containing 10 mM imidazole, stored at 4 °C.
14. Buffer A containing 150 mM imidazole, stored at 4 °C.
15. Stock buffer: 10 mM Tris-HCl (pH 7.4) containing 250 mM sucrose, stored at 4 °C.
16. Hypotonic buffer: 10 mM HEPES-KOH (pH 7.2) containing 15 mM KCl and 0.1 mM $MgCl_2$, stored at 4 °C.
17. H/KCl buffer: 25 mM HEPES-KOH (pH 7.2) containing 115 mM KCl, stored at 4 °C.
18. 0.3% trypan-blue in PBS.
19. BCA protein assay kit (Pierce).
20. D-*Erythro*-[3-^{3}H]sphingosine (20 Ci/mmol) (American Radiolabeled Chemicals) (see **Note 2**).
21. Fumonisin B_1 (Sigma), stored at −80 °C at a concentration of 2 mM.
22. UDP-glucose (Sigma), stored at −20 °C at a concentration of 50 mM.
23. GTP (sodium salt form, Sigma), neutralized with NaOH solution, stored at −80 °C at a concentration of 100 mM.
24. Dithiothreitol (Sigma), stored at −20 °C at a concentration of 100 mM.
25. Palmytoyl-CoA lithium salt (Sigma), stored at −80 °C at concentrations of 5 mM (a dense stock) or 100 μM (a diluted stock).
26. 10 × salt premix: 200 mM HEPES-KOH (pH 7.0) containing 700 mM KCl, 25 mM MgOAc, 2.5 mM GTP, 2 mM dithiothreitol, 5 mM UDP-glucose and 144 μM fumonisin B_1, stored at −80 °C.

27. ATP (sodium salt form, Sigma), neutralized with NaOH solution, stored at −80 °C at a concentration of 20 mM.
28. Creatine phosphate (Sigma), stored at −20 °C at a concentration of 200 mM.
29. Creatine phosphokinase from rabbit muscle (Sigma), stored at −80 °C at a concentration of 1000 U/mL.
30. 20 × ATP regenerating system: 1 mM ATP, 40 mM phosphocreatine, 160 U/mL creatine phosphokinase, stored at −80 °C.
31. Apyrase (Sigma), stored at −80 °C at a concentration of 1000 U/mL.
32. Chloroform/methanol (1:2, by volume), stored at room temperature.
33. Chloroform/methanol/water (65:25:4, by volume), prepared just before use.
34. Benzene/diethylether/ethylacetate/methanol/25% ammonia water (60/7.5/7.5/25/0.75, by volume), prepared just before use.
35. 0.1 M KCl, autoclaved, stored at room temperature.
36. High performance TLC (Silica gel 60, Merck), stored in a desiccator at room temperature.

III. Methods

A. Cell Culture

1. CHO-K1 is a permanent cell line derived from Chinese hamster ovary. LY-A, a CHO-K1-derived mutant cell line, is deficient in ceramide transport activity due to a mutation converting glycine 67 to glutamic acid in CERT cDNA (Fukasawa *et al.*, 1999; Hanada *et al.*, 1998; Hanada *et al.*, 2003). The LY-A cell line is available from RIKEN Cell Bank (http://www.brc.riken.jp/lab/cell/english/).
2. LY-A cells are routinely maintained in F-12 culture medium (Ham's F-12 medium supplemented with 10% newborn bovine serum, penicillin G (100 units/mL) and streptomycin sulfate (100 μg/mL)) at 33 °C in a 5% CO_2 atmosphere (see **Note 3**).

B. Preparation of LY-A Cytosol

1. LY-A cells are cultured in 25 mL of F-12 culture medium in a 15-cm dish. Two or three dishes with sub-confluent LY-A cells are necessary for the liquid culture.
2. Powder of the ES medium (4.85 g) is dissolved in 500 mL of water in a 500 mL spinner flask (Bellco Glass Inc., NewJersey, USA) by stirring, and sterilized by autoclave at 121 °C for 20 min (see **Note 1**).
3. After cooling down to room temperature, the ES medium is aseptically supplemented with 5% fetal bovine serum, 2 mM L-glutamine, 10 mM HEPES-NaOH (pH 7.4), and 0.1% $NaHCO_3$. The resultant medium is named ES culture medium. Before starting cell culture, the ES culture medium is pre-warmed overnight at 37 °C in a 5% CO_2 atmosphere.
4. Sub-confluent LY-A cells in the 15-cm dishes are trypsinized to be suspended in the F-12 culture medium (10 mL per 15-cm dish). Then, the suspension of LY-A cells are added to 500 mL of the pre-warmed ES culture medium in the spinner

flask and cultivated at 37 °C in a 5% CO_2 atmosphere with gentle stirring (80–120 r.p.m) for 3–4 days (see **Note 4**).

5. While the 500-mL scale cultivation, 1 L of ES culture medium is prepared in a 1000 mL spinner flask (Bellco Glass Inc.) as described in **Subheadings III B 2** and **III B 3**.
6. Two hundred fifty milliliters of the cell culture prepared at **Subheading III B 4.** is transferred to the spinner flask containing 1 L of ES culture medium and cultivated as described in **Subheading III B 4** to a density of ~4 × 10^5 cells/mL (see **Note 5**).
7. Cells are precipitated by centrifugation at 250 × *g* for 10 min at 4 °C. Hereafter, all manipulations are carried out at 4 °C or on ice.
8. The pellet is suspended in a 10-pellet volume of PBS, and the suspension is centrifuged at 250 × *g* for 10 min.
9. The cell pellet is suspended in a 4-pellet volume of suspension buffer (10 mM Tris-HCl (pH 7.4), 250 mM sucrose). The cell suspension is transferred to a stainless steel ball bearing homogenizer equipped with disposable 10-mL syringes (see **Note 6**). Then, the cells are broken with the homogenizer as described previously (Balch and Rothman, 1985).
10. The homogenate is centrifuged at 900 × *g* for 10 min to precipitate unbroken cells, cell debris, and nuclei.
11. The post-nuclear supernatant is centrifuged (100,000 × *g*, 1 h). The resultant supernatant fluid is centrifuged again (100,000 × *g*, 1 h) to remove the particulate fraction as much as possible.
12. The resultant supernatant fluid is collected as a cytosolic fraction. The cytosolic fraction (~1.5-mL aliquots of which are dispensed to 1.5-mL polypropylene tubes) is rapidly frozen in liquid nitrogen and stored at −80 °C until use. The protein concentration of the cytosolic fraction is ~8 mg/mL.

C. Expression and Purification of Hexahistidine (His_6)-Tagged CERT

1. BL21 (DE3) *Escherichia coli* cells transfected with a bacterial expression plasmid encoding a His_6-tagged human CERT (Hanada *et al.*, 2003) are aerobically cultured in 1 L of Luria-Bertani broth containing kanamycin (25 μg/mL) at 37 °C to an OD of 0.6 at 600 nm.
2. The culture is briefly chilled on ice. Then, after the addition of isopropyl β-D-thiogalactoside to a final concentration of 250 μM, the cells are cultured for 16 h at 25 °C.
3. Hereafter, all manipulations are done at 4 °C or on ice unless otherwise noted. The cells are harvested by centrifugation at 1400 × *g* for 15 min, and suspended in 50 mL of lysis buffer (25 mM Tris-HCl (pH 7.4), 1% Triton X-100, 1 mM sodium orthovanadate, 50 mM NaF, 5 mM sodium pyrophosphate, 270 mM sucrose, 2.5 mM 2-mercaptoethanol, and protease inhibitors.
4. The cell suspension (12.5-mL aliquots of which are dispensed to 50-mL polypropylene tubes) is frozen at −80 °C, thawed at room temperature, and

sonicated eight times for 20 s with 1-min intervals with a probe-type sonicator (model W-225R, Heat Systems-Ultrasonics Inc.) at a 24-W output.

5. The sonicated lysate is transferred to ultracentrifuge tubes, and centrifuged at 100,000 × *g* for 1 h. The supernatant fluid is collected.
6. In parallel with the centrifugation, TALON® resin, a Co^{2+}-chelated affinity resin is pre-equilibrated with buffer A (50 mM sodium phosphate buffer (pH 7.0), 300 mM NaCl).
7. The supernatant fraction (~50 mL) is incubated with 3 mL of the pre-equilibrated TALON® resin in a 50-mL tube (Corning) for 1 h with rotary shaking (see **Note 7**).
8. After centrifugation (1000 × *g*, 3 min), the precipitated resin is suspended in 45 mL of buffer A containing 10 mM imidazole and precipitated for washing. After a repeat of this washing step twice, the precipitated resin is transferred to a disposable 5-mL polypropylene column (Pierce), washed with 30 mL of buffer A containing 10 mM imidazole, and then eluted with 6 mL of buffer A containing 150 mM imidazole.
9. The elution is dialyzed against stock buffer (10 mM Tris-HCl buffer (pH 7.4), 250 mM sucrose) for 12–24 hours.
10. The dialyzed fraction as purified CERT (5–1000 μL aliquots of which are dispensed to 1.5-mL polypropylene tubes) is divided and stored at −80 °C until use. Protein concentrations are determined using the BCA protein assay kit with bovine serum albumin as the standard. Purity of CERT was more than 90% (Fig. 4), and the yield of purified CERT is ~7 mg from 1 L of bacterial culture.

D. Preparation of Perforated Cells

Perforated cells are freshly prepared on the day of the ceramide transport assay. Thus, all manipulations described in **Subheadings III D 3–III D 8**, **III E 1–III E 7**, and **III F 1–III F 5** are done in the same day.

1. LY-A cells are harvested by trypsinization, and seeded in 10 mL of F-12 culture medium at a density of 3.3 × 10^6 cells per 10-cm cell culture dish (Corning).
2. Cells are cultured at 37 °C overnight to reach sub-confluence.
3. Hereafter, all manipulations are carried out at 4 °C or on ice. The cell monolayers are washed with 5 mL of hypotonic buffer (10 mM HEPES-KOH (pH 7.2), 15 mM KCl, and 0.1 mM $MgCl_2$) twice, and incubated for 10 min in 5 mL of the hypotonic buffer.
4. The hypotonic buffer is removed and the cell monolayers are scraped in 5 mL of H/KCl buffer (25 mM HEPES-KOH (pH 7.2), 115 mM KCl) with the scraper rubber policeman (see **Note 8**).
5. The suspension is centrifuged in a 14-mL polypropylene round-bottom tube (Falcon) at 250 × *g* for 5 min.
6. The pellet as perforated cells is suspended with 5 mL of H/KCl buffer by gentle pipetting and centrifuged again (250 × *g*, 5 min).

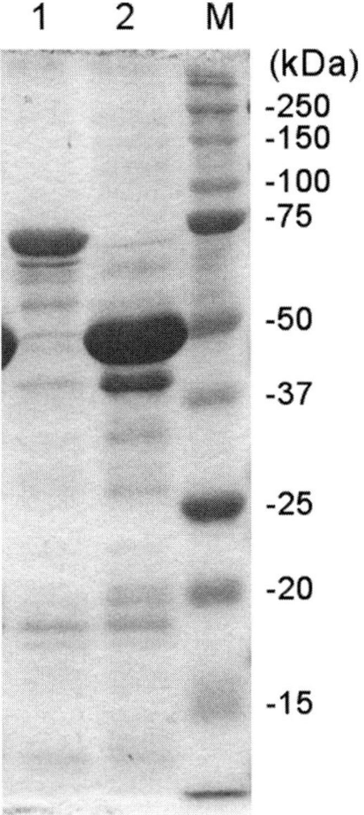

Fig. 4 SDS–PAGE analysis of purified His_6-tagged CERT. His_6-tagged CERT is expressed and purified as described (see **Subheading III C**). The proteins are analyzed by SDS–PAGE and visualized by Coomassie brilliant blue staining. Lane 1, full-length His_6-tagged CERT; lane 2, START-domain deleted His_6-tagged CERT; M, molecular mass standards.

7. The perforated cells are resuspended in H/KCl buffer (150 μL per dish). (see **Note 9**).
8. The protein concentration is determined with the BCA protein assay kit. The concentration of the perforated cells is usually 2.5–4.0 mg/mL.

E. *In vitro* Assay of Ceramide Transport From the ER to the Golgi Apparatus

1. Ethanolic solution of D-*Erythro*-[3-^{3}H]sphingosine (20 Ci/mmol) is dried under a nitrogen gas stream and dispersed in water at a concentration of 8 μM (160 μCi/mL) by vortex mixing and sonication in a bath-type sonicator. (see **Note 10**).
2. Perforated cells (1.6 mg/mL) are incubated in H/KCl buffer containing 10 μM palmitoyl-CoA and 0.8 μM [^{3}H]sphingosine (16 μCi/mL) in a 1.5-mL polypropylene tube (Eppendorf) for 30 min at 15 °C for pulse labeling of ceramide (see **Note 11**). One assay needs 25 μL (40 μg) of perforated cells.

3. After addition of fumonisin B_1 to a final concentration of 20 μM, the labeled perforated cells are incubated on ice for more than 15 min to block ceramide synthesis.
4. For one assay, purified His_6-tagged CERT (~3 ng) is diluted with 100 μg of the LY-A cytosol fraction to the final volume of 25 μL. (see **Note 12**).
5. Standard chase reactions are conducted in a final volume of 90 μL in a transport reaction mixture: 20 mM HEPES-KOH (pH 7.0), 70 mM KCl, 2.5 mM MgOAc, 250 μM GTP, 0.2 mM dithiothreitol, 0.5 mM UDP-glucose, 1 × ATP-regenerating system (see **Materials**), 20 μM fumonisin B_1, 100 μg of the cytosol fraction, and 40 μg of pre-labeled perforated cells (see **Note 13, 14**).
6. To start each transport assay, 25 μL (40 μg of protein) of the pre-labeled perforated cells are added to 65 μL of a mixture containing other components, and the resultant transport reaction mixture is incubated at 37 °C for 30 min.
7. The reaction is stopped by the addition of 700 μL of chloroform/methanol (1:2, by volume) and vortex mixing. After here, all manipulations are carried out at room temperature.

F. Extract Lipids and TLC Analysis

1. Chloroform (230 μL) and 0.1 M KCl (420 μL) are added to each sample and mixed. The mixture is centrifuged at 20,000 × *g* for 2 min for phase separation to extract lipids.
2. The lower organic solvent phase is carefully transferred to a glass tube with a glass Pasteur pipette, and dried under a nitrogen gas stream.
3. The dried lipids are dissolved in ~50 μL of chloroform/methanol (19:1, by volume), and streaked on a high performance TLC plate with a capillary glass. Also, known amounts of the standard [^{3}H]sphingosine (12, 24 and 48 nCi; 48 nCi = 2.4 pmol) are applied to the TLC plate to prepare a calibration curve.
4. The lipids on the TLC plate are developed with a solvent system of chloroform/methanol/water (65:25:4, by volume).
5. The TLC plate is dried under air. Then, TLC plate is exposed to an imaging plate (Fuji Film Inc., Tokyo, Japan) for 16–48 h.
6. Radioactive lipids are analyzed with a BAS1800 image analyzer (Fuji Film Inc., Tokyo, Japan) (Fig. 3) (see **Note 15**).
7. Giving that specific radioactivities of [^{3}H]lipids produced in the assay are the same as the activity of the standard [^{3}H]sphingosine, the amount of [^{3}H]SM produced is estimated from a calibration curve made of known amounts of [^{3}H]sphingosine.

IV. Notes

1. ES medium is a modified autoclavable Eagle's minimum essential medium enriched with amino acids, sodium pyruvate, and vitamin B_{12} to facilitate cell

proliferation and to reduce serum consumption (Koyama and Kodama, 1982). ES medium of Nissui Pharmaceutical Co. contains kanamycin, an autoclavable anti-bacterial drug. The three lids of the spinner flask are capped with pieces of aluminum foil. Make sure that at least one of the lids keeps loosened during autoclaving, cooling, and cell culturing processes.

2. D-*Erythro*-[3-^{3}H]sphingosine was sometimes degraded even when its purity was examined soon after the arrival of the compound from the manufacturer. Thus, it is recommended that the purity of [^{3}H]sphingosine is examined by TLC with a solvent system of benzene/diethylether/ethylacetate/methanol/25% ammonia water (60/7.5/7.5/25/0.75, by volume) before use (Kumagai *et al.*, 2005). Radioactive lipids are analyzed with a BAS1800 image analyzer. Sphingosine is detected at an R_f value of 0.37. We used [^{3}H]sphingosine with the purity of more than 70%.
3. CHO-K1 and LY-A cells can be cultured also at 37 °C. For F-12 culture medium, fetal bovine serum can be used in place of newborn calf serum.
4. Cell growth may be slow for the initial 1–2 days. The color of the culture medium changes from red to yellow–orange during the cell culture. When the medium is still red after 3-days cultivation, additional 1- or 2-days cultivation is required for better yields of LY-A cells.
5. We recommend that a portion of the cells adapted to the suspension culture are stored as frozen stocks in liquid nitrogen. When the adapted cells are used, the suspension culture in the spinner flask can be started without the pre-culture in dishes.
6. A ball-bearing homogenizer similar to the type developed originally by Balch and Rothmann (Balch and Rothman, 1985) might be available from EMBL precision engineering (Heidelberg, Germany) (German and Howe, 2009). For homogenization of cells to prepare the cytosol fraction, Dounce homogenizers may also work (Balch *et al.*, 1984), although we have not validated it.
7. The manufacturer's website informs that the TALON® resin binds to his-tagged proteins with higher specificity than Ni^{2+}-chelated resins (http://www.clontech.com/JP/Products/Protein_Expression_and_Purification/His-Tagged_Protein_Purification/Cobalt_Resin-Batch?sitex=10025:22372:US).
8. We do not know whether "rubber policeman" can be used as a general technical term, although the origin of this scraper is discussed (Jensen, 2008).
9. The prepared perforated cells contain a small population of unperforated intact cells, and the intact cells synthesize SM regardless of exogenous CERT. Thus, the preparation is checked by staining with 0.3% trypan-blue solution. The percentage of stained (perforated) cells should be more than 95% when observed with a hemocytometer under a phase contrast microscope.
10. Because ethanol has a severe effect on this assay, the sphingosine solution should be dried completely before dispersion in water.
11. To obtain quantitative data on perforated cell-derived activity, a portion of the perforated cells is pulse-labeled with [^{3}H]sphingosine in the absence of palmitoyl-CoA and chased as a control. The radioactivity incorporated into ceramide,

SM, and glucosylceramide in the control experiments is regarded as background activity derived from unperforated cells, and subtracted from the radioactivity of each lipid produced in the normal assay.

12. Because CERT binds to tube walls at very low protein concentrations, the cytosol of LY-A should be used for the diluent at every stage of the dilution.
13. For one assay, 4.5 μL of 20 × the ATP-regenerating system and 9.0 μL of 10 × salt solution are mixed and diluted to 40 μL with water, and then 25 μL of CERT diluted in the LY-A cytosol is added in a 2-mL polypropylene tube (Eppendorf). The transport reaction is started by adding 25 μL of prelabeled perforated cells to the 2-mL tube and mixing by gentle pipetting.
14. Dependence of the amounts of the cytosol per 40 μg of pre-labeled perforated cells on the production of [^{3}H]SM was examined previously (Funakoshi *et al.*, 2000). The cytosol fraction and perforated cells from the wild-type CHO-K1 cells can be prepared by essentially the same methods as described in **Subheadings III B** and **III D.**, respectively (Funakoshi *et al.*, 2000). Near-linearity of [^{3}H]SM production was observed from 0 to 100 μg of the CHO-K1 cytosol with 40 μg of perforated CHO-K1 cells, indicating that 100 μg of the cytosol is an appropriate amount to analyze the activity of cytosolic factors in the reconstitution system (Funakoshi *et al.*, 2000).
15. Production of the BAS image analyzer system has been stopped. Typhoon® image analyzer system (GE healthcare) may be used as an alternative to detect ^{3}H-labeled lipids separated on the TLC plate.
16. In the previous papers (Beckers *et al.*, 1987; Funakoshi *et al.*, 2000), the perforated cells are defined as semi-intact cells. However, it may be inappropriate to regard the perforated cells not having the cytosol as "semi-intact" cells. Thus, in this manuscript, semi-intact cells are described to be composed of the perforated cells, cytosol, and various supplementary factors including the ATP-regenerating system.

References

Balch, W. E., and Rothman, J. E. (1985). Characterization of protein transport between successive compartments of the Golgi apparatus: asymmetric properties of donor and acceptor activities in a cell-free system. *Arch. Biochem. Biophys.* **240**, 413–425.

Balch, W. E., Dunphy, W. G., Braell, W. A., and Rothman, J. E. (1984). Reconstitution of the transport of protein between successive compartments of the Golgi measured by the coupled incorporation of N-acetylglucosamine. *Cell* **39**, 405–416.

Beckers, C. J., Keller, D. S., and Balch, W. E. (1987). Semi-intact cells permeable to macromolecules: use in reconstitution of protein transport from the endoplasmic reticulum to the Golgi complex. *Cell* **50**, 523–534.

Fukasawa, M., Nishijima, M., and Hanada, K. (1999). Genetic evidence for ATP-dependent endoplasmic reticulum-to-Golgi apparatus trafficking of ceramide for sphingomyelin synthesis in Chinese hamster ovary cells. *J. Cell Biol.* **144**, 673–685.

Funakoshi, T., Yasuda, S., Fukasawa, M., Nishijima, M., and Hanada, K. (2000). Reconstitution of ATP- and cytosol-dependent transport of de novo synthesized ceramide to the site of sphingomyelin synthesis in semi-intact cells. *J. Biol. Chem.* **275**, 29938–29945.

Futerman, A. H., and Riezman, H. (2005). The ins and outs of sphingolipid synthesis. *Trends Cell Biol.* **15**, 312–318.

German, C. L., and Howe, C. L. (2009). Preparation of biologically active subcellular fractions using the Balch homogenizer. *Anal. Biochem.* **394**, 117–124.

Hanada, K., Hara, T., Fukasawa, M., Yamaji, A., Umeda, M., and Nishijima, M. (1998). Mammalian cell mutants resistant to a sphingomyelin-directed cytolysin. Genetic and biochemical evidence for complex formation of the LCB1 protein with the LCB2 protein for serine palmitoyltransferase. *J. Biol. Chem.* **273**, 33787–33794.

Hanada, K., Kumagai, K., Yasuda, S., Miura, Y., Kawano, M., Fukasawa, M., and Nishijima, M. (2003). Molecular machinery for non-vesicular trafficking of ceramide. *Nature* **426**, 803–809.

Hanada, K., Kumagai, K., Tomishige, N., and Yamaji, T. (2009). CERT-mediated trafficking of ceramide. *Biochim. Biophys. Acta* **1791**, 684–691.

Huitema, K., van den Dikkenberg, J., Brouwers, J. F., and Holthuis, J. C. (2004). Identification of a family of animal sphingomyelin synthases. *EMBO J.* **23**, 33–44.

Jensen, W. B. (2008). The Origin of the Rubber Policeman. *J. Chem. Educ.* **85**, 776.

Koyama, H., and Kodama, H. (1982). Adenine phosphoribosyltransferase deficiency in cultured mouse mammary tumor FM3A cells resistant to 4-carbamoylimidazolium 5-olate. *Cancer Res.* **42**, 4210–4214.

Kumagai, K., Yasuda, S., Okemoto, K., Nishijima, M., Kobayashi, S., and Hanada, K. (2005). CERT mediates intermembrane transfer of various molecular species of ceramides. *J. Biol. Chem.* **280**, 6488–6495.

Mandon, E. C., Ehses, I., Rother, J., van Echten, G., and Sandhoff, K. (1992). Subcellular localization and membrane topology of serine palmitoyltransferase, 3-dehydrosphinganine reductase, and sphinganine N-acyltransferase in mouse liver. *J. Biol. Chem.* **267**, 11144–11148.

Mizutani, Y., Kihara, A., and Igarashi, Y. (2005). Mammalian Lass6 and its related family members regulate synthesis of specific ceramides. *Biochem. J.* **390**, 263–271.

Riebeling, C., Allegood, J. C., Wang, E., Merrill Jr., A. H., and Futerman, A. H. (2003). Two mammalian longevity assurance gene (LAG1) family members, trh1 and trh4, regulate dihydroceramide synthesis using different fatty acyl-CoA donors. *J. Biol. Chem.* **278**, 43452–43459.

van Meer, G., Voelker, D. R., and Feugenson, G. W. (2008). Membrane lipids: where they are and how they behave. *Nat. Rev. Mol. Cell Biol.* **9**, 112–124.

Venkataraman, K., Riebeling, C., Bodennec, J., Riezman, H., Allegood, J. C., Sullards, M. C., Merrill Jr., A. H., and Futerman, A. H. (2002). Upstream of growth and differentiation factor 1 (uog1), a mammalian homolog of the yeast longevity assurance gene 1 (LAG1), regulates N-stearoyl-sphinganine (C18-(dihydro)ceramide) synthesis in a fumonisin B1-independent manner in mammalian cells. *J. Biol. Chem.* **277**, 35642–35649.

Yamaoka, S., Miyaji, M., Kitano, T., Umehara, H., and Okazaki, T. (2004). Expression cloning of a human cDNA restoring sphingomyelin synthesis and cell growth in sphingomyelin synthase-defective lymphoid cells. *J. Biol. Chem.* **279**, 18688–18693.

CHAPTER 7

Visualizing Mitochondrial Lipids and Fusion Events in Mammalian Cells

Huiyan Huang and Michael A. Frohman

Department of Pharmacology, Center for Developmental Genetics, Stony Brook University, Stony Brook, New York, USA

Abstract

Mitochondria are dynamic organelles that frequently undergo fusion and fission, the balance of which is critical for proper cellular functioning and viability. Most studies on mitochondrial fusion and fission mechanisms have focused on proteins thought to physically mediate the events. However, dynamic changes in membrane phospholipids also play roles in facilitating the fusion and fission events. This chapter will review the importance of lipids in mitochondrial dynamics and some

Copyright 2012, Elsevier Inc. All rights reserved.

0091-679X/10 $35.00
DOI 10.1016/B978-0-12-386487-1.00007-9

of the methods that can be used to study the function of lipids in mitochondrial fusion and fission.

I. Introduction

A. Mitochondrial Dynamics

Mitochondrial "health" requires frequent fusion of mitochondria and subsequent division (fission). The fusion process is critical for exchanging mitochondrial contents and maintaining integrity of the mitochondrial genome, whereas fission is vital for being able to distribute mitochondria to subcellular locations where ATP is most needed or during mitosis, and for facilitating turnover of damaged mitochondria via mitophagy (Braschi and McBride, 2010; Chan, 2006; Otera and Mihara, 2011). Disturbance of the mitochondrial fusion-fission balance in humans results in multiple consequences (Kane and Youle, 2010; Westermann, 2010) including peripheral neuropathy (Zuchner *et al.*, 2004), optic atrophy (Olichon *et al.*, 2006), and neonatal lethality (Waterham *et al.*, 2007).

A number of proteins critically involved in mitochondrial fusion and fission in yeast and mammals have been identified – Mitofusin 1 and 2 (Mfn1/2), Optic atrophy 1 (Opa1), human Fission 1 (hFis1), and dynamin-related protein 1 (Drp1) (Otera and Mihara, 2011). During mitochondrial fusion, the outer mitochondrial membrane proteins Mfn1/2 tether the adjacent mitochondria with their C-terminal coiled-coil domain to promote outer membrane fusion (Koshiba *et al.*, 2004), while the inner membrane protein Opa1 faces the inter membrane space and mediates inner membrane fusion (reviewed in Chen and Chan, 2010). Conversely, hFis1 resides in the outer mitochondrial membrane and has long been thought to recruit the cytosolic protein Drp1 to sites of future fission, although recently, the proposal has been made that another outer membrane protein, mitochondrial fission factor, may be the critical component that recruits Drp1 instead of hFis1 (Gandre-Babbe and van der Bliek, 2008; Otera *et al.*, 2010). Regardless of the recruitment mechanism, Drp1 self-assembles as an oligomer and wraps around the mitochondria to achieve membrane constriction and division (Mears *et al.*, 2011), analogous to the role dynamin undertakes during endocytosis (Chan, 2006). Many additional proteins have been added to the list of ones that affect mitochondrial fusion and fission as a result of either screening experiments or random discoveries, complicating our understanding of the mechanisms underlying mitochondrial fusion and fission – much remains to be uncovered and explained!

B. Lipids in Mitochondrial Dynamics

Apart from the involvement of two membranes, many aspects of the biophysical process of mitochondrial fusion and fission are analogous to the fusion and fission of cytosolic membrane vesicles as they bud from and fuse into subcellular membrane compartments. Similarly, several lipids have been discovered to be important in potentially analogous ways as well (reviewed in Furt and Moreau, 2009; Osman *et al.*, 2011). A role for the signaling lipid phosphatidic acid (PA) has been described in promoting Mfn-dependent mitochondrial fusion that is analogous to the role PA plays in SNARE protein-regulated exocytosis, although each fusion process uses a different lipid-modifying enzyme, MitoPLD and classic PLD, respectively, to locally produce PA (Choi *et al.*, 2006; Huang *et al.*, 2005; Vicogne *et al.*, 2006; Vitale *et al.*, 2001). More recently, another lipid, diacylglycerol (DAG), has been linked to mitochondrial fission events (Huang *et al.*, 2011), analogous to the role of DAG in Golgi vesiculation (Fernandez-Ulibarri *et al.*, 2007). Evidence has been presented for roles of other lipids, including cardiolipin (Khalifat *et al.*, 2008; Montessuit *et al.*, 2010) and cholesterol (Altmann and Westermann, 2005), in mitochondrial membrane morphodynamics (reviewed in Furt and Moreau, 2009; Osman *et al.*, 2011). Cardiolipin, which is primarily found in the inner membrane, has been shown to play a major role cristae dynamics (Khalifat *et al.*, 2008) and in membrane fusion through interaction with Opa1/Mgm1 (Ban *et al.*, 2010; DeVay *et al.*, 2009; Rujiviphat *et al.*, 2009). On the outer membrane, where a smaller but significant amount of cardiolipin is found (Gebert *et al.*, 2009), cardiolipin has been shown to serve as the substrate for PA production during fusion events (Choi *et al.*, 2006). In addition, Drp1-stimulation of Bax oligomerization and cytochrome c release during apoptosis requires the presence of cardiolipin in the outer membrane, and the Drp1-cardiolipin interaction is thought to trigger membrane tethering and hemifusion (Montessuit *et al.*, 2010) which might also play a role in fission in this context (Kozlovsky and Kozlov, 2003). For cholesterol, the mechanism of action is not clear, but many of the genes required for the synthesis of cholesterol have been identified as regulators of mitochondrial morphogenesis in yeast (Altmann and Westermann, 2005).

II. Imaging Mitochondrial Tubules in Overexpression or Knockdown Samples

Since directly manipulating lipids on mitochondrial membranes via introduction of liposomes or other approaches is almost impossible in living cells, controlling expression levels and/or localization of lipid-modifying enzymes is the methodology used to test the importance of individual lipids in mitochondrial morphology. The approaches used should include overexpression and inactivation (via RNAi, small molecule inhibitors, or genetic ablation) of the lipid-generating enzyme, re-targeting of enzymes that can generate the lipid but are normally found elsewhere in the cell (Komatsu *et al.*, 2010), and overexpression or re-targeting of enzymes that consume the lipid (Choi *et al.*, 2006; Huang *et al.*, 2011). Since each of these approaches has inherent caveats, use of more than one approach when possible is preferable. As well, some thought needs to be given to potential compensatory issues; there are multiple pathways for generation and consumption of most lipids, and long-term removal of a specific pathway (e.g., via genetic ablation) may have a diminished or different effect than acute removal (e.g., via a small molecule inhibitor (e.g., Su *et al.*, 2009) or chemically induced rapid re-targeting of a lipid modifying enzyme (e.g., Nishioka *et al.*, 2010)).

A. Choice of Cell Types

To assess mitochondrial morphology in mammalian cells, in theory, any cell line that is easy to culture and transfect can be used. We routinely use HeLa, NIH-3T3, or mouse embryo fibroblast (MEF) cells. However, for investigators who are interested in mitochondrial dynamics in specific cell types such as neurons or muscle cells, specialized cell types can be used as well. It is advisable to check multiple cell lines at the beginning of the intended studies, since different cell lines can have unique characteristics that result in different mitochondrial tubule lengths under basal conditions. When the intent is to study a lipid involved in mitochondrial fission, it is better for some purposes to start off with a cell line that has longer tubules such as HeLa cells in order to easily detect overexpression phenotypes, and conversely, cell lines with shorter tubules are preferred for lipids that promote fusion. However, since gain and loss-of-function experiments will yield opposite effects, the general ideal is to start with cells with intermediate length tubules.

Some cells are relatively challenging to study because of specific characteristics. For example, the OP9 and NIH3T3-L1 pre-adipocyte cell lines are relatively flat and well-spread on coverslips and easily imaged. However, upon differentiation, the cells become spherical, filled with lipid droplets, and have relatively scant cytoplasm, making it very challenging to discriminate changes in mitochondrial morphology using confocal microscopy. In some cases, however, use of stronger-adhering coverslip coatings (for example, poly-L-lysine), may elicit sufficient flattening and spreading of the cells that the imaging can be performed.

B. Mitochondrial labeling methods

There are multiple ways to label the mitochondrial tubules – mitochondrial-targeted fluorescent proteins (MitoFPs), MitoTracker dyes, and immunostaining using mitochondria-specific protein antibodies. The MitoFPs, constructed by fusing any fluorescent protein with a mitochondrial localization sequence (MLS), can be co-transfected with the lipid enzyme or co-expressed in an IRES vector (a number of such constructs are available from Clontech). However, when a specific cell line has to be used and achieving high transfection efficiency is not possible, the MitoTracker dyes and antibody staining are preferred since the mitochondria in all the cells will be labeled equally well.

There are a range of MitoTracker dyes with different wavelengths available from Invitrogen. Most dyes, such as MitoTracker Orange and MitoTracker Red, are sensitive to mitochondrial membrane potential, which is useful as an indicator of mitochondrial function, but potentially problematic if manipulation of the lipid under study might result in compromised membrane potential. In contrast, MitoTracker Green accumulates in mitochondria regardless of membrane potential. However, it should be noted that cells labeled with MitoTracker Green should be fixed with cold acetone instead of aldehyde-based fixatives.

Finally, immunostaining with mitochondria-specific antibodies requires more procedure steps and time, but is versatile, offers a broad choice of secondary antibodies, and avoids the risk of altering the morphology of mitochondria through introducing a foreign protein or dye into them. Common and abundant mitochondria-specific proteins used for imaging with commercially available antibodies include cytochrome c, cytochrome c oxidase IV, Tom 20, Tom 22, Tom 70, and porin. Worth noting, when multiple proteins of interest need to be detected at the same time, it can be preferable to use MitoFPs or MitoTracker dyes to visualize the mitochondrial population, since there are only a limited number of species that secondary antibodies are available for, limiting the number of primary antibodies that can be simultaneously detected.

C. Data Analysis

Mitochondria display diverse morphologies even within the same cell line. Therefore, it is critical to take high quality images of mitochondrial morphology for many (often one to several hundred) cells for each experimental condition (generally using confocal microscopy) and then quantify the results. Nonetheless, mitochondrial morphology is a subjective phenotype, and it can be difficult to cleanly categorize them. Therefore, before analyzing the data, it is important to become familiar with the possible mitochondrial phenotypes that can exist under different conditions, some of which are shown in Fig. 1. Mitochondria normally appear tubular, intermediate, or in a minority of cases, fragmented; however, in extreme conditions when a mitochondrial fusion protein or fission protein is overexpressed, the mitochondria can aggregate in the peri-nuclear region or fragment

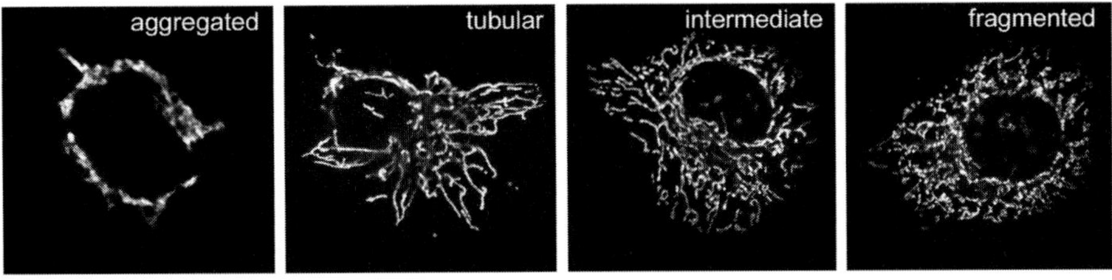

Fig. 1 Examples of mitochondrial morphology. HeLa cells display different mitochondrial morphologies, as shown for cells overexpressing MitoPLD that have aggregated mitochondria (left), or others with tubular, intermediate, or fragmented mitochondria as seen under basal conditions. Mitochondria are stained with anti-cytochrome c antibody.

into small spherical or oval mitochondria, respectively. Software analysis programs such as ImageJ (freely available from NIH) can be used to compare mitochondrial tubule lengths using the "analyze particles" function. However, the disadvantage of automated programs is that they can only distinguish very obvious phenotypes such as mitochondrial aggregation or mitochondrial fragmentation. When it comes to subtle changes in tubule length, none of the programs are as sophisticated as the trained human eye.

With sufficient numbers of images of mitochondrial morphology generated from three or more independent experiments for each treatment, the data can be analyzed (ideally by a blinded investigator) by categorizing each cell as tubular, intermediate, or fragmented, and then comparing the distribution of the categories for each sample by calculating the percentage of cells exhibiting each phenotype. Statistical analysis (*t*-test for two treatments or one-way ANOVA for more than two treatments) is then used to determine whether there is a significant difference between the experimental conditions. When the lipid has a profound effect on the mitochondrial morphology, the categories can be assigned differently to suit the extreme conditions. For example, when PA is generated by MitoPLD overexpression, it induces mitochondrial aggregation at high levels of expression and mitochondrial elongation at low levels (Choi *et al.*, 2006; Huang *et al.*, 2011). These morphological phenotypes can be tabulated as different categories. Similarly, other manipulations can cause tubulation or fragmentation (Choi *et al.*, 2006; Huang *et al.*, 2011).

III. A Quantitative Assay for Mitochondrial Fusion

Through the above experiments, the investigators should have an idea whether the lipids they are interested in help to elongate or shorten mitochondrial tubule lengths. However, the mitochondrial morphology is determined by the balance of the opposing fusion and fission events. Therefore, assessing morphological changes alone only indicates shifts in the balance of fusion and fission. Longer tubules could result either from more fusion or from fewer fission events. To determine which, mitochondrial fusion assays can be employed to assess whether the cells have abnormal rates of fusion. If the lipid is suspected to be involved in mitochondrial fission, there are other methods that have been described recently (Molina and Shirihai, 2009).

Multiple versions of mitochondrial fusion assays are available. The classic mitochondrial fusion assay introduces two different fluorescent proteins into two populations of the cells, mixes the cells and then induces cell fusion with polyethylene glycerol (PEG) or a hemagglutinating virus. The subsequent mitochondrial fusion after cell fusion is assayed several hours later by scoring the extent of coincidence of the two fluorescent proteins (Nunnari *et al.*, 1997). A second mitochondrial fusion assay was developed to take advantage of photoactivatable GFP (PA-GFP). By co-labeling the mitochondria with matrix-targeted PA-GFP and DsRed2, the fusion and fission of mitochondria can be quantitatively recorded after irradiation at regions of interest in a chosen cell (Twig *et al.*, 2006, 2008). In brief, the DsRed2 protein visualizes all of the mitochondria using red fluorescence; when the PA-GFP is activated in a restricted region of the cell, mitochondria in that area become yellow (as a combination of the DsRed and the newly fluorescent PA-GFP). As fusion subsequently occurs, mixing of matrix contents transfers the activated PA-GFP into previously red-only mitochondria, resulting in spread and dilution of the yellow color, both of which can be quantitated to describe mitochondrial dynamics (motility and fusion). Both of the above invaluable mitochondrial fusion assays are described in reviews elsewhere (Ingerman *et al.*, 2007; Molina and Shirihai, 2009). Here, we will discuss the quantitative assay for mitochondrial fusion that was newly developed by our lab (Huang *et al.*, 2010).

The quantitative assay for mitochondrial fusion works by labeling two populations of cells with split-*Renilla* luciferase (RLuc) fragments that reconstitute and emit chemiluminescence when the mitochondria fuse together following chemical induced cell fusion (Huang *et al.*, 2010). Therefore, the first steps for the assay are to choose a desired cell line and to introduce each split-RLuc construct into the cells to select for stable expressing cells, following which the assay can be performed to compare the luciferase activity between different samples.

A. Generation of Stable Cells

1. Choice of Cell Lines

In situations where small molecule inhibitors can be used, the most important consideration is the relevance of the cell type to investigator's interest in the functions that ultimately need to be assessed. In other situations, for example, where manipulation of a lipid by overexpressing or knocking-down a lipid enzyme is needed, one has to balance between the relevance of the cell type and the ease of transfection. For the assay to be valid using transient transfection approaches, at least 80% transfection efficiency needs to be achieved. If this is not feasible using standard plasmid transfection approaches, alternative ways to introduce the DNA can include methods such as viral infection. Another solution is to generate stable expression cell lines first for the lipid modifying enzyme of interest, and then introduce the split-RLuc constructs into both control and the stable-expressing cell lines.

2. Generation of Stable Cell Lines

Since the split-RLuc constructs were generated in a retroviral backbone (Huang *et al.*, 2010), generation of stable cell lines can be achieved by producing the retrovirus in a virus packaging cell line, infecting the target cells with the virus, and then selecting stable cell populations with appropriate antibiotics, in this case puromycin. The detailed protocol varies with the virus packaging cell line, each of which has its own instructions and manuals with the protocols needed to generate stable cell lines.

B. The Quantitative Assay for Mitochondrial Fusion

Once the stable cell lines are ready, two more steps that need to be carried out before performing the mitochondrial fusion assay are to determine how many cells are needed for the assay and what concentration of cycloheximide (CHX) to use during the assay. CHX is required to prevent *de novo* expression and mitochondrial targeting of reporter proteins subsequent to cell fusion, which would otherwise generate a false positive readout for the mitochondrial fusion assay (Huang *et al.*, 2010).

To determine the cell number, a range in numbers of cells from 10^5 to 10^6 should be plated in a 12-well plate and examined the following day to identify the well that has reached 80–100% confluency. If transfection of plasmid constructs or RNAi oligos is required, it will be necessary to plate a smaller number of cells, perform the transfection the following day, and wait for 1–3 days to achieve expression / knock-down of a specific mRNA and protein. The assay can be scaled up or down in size according to need.

To determine the optimal CHX concentration, the cell mixture containing each split-RLuc construct are co-plated at the cell density determined above. The next day, the cells are treated with different concentrations of CHX, fused with PEG 1500, incubated for 3 h and then assayed for luciferase activity (refer to detailed protocols below). The relative light unit (RLU) of luciferase activity will decrease with increased CHX concentration until a baseline is reached, signifying the level at which *de novo* protein production has ceased. The lowest CHX concentration that achieves baseline luciferase activity should be chosen to be used for the assay. In practice, this concentration will vary for different types of cell lines. Due to the extreme sensitivity of the luciferase activity measurement and thus the fluctuations in the readings, each sample should be measured at least three times to determine the optimal CHX concentration. Concentrations of 0, 50 μg/mL, 100 μg/mL, 200 μg/mL, and 400 μg/mL represent a range that will cover most cell types.

General protocol for mitochondrial fusion assay

1. Mix and plate cells containing each split-RLuc construct in 12-well plates with the pre-determined cell number. Number of wells to plate ($N = a \times b \times c$) depends on the number of treatments (a), number of time points (b), and number of repeats (c).
2. (Optional) The next day, transfect or infect the cells with constructs or oligos as desired.
3. When the cells reach 80–100% confluence, pre-treat with small molecule inhibitors as desired and then add CHX at the predetermined concentration 30 min before inducing cell fusion.

4. At the end of the CHX incubation period, remove the media from the wells and add 300 μL per well pre-warmed 50% PEG 1500 for exactly 60 s. Practice first to see how many samples can be handled at the same time. Divide samples into several groups if necessary.
5. After 60 s, leaving the PEG in the well, immediately add 1 mL of complete media containing cycloheximide to each well. Swirl well to get rid of the PEG and wash two more times with complete media containing cycloheximide.
6. Add complete media containing cycloheximide and then return the cells to the incubator. Collect cells at 0, 30 min, 1 h, 2 h, 3 h, 5 h, and 7 h, or more simply, 0, 1 h, and 3 h. To collect individual samples, take out the plate from incubator, add 500 μL 5 mM EDTA in PBS containing cycloheximide, wait 1 min for the cells to come off the plate and then collect the cells into an eppendorf tube. Put the plate back in the incubator. Spin down the cells at 14,000 rpm for 1 min and aspirate off the supernatant. Immediately store the cell pellets at −20 °C.
7. When all the samples are collected, lyse the cell pellets with 50 μL 1x *Renilla* luciferase assay lysis buffer on ice for 15 min. Vortex for 2 s.
8. Freshly prepare an adequate volume of luciferase substrate to perform the desired number of Renilla luciferase assays (20 μL reagent per assay sample). Add 1 volume of 100× *Renilla* luciferase substrate to 100 volumes of *Renilla* luciferase assay buffer.
9. Set up the luminometer with integration time of 5 s.
10. Add 20 μL of *Renilla* luciferase assay solution into a transparent eppendorf tube and place the tube in the luminometer.
11. Add 3 μL of the lysate sample into the solution. Pipette up and down to mix well and then initiate measurement.
12. Proceed with the next sample from step 10.

C. Data Analysis

For analysis, plot time (x) versus RLU (y) and compare the resulting curves for each experimental condition. Potential outcomes of the plots are described below (Fig. 2). Each curve should also be compared with the control experimental condition, which indicates the baseline for fusion rates for that cell type. The Luciferase signal increases over time in parallel to the progression of fusion, but then eventually begins to fall, as the reporters being to undergo degradation.

Example A: Each early time point in the experimental sample reveals a lower luciferase activity than is observed for the corresponding control sample, but the curves eventually merge, indicating that cells in the experimental condition are undergoing mitochondrial fusion at a slower rate than the cells in the control conditions.

Example B: Luciferase activity is dramatically lower in comparison to the control sample, with a peak reading at the 1 h time point instead of at 3 h, indicating that there is little on-going fusion. The residual luciferase activity is thought to come

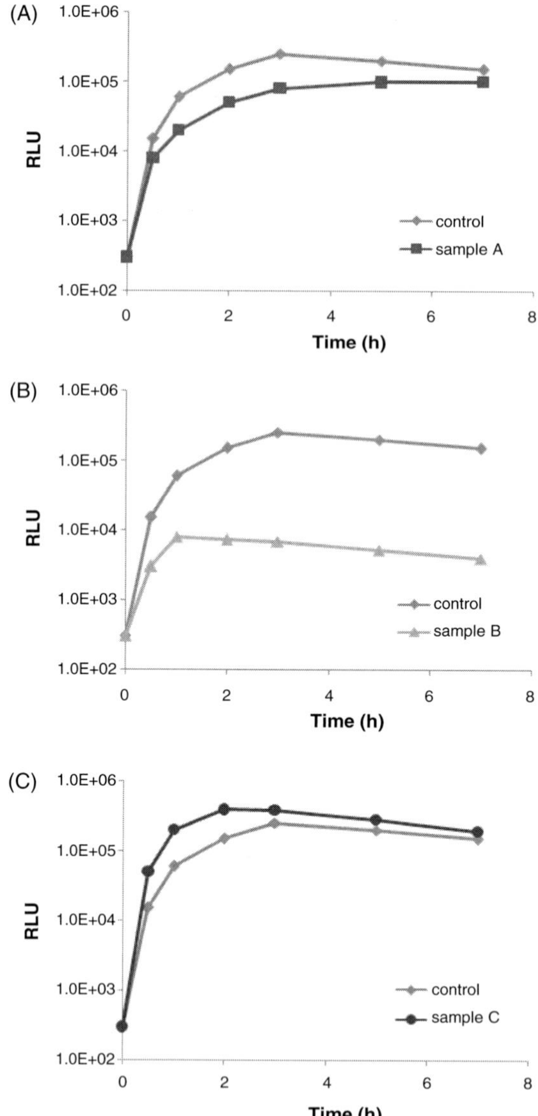

Fig. 2 Data analysis. Hypothetical examples of data representing reduced mitochondrial fusion (A), lack of mitochondrial fusion (B), and increased mitochondrial fusion (C). (For color version of this figure, the reader is referred to the web version of this book.)

from reconstitution of a small fraction of the reporter proteins in the cytoplasm, or release of the reporter proteins during apoptosis.

Example C: Each time point in the experimental sample exhibits a higher luciferase activity than that of the corresponding control sample with the peak luciferase activity occurring earlier, suggesting that the cells are undergoing mitochondrial fusion at an increased rate.

IV. Visualizing Lipids in Cells

Morphological analysis of mitochondrial length and the fusion assay described above provide functional information concerning the consequences of manipulating the levels of individual lipids on mitochondrial fusion or fission. Complementing these approaches, lipid sensors exist for a number of kinds of lipids that permit direct visualization of the lipids on the mitochondrial surface (readers should also be directed to another chapter in this volume by Sarantis and Grinstein). Sensors for PA and DAG have been successfully employed to detect these lipids on the mitochondrial surface (Choi *et al.*, 2006; Gallegos *et al.*, 2006; Huang *et al.*, 2011; Sato *et al.*, 2006).

A. Lipid Sensors

Standard lipid sensors consist of the lipid-recognizing domains of known lipid-binding proteins fused with fluorescent proteins of various colors, which enables live cell imaging of lipid dynamics to study the spatio-temporal aspects of lipid signaling. This approach has been utilized to generate sensors for a wide variety of lipids including PA (Nakanishi *et al.*, 2004; Rizzo *et al.*, 2000), DAG (Dries *et al.*, 2007; Giorgione *et al.*, 2006) and different species of phosphoinositides (Varnai and Balla, 2006), using a variety of protein domains. These sensors are very useful for visualizing the relative distribution of individual lipids within the cell. However, it should be noted that it is quite challenging to use such sensors for quantitative measurements of the lipids, in contrast to more technically challenging approaches using fluorescence resonance energy transfer (FRET)-based lipid sensors that have been developed specifically to this end. Different approaches have been developed utilizing the FRET principle to generate signals based on the lipid levels. So far, the most sophisticated method is to attach both fluorophores to the same lipid-binding domain, taking advantage of the conformational change that occurs upon lipid binding that alters the distance between the fluorophores, resulting in changes in energy transfer. Examples of such lipid sensors include FRET-based DAG (Gallegos *et al.*, 2006; Sato *et al.*, 2006), PA (Nishioka *et al.*, 2010), and PIP_3 sensors (Sato *et al.*, 2006; Tanimura *et al.*, 2004).

B. Assay Design

Most of the lipid sensors that have been described are openly available from the investigators that generated them. It is worth noting that some of the sensors appear to work better on some membrane surfaces than others (e.g., plasma membrane versus mitochondrial surface), either because the presentation of the lipid is different in the different environment due to the presence of other lipids or differences in the fatty acid composition of the lipid, or because the sensors also interact with proteins on the surface, rather than purely reacting to the target lipid. Therefore, failure to detect altered distribution of a specific lipid sensor after manipulating the lipid level on a specific membrane surface in the cell may not necessarily mean there is no change there in the lipid level. Controls that can be used for this approach include mutant lipid sensors that are unable to bind to the lipid, and inactive lipid-modifying enzymes that do not change the lipid levels. Reviews on technical and theoretical aspects of FRET measurements have been reviewed elsewhere (Sekar and Periasamy, 2003; Thaler *et al.*, 2005; van Rheenen *et al.*, 2004).

Finally, it should be noted that such sensors can only be used to probe the mitochondrial surface, since the inner leaflet of the outer membrane and both sides of the inner membrane are essentially incapable of being accessed by these protein-based sensors. Thus, some caution should be used when attempting to compare the information obtained using sensors to biochemical analyses of lipid content that assay additional leaflets and membranes.

C. Biochemical Approaches

Although the list of lipid sensors has been expanding continuously, sensors are not available for all types of lipids, and some of the available sensors are not very sensitive (readers should also be directed to another chapter in this volume by Sarantis and Grinstein). As an alternate approach, biochemical approaches such as Liquid Chromatography/Mass Spectrometry-Mass Spectrometry (LC/MS-MS) can be utilized to measure changes in concentrations of individual mitochondrial lipids (Bird *et al.*, 2011; Choi *et al.*, 2007). However, in addition to the issue cited above regarding the additional leaflets and membranes, it should also be noted that the sensors report on lipid that is accessible to binding; lipid already bound by other proteins can not be imaged by the sensor – whereas the MS approach will identify all of the lipid present. On the other hand, fairly large amounts of material are required for MS approaches (Bird *et al.*, 2011), and in some cases the lipids may undergo modification or degradation during the purification and recovery steps needed to prepare them for analysis.

V. Summary

With the discovery of the involvement of lipids such as PA, DAG, cholesterol, and cardiolipin in mitochondrial fusion and fission dynamics, there is increased awareness of the need to study other lipids for their roles in the process of mitochondrial morphodynamics.

Acknowledgment

The work is supported by NIH GM071520 and GM084251.

References

Altmann, K., and Westermann, B. (2005). Role of essential genes in mitochondrial morphogenesis in Saccharomyces cerevisiae. *Mol. Biol. Cell* **16**, 5410–5417.

Ban, T., *et al.* (2010). OPA1 disease alleles causing dominant optic atrophy have defects in cardiolipin-stimulated GTP hydrolysis and membrane tubulation. *Hum. Mol. Genet.* **19**, 2113–2122.

Bird, S. S., *et al.* (2011). Lipidomics profiling by high-resolution LC-MS and high-energy collisional dissociation fragmentation: focus on characterization of mitochondrial cardiolipins and monolysocardiolipins. *Anal. Chem.* **83**, 940–949.

Braschi, E., and McBride, H. M. (2010). Mitochondria and the culture of the Borg: understanding the integration of mitochondrial function within the reticulum, the cell, and the organism. *Bioessays* **32**, 958–966.

Chan, D. C. (2006). Mitochondrial fusion and fission in mammals. *Ann. Rev. Cell Dev. Biol.* **22**, 79–99.

Chen, H., and Chan, D. C. (2010). Physiological functions of mitochondrial fusion. *Ann. N. Y. Acad. Sci.* **1201**, 21–25.

Choi, S. Y., *et al.* (2006). A common lipid links Mfn-mediated mitochondrial fusion and SNARE-regulated exocytosis. *Nat. Cell Biol.* **8**, 1255–1262.

Choi, S. Y., *et al.* (2007). Cardiolipin deficiency releases cytochrome c from the inner mitochondrial membrane and accelerates stimuli-elicited apoptosis. *Cell Death Differ.* **14**, 597–606.

DeVay, R. M., *et al.* (2009). Coassembly of Mgm1 isoforms requires cardiolipin and mediates mitochondrial inner membrane fusion. *J. Cell Biol.* **186**, 793–803.

Dries, D. R., *et al.* (2007). A single residue in the C1 domain sensitizes novel protein kinase C isoforms to cellular diacylglycerol production. *J. Biol. Chem.* **282**, 826–830.

Fernandez-Ulibarri, I., *et al.* (2007). Diacylglycerol is required for the formation of COPI vesicles in the Golgi-to-ER transport pathway. *Mol. Biol. Cell* **18**, 3250–3263.

Furt, F., and Moreau, P. (2009). Importance of lipid metabolism for intracellular and mitochondrial membrane fusion/fission processes. *Int. J. Biochem. Cell Biol.* **41**, 1828–1836.

Gallegos, L. L., *et al.* (2006). Targeting protein kinase C activity reporter to discrete intracellular regions reveals spatiotemporal differences in agonist-dependent signaling. *J. Biol. Chem.* **281**, 30947–30956.

Gandre-Babbe, S., and van der Bliek, A. M. (2008). The novel tail-anchored membrane protein Mff controls mitochondrial and peroxisomal fission in mammalian cells. *Mol. Biol. Cell* **19**, 2402–2412.

Gebert, N., *et al.* (2009). Mitochondrial cardiolipin involved in outer-membrane protein biogenesis: implications for Barth syndrome. *Curr. Biol.* **19**, 2133–2139.

Giorgione, J. R., *et al.* (2006). Increased membrane affinity of the C1 domain of protein kinase C delta compensates for the lack of involvement of its C2 domain in membrane recruitment. *J. Biol. Chem.* **281**, 1660–1669.

Huang, H., *et al.* (2010). A quantitative assay for mitochondrial fusion using Renilla luciferase complementation. *Mitochondrion* **10**, 559–566.

Huang, H., *et al.* (2011). piRNA-associated germline nuage formation and spermatogenesis require MitoPLD pro-fusogenic mitochondrial-surface lipid signaling. *Dev. Cell* **20**, 376–387.

Huang, P., *et al.* (2005). Insulin-stimulated plasma membrane fusion of Glut4 glucose transporter-containing vesicles is regulated by phospholipase D1. *Mol. Biol. Cell* **16**, 2614–2623.

Ingerman, E., *et al.* (2007). In vitro assays for mitochondrial fusion and division. *Methods Cell Biol.* **80**, 707–720.

Kane, L. A., and Youle, R. J. (2010). Mitochondrial fission and fusion and their roles in the heart. *J. Mol. Med.* **88**, 971–979.

Khalifat, N., *et al.* (2008). Membrane deformation under local pH gradient: mimicking mitochondrial cristae dynamics. *Biophys. J.* **95**, 4924–4933.

Komatsu, T., *et al.* (2010). Organelle-specific, rapid induction of molecular activities and membrane tethering. *Nat. Methods* **7**, 206–208.

Koshiba, T., *et al.* (2004). Structural basis of mitochondrial tethering by mitofusin complexes. *Science* **305**, 858–862.

Kozlovsky, Y., and Kozlov, M. M. (2003). Membrane fission: model for intermediate structures. *Biophys. J.* **85**, 85–96.

Mears, J. A., *et al.* (2011). Conformational changes in Dnm1 support a contractile mechanism for mitochondrial fission. *Nat. Struct. Mol. Biol.* **18**, 20–26.

Molina, A. J., and Shirihai, O. S. (2009). Monitoring mitochondrial dynamics with photoactivatable [corrected] green fluorescent protein. *Methods Enzymol.* **457**, 289–304.

Montessuit, S., *et al.* (2010). Membrane remodeling induced by the dynamin-related protein Drp1 stimulates Bax oligomerization. *Cell* **142**, 889–901.

Nakanishi, H., *et al.* (2004). Positive and negative regulation of a SNARE protein by control of intracellular localization. *Mol. Biol. Cell* **15**, 1802–1815.

Nishioka, T., *et al.* (2010). Heterogeneity of phosphatidic acid levels and distribution at the plasma membrane in living cells as visualized by a Foster resonance energy transfer (FRET) biosensor. *J. Biol. Chem.* **285**, 35979–35987.

Nunnari, J., *et al.* (1997). Mitochondrial transmission during mating in Saccharomyces cerevisiae is determined by mitochondrial fusion and fission and the intramitochondrial segregation of mitochondrial DNA. *Mol. Biol. Cell* **8**, 1233–1242.

Olichon, A., *et al.* (2006). Mitochondrial dynamics and disease, *OPA1. Biochim. Biophys. Acta* **1763**, 500–509.

Osman, C., *et al.* (2011). Making heads or tails of phospholipids in mitochondria. *J. Cell Biol.* **192**, 7–16.

Otera, H., and Mihara, K. (2011). Molecular mechanisms and physiologic functions of mitochondrial dynamics. *J. Biochem.* **149**, 241–251.

Otera, H., *et al.* (2010). Mff is an essential factor for mitochondrial recruitment of Drp1 during mitochondrial fission in mammalian cells. *J. Cell Biol.* **191**, 1141–1158.

Rizzo, M. A., *et al.* (2000). The recruitment of Raf-1 to membranes is mediated by direct interaction with phosphatidic acid and is independent of association with Ras. *J. Biol. Chem.* **275**, 23911–23918.

Rujiviphat, J., *et al.* (2009). Phospholipid association is essential for dynamin-related protein Mgm1 to function in mitochondrial membrane fusion. *J. Biol. Chem.* **284**, 28682–28686.

Sato, M., *et al.* (2006). Imaging diacylglycerol dynamics at organelle membranes. *Nat. Methods* **3**, 797–799.

Sekar, R. B., and Periasamy, A. (2003). Fluorescence resonance energy transfer (FRET) microscopy imaging of live cell protein localizations. *J. Cell Biol.* **160**, 629–633.

Su, W., *et al.* (2009). FIPI, a Phospholipase D pharmacological inhibitor that alters cell spreading and inhibits chemotaxis. *Mol. Pharmacol.* **75**, 437–446.

Tanimura, A., *et al.* (2004). Fluorescent biosensor for quantitative real-time measurements of inositol 1,4,5-trisphosphate in single living cells. *J. Biol. Chem.* **279**, 38095–38098.

Thaler, C., *et al.* (2005). Quantitative multiphoton spectral imaging and its use for measuring resonance energy transfer. *Biophys. J.* **89**, 2736–2749.

Twig, G., *et al.* (2006). Tagging and tracking individual networks within a complex mitochondrial web with photoactivatable GFP. *Am. J. Physiol. Cell Physiol.* **291**, C176–C184.

Twig, G., *et al.* (2008). Fission and selective fusion govern mitochondrial segregation and elimination by autophagy. *EMBO J.* **27**, 433–446.

van Rheenen, J., *et al.* (2004). Correcting confocal acquisition to optimize imaging of fluorescence resonance energy transfer by sensitized emission. *Biophys. J.* **86**, 2517–2529.

Varnai, P., and Balla, T. (2006). Live cell imaging of phosphoinositide dynamics with fluorescent protein domains. *Biochim. Biophys. Acta* **1761**, 957–967.

Vicogne, J., *et al.* (2006). Asymmetric phospholipid distribution drives in vitro reconstituted SNARE-dependent membrane fusion. *Proc. Natl. Acad. Sci. U. S. A.* **103**, 14761–14766.

Vitale, N., *et al.* (2001). Phospholipase D1: a key factor for the exocytotic machinery in neuroendocrine cells. *EMBO J.* **20**, 2424–2434.

Waterham, H. R., *et al.* (2007). A lethal defect of mitochondrial and peroxisomal fission. *N. Engl. J. Med.* **356**, 1736–1741.

Westermann, B. (2010). Mitochondrial fusion and fission in cell life and death. *Nat. Rev. Mol. Cell Biol.* **11**, 872–884.

Zuchner, S., *et al.* (2004). Mutations in the mitochondrial GTPase mitofusin 2 cause Charcot-Marie-Tooth neuropathy type 2A. *Nat. Genet.* **36**, 449–451.

PART II

Lipid Metabolism and Signaling

CHAPTER 8

Targeted and Non-Targeted Analysis of Membrane Lipids Using Mass Spectrometry

Xue Li Guan[*,†] **and Markus R. Wenk**[*,†,‡]

[*]Department of Medical Parasitology and Infection Biology, Swiss Tropical and Public Health Institute, Basel, Switzerland

[†]University of Basel, Basel, Switzerland

[‡]Department of Biochemistry and Department of Biological Sciences, Yong Loo Lin School of Medicine, National University of Singapore, Singapore

Copyright 2012, Elsevier Inc. All rights reserved.

0091-679X/10 $35.00
DOI 10.1016/B978-0-12-386487-1.00008-0

Abstract

Mass spectrometry has gained popularity amongst cell biologists in recent years as a complementary tool to probe the metabolism and biological functions of lipids. Indeed, the technology offers unprecedented sensitivity, selectivity, and resolution to monitor lipids at the level of molecular species. This has led to further insights on how differences in fine chemical details of lipids may affect cellular processes. While different degrees of sophistication of mass spectrometric analysis exist and instruments are rapidly evolving, two general approaches, targeted and non-targeted, have been adopted by analysts. In this chapter, we describe these approaches and simple methods for rapid analyses of membrane lipids, which will serve as a starting point for future in-depth studies.

I. Introduction

The definition of lipids has undergone dramatic changes with the constant revelation of novel structures (Ito *et al.,* 2008; Korekane *et al.,* 2007) and discovery of the functions of these compounds. Indeed, the burgeoning appreciation of the critical roles of lipids in biological processes has rekindled interest in lipid research to dissect the functions of these metabolites beyond their structural and storage capacity. It is thus imperative to directly follow the entity itself to complement other techniques in cell biology in order to fully uncover the functions of the wide spectrum of lipids nature has created.

Unlike proteins or genes, which are made up of a limited number of monomeric units, lipids comprise of a structurally diverse collection of molecules that vary in physicochemical properties and dynamic range, posing a huge technical challenge to analysts. Traditional lipid analysis, such as thin layer and gas chromatography, is hampered with limited sensitivity, selectivity and resolution. Only in the last 10–15 years has lipid analysis on a systems-level scale made substantial progress, aided by advances in technologies, particularly in mass spectrometry, that afford an "-omic-centric" view of the lipid inventory of biological systems. Notably, the development of soft ionization techniques such as electrospray ionization (ESI) and matrix-assisted laser desorption/ionization (MALDI) has significantly expanded the range of lipids that can be qualitatively and quantitatively analyzed by mass spectrometry (Pulfer and Murphy, 2003). The goal of the lipidomics field, which is the systems-level scale analysis of lipids and their interacting partners, aims not only to measure the whole lipid inventories of biological systems, but to functionally annotate these physiologically relevant metabolites.

While the development of sophisticated instrumentation is desired to advance the field of lipidomics, just as important is a good understanding of the capability of available technologies and developing sensible lipidomic strategies to interrogate the specific biological questions. These take into considerations the nature of the analytes, which influences their extractability and ionization, as well as the instrumentations. Many excellent reviews on lipidomics analytical technologies are widely available

(Blanksby and Mitchell, 2010; Han *et al.,* 2012; Harkewicz and Dennis, 2011; Griffiths *et al.,* 2007; Merrill *et al.,* 2009; Myers *et al.,* 2011; Shevchenko and Simons, 2010; Wenk, 2010).Without attempting exhaustive descriptions of these aspects, we will describe in this chapter general considerations and approaches for lipid analysis by mass spectrometry for mammalian cells and tissues. The readers are referred to a recent chapter in methods in enzymology for details for yeast lipid analysis (Guan *et al.,* 2010).

II. Isolation and Purification of Membrane Lipids

In general, the choice of extraction protocol depends on the nature of the biological sample (e.g., tissue, cells or fluids) and the chemistry of the lipid of interest. Ultimately, quantitative isolation of lipids with maximal recovery and purity is desired. For monitoring recoveries as well as for quantitative purposes, a suitable cocktail of internal standards that has the same ionization properties as the analyte(s) of interest needs to be spiked into the mixture.

Glycerophospholipids, sphingolipids, and sterols are the three major classes of lipids that make up the bulk of eukaryotic cell membranes (Fig. 1). Membrane lipids can be recovered reasonably well with chloroform/methanol extraction, typically

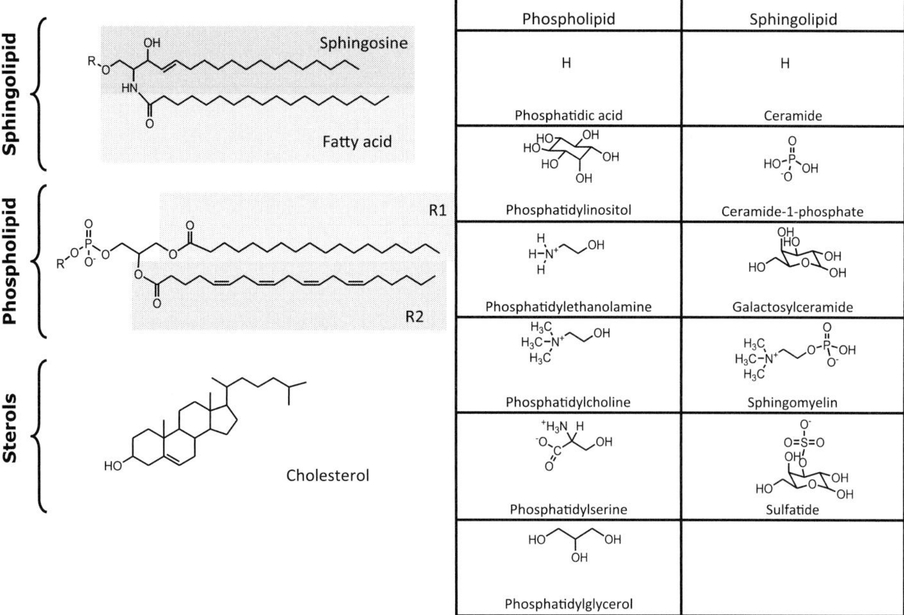

Fig. 1 Structural diversity of membrane lipids. Sphingolipids (top panel), (glycero)phospholipids (middle panel), and possible sterols (lower panel) are the major lipid classes found in eucaryotic cells. Headgroup modifications (R) of sphingolipids and (glycero)phospholipids are shown in the right hand panel.

according to Folch and co-workers (1957) or Bligh and Dyer (1959), in which they are mostly enriched in the chloroform phase. Sugar modification and phosphorylation, however, render some of these lipids highly polar, and thus may escape into the aqueous milieu during isolation. For instance, phosphoinositides, due to their polar nature and low abundance, are poorly recovered using conventional Bligh and Dyer or Folch methods of extraction, which is commonly used for bulk membrane lipids including sterols, ceramides, sphingomyelin, and major glycerophospholipids. Modifications such as acidification or use of ion pair agents as well as derivatization to aid in solubility, and therefore recovery have been reported (Gray *et al.,* 2003; Pettitt *et al.,* 2006; Wenk *et al.,* 2003).

The detection of lipids by mass spectrometry depends not only on the structure of the molecule, which entails its inherent ionization property, but is also influenced by the presence of other compounds in a complex extract which will compete for ionization or suppress signals through gas-phase reactions. This can be overcome by selective enrichment of lipids of interest. To improve the detection of sphingolipids, bulk cellular glycerophospholipids are removed after alkaline hydrolysis (Brockerhoff, 1963; Jiang *et al.,* 2007), thereby reducing signal suppression during mass spectrometric analysis.

III. Detailed Protocols for Isolation of Membrane Lipid From Mammalian Cells and Tissues

A. Lipid Standards

Phosphatidic acid, GPA (14:0/14:0), 0.06 μg
Phosphatidylcholine, GPCho (14:0/14:0), 0.8 μg
Phosphatidylethanolamine, GPEtn (14:0/14:0), 0.4 μg
Phosphatidylglycerol, GPGro(14:0/14:0), 0.04 μg
Phosphatidylserine, GPSer (14:0/14:0), 0.2 μg
Sphingomyelin, SM (12:0/dl8:l), 0.3 μg
Glucosylceramide (8:0/dl8:l), 0.08 μg
Ceramide (17:0/dl8:l) (Avanti Polar Lipids, Alabaster, AL, USA), 0.08 μg
Phosphatidylinositol, GPIns (8:0/8:0) (Echelon Biosciences, Salt Lake City, UT), 0.15 μg

Note: The above recommended amounts are based on 1×10^6 HELA cells. This should be titrated based on sample source.

B. Sample Preparation

1. For cell culture, wash and harvest 3×10^6 cells in ice-cold phosphate-buffered saline (PBS). Resuspend cells in 300 μL of PBS. Split into three aliquots for total lipid extraction, sphingolipid extraction, and protein content estimation.
2. For tissue such as rat brain, homogenize in five volumes of ice-cold PBS. Transfer aliquots of 20 mg of tissue homogenate to new tubes.

C. Total Lipid and Sphingolipid Extraction Using a Modified Bligh and Dyer Method

1. Add internal standard mixture
2. Add 900 μL of chloroform-methanol, 1:2 (volume/volume) to the cell suspension or tissue lysate
3. Vortex for 1 min
4. Stand on ice for 1 h, with intermittent vortexing
5. Add 300 μL of chloroform and 300 μL of 0.1 N HC1
6. Vortex for 1 min and incubate on ice for another 2 min
7. Centrifuge the sample for 5 min at 9000 rpm
8. Collect the lower organic phase and transfer to a new tube. Be extremely careful not to disturb and/or aspirate the intermediate disc
9. Re-extract the aqueous phase by adding 600 μL of chloroform and repeating steps 5 and 6
10. Pool organic phase
11. Dry lipid extract under vacuum or a constant stream of nitrogen gas and store at −80 °C
12. Immediately prior to analysis, reconstitute the lipid extracts in HPLC mobile phase

Note: The use of 0.1 N HC1 in Step 5 aids in recovery of acidic phospholipids. If plasmalogens are of interest, this should be replaced with deionized water.

In addition, using aliquots of the cells and tissue homogenate, sphingolipid extracts are prepared using alkaline methanolysis as described by Sullards and Merrill (2001), with slight modifications. Specifically,

1. Add 0.5 mL of methanol to the cell suspension/tissue homogenate in 13 mm × 100 mm glass test tubes.
2. Add 0.25 mL of chloroform
3. Add internal standards
4. Sonicate 30 s at room temperature and incubate overnight at 48 °C in a heat block or water bath
5. Cool and add 75 μL of 1 M methanolic KOH
6. Sonicate for about 30 s, then incubate 2 h at 37 °C
7. Neutralize solution with 6 μL of glacial acetic acid
8. To the neutralized solution, add 2 mL of chloroform and 4 mL of water, mix, and centrifuge at 4000 g for 10 min
9. Carefully collect the lower organic layer (change tip which each transfer, be careful not to aspirate the protein disc in between the two layers)
10. Dry lipid extract under vacuum or a constant stream of nitrogen gas and store at −80 °C

It should be noted that chloroform is highly toxic and should be handled with care. Alternatively, extraction of membrane lipids can be achieved by replacing chloroform with methyl-tert-butyl ether, which also offers the possibility for automation because the low density, lipid-containing organic phase forms the upper layer during phase separation (Matyash *et al.,* 2008).

IV. Mass Spectrometry-Based Approaches for Lipid Analysis

Lipid analysis by mass spectrometry can be generally classified into two approaches – targeted and non-targeted analysis. Effectively the approaches are dependent on the type of instrumentation available to the laboratories.

A. Profiling of Complex Lipid Mixtures Using Single Stage Mass Spectrometry (Non-Targeted Analysis)

Han and co-workers demonstrated the differential ionization efficiency of lipid classes based on their inherent electrical propensities and termed this separation and/or selective ionization of different lipids as "intrasource separation" (Han and Gross, 2005b). Enhanced sensitivity of microfluidics-based ionization (Han *et al.*, 2008) and ultra high resolution MS such as Fourier Transform Ion Cyclotron (FT-ICR) MS (Ivanova *et al.*, 2001; Schwudke *et al.*, 2007) have tremendously improved the separation of lipids, providing an unparalleled platform for MS-based profiling to provide a fingerprint of the lipid mixture. Information of the fine details of molecular species is indicated by the mass-to-charge ratio (m/z) and the ion intensity correlates to quantity (Zacarias *et al.*, 2002). Depending on the mass accuracy and resolution of the mass spectrometer (Table I), elemental composition of each ion detected can be obtained. However, the volume of data generated by mass spectrometry is considerable and interpretation of data was hampered during the dawn of lipidomics when there was a virtual absence of comprehensive and integrated reference databases. This situation has changed with the increasing availability of public and open source databases (Table II) as well as proprietary software tools (Table III, Taguchi and Ishikawa, 2010).

Figure 2A represents a typical mass spectrum of a mouse brain lipid extract analyzed by ESI-MS with direct infusion, that is, without prior chromatographic separation nor via loop injection using an autosampler. Instead, constant flow infusion (e.g., via a microliter infusion syringe pump or Nanomate (Advion)) (Han and Gross, 2005a, 2005b; Han *et al.*, 2008) was used. Major membrane phospholipids

Table I
Comparison of mass analysers

Instrument	Resolution	Mass Accuracy (ppm)	Sensitivity	Dynamic Range
Linear Ion Trap (LTQ)	2000	100	Femtomole	1.00E + 04
Triple quadrupole	2000	1000	Attomole	1.00E + 06
LTQ-Orbitrap	100,000	2	Femtomole	1.00E + 04
LTQ-Fourier Transform Ion Cyclon Resonance	500,000	<2	Femtomole	1.00E + 04
Quadrupole-Tof of Flight	10,000	2-5	Attomole	1.00E + 06

Table II
List of lipid-related databases

Database	Content	URL
LIPID MAPS	Largest curated lipid database containing (i) mass spectra; (ii) classifications; (iii) protocols of 10,000 lipid species.	http://www.lipidmaps.org/
LMSAD (Lipid Mass Spectrum Analysis Database)	Subset of LIPID MAPS database, encompassing structures and annotations of biologically relevant lipids. Structures of lipids in the database come from four sources: (i) LIPID MAPS Consortium's core laboratories and partners; (ii) lipids identified by LIPID MAPS experiments; (iii) computationally generated structures for appropriate lipid classes; (iv) biologically relevant lipids manually curated from LipidBank LIPIDAT and other public sources (Sud *et al.*, 2007).	http://www.lipidmaps.org/data/structure/index.html
LipidBank	A Japanese library containing >7000 lipid species, partly with experimental information including MS data and references (Taguchi *et al.*, 2007).	http://lipidbank.jp/
MassBank	A database of comprehensive high-resolution mass spectra of metabolites (Taguchi *et al.*, 2007).	http://www.massbank.jp/index-e.html
LIPIDAG	A relational database of lipid miscibility and associated information. LIPIDAG includes references to almost 1600 phase.	http://www.lipidag.ul.ie/
LIPIDAT	A central depository for information on lipid mesomorphic and polymorphic transitions and miscibility (Caffrey and Hogan, 1992).	http://www.lipidat.ul.ie/
SphingoMap	A curation of pathway map for sphingolipid biosynthesis that includes many of the known sphingolipids and glycosphingolipids arranged according to their biosynthetic origin(s).	http://sphingolab.biology.gatech.edu/
Kyoto Encyclopedia of Genes and Genomes, KEGG	A database of biological systems, consisting of genetic building blocks of genes and proteins (KEGG GENES), chemical building blocks of both endogenous and exogenous substances (KEGG LIGAND), molecular wiring diagrams of interaction and reaction networks (KEGG PATHWAY), and hierarchies and relationships of various biological objects (KEGG BRITE).	http://www.genome.jp/kegg/
Lipid library	A collection of links and information, including lipid biochemistry and analytical techniques, such as mass spectra, NMR techniques, Ag+ chromatography.	http://www.lipidlibrary.co.uk/
CyberLipids	A collection of links and information, including extraction protocols and lipid analytics.	http://www.cyberlipid.org

Table III
List of softwares for mass spectrometry spectra processing and data analysis

Software	Utility	Source	URL
LipidSearch	A high-throughput and automatic system for glycerophospholipids identification by various types of MS raw data file	Department of Metabolome, Graduate School of Medicine, the University of Tokyo (Taguchi and Ishikawa, 2010; Taguchi *et al.*, 2007)	http://metabo.umin.jp/index.htm
LipidMaps MS Tool	Find mass, number of carbons, number of double bonds, abbreviation, MS/MS product ions (neutral loss), formula, and ion based on input criteria, with links to structure and isotopic distribution.	LipidMaps (Fahy *et al.*, 2007)	http://www.lipidmaps.org/tools/index.html
Lipidlnspector	Lipid profiling and identification by multiple precursor and neutral loss scanning based in specified criteria in a data-dependent fashion	Scions (Schwudke *et al.*, 2006)	http://www.scionics.de/lipidinspector
LipidView (formerly known as Lipid Profiler)	Identification and quantification of molecular species of glycerophospholipids by automated interpretation of multiple precursor ion scan spectra, only compatible with Analyst data files	MDS Sciex (Ejsing *et al.*, 2006)	Enquiries on software can be made to MDS Sciex http://www.mdssciex.com/
Fatty Acid Analysis Tool, FAAT	Analyse high-resolution mass spectra of lipids obtained with FT-ICR instruments, includes assignment of ions from a user-defined library based on their exact mass, and comparative analysis of monoisotopic ions in complex mixture	(Leavell and Leary, 2006)	http://pubs.acs.org/doi/suppl/10.1021%2Fac0604179
Lipid Mass Spectrum Analysis, LIMSA	Peak detection and integration tool for identification (based on user defined lists)	Freeware provided by University of Helsinki (Haimi *et al.*, 2006)	http://www.helsinki.fi/science/lipids/software.html
Spectrum Extraction from Chromatographic Data, SECD	For display of LC-MS chromatograms as two-dimensional "maps" for visual inspection and extraction of mass spectra from LC-MS data	Freeware provided by University of Helsinki (Haimi *et al.*, 2006)	http://www.helsinki.fi/science/lipids/software.html

Lipid Qualitative/ Quantitative Analysis, LipidQA	Automated identification and quantification of complex lipid molecular specis in mixtures. Library of reference spectra is built based on previous fragmentation pattern of known lipid	Washington University Biomedical Mass Spectrometry Resource (Song *et al.*, 2007)	Access to LipidQA software can be obtained through Haowei Song (hsong@im.wustl.edu), John Turk (jturk@wustl.edu), or the Washington University Biomedical Mass Spectrometry Resource website
AMDMS-SL	A comprehensive freeware package designed for shotgun lipidomics	(Yang *et al.*, 2009)	www.shotgunlipidomics.com
MZmine 2	A freeware for processing, visualizing and analyzing LC-MS data	(Pluskal *et al.*, 2010)	http://mzmine.sourceforge.net
XCMS	A freeware for processing, visualizing and analyzing LC-MS data, used mainly for metabolomics	(Smith *et al.*, 2006)	http://metlin.scripps.edu/xcms/index.php
Spectromania	A software for visualization and analyzing MS data		http://www.spectromania.com/
LipidomeDB	A web application to quantify complex lipids by processing data acquired after direct infusion of a lipid-containing biological extract	(Zhou *et al.*, 2011)	

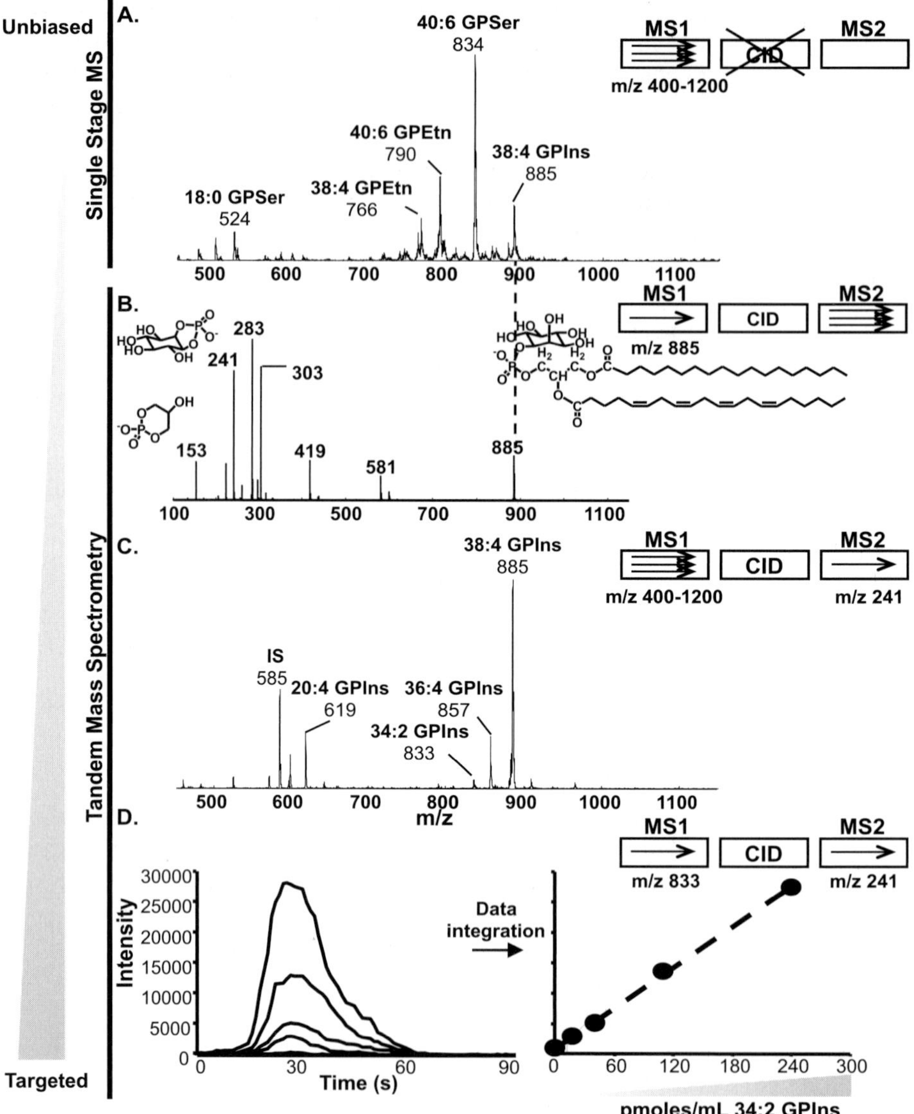

Fig. 2 Analysis of brain lipids by negative ion mode ESI-MS. (A) Single stage electrospray ionization mass spectrum (ESI-MS) in the negative ion mode. The majority of glycerophospho- and sphingo-lipids are detected in the mass range of 400–1200. The ions can be tentatively assigned by their mass-to-charge (m/z) ratio. Characterization of ions can be achieved by collision-induced dissociation (CID) and tandem mass spectrometry (MS/MS). (B) MS/MS spectra of ions with m/z 885. An ion of interest can be selected in the first mass analyzer (MS1) and after CID, the fragment ions are analyzed in the second mass analyzer (MS2). The product ions of the parent with m/z 885 (38:4 GPIns) includes m/z 153, 241, 283 and 303, which correspond to ions arising from the glycerol phosphate backbone, inositol phosphate headgroup and fatty acyls, respectively. Such information on a common fragment ion that is characteristic and specific for

and sphingolipid including ceramides and glucosylceramides deprotonate and ionize well in the negative ionization mode. Choline-containing lipids such as phosphatidylcholine and sphingomyelin although detectable, ionize better when protonated in the positive ionization mode. In addition, although cholesterol is highly abundant in the mammalian brain, due to its poor ionization efficiency in the negative ion mode, it is not detected and therefore not represented in Fig. 2A. Similarly, the signaling molecules, phosphoinositides, were not detected due to (1) poor recovery during extraction and (2) ion suppression by major lipids.

There is currently no one single method available to probe the entire cellular lipidome. The use of additives to promote ionization, alternative ionization sources such as Atmospheric Pressure Chemical Ionization (APCI) which is more suitable for less polar lipids such as cholesterol, and/or coupling to liquid chromatography (LC) represent solutions to improve the range of lipids to be measured. Specifically, using ESI-MS in positive ionization mode, with upfront separation using a C18 column, we achieved separation and detection of sterols and sterol esters in complex lipid mixtures (Fig. 3) (Shui *et al.,* 2010). Alternatively, analysis of sterol intermediates can be achieved using gas chromatography mass spectrometry (Wang *et al.,* 2009). Increased selectivity and sensitivity can be further achieved by targeted quantification by multiple reaction monitoring (please refer to section on targeted analysis and also selected references (Axelsen and Murphy, 2010; Honda *et al.,* 2010; Ikeda and Taguchi, 2010; Sullards and Merrill, 2001).

B. Differential Analysis of Lipid Profiles

High resolution MS profiling represents an attractive top-down approach to study changes in the lipidome upon perturbation. The instrument is programmed to collect data on detectable ions in order to provide a "fingerprint" for the various experimental conditions. The lipid profiles can be compared using specialized softwares (Fig. 4A and Table III) which perform functions including smoothing, peak detection, and isotopic correction (Myers *et al.,* 2011). A list of ions or identified lipids

Fig. 2 (*Cont.*) a class of lipids can be used for other MS experiments, such as multiple-reaction monitoring (MRM) and precursor ion scans. (C) Precursor ion scans for lipids containing inositol phosphate headgroup (m/z 241). The second mass analyzer is fixed at m/z 241 and the first analyzer scans the mass range of interest. Consequently, ions with the propensity to form fragment ions with m/z 241 is selectively detected. Samples can be spiked with internal standards (IS), which is typically not found naturally in the samples under investigation, to allow for semi-quantitative profiling. (D) Overlay of chromatogram (left panel) and standard curve (right panel) obtained from quantification of varying concentrations of a commercially available 34:2 GPIns by MRM. The first and second mass analyzers are fixed at the parent ion of interest and its unique fragment ion respectively and selective quantification can be attained with a reasonably good linearity. Note that 34:2 GPIns is a minor ion in the complex lipid mixture and MRM offers a selective and sensitive method for quantification.

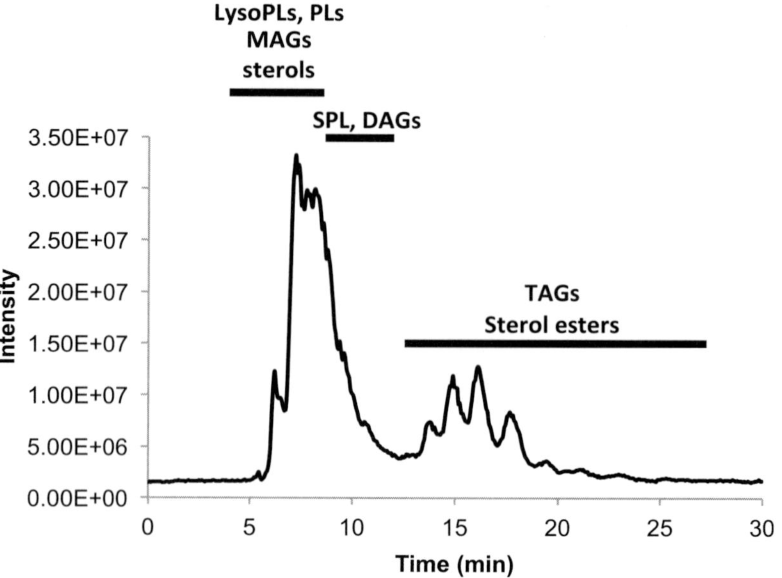

Fig. 3 LCMS analysis of cellular lipid extracts of mouse embryonic fibroblasts. Lipids were analyzed by ESI-MS with upfront liquid chromatography separation using a C18 reverse phase column (Agilent Zorbax Eclipise XDB-C18) to enhance detection of sterols in complex mixture. The HPLC system is made up of an Agilent 1100 binary pump, an Agilent 1100 thermo sampler, and an Agilent 1100 column oven. A sensitive in house method was developed using an Agilent Zorbax Eclipise XDB-C18 column. The HPLC conditions are (1) chloroform: methanol: 0.1 M Ammonium Acetate (100:100:4) as mobile phase at a flow rate of 0.25 mL. min; (2) column temperature: 25 °C; (3) injection volume: 20 μL. Abbreviations: DAGs, diacylglycerols; LysoPLs, lysophospholipids; MAGs, monoacylglycerols; PLs, glycerophospholipids; SPL, sphingolipids; TAGs, triacylglycerols.

and their abundance can be generated for further computation of differences as well as statistical analyses in paired sample sets.

Slight differences in m/z values of ions and retention time (in analysis involving upfront separation by high performance LC) between experiments are common due to small drifts in experimental conditions such as variations in temperature during time of flight (ToF) measurements and gradient elutions, and are a problem in particular when multiple spectra are to be processed. In order to perform direct analysis of the entire mass spectra obtained from high resolution mass spectrometers such as ToF instruments, and to utilize all collected data, the mass spectra should be properly aligned. Therefore, a previously developed chemometric method, which is based on correlation optimized warping (COW) (Nielsen *et al.,* 1998), was adopted by our laboratory. COW builds on piecewise stretching and compression of spectra along the m/z axis to correct for drifts (Fig. 4B) (Guan *et al.,* 2006). Importantly, such warping does not affect the peak intensities (Nielsen *et al.,* 1998). Once spectra are aligned they can be further processed arithmetically. A differential profile, which is the ratio of averaged replicate spectra from condition A, <A>, and averaged

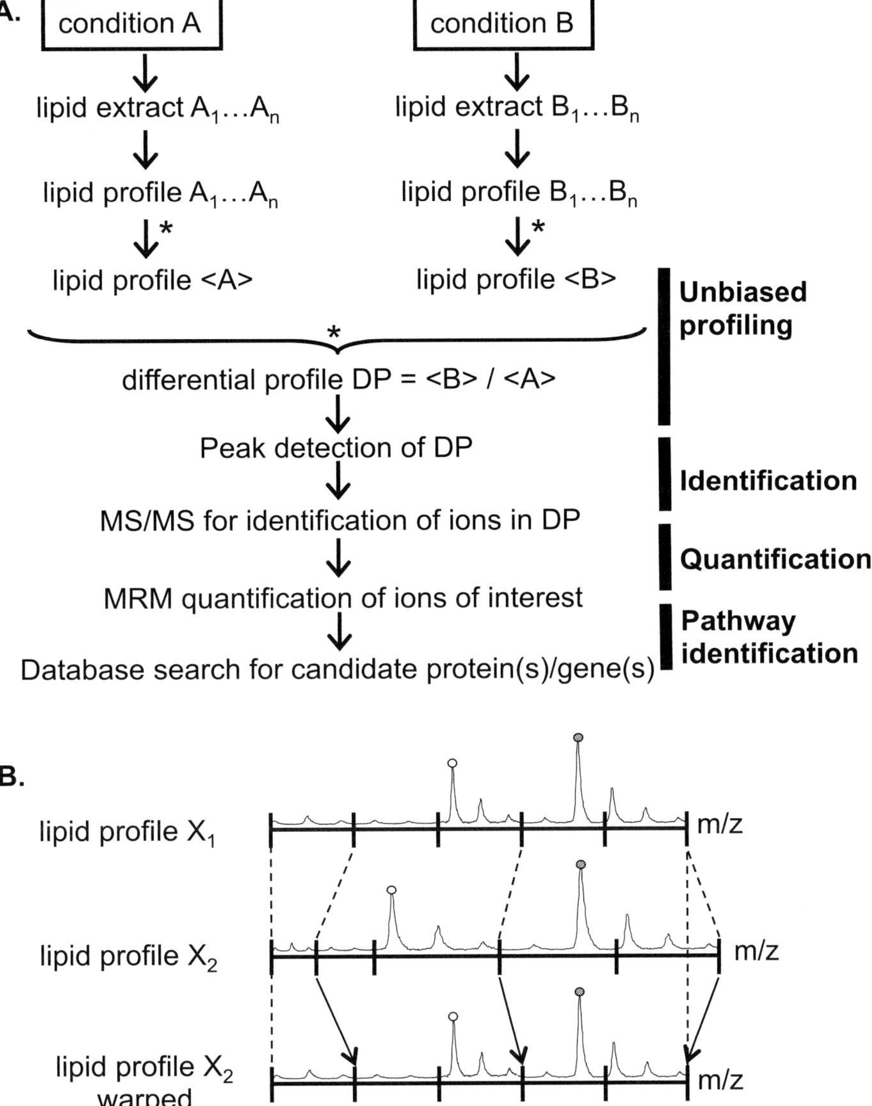

Fig. 4 Cartoon illustrating the general workflow for non-targeted and targeted identification of lipid metabolites that were altered in a paired sample system. (A) Typically two conditions for which multiple independent replicate samples (e.g., lipid extracts Al...An) are available are at the origin of the approach. Each lipid extract yields a corresponding lipid profile (e.g., lipid profile Al...An, which are single stage mass spectra in this study). These replicate spectra are next aligned using chemometric alignment based on correlation optimised warping (COW, see panel B) and subsequently the ratio of the signal intensities is computed as a function of m/z (DP). Ions of interest will be characterized by tandem mass spectrometry and selectively quantified by multiple reaction monitoring. (B) Correlation optimized warping of mass

replicate spectra from condition B, <B>, can be computed as a function of m/z. Using a peak detection algorithm, a list of peaks that are differentially regulated is generated and these candidate ions are selected for targeted analysis, which involves identification by tandem mass spectrometry (MS/MS) and quantification by multiple-reaction monitoring (MRM) (Fig. 2B, D).

Non-targeted profiling is highly appealing for unraveling previously unknown correlation between lipid metabolism and a perturbation of interest. It is highly practical at a stage whereby the inventory of the entire cellular lipidome is incomplete – as long as the lipids ionize and are detected, rapid analysis can be achieved. Another advantage is the likelihood of discovering an unexpected lipid mediator under a given condition and/or a previously uncharacterized lipid moiety. For instance, phospholipase A2 has been implicated in excitatory neurotransmission. In a model of kainite-induced excitotoxicity, in addition to identifying changes in major phospholipids with mainly polyunsaturated fatty acyl chains, which are substrates of phospholipase A2, we reported alterations in an additional cluster of ions in a mass range above 950 amu (Fig. 5). Based on tandem MS (MS2 and MS3), we speculated that this cluster of ions could potentially be N-acylated phosphatidylethanolamines, although it should be noted that unequivocal assignment can only be ascertained by more careful comparisons with the fragmentation patterns of N-acylated phosphatidylethanolamine standards (Guan *et al.*, 2006; Hansen *et al.,* 1999).

To improve resolution of lipids in a highly complex mixture, upfront LC is performed and the data obtained from LC-MS profiles can be processed in a similar fashion, taking into consideration the retention times (Shui *et al.,* 2007). Indeed, the capacity for global profiling has driven biology into a dimension, in which large volumes of complex, but inter-related data can be mined for patterns that will stimulate new hypotheses for experimental validation.

C. Targeted Lipid Analysis Using Tandem Mass Spectrometry

An alternative and more targeted approach in lipid analysis by mass spectrometry includes (i) defining the target(s) of interest to measure (here a target is defined as a specific lipid class, or a specific lipid species), (ii) detailed characterization, and (iii) development of a method by which these target metabolites are selectively measured. This is commonly achieved using triple quadrupole systems, that is, mass spectrometers with scanning mass analyzers (although reconstruction of multiple product ion spectra in ToF instruments has been used successfully in such

---

Fig. 4 (*Cont.*) spectra. COW was used as a pre-processing method for the mass spectral data. Briefly, a sample mass spectrum profile (X2) is aligned to a reference spectra (XI) by piecewise linear stretching and compression – also known as warping – of the m/z axis of the reference (XI). The asterisk denotes steps which include COW based alignment of spectra.
Adapted from Guan *et al.*, 2006.

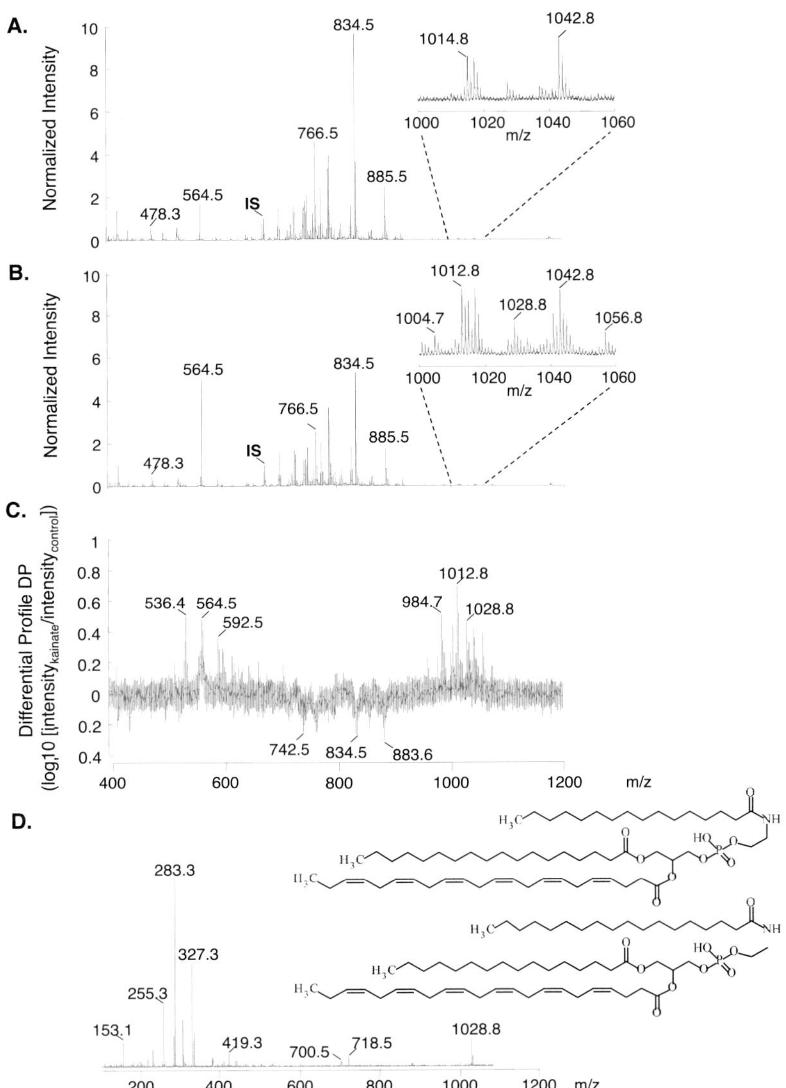

Fig. 5 Discovery of novel lipid metabolites during kainate-induced excitotoxicity. Rat hippocampal lipids were extracted using a modified Bligh and Dyer method and analyzed using negative-ion mode ESI-MS. (A) Averaged normalized spectra from saline-injected rats ($n = 3$). (B) Averaged normalized spectra from kainate-injected rats (3 days post injection, $n = 3$). (C) Differential profile of relative changes in lipid compositions following kainate exposure. Spectrum of control and kainate-treated rat hippocampal lipids are aligned and differences in ion intensities are computed using Matlab. (D) Identification of ions of interest by tandem mass spectrometry, in this case, m/z 1028. The product ion spectrum is consistent with previously published work using synthetic N-acylated phosphatidylethanolamine (NAPE) (Hansen *et al.,* 1999). Inset shows proposed structures of NAPE species with m/z 1028 (18:0 NPE (16:0/22:6) and 16:0 NPE (18:0/22:6)). Adapted from *Guan et al.*, 2006.

applications as well). The process of target characterization and measurement has been greatly facilitated by automated information-dependent acquisition, whereby the mass spectrometer is set to run a survey scan and several tandem MS events, which are usually user defined, to identify as many putative compounds as possible in a mixture within a single run (Schwudke *et al.,* 2006).

D. Lipid-Class Specific Profiling by Precursor Ion and/or Neutral Loss Scans

Tandem mass spectrometry has aided in characterization and identification of lipids (Fig. 2B). Therefore, the requirement for structural information is not an issue for targeted analysis of well-characterized lipids such as major glycerophospholipids, sphingolipids, and sterols (Table IV). Indeed, this method has been commonly applied to fingerprint cellular sublipidome, partly because of enhanced selectivity and sensitivity.

Instead of measuring all possible ions, a multi-dimensional ESI-MS approach (Ekroos *et al.,* 2002; Han and Gross, 2003) involves experimentally filtering for specific classes of lipids using precursor (PREIS) or neutral-loss (NL) scanning (also known as "focused lipidomics") (Fig. 2C and Table IV, Taguchi *et al.,* 2005). These methods are dependent on predetermination of the fragmentation pattern and

Table IV
Precursor ions, MSMS scan modes, and associated parameters for analysis of major glycerophospholipids and sphingolipids in mammalian cells and tissues

Lipid	Precursor ion	MS/MS Modes	Fragment	Mass range (m/z)	Declustering Potential (V)	Collision Energy (V)
			Glycero-phospholipid			
GPA	[M-H]-	PREIS 153	Glycerophosphate derivative	370–800	−75	−40 to −60
GPGro	[M-H]-	PREIS 153	Glycerophosphate derivative	400–800	−75	−40 to −60
GPCho	[M+H]+	PREIS 184	Phosphocholine	400–900	70	45 to 65
GPEtn	[M-H]-	PREIS 196	Glycerophosphoethanolamine derivative	400–800	−75	−40 to −65
	[M+H]+	NL 141	Phosphoethanolamine			
GPSer	[M-H]-	NL 87	Serine	400–800	−75	−25
	[M-H]-	PREIS 153	Glycerophosphate derivative	400–800	−75	−40 to −60
GPIns	[M-H]-	PREIS 241	Cyclic inositol phosphate	450–900	−90	−45 to −60
	[M-H]-	PREIS 153	Glycerophosphate derivative	450–900	−75	−45 to −65
			Sphingolipid			
Ceramide	[M+H]+	PREIS 264	Double dehydration product of d18:1 sphingoid base	450–700	60	40 to 50
Sphingomyelin	[M+H]+	PREIS 184	Phosphocholine	600–900	80	45 to 60
Hexosylceramide	[M+H]+	PREIS 264	Double dehydration product of d18:1 sphingoid base	650–900	60	45 to 60

Abbreviations: PREIS, Precursor ion scan; NL, Neutral loss scan structure as it is prudent to find a product ion that is unique to the structure. Specifically, ions of interest are selected by the first quadrupole where they will undergo collision-induced dissociation and the fragments are passed through the third quadrupole and a fragmentation profile is generated (Fig. 2B). In this case, the ion with m/z 885, is a 38:4 phosphatidylinositol with the fragment ions, m/z 241 (dehydrated inositol phosphate), 153 (dehydrated glycerophosphate), 283 (stearic acid), and 303 (arachidonic acid).

Using the diagnostic fragment of inositol phosphate and glycerophosphate, the mass spectrometer can be set to selectively generate the phosphatidylinositol profile in the precursor ion scan mode (Fig. 2C). In the PREIS scan mode, the first quadrupole scans over a selected mass range and the ions sequentially enter the collision cell where collision energy is applied to induce fragmentation. The third quadrupole is set to transmit a selected single product ion, such as inositol phosphate (m/z 241). A profile comprising ions with the propensity to form fragment ions with m/z 241 is thus generated.

Conversely, a neutral loss scan is used to profile several classes of glycerophospholipids when the charge of the headgroup does not localize to the lipid headgroup after fragmentation. In this mode, the first and third quadrupoles are linked and scanned at the same rate over the same mass range with a constant mass difference, for instance 87 amu in negative mode for the loss of serine. The resultant mass spectrum contains all the precursor ions that lose a neutral species of selected mass, in the case of a neutral loss of 87 amu, a phosphatidylserine profile is generated.

E. Quantification by Multiple Reaction Monitoring

While triple quadrupole instruments tend to be limited in mass resolution, high selectivity for quantification of specific lipid molecular species can be achieved by monitoring pairs of parent ions and the diagnostic fragment/daughter ions in tandem. This method is known as single reaction monitoring (SRM) or multiple reactions monitoring (MRM), depending on the number of metabolites measured. With the information of parent and fragment ions obtained from single stage profiling and characterization by collision induced dissociation, a list of MRM transitions can be generated for individual molecular species of lipids. Specifically, the first quadrupole is set to monitor specific parent ions for example, 833.6 in the negative ion mode for 34:2 phosphatidylinositol, which will undergo collision-induced dissociation in the collision cell, and the fragment ions passes through the third quadrupole which is set to selectively monitor the diagnostic ions of the inositol phosphate, in this case, m/z 241 (Fig. 2D). Each individual ion dissociation pathway is optimized with regard to collision energy and declustering potential (terminology specific for an ABI Sciex triple quadrupole) to minimize variations in relative ion abundance due to differences in rates of dissociation. For quantification, the signal intensities are normalized to the relevant internal standards, which are typically spiked before lipid extraction to control for recovery. Tables V and VI summarize the MRM transitions for major diacyl glycerophospholipids and sphingolipids.

Table V

Parent (Ql)/ Daughter (Q3) ions transitions for multiple reaction monitoring of acyl-containing glycerophospholipids. Internal standards used are 14:0/14:0 GPA, GPGro, GPEtn, GPSer, GPCho, and 8:0/8:0 GPIns

Molecular Species	GPA		GPGro		GPEtn		GPSer		GPIns		GPCho	
	Q1 $[M-H]^-$	Q3	Q1 $[M-H]^-$	Q3	Q1 $[M-H]^-$	Q3	Q1 $[M-H]^-$	Q3	Q1 $[M-H]^-$	Q3	Q1 $[M+H]^+$	Q3
16:1 (Lyso)	407.2	153.1	481.3	153.1	450.3	196.1	494.3	407.3	569.3	241.1	494.3	184.1
16:0 (Lyso)	409.2	153.1	483.3	153.1	452.3	196.1	496.3	409.3	571.3	241.1	496.3	184.1
18:1 (Lyso)	435.3	153.1	509.3	153.1	478.3	196.1	522.3	435.3	597.3	241.1	522.3	184.1
18:0 (Lyso)	437.3	153.1	511.3	153.1	480.3	196.1	524.3	437.3	599.3	241.1	524.3	184.1
32:2	643.4	153.1	717.5	153.1	686.5	196.1	730.5	643.5	805.5	241.1	730.5	184.1
32:1	645.4	153.1	719.5	153.1	688.5	196.1	732.5	645.5	807.5	241.1	732.5	184.1
32:0	647.4	153.1	721.5	153.1	690.5	196.1	734.5	647.5	809.5	241.1	734.5	184.1
34:2	671.5	153.1	745.5	153.1	714.5	196.1	758.5	671.5	833.5	241.1	758.6	184.1
34:1	673.5	153.1	747.5	153.1	716.5	196.1	760.5	673.5	835.5	241.1	760.6	184.1
34:0	675.5	153.1	749.5	153.1	718.5	196.1	762.5	675.5	837.5	241.1	762.6	184.1
36:4	695.5	153.1	769.5	153.1	738.5	196.1	782.5	695.5	857.6	241.1	782.6	184.1
36:3	697.5	153.1	771.5	153.1	740.5	196.1	784.5	697.5	859.6	241.1	784.6	184.1
36:2	699.5	153.1	773.5	153.1	742.5	196.1	786.5	699.5	861.6	241.1	786.6	184.1
36:1	701.5	153.1	775.6	153.1	744.6	196.1	788.6	701.6	863.6	241.1	788.6	184.1
36:0	703.5	153.1	777.6	153.1	746.6	196.1	790.6	703.6	865.6	241.1	790.6	184.1
38:5	721.5	153.1	795.6	153.1	764.6	196.1	808.6	721.6	883.6	241.1	808.6	184.1
38:4	723.5	153.1	797.6	153.1	766.6	196.1	810.6	723.6	885.6	241.1	810.6	184.1
38:3	725.5	153.1	799.6	153.1	768.6	196.1	812.6	725.6	887.6	241.1	812.6	184.1
38:2	727.5	153.1	801.6	153.1	770.6	196.1	814.6	727.6	889.6	241.1	814.6	184.1
38:1	729.6	153.1	803.6	153.1	772.6	196.1	816.6	729.6	891.6	241.1	816.6	184.1
38:0	731.6	153.1	805.6	153.1	774.6	196.1	818.6	731.6	893.6	241.1	818.6	184.1
40:6	749.6	153.1	821.6	153.1	790.6	196.1	834.6	747.6	909.6	241.1	834.7	184.1
40:5	751.6	153.1	823.6	153.1	792.6	196.1	836.6	749.6	911.6	241.1	836.7	184.1
40:4	753.6	153.1	825.6	153.1	794.6	196.1	838.6	751.6	913.6	241.1	838.7	184.1
40:3	755.6	153.1	827.6	153.1	796.6	196.1	840.6	753.6	915.6	241.1	840.7	184.1
40:2	757.7	153.1	829.6	153.1	798.6	196.1	842.6	755.6	917.6	241.1	842.7	184.1
40:1	759.7	153.1	831.6	153.1	800.6	196.1	844.6	757.6	919.6	241.1	844.7	184.1
40:0	777.7	153.1	833.6	153.1	802.6	196.1	846.6	759.6	921.6	241.1	846.7	184.1
Internal Standards	591.4	153.1	665.5	153.1	634.5	196.1	678.5	591.5	585.3	241.1	678.5	184.1

Table VI
Parent (Ql)/ Daughter (Q3) ions transitions for multiple reaction monitoring of sphingolipids Internal standards used are 17:0/dl8:l Ceramide, 12:0/d 18:1 Sphingomyelin and 8:0/d 18:1 Glucosylceramide

Molecular Species	Ceramide		Sphingomyelin		Hexosylceramide	
	Q1 $[M+H]^+$	Q3	Q1 $[M+H]^+$	Q3	Q1 $[M+H]^+$	Q3
16:1/d18:1	536.5	264.2	701.6	184.1	698.6	264.2
16:0/d18:1	538.5	264.2	703.6	184.1	700.6	264.2
16:1/d18:0	538.5	266.2	703.6	184.1	700.6	266.2
16:0/d18:0	540.5	266.2	705.6	184.1	702.6	266.2
18:1/d18:1	564.5	264.2	729.6	184.1	726.6	264.2
18:0/d18:1	566.5	264.2	731.6	184.1	728.6	264.2
18:1/d18:0	566.5	266.2	731.6	184.1	728.6	266.2
18:0/d18:0	568.5	266.2	733.6	184.1	730.6	266.2
20:1/d18:1	592.5	264.2	757.6	184.1	754.6	264.2
20:0/d18:1	594.5	264.2	759.6	184.1	756.6	264.2
20:1/d18:0	594.5	266.2	759.6	184.1	756.6	266.2
20:0/d18:0	596.5	266.2	761.6	184.1	758.6	266.2
22:1/d18:1	620.5	264.2	785.6	184.1	782.6	264.2
22:0/d18:1	622.5	264.2	787.6	184.1	784.6	264.2
22:1/d18:0	622.5	266.2	787.6	184.1	784.6	266.2
22:0/d18:0	624.5	266.2	789.6	184.1	786.6	266.2
24:1/d18:1	648.6	264.2	813.7	184.1	810.7	264.2
24:0/d18:1	650.6	264.2	815.7	184.1	812.7	264.2
24:1/d18:0	650.6	266.2	815.7	184.1	812.7	266.2
24:0/d18:0	652.6	266.2	817.7	184.1	814.7	266.2
26:1/d18:1	676.6	264.2	841.7	184.1	838.7	264.2
26:0/d18:1	678.6	264.2	843.7	184.1	840.7	264.2
26:1/d18:0	678.6	266.2	843.7	184.1	840.7	266.2
26:0/d18:0	680.6	266.2	845.7	184.1	842.7	266.2
Internal Standards	552.5	264.2	647.5	184.1	588.5	264.2

Targeted analysis offers the advantages of enhanced sensitivity and also, the inherent redundancy in data collection and analysis, particularly in applications with a working hypothesis, is reduced. Furthermore, the quantitative information gathered is well-suited for downstream applications such as pathway modeling and simulations. One of the limitations of this approach is the general applicability due the diversity and difference in lipid catalogues between various systems. For instance, a list of MRM transitions for quantification of the lipidome of a neuronal cell line may not be fully applicable to a red blood cell, not to mention the transferability of methods between different tissues, organs, and obviously organisms (Andreyev *et al.,* 2010; Ejsing *et al.,* 2009; Quehenberger *et al.,* 2010; Sartain *et al.,* 2011). Indeed, both non-targeted and targeted analyses have their own strengths and weaknesses, and thus these approaches are highly complementary and their combination is a powerful discovery tool for lipidomics.

V. Details of Lipidomics Analysis

A. Reagents

LC-MS grade chloroform, methanol.

Note: High quality solvents are required to minimize background noise and contamination.

B. Non-Targeted Profiling

1. ESI-MS and MS/MS analyses are performed on a Waters Micromass quadrupole Time-of-Flight (qToF) micro mass spectrometer, coupled to a Waters CapLC liquid chromatography system (Waters Corp., Milford, MA, USA).
2. 2|iL of sample is directly introduced into the mass spectrometer by loop injection. We recommend starting with a final concentration of 10 μM.
3. Chloroform-methanol (1:1, v/v) is used as the mobile phase at a flow rate of 15 μL/min.
4. For full scan MS, maintain the capillary and cone voltages at 3.0 kV and 50 V, respectively. The source and desolvation temperatures are set at 80 and 250 °C respectively. Mass spectra are acquired in the negative ion mode from m/z 400 to 1000, with a frequency of 1 scan per second and an acquisition time of 3 min.
5. For MS/MS, collision energy ranges from 25 to 80 eV.
6. For data processing, chromatograms are combined to generate combined spectra and a corresponding spectrum list using MassLynx 4.0 (Waters Corp.).
7. The data are next migrated to MatLAB (MathWorks, Inc., Natick, MA) for alignment of spectra using correlation-optimized warping (COW) (Fig. 4) and differences in signal intensities between control and treatment are computed to generate a differential profile as described by Guan *et al.* (2006).

C. Targeted Quantification

1. Quantification of individual lipid molecular species is performed using multiple reaction monitoring (MRM) with an Applied Biosystems 4000 Q-Trap mass spectrometer (ABISciex, Foster City, CA) coupled to a high performance liquid chromatography system (Agilent Technologies, Santa Clara, CA).
2. Sample (20 μL) is directly introduced into the mass spectrometer by loop injection.
3. Chloroform:methanol (1:1) is used as a mobile phase at a flow rate of 250 μL/min.
4. Mass spectrometry is recorded in both negative and positive ESI modes. ESI conditions are: turbo spray source voltage, 4500 V; source temperature, 250 °C; GS1: 40.00, GS2: 30.00, curtain gas: 25.
5. Precursor ion and neutral loss scans for major glycerophospholipids and sphingolipids can be performed as summarized in Table IV. This step is particularly

useful for molecular species determination for each lipid classes to ensure a comprehensive coverage for subsequent MRM quantification.

6. In an MRM experiment, the first quadrupole, Ql, is set to pass the precursor ion of interest to the collision cell, Q2, where it undergoes collision-induced dissociation. The third quadruple, Q3, is set to pass the structure specific product ion characteristic of the precursor lipid of interest. These transition pairs for major glycerophospholipids and sphingolipids are summarized in Tables V and VI. As discussed above, each individual ion dissociation pathway is optimized with regard to collision energy and declustering potential to minimize variations in relative ion abundance due to differences in rates of dissociation.
7. For quantification, signals of individual ions are normalized to the respective internal standards, which are spiked according to the starting amounts of material.

References

Andreyev, A. Y., *et al.* (2010). Subcellular organelle lipidomics in TLR-4-activated macrophages. *J. Lipid Res.* **51**, 2785–2797.

Axelsen, P. H., and Murphy, R. C. (2010). Quantitative analysis of phospholipids containing arachidonate and docosahexaenoate chains in microdissected regions of mouse brain. *J. Lipid Res.* **57**, 660–671.

Blanksby, S. J., and Mitchell, T. W. (2010). Advances in mass spectrometry for lipidomics. *Annu. Rev. Anal. Chem. (Palo. Alto. Calif.)* **3**, 433–465.

Bligh, G., and Dyer, W. J. (1959). A rapid method of total lipid extraction and purification. *Can. J. Biochem. Physiol.* **37**, 911–917.

Brockerhoff, H. (1963). Breakdown of phospholipids in mild alkaline hydrolysis. *J. Lipid Res.* **35**, 96–99.

Caffrey, M., and Hogan, J. (1992). LIPID AT: a database of lipid phase transition temperatures and enthalpy changes. DMPC data subset analysis. *Chem. Phys. Lipids* **61**, 1–109.

Ejsing, C. S., Duchoslav, E., Sampaio, J., Simons, K., Bonner, R., Thiele, C., Ekroos, K., and Shevchenko, A. (2006). Automated identification and quantification of glycerophospholipid molecular species by multiple precursor ion scanning. *Anal. Chem.* **78**, 6202–6214.

Ejsing, C. S., Sampaio, J. L., Surendranath, V., Duchoslav, E., Ekroos, K., Klemm, R. W., Simons, K., and Shevchenko, A. (2009). Global analysis of the yeast lipidome by quantitative shotgun mass spectrometry. *Proc. Natl. Acad. Sci. U. S. A* **106**, 2136–2141.

Ekroos, K., Chernushevich, V., Simons, K., and Shevchenko, A. (2002). Quantitative profiling of phospholipids by multiple precursor ion scanning on a hybrid quadrupole time-of-flight mass spectrometer. *Anal. Chem.* **74**, 941–949.

Fahy, E., Sud, M., Cotter, D., and Subramaniam, S. (2007). LIPID MAPS online tools for lipid research. *Nucleic Acids Res.* **35**, W606–W612.

Folch, J., Lees, M., and Sloane Stanley, G. H. (1957). A simple method for the isolation and purification of total lipides from animal tissues. *J. Biol. Chem.* **226**, 497–509.

Gray, A., Olsson, H., Batty, I. H., Priganica, L., and Peter, D. C. (2003). Nonradioactive methods for the assay of phosphoinositide 3-kinases and phosphoinositide phosphatases and selective detection of signaling lipids in cell and tissue extracts. *Anal. Biochem.* **313**, 234–245.

Griffiths, W. J., Karu, K., Hornshaw, M., Woffendin, G., and Wang, Y. (2007). Metabolomics and metabolite profiling: past heroes and future developments. *Eur. J. Mass Spectrom. (Chichester, Eng)* **13**, 45–50.

Guan, X. L., He, X., Ong, W. Y., Yeo, W. K., Shui, G., and Wenk, M. R. (2006). Non-targeted profiling of lipids during kainate-induced neuronal injury. *FASEB J.* **20**, 1152–1161.

Guan, X. L., Riezman, I., Wenk, M. R., and Riezman, H. (2010). Yeast lipid analysis and quantification by mass spectrometry. *Meth. Enzymol.* **470**, 369–391.

Haimi, P., Uphoff, A., Hermansson, M., and Somerharju, P. (2006). Software tools for analysis of mass spectrometric lipidome data. *Anal. Chem.* **78**, 8324–8331.

Han, X., and Gross, R. W. (2005a). Shotgun lipidomics: multidimensional MS analysis of cellular lipidomes. *Expert Rev. Proteomics* **2**, 253–264.

Han, X., and Gross, R. W. (2005b). Shotgun lipidomics: electrospray ionization mass spectrometric analysis and quantitation of cellular lipidomes directly from crude extracts of biological samples. *Mass Spectrom. Rev* **24**, 367–412.

Han, X., and Gross, R. W. (2003). Global analyses of cellular lipidomes directly from crude extracts of biological samples by ESI mass spectrometry: a bridge to lipidomics. *J. Lipid Res.* **44**, 1071–1079.

Han, X., Yang, K., and Gross, R. W. (2008). Microfluidics-based electrospray ionization enhances the intrasource separation of lipid classes and extends identification of individual molecular species through multi-dimensional mass spectrometry: development of an automated high-throughput platform for shotgun lipidomics. *Rapid Commun. Mass Spectrom.* **22**, 2115–2124.

Han, X., Yang, K., and Gross, R. W. (2012). Multi-dimensional mass spectrometry-based shotgun lipidomics and novel strategies for lipidomic analyses. *Mass Spectrom. Rev.* **31**(1), 134–178.

Hansen, H. H., Hansen, S. H., Bjornsdottir, I., and Hansen, H. S. (1999). Electrospray ionization mass spectrometric method for the determination of cannabinoid precursors: N-acylethanolamine phospholipids (NAPEs). *J. Mass Spectrom.* **34**, 761–767.

Harkewicz, R., and Dennis, E. A. (2011). Applications of mass spectrometry to lipids and membranes. *Annu. Rev. Biochem.* **80**, 301–325.

Honda, A., Miyazaki, T., Ikegami, T., Iwamoto, J., Yamashita, K., Numazawa, M., and Matsuzaki, Y. (2010). Highly sensitive and specific analysis of sterol profiles in biological samples by HPLC-ESI-MS/MS. *J. Steroid Biochem. Mol. Biol.* **121**, 556–564.

Ikeda, K., and Taguchi, R. (2010). Highly sensitive localization analysis of gangliosides and sulfatides including structural isomers in mouse cerebellum sections by combination of laser microdissection and hydrophilic interaction liquid chromatography/electrospray ionization mass spectrometry with theoretically expanded multiple reaction monitoring. *Rapid Commun. Mass Spectrom.* **24**, 2957–2965.

Ito, S., Nabetami, T., Shinoda, Y., Nagatsuka, Y., and Hirabayashi, Y. (2008). Quantitative analysis of a novel glucosylate phospholipid by liquid chromatography-mass spectrometry. *Anal. Biochem.* **376**, 252–257.

Ivanova, P. T., Cerda, B. A., Horn, D. M., Cohen, J. S., McLafferty, F. W., and Brown, H. A. (2001). Electrospray ionization mass spectrometry analysis of changes in phospholipids in RBL-2H3 mastocytoma cells during degranulation. *Proc. Natl. Acad. Sci. U. S. A.* **98**, 7152–7157.

Jiang, X., Cheng, H., Yang, K., Gross, R. W., and Han, X. (2007). Alkaline methanolysis of lipid extracts extends shotgun lipidomics analyses to the low-abundance regime of cellular sphingolipids. *Anal. Biochem.* **371**, 135–145.

Korekane, H., Tsuji, S., Noura, S., Ohue, M., Sasaki, Y., Imaoka, S., and Miyamoto, Y. (2007). Novel fucogangliosides found in human colon adenocarcinoma tissues by means of glycomic analysis. *Anal. Biochem.* **364**, 37–50.

Leavell, M. D., and Leary, J. A. (2006). Fatty acid analysis tool (FAAT): An FT-ICRMS lipid analysis algorithm. *Anal. Chem.* **78**, 5497–5503.

Matyash, V., Liebisch, G., Kurzchalia, T. V., Shevchenko, A., and Schwudke, D. (2008). Lipid extraction by methyl-tert-butyl ether for high-throughput lipidomics. *J. Lipid Res.* **49**, 1137–1146.

Merrill Jr., A. H., Stokes, T. H., Momin, A., Park, H., Portz, B. J., Kelly, S., Wang, E., Sullards, M. C., and Wang, M. D. (2009). Sphingolipidomics: a valuable tool for understanding the roles of sphingolipids in biology and disease. *J. Lipid Res.* **50**(Suppl), S97–S102.

Myers, D. S., Ivanova, P. T., Milne, S. B., and Brown, H. A. (2011). Quantitative analysis of glycerophospholipids by LC-MS: Acquisition, data handling, and interpretation. *Biochim. Biophys. Acta.* **1811** (11), 748–757.

Nielsen, N. V., Carstensen, J. M., and Smedsgaard, J. (1998). Alignment of single and multiple wavelength chromatographic profiles for chemometric data analysis using correlation optimised warping. *J. Chromatogr. A* **805**, 17–35.

Pettitt, T. R., Dove, S. K., Lubben, A., Calaminus, S. D., and Wakelam, M. J. (2006). Analysis of intact phosphoinositides in biological samples. *J. Lipid Res.* **47**, 1588–1596.

Pluskal, T., Castillo, S., Villar-Briones, A., and Oresic, M. (2010). MZmine 2: modular framework for processing, visualizing, and analyzing mass spectrometry-based molecular profile data. *BMC. Bioinformatics* **11**, 395.

Pulfer, M., and Murphy, R. C. (2003). Electrospray mass spectrometry of phospholipids. *Mass Spectrom. Rev.* **22**, 332–364.

Quehenberger, O., *et al.* (2010). Lipidomics reveals a remarkable diversity of lipids in human plasma. *J. Lipid Res.* **51**, 3299–3305.

Sartain, M. J., Dick, D. L., Rithner, C. D., Crick, D. C., and Belisle, J. T. (2011). Lipidomic analyses of Mycobacterium tuberculosis based on accurate mass measurements and the novel "Mtb LipidDB". *J. Lipid Res.* **52**, 861–872.

Schwudke, D., Hannich, J. T., Surendranath, V., Grimard, V., Moehring, T., Burton, L., Kurzchalia, T., and Shevchenko, A. (2007). Top-down lipidomic screens by multivariate analysis of high-resolution survey mass spectra. *Anal. Chem.* **79**, 4083–4093.

Schwudke, D., Oegema, J., Burton, L., Entchev, E., Hannich, J. T., Ejsing, C. S., Kurzchalia, T., and Shevchenko, A. (2006). Lipid profiling by multiple precursor and neutral loss scanning driven by the data-dependent acquisition. *Anal. Chem.* **78**, 585–595.

Shevchenko, A., and Simons, K. (2010). Lipidomics: coming to grips with lipid diversity. *Nat. Rev. Mol. Cell Biol.* **11**, 593–598.

Shui, G., Bendt, A. K., Pethe, K., Dick, T., and Wenk, M. R. (2007). Sensitive profiling of chemically diverse bioactive lipids. *J Lipid Res.* **48**, 1976–1984.

Shui, G., Guan, X. L., Low, C. P., Chua, G. H., Goh, J. S., Yang, H., and Wenk, M. R. (2010). Toward one step analysis of cellular lipidomes using liquid chromatography coupled with mass spectrometry: application to Saccharomyces cerevisiae and Schizosaccharomyces pombe lipidomics. *Mol. Biosyst.* **6**, 1008–1017.

Smith, C. A., Want, E. J., O'Maille, G., Abagyan, R., and Siuzdak, G. (2006). XCMS: processing mass spectrometry data for metabolite profiling using nonlinear peak alignment, matching, and identification. *Anal. Chem.* **78**, 779–787.

Song, H., Hsu, F. F., Ladenson, J., and Turk, J. (2007). Algorithm for processing raw mass spectrometric data to identify and quantitate complex lipid molecular species in mixtures by data-dependent scanning and fragment ion database searching. *J. Am. Soc. Mass Spectrom.* **18**, 1848–1858.

Sud, M., *et al.* (2007). LMSD: LIPID MAPS structure database. *Nucleic Acids Res.* **35**, D527–D532.

Sullards, M. C., and Merrill Jr., A. H. (2001). Analysis of sphingosine 1-phosphate, ceramides, and other bioactive sphingolipids by high-performance liquid chromatography-tandem mass spectrometry. *Sci STKE* **2001**(67), pl1.

Taguchi, R., Houjou, T., Nakanishi, H., Yamazaki, T., Ishida, M., Imagawa, M., and Shimizu, T. (2005). Focused lipidomics by tandem mass spectrometry. *J. Chromatogr. B Analyt. Technol. Biomed. Life Sci.* **823**, 26–36.

Taguchi, R., and Ishikawa, M. (2010). Precise and global identification of phospholipid molecular species by an Orbitrap mass spectrometer and automated search engine Lipid Search. *J. Chromatogr. A* **1217**, 4229–4239.

Taguchi, R., Nishijima, M., and Shimizu, T. (2007). Basic analytical systems for lipidomics by mass spectrometry in Japan. *Meth. Enzymol.* **432**, 185–211.

Wang, Y., *et al.* (2009). Targeted lipidomic analysis of oxysterols in the embryonic central nervous system. *Mol. Biosyst.* **5**, 529–541.

Wenk, M. R. (2010). Lipidomics: new tools and applications. *Cell* **143**, 888–895.

Wenk, M. R., Lucast, L., Di Paolo, G., Romanelli, A. J., Suchy, S. F., Nussbaum, R. L., Cline, G. W., Shulman, G. L., McMurray, W., and De Camilli, P. (2003). Phosphoinositide profiling in complex lipid mixtures using electrospray ionization mass spectrometry. *Nat. Biotechnol.* **21**, 813–817.

Yang, K., Cheng, H., Gross, R. W., and Han, X. (2009). Automated lipid identification and quantification by multidimensional mass spectrometry-based shotgun lipidomics. *Anal. Chem.* **81**, 4356–4368.

Zacarias, A., Bolanowski, D., and Bhatnagar, A. (2002). Comparative measurements of multicomponent phospholipid mixtures by electrospray mass spectroscopy: relating ion intensity to concentration. *Anal. Biochem.* **308**, 152–159.

Zhou, Z., Marepally, S. R., Nune, D. S., Pallakollu, P., Ragan, G., Roth, M. R., Wang, L., Lushington, G. H., Visvanathan, M., and Welti, R. (2011). LipidomeDB Data Calculation Environment: Online Processing of Direct-Infusion Mass Spectral Data for Lipid Profiles. *Lipids* **46**, 879–884.

CHAPTER 9

Modulation of Host Phosphoinositide Metabolism During *Salmonella* Invasion by the Type III Secreted Effector SopB

Bernhard Roppenser[*], **Sergio Grinstein**[*,†,‡] **and John H. Brumell**[*,‡,§]

[*]Cell Biology Program, Hospital for Sick Children, Toronto, Ontario, Canada

[†]Department of Biochemistry, University of Toronto, Ontario, Canada

[‡]Institute of Medical Science, University of Toronto, Ontario, Canada

[§]Department of Molecular Genetics, University of Toronto, Ontario, Canada

Abstract

Phosphoinositides (PI) play an important role in many different cellular processes. Their generation and functions, however, are very dynamic, and the detection of localized events usually requires very precise imaging techniques. Recent advances

Copyright 2012, Elsevier Inc. All rights reserved.

0091-679X/10 $35.00
DOI 10.1016/B978-0-12-386487-1.00009-2

in lipid research raised the possibility of designing molecular probes to specifically detect lipids in different subcellular compartments and have provided new tools to directly image PI dynamics in living cells. *Salmonella* is a pathogenic bacterium that has the ability to invade host cells and grow intracellularly. To this end, they secrete specialized virulence proteins (effectors) directly into the cytosol of host cells. These effectors modulate signaling pathways to initiate bacterial uptake and promote intracellular survival. SopB, one of the many effector proteins that are translocated into host cells, has PI phosphatase activity and directly modulates PI metabolism. In this chapter, we describe a method to transfect PI-binding domains fused to fluorescent proteins as probes to monitor lipid dynamics during a *Salmonella* invasion in living cells using a spinning-disk confocal microscope.

I. Introduction

Phosphoinositides (PI) comprise only a small fraction of phospholipids in biological membranes but undoubtedly play a major role in a variety of cellular processes such as cell signaling, survival, membrane trafficking, membrane structure, and cytoskeletal dynamics (Di Paolo and De Camilli, 2006). PIs can be phosphorylated at the hydroxyl residues at positions 3, 4, or 5 of the inositol ring to produce seven different species. Given the importance of PIs in cellular processes, their metabolism and spatial and temporal distribution are tightly regulated by a large number of kinases and phosphatases (Sasaki *et al.*, 2009).

The phosphorylated head groups of the inositol ring of PIs are exposed to the cytosolic side of membranes where they are recognized by host proteins. These proteins associate specifically with individual PI species by defined domains including the PH (pleckstrin homology), FYVE (Fab1, YOTB, Vac1, EEA1), or PX (Phox homology) domain (See Table I). When expressed separately from the rest of the protein, these modules retain their capability to recognize and bind to the target lipid; when such modules are fused to a fluorescent marker, such as green or red fluorescent protein (GFP or RFP), live imaging of lipid dynamics becomes possible. This technique has become very popular in recent years and has emerged as a valuable tool in studying PI metabolism or lipid distribution (Varnai and Balla, 2006).

Bacterial pathogens have evolved various mechanisms to interfere with the machinery of their host cells, enabling them to gain entry into and survive intracellularly (Cossart and Sansonetti, 2004; Flannagan *et al.*, 2009). Considering the importance of PIs in phagocytosis, actin dynamics, and membrane trafficking, it is not surprising that pathogenic bacteria have developed means to exploit PI metabolism (Hilbi, 2006; Pizarro-Cerda and Cossart, 2004).

Salmonella enterica serovar Typhimurium (*S.* Typhimurium) is a facultative intracellular pathogen that can invade a variety of non-phagocytic cells. To accomplish this, they use a type III secretion system (T3SS), a needle-like structure on the bacterial surface, to directly inject effector proteins into host cells (Galan and Wolf-Watz, 2006). These translocated proteins modulate signal transduction pathways inducing bacterial uptake and allowing intracellular survival (Knodler and

Table I

Common lipid-binding probes that are used to study phosphoinositides during *Salmonella* invasion

Lipid	Probe	References
PI(3)P	FYVE	[a]
	PX-p40phox	[b]
$PI(4,5)P_2$	PH-PLCδ	[c]
$PI(3,4)P_2/PI(3,4,5)P_3$	PH-Akt	[d]
	PH-PDK1	[e]

[a] Raiborg, C., Bremnes, B., Mehlum, A., Gillooly, D. J., D'Arrigo, A., Stang, E., and Stenmark, H. (2001). FYVE and coiled-coil domains determine the specific localisation of Hrs to early endosomes. *J. Cell Sci.* **114,** 2255–2263.

[b] Ellson, C. D., Gobert-Gosse, S., Anderson, K. E., Davidson, K., Erdjument-Bromage, H., Tempst, P., Thuring, J. W., Cooper, M. A., Lim, Z. Y., Holmes, A. B., Gaffney, P. R., Coadwell, J., Chilvers, E. R., Hawkins, P. T., and Stephens, L. R. (2001). PtdIns(3)P regulates the neutrophil oxidase complex by binding to the PX domain of p40(phox). *Nat. Cell Biol.* **3,** 679–682.

[c] Stauffer, T. P., Ahn, S., and Meyer, T. (1998). Receptor-induced transient reduction in plasma membrane PtdIns(4,5)P_2 concentration monitored in living cells. *Curr. Biol.* **8,** 343–346.

[d] Servant, G., Weiner, O. D., Herzmark, P., Balla, T., Sedat, J. W., and Bourne, H. R. (2000). Polarization of chemoattractant receptor signaling during neutrophil chemotaxis. *Science* **287,** 1037–1040.

[e] Komander, D., Fairservice, A., Deak, M., Kular, G. S., Prescott, A. R., Peter Downes, C., Safrany, S. T., Alessi, D. R., and van Aalten, D. M. (2004). Structural insights into the regulation of PDK1 by phosphoinositides and inositol phosphates. *EMBO J.* **23,** 3918–3928.

Steele-Mortimer, 2003). *S.* Typhimurium's T3SS effectors play a key role in the initial step of the bacterial invasion, triggering actin rearrangements and membrane ruffling by either activating RhoGTPases or by directly interacting with actin, leading to internalization (Schlumberger and Hardt, 2006).

One effector, SopB (also known as SigD), has been shown to play an important role in both bacterial invasion and maturation of the *Salmonella*-containing vacuole (SCV) (Bakowski *et al.*, 2010; Wasylnka *et al.*, 2008; Zhou *et al.*, 2001). SopB is a phosphoinositide phosphatase, which shows homology to mammalian inositol polyphosphate 4-phosphatases and IpgD, an effector secreted by *Shigella flexneri*. SopB contains a catalytic domain with a conserved cysteine residue that is essential for all of its biological effects (Marcus *et al.*, 2001; Niebuhr *et al.*, 2000; Norris *et al.*, 1998; Zhou *et al.*, 2001). SopB also has a C-terminal region with homology to the mammalian phosphatase synaptojanin. *In vitro*, SopB hydrolyzes $PI(3,4)P_2$, $PI(3,5)P_2$, and $PI(3,4,5)P_3$, but other substrates such as $PI(4,5)P_2$ and inositol polyphosphates are also dephosphorylated with high activity (Marcus *et al.*, 2001; Norris *et al.*, 1998; Zhou *et al.*, 2001).

The *in vivo* substrate specificity of SopB remains unclear. A recent study showed that SopB promotes disappearance of $PI(4,5)P_2$ at the invaginating regions of the plasma membrane, which seems to be responsible for sealing of the membrane and SCV formation (Terebiznik *et al.*, 2002). In HeLa cells that were infected with a *S.* Typhimurium strain lacking SopB, vesicular fission was significantly delayed. The PH domain of PLCδ coupled to GFP was used in this study to monitor $PI(4,5)P_2$ levels during the invasion (Terebiznik *et al.*, 2002) (see Figs. 1, 2).

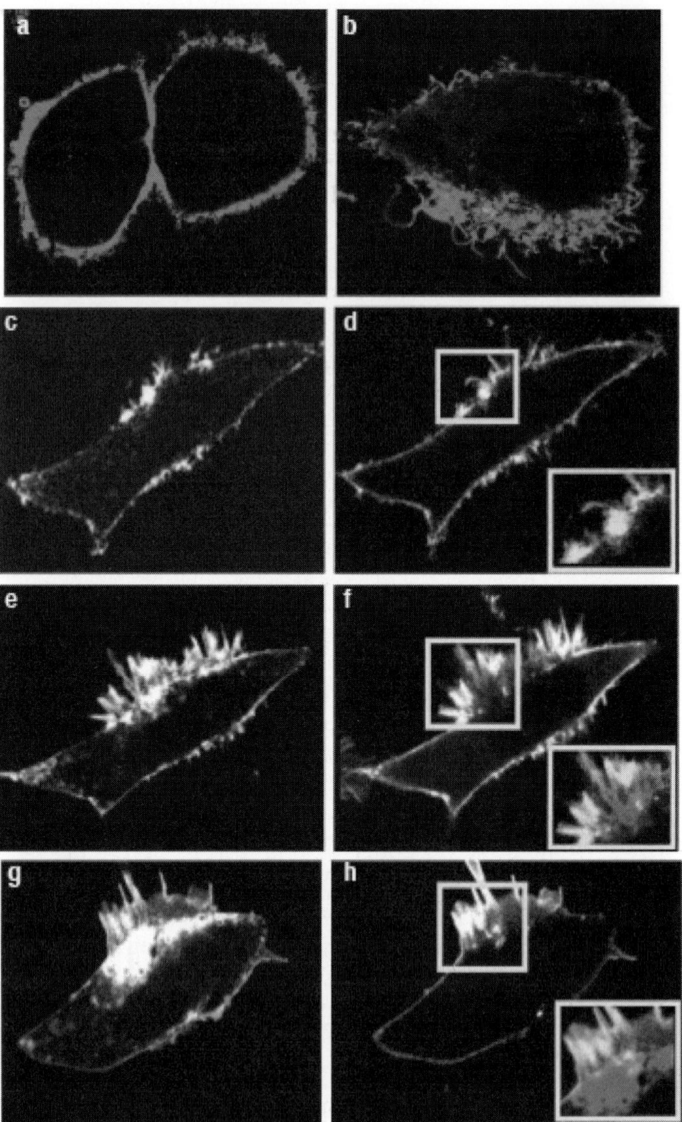

Fig. 1 Disappearance of $PI(4,5)P_2$ during *Salmonella* invasion. HeLa cells were transfected with PLCδ-PH-GFP and (a) left uninfected or (b) infected with WT RFP-expressing *Salmonella* for 10 min. Cells were transfected with both PM-CFP to visualize the plasma membrane and PLCδ-PH-YFP and then infected with WT *Salmonella*. Pictures were acquired after (c and d) 2.5 min, (e and f) 5 min, and (g and h) 20 min. Insets show magnifications of boxed regions. PM-CFP is shown in red and PLCδ-PH-YFP in green. Extensive co-localization between the two probes is visible near the tip of the ruffles. However, the invaginating region below the invasion ruffle is mostly lacking $PI(4,5)P_2$ but still contains PM-CFP. Images are reproduced from (Terebiznik *et al.*, 2002). (See color plate.)

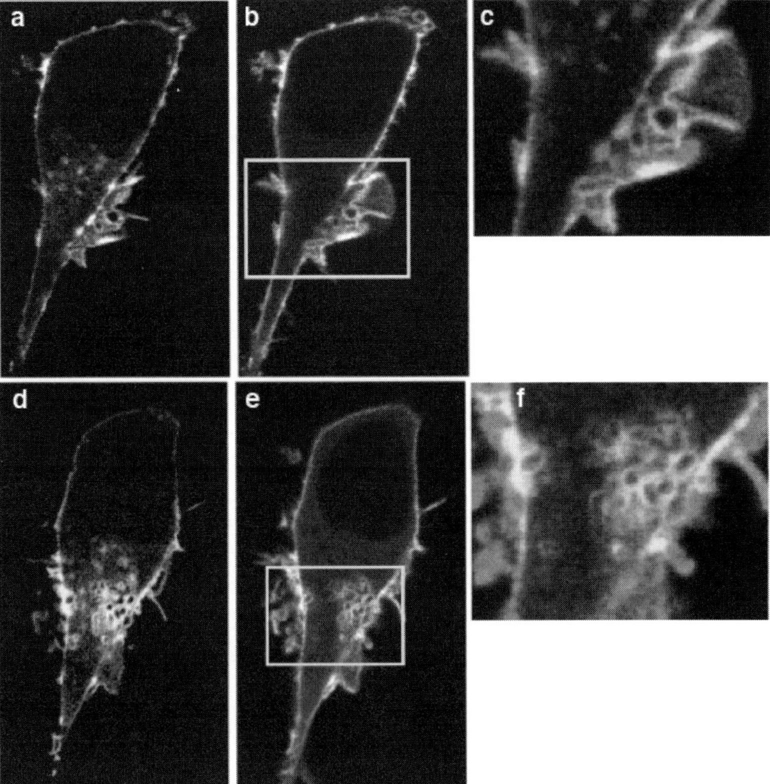

Fig. 2 Effect of SopB-deficient *Salmonella* on $PI(4,5)P_2$. HeLa cells were co-transfected with PLCδ-PH-GFP and PM-CFP and then infected with SopB-deficient *Salmonella*. Images were acquired after 5 min (a–c) or 30 min (d–f). c and f show magnifications of boxed regions. PLCδ-PH-GFP is shown in green and PM-CFP in red. Disappearance of $PI(4,5)P_2$ at the invaginating regions is greatly diminished. Images are reproduced from (Terebiznik *et al.*, 2002). (See color plate.)

Lipid extraction and high-performance liquid chromatography (HPLC) was also used to show that SopB is responsible for the generation of PI(5)P by dephosphorylating $PI(4,5)P_2$ (Mallo *et al.*, 2008; Mason *et al.*, 2007). Similarly, the *Shigella* homolog IpgD specifically hydrolyzes $PI(4,5)P_2$ at the plasma membrane to yield PI(5)P (Niebuhr *et al.*, 2002). Dephosphorylation of $PI(4,5)P_2$ by SopB also contributes to changes in membrane surface charge, resulting in inhibition of SCV-lysosome fusion (Bakowski *et al.*, 2010).

SopB can activate the pro-survival kinase Akt, preventing apoptosis in infected cells (Knodler *et al.*, 2005; Steele-Mortimer *et al.*, 2000). Canonical Akt activation occurs upon stimulation of various receptors that activate class I PI3-kinase, which in turn leads to production of $PI(3,4)P_2$ and $PI(3,4,5)P_3$ and recruitment of Akt from the cytosol to the plasma membrane where it is activated (Franke, 2008). How SopB activates Akt is unclear, given the fact that SopB is able to dephosphorylate $PI(3,4)P_2$ and $PI(3,4,5)P_3$ *in vitro*, and that both of these PIs are required to activate Akt. In the case of the *Shigella* homolog IpgD, Pendaries *et al.* (2006) suggested a mechanism whereby PI(5)P generated by the phosphatase activates a class I PI3-kinase. Whether a similar mechanism applies in *Salmonella*-infected cells needs further investigation.

At the same time that SopB hydrolyzes $PI(4,5)P_2$ in host cells to generate PI(5)P, elevated levels of $PI(3,4)P_2$ and $PI(3,4,5)P_3$ can be detected in lipid extracts by HPLC or visualized at the ruffles by transfection of the PH-domain of Akt. Remarkably, the accumulation of $PI(3,4)P_2$ and $PI(3,4,5)P_3$ was found to be resistant to the PI3-kinase inhibitor LY294002 (Mallo *et al.*, 2008) (see Fig. 3). Whether this localized increase in $PI(3,4)P_2$ and $PI(3,4,5)P_3$ formation at the ruffles contributes to Akt activation remains to be elucidated, but class I or class III PI3-kinases do not seem to play a role, since the appearance of the 3'-phosphorylated inositides is LY294002 resistant.

SopB has been suggested to promote PI(3)P formation on nascent SCVs through the hydrolysis of $PI(3,4)P_2$ and $PI(3,4,5)P_3$ at this compartment (Hernandez *et al.*, 2004). A study from our laboratory, however, could demonstrate that Vps34, which is recruited via Rab5, is essential for PI(3)P localization on SCVs and this formation was blocked by PI3-kinase inhibitors such as wortmannin or LY294002 and by specific knock-down of Vps34 with siRNA (Mallo *et al.*, 2008) (see Fig. 4). Therefore, SopB initiates two PI3-kinase signaling events during invasion: (i) formation of $PI(3,4)P_2$ and $PI(3,4,5)P_3$ at invasion ruffles by a wortmannin/LY294002-insensitive mechanism and (ii) production of PI(3)P on nascent SCVs by promoting delivery of Rab5 and Vps34 to this compartment.

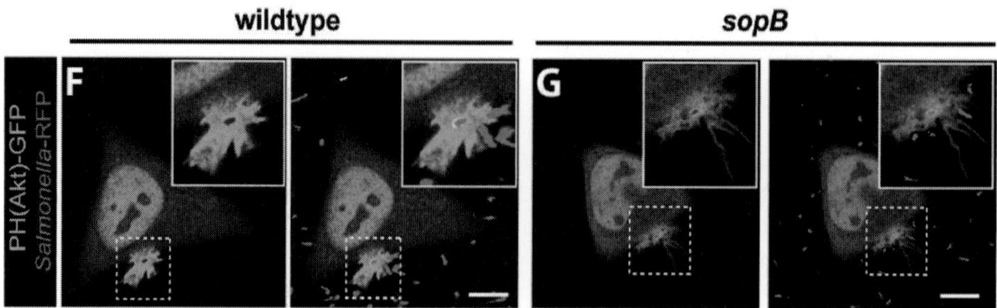

Fig. 3 SopB mediates $PI(3,4)P_2$ and $PI(3,4,5)P_3$ production at the plasma membrane. HeLa cells were transfected with PH-Akt-GFP and infected with either RFP-expressing WT or SopB-deficient *Salmonella* for <15 min. Pictures were acquired with a spinning disk confocal microscope. Insets show magnifications of dashed boxes. *Salmonella* initiates massive accumulation of $PI(3,4)P_2$ and $PI(3,4,5)P_3$ at the invasion ruffle, which was dependent on SopB. © Rockefeller University Press, 2008. Originally published in *J. Cell Biol.* doi: 10.1083/jcb.200804131. (See color plate.)

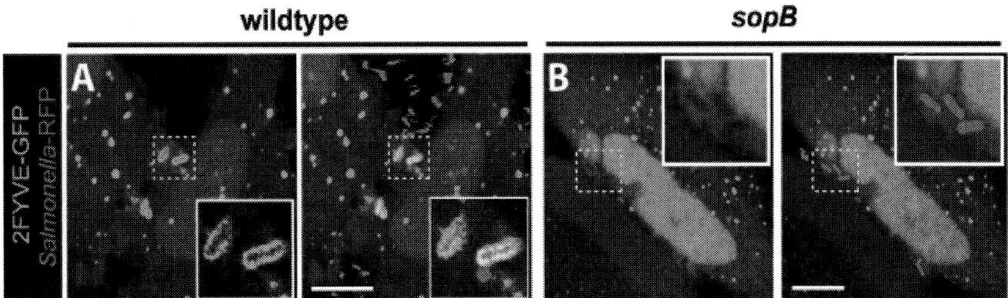

Fig. 4 SopB is required for PI(3)P localization on SCVs. HeLa cells were transfected with 2FYVE-GFP and infected with either RFP-expressing WT or SopB-deficient *Salmonella* for <15 min. Pictures were taken with a spinning disk confocal microscope. Insets show magnifications of dashed boxes. 2FYVE-GFP is recruited to the SCV in a SopB-dependent manner. © Rockefeller University Press, 2008. Originally published in *J. Cell Biol.* doi: 10.1083/jcb.200804131. (See color plate.)

II. Rationale

PI-binding modules fused to fluorescent proteins have gained popularity in studying lipid metabolism and this method has become an essential tool for dynamic imaging of lipids in living cells. Due to its ability to modulate lipid-signaling pathways during an invasion, *Salmonella* provides a powerful model to study acute changes in PI metabolism. The method described below combines the advantages of PI-binding domains as fluorescent molecular probes and of *Salmonella* as a model organism to study PI modulation during an infection (Fig. 5).

III. Materials and Media

A. Cell Culture

- HeLa cells, human cervical cancer cell line (ATCC®, Manassas, VA)
- Dulbecco's modified Eagle's medium, high glucose (DMEM/High Glucose) (Fisher Scientific, Ottawa, ON, Canada) supplemented with 10% heat-inactivated fetal bovine serum (FBS) (Wisent, Mississauga, ON, Canada)
- RPMI-1640 supplemented with L-glutamine and HEPES (Fisher Scientific)
- 0.05% Trypsin with 0.53 mM EDTA (Wisent)
- Sterile phosphate-buffered saline (PBS) (Fisher Scientific)
- Tissue culture flasks, T-75 (Becton Dickinson, Mississauga, ON, Canada)
- Six-well tissue culture plates (Becton Dickinson)
- Round glass cover slips, 25 mm diameter (Fisher Scientific)

B. Cell Transfection

- Genejuice® transfection reagent (EMD Chemicals, Mississauga, ON, Canada)
- DMEM without serum (Fisher Scientific)

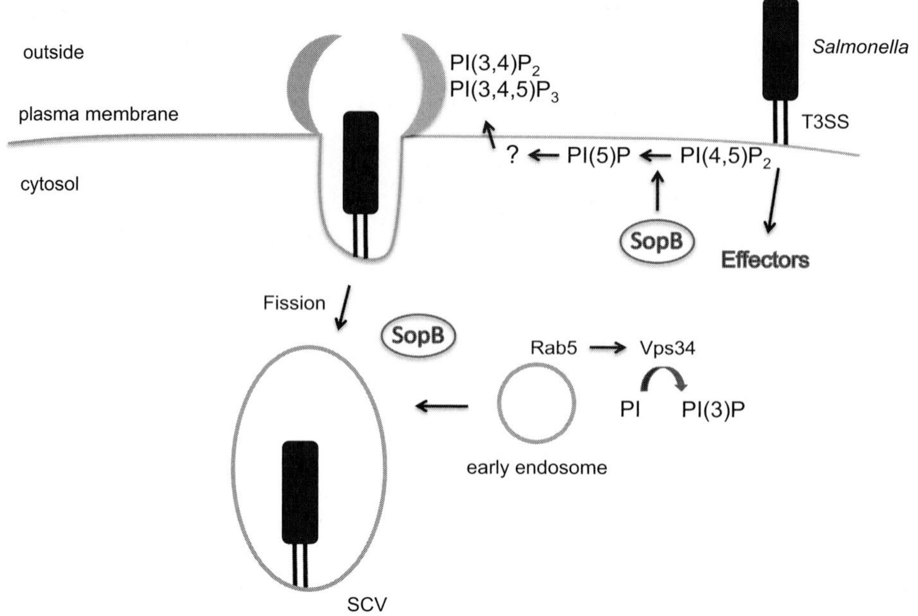

Fig. 5 Model of PI modulation during *Salmonella* invasion. SopB dephosphorylates PI(4,5)P_2 to produce PI(5)P and generates PI(3,4)P_2 and PI(3,4,5)P_3 at the invasion ruffle by an unknown mechanism. Rab5-containing endosomes are recruited to the SCV in a SopB-dependent manner. Rab5 recruits/activates Vps34 to mediate PI(3)P formation on SCVs via phosphorylation of PI. (For color version of this figure, the reader is referred to the web version of this book.)

C. Bacterial Invasion

- *Salmonella enterica* serovar Typhimurium (*S.* typhimurium), SL1344 WT (Hoiseth and Stocker, 1981)
- *S.* Typhimurium, SL1344 ΔSopB (Steele-Mortimer *et al.*, 2000)
- *S.* Typhimurium SL1344-RFP, WT *Salmonella* expressing mRFP (Birmingham *et al.*, 2006)
- *S.* Typhimurium SL1344 ΔSopB-RFP, mutant SopB *Salmonella* expressing mRFP (Bakowski *et al.*, 2008)
- Luria-Bertani broth (LB) (BioShop, Burlington, ON, Canada)
- Antibiotics for selection: Streptomycin, Chloramphenicol, Ampicillin, Kanamycin (Sigma-Aldrich, Oakville, ON, Canada)
- AlexaFluor® 647 carboxylic acid, succinimidyl ester (NHS-647) (Invitrogen, Burlington, ON, Canada)

D. Microscope

- A Leica DMIRE2 inverted fluorescence microscope with a spinning disk confocal scan-head, equipped with four separate diode-pumped solid-state laser lines (405,

491, 561, and 638 nm) (Spectral Applied Research, Richmond Hill, ON, Canada), an ASI-motorized XY-stage, and a Piezo Focus Drive (Quorum Technologies, Guelph, ON, Canada). Images were acquired with a back-thinned EM-CCD camera (Hamamatsu) and Volocity acquisition software (Perkin Elmer, Waltham, MA)

- Attofluor® cell chamber for live-cell microscopy for 25-mm cover slips (Invitrogen, Burlington, Canada)
- Tempcontrol 37 with heating insert P (Pecon, Germany)

IV. Methods

Although the protocol described here is performed using HeLa cells, the same procedure can be applied to any other cell type that is susceptible to *Salmonella* invasion. The protocol focuses on live-cell imaging of the bacterial invasion. However, it can also be utilized to examine fixed cells. To this end, cells need to be fixed at the desired time-points during the invasion with paraformaldehyde (usually 2.5% final) and can then be analyzed on the same microscope that is used for live-cell imaging. Alternatively, if quantifications for statistical analysis need to be performed, an epifluorescence microscope with compatible light sources is usually sufficient.

To visualize the bacteria during live-cell imaging, we routinely use *Salmonella* strains that express RFP. This, however, is only feasible when no other red fluorescence-emitting proteins are expressed. If both GFP- and RFP-tagged probes against PIs are used, the bacteria can be stained with NHS-647 prior to the invasion. NHS-647 is an isomer-free, amine-reactive dye that is spectrally identical, yet significantly brighter than the Cy5 dye.

A. Cell Culture

- HeLa cells are routinely grown in 10 mL DMEM/High Glucose supplemented with 10% FBS at 37 °C and 5% CO_2 without antibiotics in T-75 tissue culture flasks. Cells can usually be used for experiments when they reach 60–80% confluence, depending on the number of invasions that need to be performed. A confluent T-75 flask of HeLa cells normally holds 10 x 10^6 to 15 x 10^6 cells; about 10^6 HeLa cells are needed for six invasion experiments.

B. Cell Transfection

- One day prior to transfection, seed the cells on 25 mm cover slips in six-well tissue culture plates at a density of 16 x 10^4 per well in 4 mL DMEM.
- The next day, remove DMEM and add 2 mL of fresh medium to the six wells. We typically use 1 μg DNA and 3 μL Genejuice® per well for transfections.

- To prepare transfection complexes, add 3 μL of Genejuice® to 100 μL serum-free DMEM, incubate for 5 min at room temperature, and then mix with 1 μg of plasmid DNA.
- After incubation for 15 min at room temperature the mixture is added drop-wise to a single well.
- Cells are then kept for another 12–16 h at 37 °C to allow expression of the plasmid.
- For co-transfections of two different plasmids, a total amount of 1 μg plasmid DNA is used. The ratio of DNA and transfection reagent may vary with each plasmid and might have to be determined for each plasmid individually. Furthermore, transfection reagents other than Genejuice® might be more suitable for different cell lines or plasmids. For some plasmids it is also important to use a high concentration of DNA stock (e.g., 2 mg/mL).

C. Bacterial Invasion and Microscopy

- Overnight cultures of bacterial strains are prepared on the same day the cells are transfected. In our laboratory, we prepare 2 mL of LB-medium with appropriate antibiotic selection and inoculate with a single bacterial colony from an agar plate with freshly streaked bacteria. Bacterial cultures are grown for at least 12 h shaking in a 37 °C incubator.
- On the next day, 300 μL of the bacteria are subcultured in 10 mL of LB in a 125 mL Erlenmeyer flask without any antibiotic for 3 h shaking at 37 °C until they reach late-log phase. During this growth phase, *Salmonella* displays optimal invasion of HeLa cells (Steele-Mortimer *et al.*, 1999).
- If RFP-expressing *Salmonella* are used, the bacterial inoculum is prepared by pelleting 1 mL of the subculture at 10,000 x g for 2 min. Then wash once with PBS and resuspend in 250 μL of RPMI-1640. The bacteria can now be used for infection experiments.
- For bacterial staining with NHS-647, centrifuge 1 mL of the subculture at 10,000 x g for 2 min, wash three times with PBS, and resuspend in 250 μL of PBS. Add 5 μL of the NHS-647 reagent (10 mg/mL) to the bacterial suspension for a final concentration of 0.2 mg/mL. After shaking the suspension for 5 min in a 37 °C incubator, centrifuge bacteria again at 10,000 x g for 2 min, wash once with 1 mL of PBS, and resuspend in 250 μL RPMI-1640. The bacteria are ready to use now.
- The microscope should be turned on at least one hour prior to infection and the digital heating device set to 37 °C.
- The microscope stage should be calibrated to ensure correct movement of the stage.
- Quickly transfer a 25 mm cover slip from a well containing transfected cells and place it in the Attofluor® cell chamber. Aspirate any remaining medium and add 500 μL of warm RPMI-1640. The switch to HEPES-buffered RPMI medium is necessary when the cells are not kept under CO_2 during microscopy.
- Transfer the cell chamber to the spinning disk microscope; focus on one transfected cell, and set top and bottom of the cell with the Piezo Focus Drive. This

should be done as quickly as possible and with low laser power to avoid any photobleaching.

- Acquisition of multiple cells will help enhance the possibility of observing PI dynamics in infected cells (not all cells are infected by bacteria).
- Add 100 μL of above-prepared bacteria and start imaging Z-stacks with Z-steps between 0.3 and 0.5 μm. Great care should be taken when adding the bacteria to the cells, so the focus on the cell is not disturbed. Exposure times should be kept as short as possible to minimize photobleaching. Binning (2x) should be considered as a trade-off of resolution for decreased photobleaching.

  For short-term acquisition (up to 10 min), one frame every 5–10 s is sufficient. For longer acquisition times (30–60 min) one image every minute is usually sufficient. Acquisition time depends on the PI probe of interest that was transfected and should not exceed 60 min. Longer imaging is not recommended without an incubation chamber that provides humidification.

D. Analysis

In our laboratory, we routinely use Volocity software for analysis of acquired images. Depending on various factors, such as fluorescence background or transfection efficiency, deconvolution may be recommended to obtain a clearer image. We usually employ the fast iterative deconvolution tool of Volocity, which is a software algorithm that can be applied to acquired images to reverse the optical distortion of the microscope resulting in a sharper image. A minimum of three Z-slices is required for deconvolution of images.

Thorough visual analysis will provide a good general impression of any changes in lipid distribution and might be enough to draw conclusions about lipid dynamics. If no quantification is needed, a series of images or a movie is usually sufficient. However, if fluorescence redistribution is not obvious or needs to be compared to other samples or experiments, proper quantitative analysis should be carried out. A common way to do this is by measuring mean or total fluorescence intensity of a defined region of interest (ROI) and normalizing it to a region in the cell where no lipid changes are expected (i.e., the cytosol). The ratio of these two intensities can then be compared to other cells.

A different tool provided by Volocity is “Measure Line Profile”, which measures fluorescence intensity values along a line drawn through a ROI. The software generates an intensity histogram that can be compared to similar structures in different cells or experiments.

For a convenient way to carry out statistical analysis, the cells are fixed after invasion and mounted on glass slides. The fixed samples can then easily be analyzed by either confocal or epifluorescence microscopy by counting an appropriate number of structures of interest.

Sometimes 3D-reconstruction of acquired Z-stacks also helps to gain more information on lipid localization or redistribution, especially when dealing with structures that protrude from the cell surface, like *Salmonella* invasion ruffles. In these instances it is often recommended to make rotational movies, with different channels visualized individually and then "layered on" the other channels. Use of the "transparency" feature also allows better 3D visualization of different PIs in subcellular structures.

As a good control, it is recommended to co-transfect the cells with a soluble fluorescent protein alone (not attached to a lipid-binding domain) or to use a nonspecific membrane marker such as the FM dyes. This assures that increased fluorescence intensity is not just due to folding of membranes resulting in enrichment of fluorescent protein. In addition, transfection of a mutant version of the PI-binding domain lacking the ability to interact with lipids could be helpful to serve as a control for specificity.

V. Summary and Conclusions

In this chapter, we have presented a general technique to monitor the lipid dynamics during a *Salmonella* invasion using confocal microscopy. This method provides many advantages to conventional methods such as HPLC or thin layer chromatography (TLC), which only measure total cellular lipid content without providing information on subcellular localization. Some cellular processes such as phagocytosis, however, are very rapid and the lipids that are involved persist only transiently and locally. To overcome these problems, PI-binding domains as molecular probes present a good alternative as molecular probes to investigate immediate changes in subcellular lipid distribution. Although they have become an indispensable tool in studying inositol lipids, great care must be taken in interpreting the data obtained using these probes and some considerations should be kept in mind (Balla *et al.*, 2000). The fluorescent protein used as a beacon molecule may inhibit binding or alter the characteristics of the probe. Also, some PI-binding domains have restricted access to only a subset of different lipid pools in the cell; in addition, they may bind to soluble inositol phosphates, reducing their ability to detect membrane-associated lipids. Additionally, when expressed at sufficiently high concentrations, the molecular reporters might compete with endogenous lipid-binding effectors. Yet, with the right set of controls, lipid-binding probes can be a very reliable and valuable tool to learn much about the function of bacterial PI-metabolizing enzymes as well as to gain knowledge about fundamental cellular processes.

Acknowledgments

John H. Brumell holds an Investigators in Pathogenesis of Infectious Disease Award from the Burroughs Wellcome Fund. Infrastructure for the Brumell Laboratory was provided by a New Opportunities Fund from the Canadian Foundation for Innovation and the Ontario Innovation Trust. This work was supported by operating grant from the Canadian Institutes of Health Research (MOP# 93634). Bernhard Roppenser is supported by the Deutsche Forschungsgemeinschaft.

References

Bakowski, M. A., Braun, V., and Brumell, J. H. (2008). Salmonella-containing vacuoles: directing traffic and nesting to grow. *Traffic* **9**, 2022–2031.

Bakowski, M. A., Braun, V., Lam, G. Y., Yeung, T., Do Heo, W., Meyer, T., Finlay, B. B., Grinstein, S., and Brumell, J. H. (2010). The phosphoinositide phosphatase SopB manipulates membrane surface charge and trafficking of the Salmonella-containing vacuole. *Cell Host Microbe* **7**, 453–462.

Balla, T., Bondeva, T., and Varnai, P. (2000). How accurately can we image inositol lipids in living cells? *Trends Pharmacol. Sci.* **21**, 238–241.

Birmingham, C. L., Smith, A. C., Bakowski, M. A., Yoshimori, T., and Brumell, J. H. (2006). Autophagy controls Salmonella infection in response to damage to the Salmonella-containing vacuole. *J. Biol. Chem.* **281**, 11374–11383.

Cossart, P., and Sansonetti, P. J. (2004). Bacterial invasion: the paradigms of enteroinvasive pathogens. *Science* **304**, 242–248.

Di Paolo, G., and De Camilli, P. (2006). Phosphoinositides in cell regulation and membrane dynamics. *Nature* **443**, 651–657.

Flannagan, R. S., Cosio, G., and Grinstein, S. (2009). Antimicrobial mechanisms of phagocytes and bacterial evasion strategies. *Nat. Rev. Microbiol.* **7**, 355–366.

Franke, T. F. (2008). PI3K/Akt: getting it right matters. *Oncogene* **27**, 6473–6488.

Galan, J. E., and Wolf-Watz, H. (2006). Protein delivery into eukaryotic cells by type III secretion machines. *Nature* **444**, 567–573.

Hernandez, L. D., Hueffer, K., Wenk, M. R., and Galan, J. E. (2004). Salmonella modulates vesicular traffic by altering phosphoinositide metabolism. *Science* **304**, 1805–1807.

Hilbi, H. (2006). Modulation of phosphoinositide metabolism by pathogenic bacteria. *Cell Microbiol.* **8**, 1697–1706.

Hoiseth, S. K., and Stocker, B. A. (1981). Aromatic-dependent Salmonella typhimurium are non-virulent and effective as live vaccines. *Nature* **291**, 238–239.

Knodler, L. A., Finlay, B. B., and Steele-Mortimer, O. (2005). The Salmonella effector protein SopB protects epithelial cells from apoptosis by sustained activation of Akt. *J. Biol. Chem.* **280**, 9058–9064.

Knodler, L. A., and Steele-Mortimer, O. (2003). Taking possession: biogenesis of the Salmonella-containing vacuole. *Traffic* **4**, 587–599.

Mallo, G. V., Espina, M., Smith, A. C., Terebiznik, M. R., Aleman, A., Finlay, B. B., Rameh, L. E., Grinstein, S., and Brumell, J. H. (2008). SopB promotes phosphatidylinositol 3-phosphate formation on Salmonella vacuoles by recruiting Rab5 and Vps34. *J. Cell Biol.* **182**, 741–752.

Marcus, S. L., Wenk, M. R., Steele-Mortimer, O., and Finlay, B. B. (2001). A synaptojanin-homologous region of Salmonella typhimurium SigD is essential for inositol phosphatase activity and Akt activation. *FEBS Lett.* **494**, 201–207.

Mason, D., Mallo, G. V., Terebiznik, M. R., Payrastre, B., Finlay, B. B., Brumell, J. H., Rameh, L., and Grinstein, S. (2007). Alteration of epithelial structure and function associated with PtdIns(4,5)P2 degradation by a bacterial phosphatase. *J. Gen. Physiol.* **129**, 267–283.

Niebuhr, K., Giuriato, S., Pedron, T., Philpott, D. J., Gaits, F., Sable, J., Sheetz, M. P., Parsot, C., Sansonetti, P. J., and Payrastre, B. (2002). Conversion of PtdIns(4,5)P(2) into PtdIns(5)P by the S.flexneri effector IpgD reorganizes host cell morphology. *EMBO J.* **21**, 5069–5078.

Niebuhr, K., Jouihri, N., Allaoui, A., Gounon, P., Sansonetti, P. J., and Parsot, C. (2000). IpgD, a protein secreted by the type III secretion machinery of Shigella flexneri, is chaperoned by IpgE and implicated in entry focus formation. *Mol. Microbiol.* **38**, 8–19.

Norris, F. A., Wilson, M. P., Wallis, T. S., Galyov, E. E., and Majerus, P. W. (1998). SopB, a protein required for virulence of Salmonella dublin, is an inositol phosphate phosphatase. *Proc. Natl. Acad. Sci. U. S. A.* **95**, 14057–14059.

Pendaries, C., Tronchere, H., Arbibe, L., Mounier, J., Gozani, O., Cantley, L., Fry, M. J., Gaits-Iacovoni, F., Sansonetti, P. J., and Payrastre, B. (2006). PtdIns5P activates the host cell PI3-kinase/Akt pathway during Shigella flexneri infection. *EMBO J.* **25**, 1024–1034.

Pizarro-Cerda, J., and Cossart, P. (2004). Subversion of phosphoinositide metabolism by intracellular bacterial pathogens. *Nat. Cell Biol.* **6**, 1026–1033.

Sasaki, T., Takasuga, S., Sasaki, J., Kofuji, S., Eguchi, S., Yamazaki, M., and Suzuki, A. (2009). Mammalian phosphoinositide kinases and phosphatases. *Prog. Lipid Res.* **48**, 307–343.

Schlumberger, M. C., and Hardt, W. D. (2006). Salmonella type III secretion effectors: pulling the host cell's strings. *Curr. Opin. Microbiol.* **9**, 46–54.

Steele-Mortimer, O., Knodler, L. A., Marcus, S. L., Scheid, M. P., Goh, B., Pfeifer, C. G., Duronio, V., and Finlay, B. B. (2000). Activation of Akt/protein kinase B in epithelial cells by the Salmonella typhimurium effector sigD. *J. Biol. Chem.* **275**, 37718–37724.

Steele-Mortimer, O., Meresse, S., Gorvel, J. P., Toh, B. H., and Finlay, B. B. (1999). Biogenesis of Salmonella typhimurium-containing vacuoles in epithelial cells involves interactions with the early endocytic pathway. *Cell Microbiol.* **1**, 33–49.

Terebiznik, M. R., Vieira, O. V., Marcus, S. L., Slade, A., Yip, C. M., Trimble, W. S., Meyer, T., Finlay, B. B., and Grinstein, S. (2002). Elimination of host cell PtdIns(4,5)P(2) by bacterial SigD promotes membrane fission during invasion by Salmonella. *Nat. Cell Biol.* **4**, 766–773.

Varnai, P., and Balla, T. (2006). Live cell imaging of phosphoinositide dynamics with fluorescent protein domains. *Biochim. Biophys. Acta* **1761**, 957–967.

Wasylnka, J. A., Bakowski, M. A., Szeto, J., Ohlson, M. B., Trimble, W. S., Miller, S. I., and Brumell, J. H. (2008). Role for myosin II in regulating positioning of Salmonella-containing vacuoles and intracellular replication. *Infect. Immun.* **76**, 2722–2735.

Zhou, D., Chen, L. M., Hernandez, L., Shears, S. B., and Galan, J. E. (2001). A Salmonella inositol polyphosphatase acts in conjunction with other bacterial effectors to promote host cell actin cytoskeleton rearrangements and bacterial internalization. *Mol. Microbiol.* **39**, 248–259.

CHAPTER 10

Acute Manipulation of Phosphoinositide Levels in Cells

Belle Chang-Ileto, Samuel G. Frere and Gilbert Di Paolo

Department of Pathology and Cell Biology, Taub Institute for Research on Alzheimer's Disease and the Aging Brain, Columbia University Medical Center, New York, USA

Abstract

Phosphoinositides are membrane-bound signaling phospholipids that function in a myriad of cellular processes, including membrane trafficking, cytoskeletal dynamics, ion channel and transporter function, and signal transduction. In order to better

Copyright 2012, Elsevier Inc. All rights reserved.

0091-679X/10 $35.00
DOI 10.1016/B978-0-12-386487-1.00010-9

understand the role of phosphoinositides in cellular processes, different approaches to study the effects of the presence or absence of these lipids must be devised. Conventional approaches of manipulating phosphoinositide levels such as over-expression or genetic ablation of lipid enzymes cause prolonged exposure of the cells to changes in lipid levels that could result in compensatory actions by the cell or downstream alterations in cell physiology. In this chapter we present an approach used recently by various laboratories, including our own, to acutely manipulate phosphoinositide levels at target locations using chemically induced dimerization (CID) that can be spatially and temporally controlled. We discuss considerations when designing expression constructs for targeting specific cellular compartment membranes and present examples from the literature on different ways of perturbing phosphoinositide levels at particular organelle membranes using CID. In addition, we provide details on image acquisition, data collection, and data interpretation. CID technology can be applied to many lipid enzymes to broaden the understanding of the role lipid signaling plays in cell physiology.

I. Introduction

A. Phosphoinositides, Membrane Dynamics and Intracellular Signaling

Phosphoinositides play prominent roles in the regulation of numerous cellular functions from membrane trafficking, cytoskeletal dynamics, ion channel and transporter function, to signal transduction (Balla *et al.*, 2009; Di Paolo and De Camilli, 2006; Falkenburger *et al.*, 2010; Hilgemann *et al.*, 2001; Suh and Hille, 2005; Yin and Janmey, 2003). Phosphatidylinositol (PI) serves as the precursor of seven phosphoinositide species, which are differentially phosphorylated at the 3, 4, and/or 5 position(s) of the inositol ring. While PI makes up less than 15% of phospholipids found in eukaryotic cells, its phosphorylated derivatives are found at even lower levels, with PI(4,5)P_2 and PI4P as the most abundant species of the phosphoinositides (Di Paolo and De Camilli, 2006). Despite their low abundance, phosphoinositides play critical functions in many cellular processes, in part due to their high turnover and the eclectic nature of their phosphorylated head groups.

Phosphoinositides are typically found concentrated at the cytoplasmic face of cellular membranes with their inositol ring, or headgroup, exposed to the cytosolic milieu, available to interact with cytosolic proteins or membrane protein cytodomains. These molecules therefore play an important role in controlling the membrane-cytosol interface. Phosphoinositides can regulate the function of integral membrane proteins at the plasma membrane (PM), such as ion channels and ion transporters (Suh and Hille, 2008). They can also serve as scaffolds that bring together the cytoskeleton or coat proteins with the cytoplasmic membrane surface (Di Paolo and De Camilli, 2006; Haucke, 2005). For example, PI(4,5)P_2 participates in the regulation of actin polymerization by binding N-WASP. This causes a conformational change that allows the recruitment and activation of the ARP2/3 complex resulting in the nucleation of actin filaments (Logan and Mandato, 2006; Mao and Yin, 2007; Pollard and Borisy, 2003; Rohatgi *et al.*, 2000).

B. Conventional Methods of Manipulating PI Levels

Conventional methods of manipulating phosphoinositide levels employed to investigate the role of these lipids in cells include exogenous application of lipids, genetic manipulation, and use of pharmacological inhibitors. The most basic of these methods are the exogenous application of phosphoinositides using polyamine carriers which "shuttle" the lipids into the cell in order to reach intracellular membranes (Ozaki *et al.*, 2000) and direct microinjection of lipid micelles into the cell (Golebiewska *et al.*, 2008). A more sophisticated approach is the recently developed membrane-permeant "caged" phosphoinositides that allow release or exposure of the "caged" lipid upon photoactivation (Subramanian *et al.*, 2010). Additionally, metabolically stabilized variants of phosphoinositides have been developed in order to dissect the effects of these lipids from those of their metabolites (Huang *et al.*, 2007; Xu *et al.*, 2006; Zhang *et al.*, 2006a, 2006b). Such methods result in the rapid entry of phosphoinositides into the cell but do not allow for the facile control of lipid concentrations or destination. A more common approach is to use genetic manipulation to over-express inositide kinases or phosphatases (Kahlfeldt *et al.*, 2010; Kim *et al.*, 2006; Krauss *et al.*, 2003), to silence specific genes through RNAi (Choudhury *et al.*, 2005; Prasad and Decker, 2005; Wang *et al.*, 2004), or to engineer organisms deficient in a gene of interest (for example, see references (Cremona *et al.*, 1999; Di Cristofano *et al.*, 1998; Di Paolo *et al.*, 2004; Gary *et al.*, 1998; Harris *et al.*, 2000; Schu *et al.*, 1993; Stambolic *et al.*, 1998; Verstreken *et al.*, 2003; Zhou *et al.*, 2010). Modulation of the target phosphoinositide concentration at the desired membrane compartment is achieved using these genetic techniques; however, the effects of prolonged exposure to these changes in lipid levels may result in downstream alterations in cellular processes that are difficult to experimentally resolve from the function directly controlled by the lipid. A more selective system of perturbing lipid levels would be the use of pharmacological inhibitors of lipid enzymes that allow for well-defined suppression of a specific enzyme(s). Inhibitors for the PI 3-kinase family have been well studied with the broadly acting small molecules wortmannin and LY249002 and the more recently developed isoform-specific agents (Knight *et al.*, 2006). Wortmannin and LY294002 (used at higher concentrations) as well as phenylarsine oxide have been used to inhibit PI 4-kinases (Downing *et al.*, 1996; Nakanishi *et al.*, 1995; Sorensen *et al.*, 1998; Varnai and Balla, 1998; Wiedemann *et al.*, 1996). However, well-defined, specific pharmacological drugs for other lipid kinases and phosphatases have not been discovered.

C. Chemical Inducers of Dimerization (CIDs)

The approach of chemically induced dimerization (CID) has been used in several recent phosphoinositide studies investigating various cellular processes, such as endocytosis, endosomal morphology and cargo sorting, ion channel function and cytoskeletal dynamics (see Table I). The use of CID allows for the acute and localized regulation of phosphoinositide synthesis or degradation and circumvents issues that arise with standard genetic and molecular approaches, which typically cause prolonged perturbations in phosphoinositide levels. Schreiber, Crabtree and their colleagues first introduced the use of CID in a landmark study on T-cell receptor-mediated signaling pathways in 1993 (Spencer *et al.*, 1993). In their approach, the immunosuppressant molecule FK506, which binds both FK-binding protein 12 (FKBP12) and calcineurin (Liu *et al.*, 1991), was synthetically dimerized (and named FK1012) not only to inactivate its immunosuppressive properties (*i.e.*, block its calcineurin-binding site) but in order to create the ability to bind two molecules of FKBP12 resulting in homodimerization of this protein. The ability of FK1012 to homodimerize the cytoplasmic domains of the T-cell receptor zeta chain fused to FKBP12 was shown to successfully induce downstream signaling cascades elicited normally by intact T-cell receptor dimerization.

A common variant to this original approach is the use of another immunosuppressive agent, rapamycin. Like FK506, this molecule also binds FKBP12 but instead of calcineurin, its second binding target is the FKBP-rapamycin binding (FRB) domain of mammalian target of rapamycin (mTOR) (Chen *et al.*, 1995). Rapamycin [or its chemical analogs, such as iRAP (Inoue *et al.*, 2005) or AP21967 (Chong *et al.*, 2002)] in CID can be used to induce heterodimerization of two distinct proteins/domains, engineered as a fusion partner with either FKBP12 or FRB.

CID can be used to induce complexes of the same protein (homodimerization) as well as complexes of two distinct proteins (heterodimerization) depending on the dimerizer compound (such as FK1012 and rapamycin) and the protein fusions used (FKBP12 alone or FKBP12 in conjunction with FRB). Variants of dimerizers and of the protein modules FRB and FKBP have been developed to overcome undesirable interactions with endogenous molecules as well as to create the ability to do orthogonal studies in which control of multiple enzymes is desired (Bayle *et al.*, 2006; Belshaw *et al.*, 1996; Choi *et al.*, 1996; Chong *et al.*, 2002; Clackson *et al.*, 1998; Inoue *et al.*, 2005; Liberles *et al.*, 1997). Native interactions between proteins as well as novel interactions and functions can be studied using such reagents. A reverse dimerization strategy uses a self-dimerizing FKBP mutant so that the initial condition of fusion protein partners is an aggregated complex that can be disassembled upon addition of a synthetic monomeric FKBP ligand (Rollins *et al.*, 2000). Localization and mislocalization of protein targets can be achieved using appropriate or inappropriate targeting signals or protein pairs. Protein stability can also be controlled using CID (Banaszynski *et al.*, 2006; Stankunas *et al.*, 2003, 2007).

D. CID and Phosphoinositides

Many of the phosphoinositide studies employing CID have focused on the consequences of degradation of the $PI(4,5)P_2$ population at the PM (see Table I). In these

Table I
Examples in the literature of CID use to manipulate phosphoinositide levels

Lipid Manipulated	Enzymatic Tool	Targeted Membrane/ Compartment; Targeting Sequence/Protein	Dimerizer Used	Phenotypes	Reference
PI(4 5)P_2	type IV 5-phosphatase domain	PM; palmitoylation sequence of GAP43	rapamycin, AP21967	Termination of ATP-induced Ca^{2+} signal, inactivation of TRPM8 channels, blockade of TfnR and EGFR endocytosis	(Varnai *et al.*, 2006)
PI(4,5)P_2	Inp54p phosphatase	PM; Lyn_{11}	iRAP	Suppression of KCNQ ion channel current	(Suh *et al.*, 2006)
PI(4,5)P_2	Inp54p phosphatase	PM; Lyn_{11}	iRAP	Dissociation of proteins with polybasic clusters from the PM (concomitant with the loss of PI(3,4,5)P_3)	(Heo *et al.*, 2006)
PI3P	myotubularin 1	endosomes; Rab5a	AP21967	TfnR accumulation in Rab5apositive endosomes, endosome tubulation	(Fili *et al.*, 2006)
PI(4,5)P_2	type IV 5-phosphatase domain	PM; palmitoylation sequence of GAP43	iRAP	Loss of endocytic CCPs, dissociation of endocytic adaptors, dynamin and Arp2/3 complex from PM	(Zoncu *et al.*, 2007)
PI(4,5)P_2	Inp54p phosphatase domain	PM; Lyn_{11}	iRAP	Decreased TfnR endocytosis, increased TfnR concentration on PM, dissociation of AP-2 from PM	(Abe *et al.*, 2008)
PI(4,5)P_2	p110 catalytic subunit of type I PI 3-kinase	PM; palmitoylation sequence of GAP43	iRAP	PIP_3 production, membrane ruffles, accelerated CCP maturation and turnover	(Nakatsu *et al.*, 2010)
PI4P	Sac1 phosphatase	Golgi; type I Golgi protein	rapamycin	Loss of clathrin adaptors from Golgi membrane and suppression of cargo exit from Golgi; delay of PI(4,5)P_2 replenishment at the PM during PLC activation	(Szentpetery *et al.*, 2010)
PI(4,5)P_2	5-phosphatase domain of Synj1	PM; endophilin N-BAR domain	rapamycin, AP21967	Membrane tubule fission	(Chang-Ileto *et al.*, 2011)

CCP: clathrin-coated pit; EGFR: epidermal growth factor receptor; PLC: phospholipase C; PM: plasma membrane; Synj1: synaptojanin 1; Tfn: transferrin; TfnR: transferrin receptor; TRPM8: methanol-activated transient receptor potential melastatin 8.

studies, a $PI(4,5)P_2$ 5-phosphatase domain fused to FKBP is recruited to the PM via PM-anchored peptides or proteins fused to FRB upon the addition of rapamycin or one of its analogs resulting in the rapid elimination of $PI(4,5)P_2$, as first used by the Balla (Varnai *et al.*, 2006) and Hille groups (Suh *et al.*, 2006). Using the palmitoylation sequence of GAP43 as the PM-anchor and the phosphatase domain of type IV 5-phosphatase, Varnai *et al.* (2006) showed that upon dephosphorylation of $PI(4,5)P_2$ at the PM multiple processes were affected including Ca^{2+} signaling, TRPM8 channel function and receptor internalization/endocytosis. Using this same combination of phosphatase and PM-anchor, Zoncu *et al.* (2007) investigated the effect of $PI(4,5)P_2$ loss on clathrin coat dynamics. $PI(4,5)P_2$ depletion resulted in the disappearance of clathrin-coated pits (CCPs) at the cells surface along with various endocytic adaptors, such as AP-2 and epsin, as well as the dissociation of dynamin and Arp2/3 from the membrane indicating a role for $PI(4,5)P_2$ not only in endocytosis, but also in actin nucleation and actin-mediated motility at the cell edge. The use of Lyn_{11} as the PM-anchor and Inp54p to degrade $PI(4,5)P_2$ was used to demonstrate that KCNQ ion channel function is dependent on the presence of this phosphoinositide (Suh *et al.*, 2006) and that transferrin receptor (TfnR) endocytosis was more sensitive to changes in $PI(4,5)P_2$ levels than TfnR recycling (Abe *et al.*, 2008). In order to study the role of $PI(4,5)P_2$ in targeting proteins with polybasic clusters to the PM, Heo *et al.* (2006) used the Lyn_{11}/Inp54p combination to selectively decrease $PI(4,5)P_2$ levels and showed that this lipid in concert with $PI(3,4,5)P_3$ was responsible for targeting of proteins with polybasic clusters to the PM. We used the CID approach to demonstrate that recruitment of the 5-phosphatase domain of synaptojanin 1 to endophilin-induced tubules at the cell surface and the resulting $PI(4,5)P_2$ turnover at these structures resulted in membrane fission (Chang-Ileto *et al.*, 2011).

CID has also been used to investigate the role of other phosphoinositides in the cell. In a study on the role of $PI(3,4,5)P_3$ and the inositol 5-phosphatase SHIP2 on CCP dynamics, recruitment of the p110 catalytic subunit of type I PI 3-kinase to the PM was used to induce acute phosphorylation of $PI(4,5)P_2$ and $PI(3,4,5)P_3$ production that resulted in the acceleration of CCP assembly and turnover (Nakatsu *et al.*, 2010). Since SHIP2 has been shown to be a negative regulator of $PI(3,4,5)P_3$-dependent signaling, this result was in agreement with SHIP2's role in dampening insulin signaling through its regulation of CCP dynamics via $PI(3,4,5)P_3$ (and $PI(4,5)P_2$) turnover. An examination of the role of PI3P on endosomal function and cargo traffic used Rab5 to localize the inositol lipid phosphatase myotubularin 1 to early endosomal membranes in order to decrease PI3P (and $PI(3,5)P_2$) in this compartment (Fili *et al.*, 2006). This study revealed that the phosphoinositide PI3P had an important influence on endosomal morphology and cargo passage through these Rab5-positive endosomes. Finally, manipulation of PI4P on Golgi membrane revealed that cargo and some clathrin adaptors depend on this lipid for proper trafficking or localization and that this lipid contributes to the maintenance of $PI(4,5)P_2$ at the PM (Szentpetery *et al.*, 2010).

E. Targeting of CID Protein Modules

Because of their heterogeneous subcellular distribution, phosphoinositides can serve as markers for different organelles and control the targeting of proteins with phospholipid-binding domains to their destinations (Behnia and Munro, 2005; Di Paolo and De Camilli, 2006). A map of cellular membranes can be drawn according to the distribution of phosphoinositides to illustrate the role of these lipids in organelle identity (Fig. 1). Several different approaches can be used to localize your protein(s) of interest to the desired cellular membrane/organelle (Varnai and Balla, 2007). Many proteins that interact with phosphoinositides possess specific phospholipid-binding domains that mediate the recruitment of these proteins to specific membrane domains in the cell (Lemmon, 2008). For example, since the pleckstrin homology domain (PH) of phospholipase Cδ1 (PLCδ1) recognizes PI(4,5)P_2, a phosphoinositide concentrated at the PM, the PH domain targets PLCδ1 to the PM (Garcia *et al.*, 1995; Lemmon *et al.*, 1995). Meanwhile, the PH domain of the four-phosphate-adaptor protein 1 (FAPP1) is specific for PI4P thereby targeting this protein to the *trans*-Golgi network where PI4P is predominant (Godi *et al.*, 2004). Another phospholipid-binding domain, FYVE (Fab1, YOTB, Vac1, EEA1) specifically recognizes PI3P and can be found

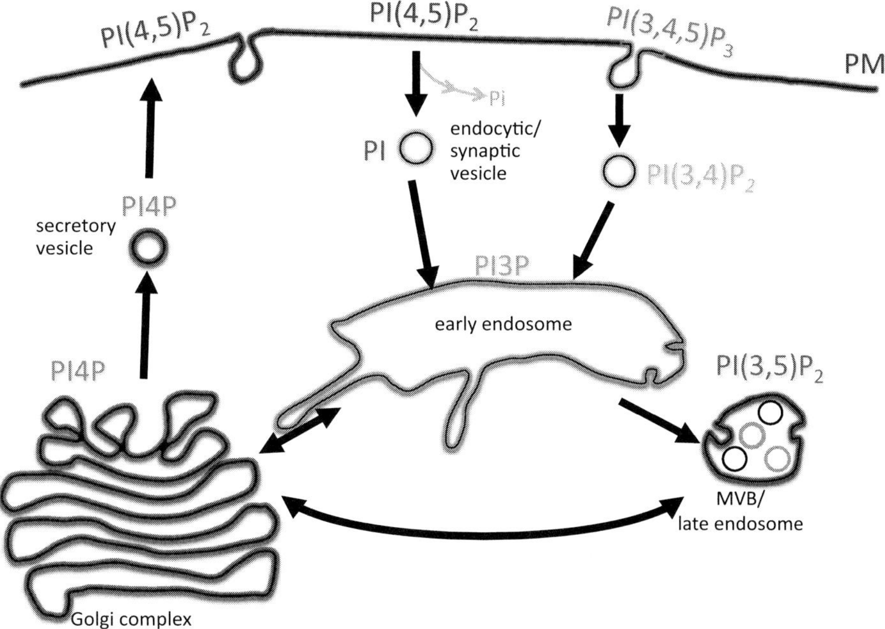

Fig. 1 Phosphoinositides and organelle identity. Phosphoinositides are heterogeneously distributed throughout the cell and can be used as markers for different organelles as illustrated by their discrete subcellular distribution. Arrows indicate the flow of membrane in membrane trafficking routes. PI5P localization in cells is unclear. (For color version of this figure, the reader is referred to the web version of this book.)

in endosomal proteins such as Hrs (hepatocyte growth factor regulated tyrosine kinase substrate) (Gillooly *et al.*, 2001) and early endosome antigen 1 (EEA1).

Organelle targeting motifs for fusion to either FRB or FKBP were evaluated by the Inoue group for the Golgi using a peptide from the giantin protein; for the mitochondria using a peptide from either Tom20 or monoamine oxidase; for the ER using cytochrome b5; and for lysosomes using LAMP1 (Komatsu *et al.*, 2010). Additionally, other validated organelle targeting motifs include the ER localization sequence of human Sac1 phosphatase for FRB targeting to the ER (Varnai *et al.*, 2007), AKAP sequences for targeting mitochondria (Csordas *et al.*, 2010; Korzeniowski *et al.*, 2010) and the type I Golgi protein Tgn38 to target the Golgi (Szentpetery *et al.*, 2010). A potential confounding factor in using targeting motifs that recognize a particular phosphoinositide(s) is that use of the CID approach may eliminate that very phosphoinositide(s) resulting in dislocalization of the targeting motif. Alternative targeting motifs that do not depend on phosphoinostide recognition should be considered, such as transmembrane proteins or membrane anchor sequences, as highlighted above.

II. Rationale

Phosphoinositides reside in the lipid bilayer and not only contribute to the chemical properties of the membrane bilayer but play an essential role in regulating protein recruitment and activity at cellular membranes. Perturbations to the levels of these lipids in the membrane can provide valuable insights into the functions of phosphoinositides in cellular processes. However, prolonged disturbance of phosphoinositide levels via standard genetic or molecular genetic approaches can provide an inaccurate picture of singular processes since it may give rise to compensatory responses either through adjustments of lipid metabolism or through the modulation of phosphoinositide-binding effector complexes in the cells' attempt to correct the initial defect. Consequently, phenotypes obtained through these prolonged manipulations of phosphoinositide levels may be indirect and difficult to interpret. Additionally, studies suggest that phosphoinositide conversion (*i.e.*, the phosphorylation/dephosphorylation of one species into another) can be "catastrophic" in nature, in that it can occur within time scales of seconds or minutes. While this concept is well-established in signal transduction areas (where the $PI(4,5)P_2$-to-$PI(3,4,5)P_3$ conversion is known to acutely turn on signaling pathways), it is less established in the area of membrane trafficking. However, recent work from a variety of groups, including our own, suggests that acute phosphoinositide conversions can govern key membrane trafficking processes (Chang-Ileto *et al.*, 2011; Rusk *et al.*, 2003; Szentpetery *et al.*, 2010; Terebiznik *et al.*, 2002). It has thus become clear in the past few years that the field was in need of novel approaches allowing for acute manipulation of phosphoinositide levels in cells. Based on the many precedents utilizing CID in various fields, this methodology became the approach of choice for the direct spatial and temporal control of phosphoinositide level manipulation (Table I).

In this chapter, we describe a variant of the CID approach to study changes in $PI(4,5)P_2$ levels specifically on endocytic membranes and their effect on membrane fission.

III. Preparation of Expression Constructs

Dimerizers can be classified as either homodimerizers or heterodimerizers. Homodimerizers are symmetric dimerizers that bind and bring together two identical ligand binding domains. There can be a single protein/protein domain of interest designed as a fusion partner to the ligand binding domain or multiple protein/protein domains. Thus, the homodimerization system can be used to oligomerize a single fusion protein or multiple ones. However, in the latter case, when the homodimerizer is added to a system of two different fusion proteins, there will be multiple complexes formed with approximately half of the complexes made up of the two different fusion proteins and a quarter of each dimer of the individual fusion proteins. Use of heterodimerizers would ensure the induction of complexes of two fusion proteins without the side product of homodimers. In this chapter, we describe the use of rapamycin and a rapamycin analog, AP21967 (Clackson, 2006), to induce heterodimerization of the ligand binding domains FK506 binding protein (FKBP) and the rapamycin binding fragment (FRB) of mTOR (Fig. 2A). The use of AP21967 presents an important control of phenotypes observed using rapamycin since this rapalog can only be recognized by an engineered FRB molecule (with a T2098L mutation) and not by endogenous FRB (Clackson, 2006). This mutant FRB molecule, however, is still able to interact with rapamycin. It is of note that this analog must be added in such high concentrations in order to induce rapid changes that its selectivity of mutant over wild type FRB may be questionable. In addition, iRAP preparations have been shown to be contaminated with rapamycin and thus the effectiveness of the analog is difficult to separate from that of rapamycin itself (Edwards and Wandless, 2007).

PI(4,5)P_2 plays a fundamental role in clathrin-mediated endocytosis. This lipid has been shown to be critical for the recruitment of key endocytic adaptor and accessory proteins to the PM and is thus important for both the initiation and the progression of endocytosis. In order to dissect out a potential role of PI(4,5)P_2 in membrane fission, we used endophilin-induced membrane tubules to model endocytic structures and more specifically bud necks and CID to induce an acute PI(4,5)P_2-to-PI4P conversion at these membrane sites and analyzed ensuing changes in membrane dynamics using fluorescence microscopy. We were interested in whether membrane fission is modulated by the interaction between the BAR domain-containing protein endophilin and the inositol 5-phosphatase synaptojanin and the PI(4,5)P_2 turnover mediated by synaptojanin. Using expression constructs originally available through Ariad Pharmaceuticals (and now Clontech), the 5-phosphatase domain of synaptojanin 1 was fused to the FKBP domain along with the fluorescent tag mRFP while endophilin, without its SH3 domain, was fused to the mutant FRB domain along with the fluorescent tag EGFP (Figs. 2B and 3A). The protein domains responsible for controlling the interaction between these two endocytic proteins (PRD for synaptojanin and SH3 for endophilin) were omitted from these constructs so that the interaction between these proteins could be precisely controlled by the addition of rapamycin or the rapalog AP21967. Overexpression of endophilin, and in this case, endophilin without its SH3 domain, in COS7 cells results in the massive tubulation of

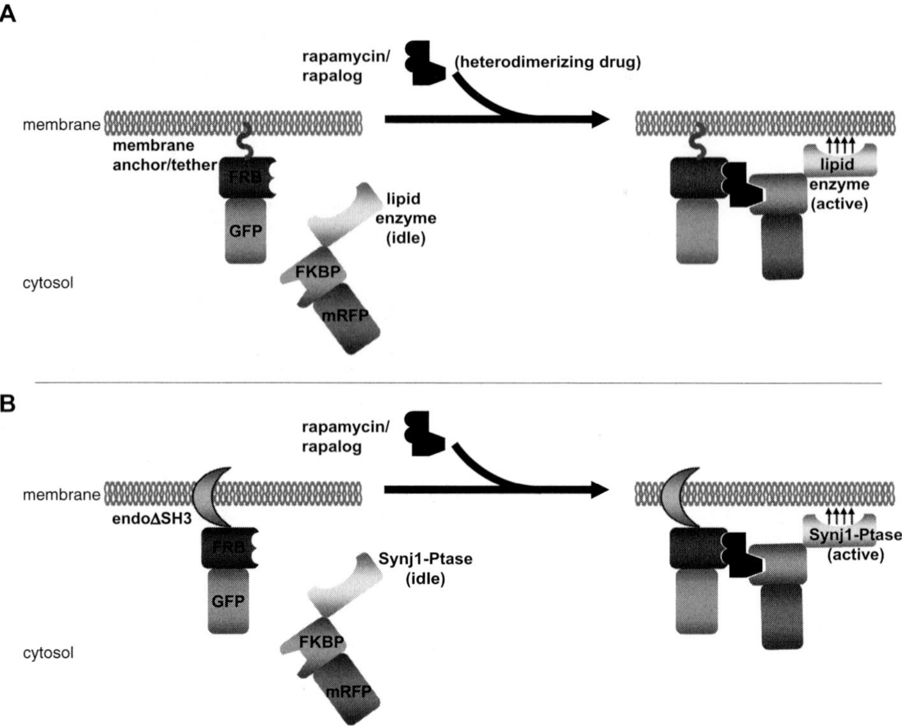

Fig. 2 Chemically induced dimerization strategy. (A) A general schematic of CID using the protein modules FRB and FKBP and heterodimerizer rapamycin or one of its chemical analogs (rapalog). The FRB module is fused to a fluorescent tag (GFP in this example) and is targeted to a membrane compartment by a membrane anchor or tether. This membrane anchor/tether can be either a peptide sequence such as a myristoylation signal or a protein/protein domain that either spans or binds to a target membrane. The FKBP module is fused to another fluorescent tag (RFP in this example) and a catalytic domain of a lipid enzyme without any targeting sequences that the full length enzyme may contain. In the pre-rapamycin state (left side), the FKBP-lipid enzyme is soluble. Upon application of rapamycin or a rapalog, this heterodimerizers bring together the FRB and FKBP domains and the FKBP chimera is recruited to the membrane at which the FRB chimera is located. This brings the lipid enzyme into close proximity of the target membrane, allowing for synthesis or turnover of the desired phosphoinositide located at the target membrane. B) The CID strategy used in our study to analyze the effects of $PI(4,5)P_2$ turnover by the endophilin-Synj1 partnership on membrane fission. In this study, the endophilin N-BAR domain was used as the membrane tether. This endophilin domain is also able to induce membrane tubulation similar to full length protein. The SH3 domain is omitted in order to avoid any recruitment of endogenous synaptojanin protein and other PRD domain-containing partners. The inositol 5-phosphatase domain from Synj1 was used to hydrolyze at $PI(4,5)P_2$ at the endophilin-induced membrane tubules upon recruitment to these membranes by the addition of rapamycin/rapalog. The Sac1 domain of Synj1 was omitted in this study in order to study only the enzyme's ability to dephosphorylate $PI(4,5)P_2$ on the 5' position since the Sac1 domain has been reported to dephosphorylate several different phosphoinositide substrates *in vitro* (Guo *et al.*, 1999). As with endophilin, the PRD domain of Synj1 was omitted in order to avoid any recruitment of Synj1 to membranes via interaction with SH3 domain-containing proteins. (For color version of this figure, the reader is referred to the web version of this book.)

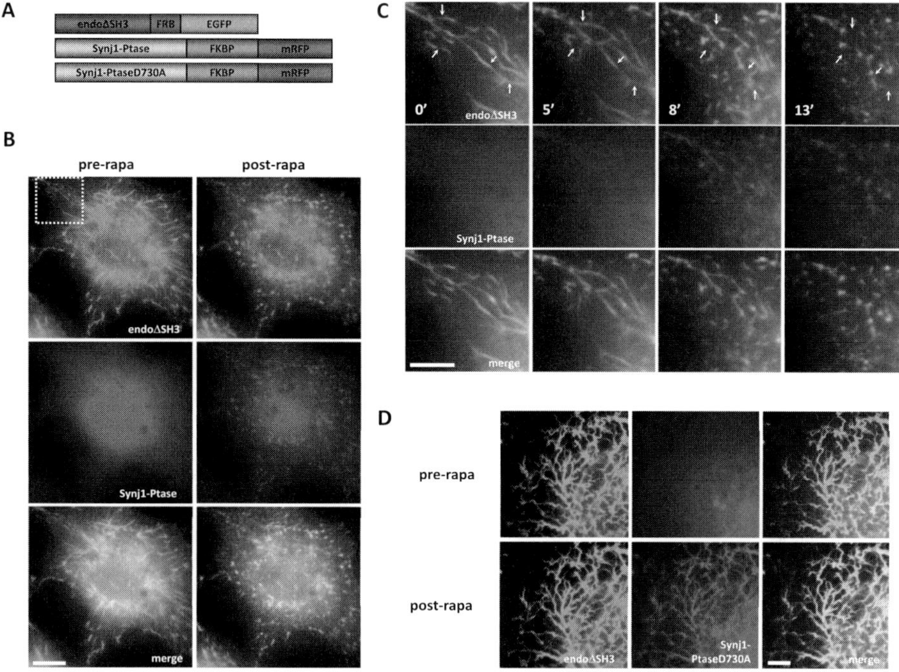

Fig. 3 Acute recruitment of the inositol 5-phosphatase domain of Synj1 to endophilin-induced tubules results in membrane fragmentation and condensation. (A) Diagram of the heterodimerization constructs: rat endophilin1 without the SH3 domain (endoΔSH3) is fused to the NH_2-terminal side of the FRB domain which is followed by EGFP; the inositol 5-phosphatase domain of Synj1, either wild type or catalytically dead mutant D730A (Synj1-Ptase, Synj1-PtaseD730A respectively), was fused to the NH_2-terminal side of two FKBP domains which is followed by mRFP. (B) In transfected COS-7 cells, Synj1-Ptase shows a diffuse signal before rapalog treatment (pre-rapa). Rapalog treatment results in Synj1-Ptase recruitment to endophilin-induced membrane tubules and fragmentation and condensation of the membrane tubules take place (post-rapa). (C) A time-lapse view of the fragmentation and condensation events upon rapalog treatment from a magnified field of the cell seen in (B) as indicated by the dotted square. Example fragmentation sites are indicated by the arrows. (D) In transfected COS-7 cells, Synj1-PtaseD730A showed diffuse signal before rapalog treatment (pre-rapa). Rapalog addition results in Synj1-PtaseD730A recruitment to endophilin-induced membrane tubules but no fragmentation/condensation events are observed (post-rapa). Scale bars represent (B) 10 μm, (C) 5 μm, and (D) 5 μm.
[Reprinted from *Developmental Cell*, Vol. 20, Chang-Ileto, B., Frere, S.G., Chan, R.B., Voronov, S.V., Roux, A., and Di Paolo, G., Synaptojanin 1-mediated PI(4,5)P_2 hydrolysis is modulated by membrane curvature and facilitates membrane fission, Pages No. 206–218, copyright 2011, with permission from Elsevier.] (See color plate.)

the PM (Fig. 3, pre-rapa state). In our system, we were able to take advantage of this phenomenon for two purposes. The first purpose was to use endophilin to localize the FRB domain to the PM where PI(4,5)P_2 is enriched and thereby using this construct to localize synaptojanin to these tubular membranes once rapamycin/AP21967 was added. The second purpose for using endophilin overexpression was to create the tubular membranes that would serve as a model for endocytic bud necks, the site at which membrane fission occurs during the endocytic process.

A. Factors to Consider When Designing Expression Constructs

1. *Membrane localization*: The final heterodimerized complex will need to be localized to the target membrane. How will this be achieved? One of the proteins of interest may have the intrinsic ability to localize to the target membrane/organelle. In our case, endophilin could be used as the targeting signal to the PM. Another possibility is the use of a localization peptide fused to either the FKBP or FRB construct, such as the myristoylation signal included in the pC_4M-F2E construct encoding the FKBP, that targets its fusion partners to the cytoplasmic face of cellular membranes.
2. *Position of fusion proteins in the expression constructs*: The organization of the fusion proteins (protein/domain of interest, FKBP/FRB, fluorescent or reporter tag) must be considered with respect to the accessibility of the different domains and potential effects on structure and/or activity of the proteins/domains. For example, if an enzyme or enzyme domain is being studied, constructs with different arrangement of the fusion domains may have to be engineered to assess whether placing another protein on its NH_2-terminus or COOH-terminus affects the catalytic activity. In our case, we replaced the SH3 domain of endophilin and the proline-rich tail of synaptojanin 1 with the FRB and FKBP domains, respectively, so as to preserve the targeting modules at the COOH terminus of these proteins.
3. *Fluorescent tags*: Which proteins will be tracked using a fluorescence tag and any fluorescence tags needed for controls must all be carefully considered and coordinated. Generally, monomeric RFP can be combined with eGFP, but CFP and YFP pairs are also possible. In addition, various other color fluorescent proteins, including blue, have been developed (Day and Davidson, 2009) and offer even more possibilities for color pairing of proteins.
4. *Controls and probes*: For every lipid enzymes utilized in the CID approach, it is essential to use a catalytically inactive counterpart in order to make sure the observed effects are dependent on the catalytic activity. Generally, this is achieved by generating point mutant versions of the wild-type construct. Such mutants are often reported in the literature for most lipid enzymes, and if this is not the case, appropriate enzymatic assays should be conducted to verify that the mutations abolish the enzymatic activity. Aside from following the protein(s) of interest, the phosphoinositide and/or membrane should also be monitored as controls for the experiment. For example, we tracked the presence or disappearance of $PI(4,5)P_2$ from the PM using the PH domain from the enzyme PLCδ1. Membrane markers used for the plasma include exogenously added styryl dyes

(*e.g.*, FM4-64, FM1-43) and fluorophore labeled-wheat germ agglutinin (WGA). Additionally, fluid phase tracers, such as fluorophore labeled dextrans could be used to label the tubules, although they are not membrane probes *per se*, unlike styryl dyes and WGA.

IV. Expression of Fusion Proteins and Cell Maintenance

A. Expression of Fusion Proteins

When more than one construct is used to encode for various recombinant proteins, issues of stoichiometry between the various proteins may arise. For example, in initial co-transfections of our endophilin and synaptojanin contructs at equal concentrations, we did not obtain membrane tubulation. By varying the ratio of endophilin and synaptojanin DNA constructs used for transfections, we were able to find an acceptable range of DNA used that resulted in membrane tubulation and detectable levels of synaptojanin. Specifically, we lowered the amount of synaptojanin construct used alongside the endophilin construct until we were able to produce the membrane tubulation. Presumably the cause of the lack of membrane tubulation in the initial experiments was due to the high concentration of synaptojanin that could either catabolize a significant population of the $PI(4,5)P_2$ molecules so that endophilin no longer had a high affinity for the PM or the excess synaptojanin was able to hydrolyze enough $PI(4,5)P_{2,}$ so that the fission process in many cases probably already occurred.

B. Cells to Image Membrane Dynamics

COS7 cells are commonly used for the analysis of membrane tubulation caused by BAR domain-containing proteins (Gallop *et al*., 2006; Itoh *et al*., 2005; Peter *et al*., 2004). These cells have a relatively flat morphology that allows for the visualization of membrane tubulation upon BAR-protein overexpression since a large number of tubular structures derived from the PM are in the same plane, particularly in the peripheral regions of the cell. Visualization of membrane tubules in other cell types, such as CHO and 293, was difficult if not unsuccessful.

C. Cell Culture Conditions

COS-7 cells were grown at 37 °C at 5% CO_2. All media preparations, cell passaging, and cell plating were performed in a sterile laminar flow hood. Media was stored at 4 °C in the dark. COS-7 cells were maintained in DMEM Medium supplemented with 10% fetal bovine serum, 1% penicillin/streptomycin, and 1% Glutamine or GlutaMAX. Care was taken while maintaining COS-7 cells to prevent overcrowding, which often led to decreased post-transfection cell imaging quality and survival.

Glass coverslips (Warner Instruments) used for growing transfected COS-7 cells were stored in EtOH. Prior to transfection, the coverslips were removed from the EtOH solution and excess alcohol was removed by passing each coverslip quickly through a flame. The coverslips were placed in a tissue culture dish and allowed to cool. COS-7 cells were transfected using a modified reverse transfection technique (Invitrogen) in which the cells were transfected and plated onto the glass coverslips at the same time. COS-7 cells were grown on the glass coverslips for 14–21 h post-transfection. Time-lapse recordings were performed within this 14–21 h period since the tubules become very stable and are modestly shortened by the recruitment of Synj1 and the hydrolysis of $PI(4,5)P_2$ after this period.

V. Microscopy

A. Choice of Imaging Chamber and Conditions

For live imaging, transfected coverslips were placed in a Chamlide recording chamber (Life Cell Instruments, South Korea) filled with 300 μL HBS solution (10 mM HEPES [pH 7.4], 136 mM NaCl, 2.5 mM KCl, 2 mM $CaCl_2$, 1.3 mM $MgCl_2$, 10 mM glucose), and imaging was performed at 37 °C using the Olympus IX-81 microscope equipped with a heating incubator (MIU-IBC-IF), connected to a Hamamatsu CCD camera (Hamamatsu Photonics) and mounted with a 60x oil-immersed UIS2 objective (Olympus).

B. Application of CIDs

Cells were treated with either 5 μM rapalog (200X stock in EtOH, as 200X) or 100 nM rapamycin (200X stock in DMSO) (Spencer *et al.*, 1993). 1.5 μL of the rapamycin/rapalog was added directly to the HBS solution overlaid on the coverslip while in the recording chamber.

C. Imaging

Images were collected using Slidebook 5.0 software (Olympus). For FM 4-64 (Invitrogen) staining, cells were incubated with the dye in HBS. See Fig. 3 for examples of the membrane tubules in cells before and after rapamycin/AP21967 application with either WT Synj1-Ptase or the catalytically inactive Synj1-PtaseD730A.

D. Analysis

Tubule length before and after rapamycin or rapalog treatment was measured using the ImageJ software (NIH, USA). For each cell recorded, the initial length of tubules before the application of rapamycin/rapalog was measured by drawing a segmented thin line along the tubules (Li). The residual length of the same tubules was measured similarly from the image obtained 2 min after the application of rapamycin (Lr) (Fig. 4A, B). The lengths of the tubules before (SLi) and after (SLr) were summated and the difference (SLi-SLr) was normalized to the total lengths of the tubules before rapamycin treatment (SLi) (Fig. 4C). Tubule length

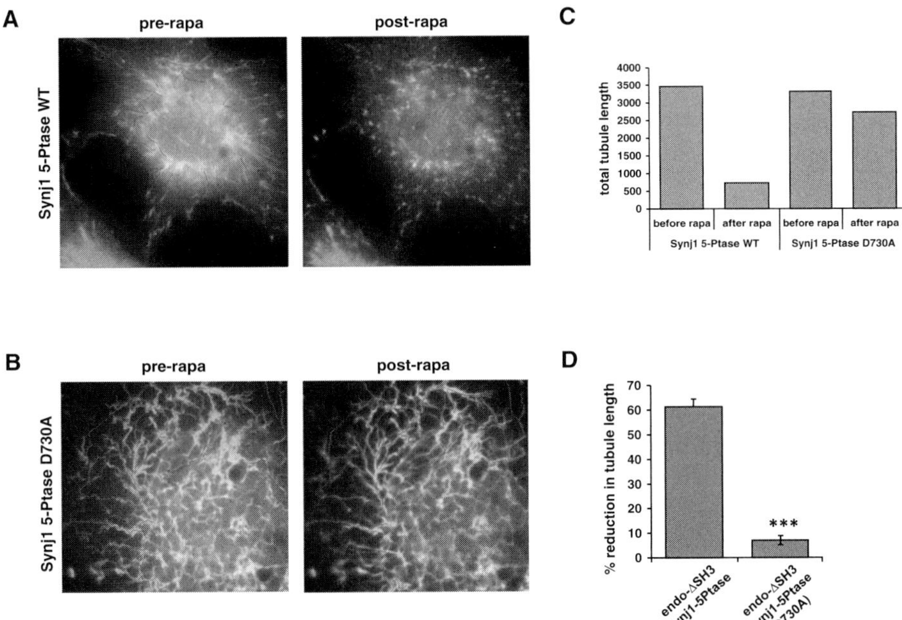

Fig. 4 Quantification of membrane fission and condensation. (A) A COS-7 cell expressing endoΔSH3-FRB-GFP and Synj1-Ptase-FKBP-RFP is shown before and after rapamycin treatment. Using ImageJ, lines were drawn to follow the endophiilin-induced tubules before application of rapamycin. The evolution of these tubules was then followed after rapamycin treatment to select the broken tubules resulting from the recruitment of Synj1-Ptase-FKBP-RFP to these tubules. The final picture was used to draw the resulting broken tubules. (B) The same method was used to draw tubules for a COS-7 cell expressing endoΔSH3-FRB-GFP and Synj1-PtaseD730A-FKBP-RFP before and after rapamycin treatment. (C) Quantification of tubule length of the cells shown in (A) and (B) using ImageJ shows the dramatic decrease in total tubule length in the cell expressing wild type Synj1-Ptase after rapamycin treatment in comparison to the small decrease experienced by the cell expressing the catalytically inactive Synj1-PtaseD730A. (D) Comparison of the % reduction in tubule length in cells expressing either WT or catalytically inactive Synj1-Ptase with a $n = 24$ for WT or $n = 9$ for D730A. Data are represented as mean ± SEM. ***$p < 0.0001$.
[Panel E is reprinted from *Developmental Cell*, Vol. 20, Chang-Ileto, B., Frere, S.G., Chan, R.B., Voronov, S.V., Roux, A., and Di Paolo, G., Synaptojanin 1-mediated PI(4,5)P_2 hydrolysis is modulated by membrane curvature and facilitates membrane fission, Pages No. 206–218, copyright 2011, with permission from Elsevier.] (See color plate.)

inversely correlated with the extent of fission (Fig. 4D), although a more accurate approach to quantify tubular fission would involve a scoring of the breakpoints per unit of tubule length. This was difficult to achieve because of the intricate network of tubules obtained in these experiments, but would be easier to achieve in a system with low tubule density (for instance, in a cell free assay). We also observed that newly formed and short tubules did not tend to break but had a greater propensity to fully condense.

VI. Considerations

There are several important considerations when using the CID approach to study the role of a phosphoinositide. Above all, the conversion of a phosphoinositide into another species as a result of CID has two consequences: (i) a reduction in the levels of the substrate; and (ii) an increase in the levels of the product. While these reactions can be monitored using genetically encoded probes to both the substrate and the product, the biological effects can be potentially caused by either phenomenon or even a combination of both. We thus recommend that whenever possible, alternate manipulations may be conducted to target the same lipid. For instance, PI(4,5)P_2 may be dephosphorylated by 5-phosphatases or cleaved by PLC into diacylglycerol and inositol-trisphosphate. If both types of hydrolysis produce the same biological effect, the latter is most likely caused by the loss of PI(4,5)P_2 (rather than the generation of PI(4,5)P_2 metabolites).

Another important point to consider is how to follow the affected membrane or target phosphoinositide without perturbing the system being studied. Protein domains that recognize phosphoinositides have been widely used as reporters for their target lipids (Lemmon, 2008; Varnai and Balla, 2007). However, the binding of these domains to phosphoinositides may affect the availability of these lipids as substrates for the lipid enzymes or the functions being studied. Indeed, over-expression of the PH domain of PLCδ1, commonly used to visualize the localization of PI(4,5)P_2 in cells, has been reported to affect cellular functions (Varnai *et al.*, 2005). In our hands, over-expression of PLCδ1-PH fused to a fluorescent tag alongside our endophilin constructs resulted in blockade of endophilin-induced tubule formation. Diluting the concentration of PLCδ1-PH used in transfection remedied this effect and allowed the visualization of PI(4,5)P_2-containing membrane and formation of membrane tubules. In addition, fluorescently-labeled membrane and organelle markers, such as FM 4-64 and wheat germ agglutinin for the PM, may also distort experimental results by masking the lipid substrate or protein(s) essential in the cellular process being studied, or may possibly make the membrane too rigid or more stable for fission.

Finally, the FRB protein module has been reported to be subject to a high rate of turnover along with any accompanying fusion partners (Edwards and Wandless, 2007; Stankunas *et al.*, 2003). This protein instability is reversed upon the addition of ligand (*i.e.*, rapamycin or one of its analogs). These features offer the disadvantage of potentially low levels of protein in the target cell particularly if the fusion-FRB chimera is particularly unstable. At the same time, this feature can also be seen as an advantage in that actions of the FRB fusion partner have a lower chance of confounding results until the heterodimerizer is added to the system.

VII. Summary and Conclusion

The use of CID in studying the roles that phosphoinositides play in cellular functions offers many advantages over the existing conventional approaches of exogenous application of lipids, genetic manipulation and use of pharmacological inhibitors. CID allows for the targeted synthesis or elimination of lipids in a spatially- and temporally-controllable manner. Whereas application of lipids to cells and over-expression of lipid enzymes in cells lead to gross or prolonged changes in a cell, CID offers a more sophisticated approach to controlling phosphoinositide levels to reveal their roles in cellular physiology. This method is amenable to a broad variety of lipid enzymes and lipid modifications (*i.e.*, by such enzymes as phospholipases and phosphatidic acid phosphatases) as long as the catalytic domains/regions of these lipid enzymes are not already embedded in the membrane (as is the case for lipid phosphate phosphatases (Sigal *et al.*, 2005)). This techniques is also amenable to all membranous organelles in the cell provided targeting domains are available and lipid substrates are present on these membranes. In addition, this method can be used to deplete the cytosol of soluble lipid enzymes by artificially mistargeting them to cellular compartments where the target substrate are not present in systems operating in a knockdown or null background of the lipid enzyme. As can be seen in phosphoinositide studies using the CID technology, much progress can be gained in the understanding the role of phosphoinositides in the cell. However, there is still a necessity for the identification of pharmacological activators and inhibitors of lipid enzymes and better probes for some phosphoinositides in order to gain even greater strides in our knowledge of the role of phosphoinositides and their lipid enzymes.

Acknowledgments

We thank Robin B. Chan and Aurélien Roux for their work in our original research manuscript (Chang-Ileto *et al.*, 2011). We are also grateful for Dr. Chan for his critical reading of this manuscript. Work on phosphoinositides in the Di Paolo lab is funded by NIH grants R01 NS056049 and R01 HD05547 and by the McKnight Endowment Fund. Belle Chang-Ileto is funded by the NIH (F31 NS058096).

References

Abe, N., *et al.* (2008). Dissecting the role of PtdIns(4,5)P2 in endocytosis and recycling of the transferrin receptor. *J. Cell Sci.* **121**, 1488–1494.

Balla, T., *et al.* (2009). Phosphoinositide signaling: new tools and insights. *Physiology (Bethesda)* **24**, 231–244.

Banaszynski, L. A., *et al.* (2006). A rapid, reversible, and tunable method to regulate protein function in living cells using synthetic small molecules. *Cell* **126**, 995–1004.

Bayle, J. H., *et al.* (2006). Rapamycin analogs with differential binding specificity permit orthogonal control of protein activity. *Chem. Biol.* **13**, 99–107.

Behnia, R., and Munro, S. (2005). Organelle identity and the signposts for membrane traffic. *Nature* **438**, 597–604.

Belshaw, P. J., *et al.* (1996). Controlling protein association and subcellular localization with a synthetic ligand that induces heterodimerization of proteins. *Proc. Natl. Acad. Sci. U. S. A.* **93**, 4604–4607.

Chang-Ileto, B., *et al.* (2011). Synaptojanin 1-mediated PI(4,5)P2 hydrolysis is modulated by membrane curvature and facilitates membrane fission. *Dev. Cell* **20**, 206–218.

Chen, J., *et al.* (1995). Identification of an 11-kDa FKBP12-rapamycin-binding domain within the 289-kDa FKBP12-rapamycin-associated protein and characterization of a critical serine residue. *Proc. Natl. Acad. Sci. U. S. A.* **92**, 4947–4951.

Choi, J. W., *et al.* (1996). Structure of the FKBP12-rapamycin complex interacting with the binding domain of human FRAP. *Science* **273**, 239–242.

Chong, H., *et al.* (2002). A system for small-molecule control of conditionally replication-competent adenoviral vectors. *Mol. Ther.* **5**, 195–203.

Choudhury, R., *et al.* (2005). Lowe syndrome protein OCRL1 interacts with clathrin and regulates protein trafficking between endosomes and the trans-Golgi network. *Mol. Biol. Cell* **16**, 3467–3479.

Clackson, T. (2006). Dissecting the functions of proteins and pathways using chemically induced dimerization. *Chem. Biol. Drug Des.* **67**, 440–442.

Clackson, T., *et al.* (1998). Redesigning an FKBP-ligand interface to generate chemical dimerizers with novel specificity. *Proc. Natl. Acad. Sci. U. S. A.* **95**, 10437–10442.

Cremona, O., *et al.* (1999). Essential role of phosphoinositide metabolism in synaptic vesicle recycling. *Cell* **99**, 179–188.

Csordas, G., *et al.* (2010). Imaging interorganelle contacts and local calcium dynamics at the ER-mitochondrial interface. *Mol. Cell* **39**, 121–132.

Day, R. N., and Davidson, M. W. (2009). The fluorescent protein palette: tools for cellular imaging. *Chem. Soc. Rev.* **38**, 2887–2921.

Di Cristofano, A., *et al.* (1998). Pten is essential for embryonic development and tumour suppression. *Nat. Genet.* **19**, 348–355.

Di Paolo, G., and De Camilli, P. (2006). Phosphoinositides in cell regulation and membrane dynamics. *Nature* **443**, 651–657.

Di Paolo, G., *et al.* (2004). Impaired PtdIns(4,5)P2 synthesis in nerve terminals produces defects in synaptic vesicle trafficking. *Nature* **431**, 415–422.

Downing, G. J., *et al.* (1996). Characterization of a soluble adrenal phosphatidylinositol 4-kinase reveals wortmannin sensitivity of type III phosphatidylinositol kinases. *Biochemistry* **35**, 3587–3594.

Edwards, S. R., and Wandless, T. J. (2007). The rapamycin-binding domain of the protein kinase mammalian target of rapamycin is a destabilizing domain. *J. Biol. Chem.* **282**, 13395–13401.

Falkenburger, B. H., *et al.* (2010). Phosphoinositides: lipid regulators of membrane proteins. *J. Physiol.* **588**, 3179–3185.

Fili, N., *et al.* (2006). Compartmental signal modulation: Endosomal phosphatidylinositol 3-phosphate controls endosome morphology and selective cargo sorting. *Proc. Natl. Acad. Sci. U. S. A.* **103**, 15473–15478.

Gallop, J. L., *et al.* (2006). Mechanism of endophilin N-BAR domain-mediated membrane curvature. *EMBO J.* **25**, 2898–2910.

Garcia, P., *et al.* (1995). The pleckstrin homology domain of phospholipase C-delta1 binds with high affinity to phosphatidylinositol 4,5-bisphosphate in bilayer membranes. *Biochemistry* **34**, 16228–16234.

Gary, J. D., *et al.* (1998). Fab1p is essential for PtdIns(3)P 5-kinase activity and the maintenance of vacuolar size and membrane homeostasis. *J. Cell Biol.* **143**, 65–79.

Gillooly, D. J., *et al.* (2001). Cellular functions of phosphatidylinositol 3-phosphate and FYVE domain proteins. *Biochem. J.* **355**, 249–258.

Godi, A., *et al.* (2004). FAPPs control Golgi-to-cell-surface membrane traffic by binding to ARF and PtdIns(4)P. *Nat. Cell Biol.* **6**, 393–404.

Golebiewska, U., *et al.* (2008). Diffusion coefficient of fluorescent phosphatidylinositol 4,5-bisphosphate in the plasma membrane of cells. *Mol. Biol. Cell* **19**, 1663–1669.

Guo, S., *et al.* (1999). SAC1-like domains of yeast SAC1, INP52, and INP53 and of human synaptojanin encode polyphosphoinositide phosphatases. *J. Biol. Chem.* **274**, 12990–12995.

Harris, T. W., *et al.* (2000). Mutations in synaptojanin disrupt synaptic vesicle recycling. *J. Cell Biol.* **150**, 589–600.

Haucke, V. (2005). Phosphoinositide regulation of clathrin-mediated endocytosis. *Biochem. Soc. Trans.* **33**, 1285–1289.

Heo, W. D., *et al.* (2006). PI(3,4,5)P3 and PI(4,5)P2 lipids target proteins with polybasic clusters to the plasma membrane. *Science* **314**, 1458–1461.

Hilgemann, D. W., *et al.* (2001). The complex and intriguing lives of PIP2 with ion channels and transporters. *Sci. STKE* **2001**, RE19.

Huang, W., *et al.* (2007). Stabilized phosphatidylinositol-5-phosphate analogues as ligands for the nuclear protein ING2: chemistry, biology, and molecular modeling. *J. Am. Chem. Soc.* **129**, 6498–6506.

Inoue, T., *et al.* (2005). An inducible translocation strategy to rapidly activate and inhibit small GTPase signaling pathways. *Nat. Methods* **2**, 415–418.

Itoh, T., *et al.* (2005). Dynamin and the actin cytoskeleton cooperatively regulate plasma membrane invagination by BAR and F-BAR proteins. *Dev. Cell* **9**, 791–804.

Kahlfeldt, N., *et al.* (2010). Molecular basis for association of PIPKI gamma-p90 with clathrin adaptor AP-2. *J. Biol. Chem.* **285**, 2734–2749.

Kim, S., *et al.* (2006). Regulation of transferrin recycling kinetics by PtdIns[4,5]P2 availability. *Faseb J.* **20**, 2399–2401.

Knight, Z. A., *et al.* (2006). A pharmacological map of the PI3-K family defines a role for p110alpha in insulin signaling. *Cell* **125**, 733–747.

Komatsu, T., *et al.* (2010). Organelle-specific, rapid induction of molecular activities and membrane tethering. *Nat. Methods* **7**, 206–208.

Korzeniowski, M. K., *et al.* (2010). Activation of STIM1-Orai1 involves an intramolecular switching mechanism. *Sci. Signal* **3**, ra82.

Krauss, M., *et al.* (2003). ARF6 stimulates clathrin/AP-2 recruitment to synaptic membranes by activating phosphatidylinositol phosphate kinase type Igamma. *J. Cell Biol.* **162**, 113–124.

Lemmon, M. A. (2008). Membrane recognition by phospholipid-binding domains. *Nat. Rev. Mol. Cell Biol.* **9**, 99–111.

Lemmon, M. A., *et al.* (1995). Specific and high-affinity binding of inositol phosphates to an isolated pleckstrin homology domain. *Proc. Natl. Acad. Sci. U. S. A.* **92**, 10472–10476.

Liberles, S. D., *et al.* (1997). Inducible gene expression and protein translocation using nontoxic ligands identified by a mammalian three-hybrid screen. *Proc. Natl. Acad. Sci. U. S. A.* **94**, 7825–7830.

Liu, J., *et al.* (1991). Calcineurin is a common target of cyclophilin-cyclosporin A and FKBP-FK506 complexes. *Cell* **66**, 807–815.

Logan, M. R., and Mandato, C. A. (2006). Regulation of the actin cytoskeleton by PIP2 in cytokinesis. *Biol. Cell* **98**, 377–388.

Mao, Y. S., and Yin, H. L. (2007). Regulation of the actin cytoskeleton by phosphatidylinositol 4-phosphate 5 kinases. *Pflugers Arch.* **455**, 5–18.

Nakanishi, S., *et al.* (1995). A wortmannin-sensitive phosphatidylinositol 4-kinase that regulates hormone-sensitive pools of inositolphospholipids. *Proc. Natl. Acad. Sci. U. S. A.* **92**, 5317–5321.

Nakatsu, F., *et al.* (2010). The inositol 5-phosphatase SHIP2 regulates endocytic clathrin-coated pit dynamics. *J. Cell Biol.* **190**, 307–315.

Ozaki, S., *et al.* (2000). Intracellular delivery of phosphoinositides and inositol phosphates using polyamine carriers. *Proc. Natl. Acad. Sci. U. S. A.* **97**, 11286–11291.

Peter, B. J., *et al.* (2004). BAR domains as sensors of membrane curvature: the amphiphysin BAR structure. *Science* **303**, 495–499.

Pollard, T. D., and Borisy, G. G. (2003). Cellular motility driven by assembly and disassembly of actin filaments. *Cell* **112**, 453–465.

Prasad, N. K., and Decker, S. J. (2005). SH2-containing 5'-inositol phosphatase, SHIP2, regulates cytoskeleton organization and ligand-dependent down-regulation of the epidermal growth factor receptor. *J. Biol. Chem.* **280**, 13129–13136.

Rohatgi, R., *et al.* (2000). Mechanism of N-WASP activation by CDC42 and phosphatidylinositol 4, 5-bisphosphate. *J. Cell Biol.* **150**, 1299–1310.

Rollins, C. T., *et al.* (2000). A ligand-reversible dimerization system for controlling protein-protein interactions. *Proc. Natl. Acad. Sci. U. S. A.* **97**, 7096–7101.

Rusk, N., *et al.* (2003). Synaptojanin 2 functions at an early step of clathrin-mediated endocytosis. *Curr. Biol.* **13**, 659–663.

Schu, P. V., *et al.* (1993). Phosphatidylinositol 3-kinase encoded by yeast VPS34 gene essential for protein sorting. *Science* **260**, 88–91.

Sigal, Y. J., *et al.* (2005). Integral membrane lipid phosphatases/phosphotransferases: common structure and diverse functions. *Biochem. J.* **387**, 281–293.

Sorensen, S. D., *et al.* (1998). A role for a wortmannin-sensitive phosphatidylinositol-4-kinase in the endocytosis of muscarinic cholinergic receptors. *Mol. Pharmacol.* **53**, 827–836.

Spencer, D. M., *et al.* (1993). Controlling signal transduction with synthetic ligands. *Science* **262**, 1019–1024.

Stambolic, V., *et al.* (1998). Negative regulation of PKB/Akt-dependent cell survival by the tumor suppressor PTEN. *Cell* **95**, 29–39.

Stankunas, K., *et al.* (2003). Conditional protein alleles using knockin mice and a chemical inducer of dimerization. *Mol. Cell* **12**, 1615–1624.

Stankunas, K., *et al.* (2007). Rescue of degradation-prone mutants of the FK506-rapamycin binding (FRB) protein with chemical ligands. *Chembiochem* **8**, 1162–1169.

Subramanian, D., *et al.* (2010). Activation of membrane-permeant caged PtdIns(3)P induces endosomal fusion in cells. *Nat. Chem. Biol.* **6**, 324–326.

Suh, B. C., and Hille, B. (2005). Regulation of ion channels by phosphatidylinositol 4,5-bisphosphate. *Curr. Opin. Neurobiol.* **15**, 370–378.

Suh, B. C., and Hille, B. (2008). PIP2 is a necessary cofactor for ion channel function: how and why? *Annu. Rev. Biophys.* **37**, 175–195.

Suh, B. C., *et al.* (2006). Rapid chemically induced changes of PtdIns(4,5)P2 gate KCNQ ion channels. *Science* **314**, 1454–1457.

Szentpetery, Z., *et al.* (2010). Acute manipulation of Golgi phosphoinositides to assess their importance in cellular trafficking and signaling. *Proc. Natl. Acad. Sci. U. S. A.* **107**, 8225–8230.

Terebiznik, M. R., *et al.* (2002). Elimination of host cell PtdIns(4,5)P(2) by bacterial SigD promotes membrane fission during invasion by Salmonella. *Nat. Cell Biol.* **4**, 766–773.

Varnai, P., and Balla, T. (1998). Visualization of phosphoinositides that bind pleckstrin homology domains: calcium- and agonist-induced dynamic changes and relationship to myo-[3H]inositol-labeled phosphoinositide pools. *J. Cell Biol.* **143**, 501–510.

Varnai, P., and Balla, T. (2007). Visualization and manipulation of phosphoinositide dynamics in live cells using engineered protein domains. *Pflugers Arch.* **455**, 69–82.

Varnai, P., *et al.* (2005). Selective cellular effects of overexpressed pleckstrin-homology domains that recognize PtdIns(3,4,5)P3 suggest their interaction with protein binding partners. *J. Cell Sci.* **118**, 4879–4888.

Varnai, P., *et al.* (2006). Rapidly inducible changes in phosphatidylinositol 4,5-bisphosphate levels influence multiple regulatory functions of the lipid in intact living cells. *J. Cell Biol.* **175**, 377–382.

Varnai, P., *et al.* (2007). Visualization and manipulation of plasma membrane-endoplasmic reticulum contact sites indicates the presence of additional molecular components within the STIM1-Orai1 Complex. *J. Biol. Chem.* **282**, 29678–29690.

Verstreken, P., *et al.* (2003). Synaptojanin is recruited by endophilin to promote synaptic vesicle uncoating. *Neuron* **40**, 733–748.

Wang, Y. J., *et al.* (2004). Critical role of PIP5KI{gamma}87 in InsP3-mediated Ca(2+) signaling. *J. Cell Biol.* **167**, 1005–1010.

Wiedemann, C., *et al.* (1996). Chromaffin granule-associated phosphatidylinositol 4-kinase activity is required for stimulated secretion. *EMBO J.* **15**, 2094–2101.

Xu, Y., *et al.* (2006). Chemical synthesis and molecular recognition of phosphatase-resistant analogues of phosphatidylinositol-3-phosphate. *J. Am. Chem. Soc.* **128**, 885–897.

Yin, H. L., and Janmey, P. A. (2003). Phosphoinositide regulation of the actin cytoskeleton. *Annu. Rev. Physiol.* **65**, 761–789.

Zhang, H., *et al.* (2006a). Synthesis and biological activity of PTEN-resistant analogues of phosphatidylinositol 3,4,5-trisphosphate. *J. Am. Chem. Soc.* **128**, 16464–16465.

Zhang, H., *et al.* (2006b). Synthesis and biological activity of phospholipase C-resistant analogues of phosphatidylinositol 4,5-bisphosphate. *J. Am. Chem. Soc.* **128**, 5642–5643.

Zhou, X., *et al.* (2010). Deletion of PIK3C3/Vps34 in sensory neurons causes rapid neurodegeneration by disrupting the endosomal but not the autophagic pathway. *Proc. Natl. Acad. Sci. U. S. A.* **107**, 9424–9429.

Zoncu, R., *et al.* (2007). Loss of endocytic clathrin-coated pits upon acute depletion of phosphatidylinositol 4,5-bisphosphate. *Proc. Natl. Acad. Sci. U. S. A.* **104**, 3793–3798.

CHAPTER 11

Regulation of Phosphoinositide-Metabolizing Enzymes by Clathrin Coat Proteins

Marnix Wieffer*, Volker Haucke*,† and Michael Krauß*

*Institute of Chemistry and Biochemistry, Freie Universität Berlin, Takustraße, Berlin, Germany

†Leibniz-Institut für Molekulare Pharmakologie (FMP), Robert-Rössle-Straße, Berlin, Germany

Abstract

Clathrin plays key roles in endocytic and endo-lysosomal membrane dynamics by facilitating the formation of coated vesicles at the plasma membrane and at the trans-Golgi network (TGN)/endosomal boundary. Assembly of the clathrin lattice

Copyright 2012, Elsevier Inc. All rights reserved.

0091-679X/10 $35.00
DOI 10.1016/B978-0-12-386487-1.00011-0

critically depends on adaptor proteins and accessory proteins, which connect the clathrin scaffold to the membrane and to transmembrane cargo including receptors, transporters, channels, and SNARE proteins. The recruitment of adaptor proteins to membrane surfaces is triggered by coincidence-detection mechanisms involving phosphoinositides (PIs), cargo proteins, and in many cases small GTPases. To tightly regulate coat formation, there is extensive cross-talk between PI-metabolizing enzymes and adaptor proteins. One of the best studied examples is the endocytic clathrin adaptor complex AP-2, which binds plasma membrane-enriched $PI(4,5)P_2$. In neurons, $PI(4,5)P_2$ is synthesized from PI(4)P primarily by the γ-isoform of the type I phosphatidylinositol 4-phosphate 5-kinase family (PIPKIγ), whose enzymatic activity is regulated by direct binding to, amongst others, the small GTPase Arf6 and AP-2. Cargo-bound AP-2 potently stimulates PIPK1γ activity and thereby drives AP-2-membrane interactions. This feed-forward loop is thought to facilitate membrane translocation of additional AP-2 molecules and concomitantly clathrin, but also of endocytic accessory proteins, many of which directly associate with $PI(4,5)P_2$. It is likely that similar mechanisms support the formation of coated vesicles at the trans-Golgi network (TGN) and on endosomes, involving PI(4)P and PI(3)P respectively, but detailed knowledge is lacking to date. To explore how coat proteins regulate kinase activity, assays are needed to sensitively detect subtle changes in PI synthesis and to discriminate between the various PI species. Here we describe a sensitive and specific radioactivity-based assay to measure PI kinase activity.

I. Introduction

A. Phosphoinositides Regulate Intracellular Membrane Traffic

Intracellular membrane traffic depends on transport vesicles and tubules, which shuttle proteins and lipids between compartments. The formation of such transport carriers requires the assembly of proteinaceous coats that couple cargo selection and recruitment of accessory proteins to the generation of membrane curvature (Bonifacino and Glick, 2004; McMahon and Mills, 2004). To ensure proper sorting of cargo proteins (e.g., signaling receptors, channels, transporters, cell adhesion molecules, SNAREs, lysosomal enzymes, etc.), coat formation is tightly regulated in space and time via the collaborative activity of multiple factors, including small GTPases and PIs (Di Paolo and De Camilli, 2006).

PIs represent a subclass of phospholipids, which can be reversibly phosphorylated at the 3-, 4-, and 5-positions of their inositol headgroup, giving rise to seven different PI subspecies. These display distinct charge and space filling properties, and are thus able to recruit select subsets of proteins containing PI-selective lipid-binding domains, including adaptor proteins that form part of vesicle coats. Phosphorylation and dephosphorylation of PIs is mediated by substrate-specific PI kinases and phosphatases, respectively, which display defined subcellular localizations. As a consequence, PI-metabolizing enzymes and their products are considered part of a membrane subcompartment identity code (Behnia and Munro, 2005; Krauss and Haucke, 2007). Indeed, PI(4)P is predominantly found at the TGN, and perhaps within recycling endosomes, whereas early endosomes are enriched in PI(3)P. The predominant PI formed at the plasma membrane is PI(4,5)P_2, hereafter referred to as PIP_2.

Importantly, PIs are not evenly distributed across intracellular membrane surfaces, but rather synthesized *de novo* within membrane subdomains of distinct functionalities, for example coat-dependent membrane deformation. Budding of transport carriers leads to the rapid degradation of PIs, causing vesicle coats to dissociate to allow for subsequent fusion with the appropriate target organelle. Hence, the recruitment and activation of PI-metabolizing enzymes has to be tightly regulated in space and time. Such spatiotemporal control involves PI synthesis through PI kinases at sites, where adaptors of vesicle coats recognize cargo, and the activation of PI-phosphatases on highly curved budding vesicles during membrane fission or uncoating (Chang-Ileto *et al.*, 2011).

B. Clathrin-mediated Endocytosis Depends on PIP_2

Many proteins involved in clathrin-mediated endocytosis (CME) harbor defined PIP_2 recognition domains or more cryptic patches of positively charged amino acids. The endocytic clathrin adaptor complex AP-2, which serves as a central hub within a complex interaction network (Wieffer *et al.*, 2009), carries multiple basic residues in three of its four subunits (Gaidarov and Keen, 1999; Jackson *et al.*, 2010; Rohde *et al.*, 2002). Initial recruitment to the plasma membrane likely involves binding of AP-2α to PIP_2 (Gaidarov and Keen, 1999; Rohde *et al.*, 2002). Phosphorylation of AP-2 μ, perhaps in conjunction with GTPases, then enables a large-scale conformational change of AP-2, during which the C-terminal domain of AP-2 μ relocates to an orthogonal face of the complex, simultaneously unblocking its cargo-binding sites. This structural reorganization results in a coplanar arrangement of its membrane interaction surfaces that enables four PIP_2 molecules to bind to AP-2 simultaneously (one to α, two to $\mu 2$, and one to $\beta 2$) along with endocytic motif-containing cargo. Thus, endocytic cargo binding of AP-2 is driven by its interaction with PIP_2-containing membranes (Jackson *et al.*, 2010).

Likewise, membrane recruitment of several other endocytic proteins including CALM/AP180, epsin, Dab2, ARH, and dynamin is also regulated by local PIP_2 synthesis (Haucke, 2005). Accordingly, acute or chronic depletion of plasmalemmal PIP_2 abrogates CME by abolishing proper targeting of the endocytic machinery (Krauss *et al.*, 2003; Varnai *et al.*, 2006; Zoncu *et al.*, 2007).

The transient nature of CME requires strong spatio-temporal enzymatic regulation. We therefore previously envisaged that the formation of AP-2/clathrin coats at the plasma membrane should be accompanied by the concomitant recruitment and/ or activation of PIPKI-isozymes. In the course of our studies, we and others identified AP-2 as a binding partner of PIPKIγ (Kahlfeldt *et al.*, 2010; Krauss *et al.*, 2006; Nakano-Kobayashi *et al.*, 2007; Thieman *et al.*, 2009), the major PIP_2-synthesizing enzyme at neuronal synapses (Wenk *et al.*, 2001).

Both, AP-2 μ- and β-subunits are involved in complex formation, and several interaction sites within PIPKIγ have been mapped including the kinase core domain and short peptides reminiscent of sorting signals within its C-terminal tail domain (Kahlfeldt *et al.*, 2010). Binding of tyrosine motif-containing cargo or of the PIPKIγ tail domain to AP-2 μ potently stimulates kinase activity, thereby facilitating the generation of local pools of PIP_2 at sites of endocytic vesicle formation. Such feed-forward loops might be used to create metastable membrane sites for the recruitment of PIP_2-dependent endocytic proteins, a process facilitated by additional factors such as the small GTPase Arf6, which in its GTP-bound state can bind to PIPK1γ and activate its enzymatic activity (Krauss *et al.*, 2003).

It is likely that feed-forward loops similar to AP-2/PIPK1γ also exist at other locations including endosomes and the TGN. In support of this view, the TGN localized-adaptor proteins GGA1-3 and AP-1 specifically bind to PI(4)P, and the adaptor protein AP-3 has been claimed to interact with the PI(4)P-synthesizing enzyme PI4K2α on a subpopulation of endosomes (Craige *et al.*, 2008; Wang *et al.*, 2003, 2007). Whether these coat components facilitate PI metabolism in a manner similar to the regulation of PIPKIγ by AP-2 remains to be investigated. The TGN and endosomes represent major protein sorting stations and regulate the biogenesis of important organelles like signaling endosomes, synaptic vesicles, and lytic granules (De Matteis and Luini, 2008). Hence, this pathway remains a fruitful territory for future studies.

We now turn to the detailed description of a biochemical assay for analyzing the effect of coat proteins on PI-metabolizing enzymes.

C. Analyzing Enzymatic Activity

Various methods are available to study the activity of PI-metabolizing enzymes. Lipid mass spectrometry can be used to analyze complex product mixtures, but this approach requires high-end equipment and specialized expertise (Wenk, 2010). PI-specific antibodies or PI-specific binding domains can be used in liposome sedimentation assays and/or for microscopy-based analysis, but tend to be partly non-specific or lack sensitivity (Varnai and Balla, 2006). Thus, for analyzing kinase-activation by accessory proteins, we prefer to use kinase activity assays involving radiolabels (Fig. 1) followed by thin-layer chromatography (TLC; described in detail below) or High Performance Liquid Chromatography (HPLC) analysis (see i.e., Nasuhoglu *et al.*, 2002 for details).

Here, we combine PI-containing liposomes or detergent-containing micelles, kinase (purified or present in cytosol derived from overexpressing cells) and [γ-^{32}P]-ATP. After a defined reaction time lipids are extracted, separated by TLC chromatography, and detected with a radiation-sensitive screen or film. This method

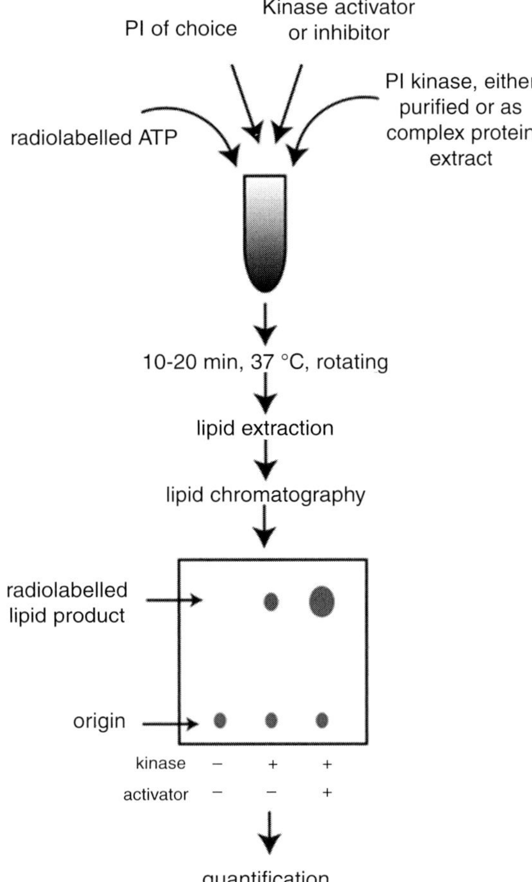

Fig. 1 Radioactive kinase assay workflow. For kinase assays, the following components are mixed in assay buffer (1) PI of choice, either as detergent solubilized PI or as PI containing-liposomes, (2) radiolabeled ATP [γ^{-32P}] (3), PI kinase of choice, either added as a complex mixture or purified (antibody or binding partner), (4) and optionally an activator or inhibitor of kinase activity. The mixture is incubated for 10–20 min at 37 °C and the lipids are extracted with methanol/chloroform. Subsequently, lipids are separated by TLC chromatography, and radioactive PI products are visualized by a radiation-sensitive film or screen.

combines the superior sensitivity of radioisotopes with a high degree of specificity by the use of specific substrates and lipid product separation. In addition, the method is economic and can be used in a standard biochemical laboratory with low-cost equipment. The use of radiosensitive screens and compatible imaging equipment further enables a broad dynamic range of detection and, hence, allows for reliable quantification.

D. Sources of PI Kinases

The kinase used in the assay can be purified from various eukaryotic sources, including insect or mammalian cell systems and tissue extracts. Some kinases can also be purified as recombinant proteins from bacteria. However, in our experience, most full-length enzymes recombinantly expressed in bacteria are either poorly soluble or display severely reduced activity. Possible reasons could be the lack of appropriate post-translational modifications (such as lipidation, phosphorylation), improper folding, or the absence of co-factors essential for physiological kinase activity.

Most of the known PI-metabolizing enzymes are expressed at low levels in tissue culture cells, often necessitating the use of kinases from eukaryotic over-expression system. An advantage of this setup is that mutant (kinase-inactive, binding-deficient) enzymes can be analyzed.

However, as many PI kinases and phosphatases cannot be constitutively over-expressed due to toxic side effects, we prefer to work with stable inducible cell lines.

II. Inducible Expression of PI-Metabolizing Enzymes

A. Generation of Flp-In/tetR Cells

Preparing stable lines by classical transfection, selection, and random integration can be time consuming and has a low success rate. However, many alternative approaches are available. Our system of choice is the Flp-In system developed by Invitrogen®. Flp-In cells have a defined Flp recombinase target (FRT) site, which is used for an efficient and targeted integration of the gene of interest. Various cell lines can be purchased from Invitrogen®, or can be generated by transfecting cells with FRT site-containing plasmid (pFRT/LacZEO). Moreover, expression from Flp-In cells can be made inducible by stably transfecting cells with a plasmid encoding the tet-repressor. For kinase assays we prefer the use of HEK293T Flp-In cells (Invitrogen®), as they express high levels of exogenous protein. In addition, they grow to high density, making it easy to pick single clones following antibiotic selection. HEK293T Flp-In cells are cultured in DMEM (4.5 g/L glucose) containing 10% FCS, penicillin, streptomycin, L-glutamine and 100 μg/mL zeocin (FRT plasmid selectable marker). We stably transfect these cells with a tet-repressor plasmid (pcDNA6/TR) and select for stable clones using zeocin and blasticidin (10 μg/mL) generating HEK293T Flp-In/tetR cells (Fig. 2a).

B. Stable Integration of the Gene of Interest in Flp-In/tetR Cells

Next, cDNA encoding the kinase of interest is cloned into the pcDNA5/FRT/TO plasmid, transfected into HEK293T Flp-In/tetR cells, together with the pOG44 plasmid, from which Flp recombinase is expressed (Fig. 2a). The recombinase will ensure efficient recombination between plasmid-encoded and chromosomal FRT

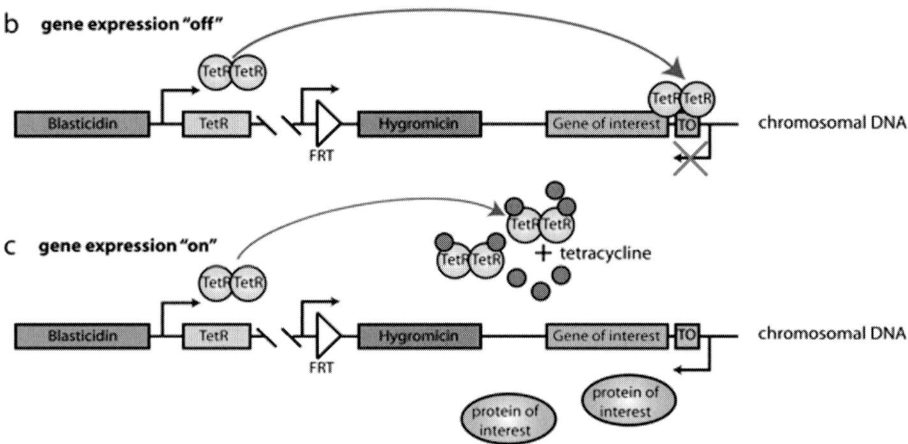

Fig. 2 Generation of inducible stable Flp-In cell lines**.** Long-term overexpression of PI-metabolizing enzymes is often toxic to tissue culture cells. This problem can be avoided by the use of stable lines, in which protein expression can be induced for short times. Here, we favor the Invitrogen® Flp-In system. (a) Flp-In cells can be obtained from Invitrogen®, or can be generated by stable transfection or the pFRT/LacZEO plasmid. Next, Flp-In/tetR cells are produced by stable integration of the tet-repressor (tetR) plasmid (pcDNA6/TR). To generate Flp-In cells expressing the gene of interest, the target gene has to be cloned into the pcDNA5/FRT/TO plasmid. The plasmid is then transfected into Flp-In/tetR cells, together with a plasmids encoding an appropriate recombinase (pOG44), ensuring recombination between the chromosomal and plasmid DNA, and targeted integration. (b) Tet repressor (TetR) binds to the tet operator site (TO), blocking transcription of the gene of interest. (c) Tetracycline, or derivates such as doxycycline, binds TetR dimers and prevent binding to TO, allowing gene expression to start. (For color version of this figure, the reader is referred to the web version of this book.)

sites and will thus allow for stable integration of the plasmid into the chromosomal DNA (Fig. 2b). We use six-well dishes for Lipofectamine 2000-based transfection with cells grown to about 80% confluence. Four hours post-transfection the medium is replaced by fresh, antibiotic-free medium, and 8 h post-transfection cells are split at a ratio of 1:20 into four 10 cm-dishes. As the zeocin-resistance is disrupted upon integration of the plasmid, the medium is replaced by DMEM containing 100 μg/mL hygromycin, the selection marker conferred by the pcDNA5/FRT/TO-vector, 24 h post-transfection. Non-transfected cells will start to die and detach from the dish after 4 days of incubation; from this time point on the medium is exchanged every other day. As soon as confluence has declined below 10% the medium is collected in Falcon tubes and centrifuged for 2 min at 3000 x *g*; the resulting supernatant is mixed one to one with fresh medium containing hygromycin and re-applied to the cells. This step can be repeated as long as the amount of floating cells and remnants steadily increases. Individual colonies are finally allowed to expand until stable clones have grown large enough for isolation. To isolate single colonies, the medium is exchanged against PBS; individual colonies are carefully detached with a 200 μL pipette and transferred into a 24-well dish containing 100 μL trypsin/EDTA. The dish is incubated for 5 min at 37 °C, and subsequently the cells are carefully resuspended in 500 μL full medium containing antibiotics. As soon as cells have grown to confluence they can be split as follows: One fifth of the cells is transferred onto two matrigel-coated glass coverslips in a 24-well plate, the residual suspension is transferred into a six-well dish.

When cells in the 24-well dish have grown to about 60% confluence, 1 μg/mL doxycycline is added to one of these wells. Cells are grown for 12 h, fixed in 4% PFA, and analyzed for protein expression using fluorescence microscopy. Clones expressing the protein of interest in fewer than 80% of the cells, or expressing the enzyme constitutively (i.e., in the absence of doxycycline) are discarded, whereas the residual clones are expanded. Note that the Flp-In system provides a high probability for insertion at a specific site within the genome. Hence, most individual clones will express the protein of interest at similar levels (Fig. 2c).

The total time required until the first clones can be picked is about 3 weeks. It takes about two additional weeks until first cryo-stocks can be prepared for long-term storage.

III. Radioactive Kinase Activity Assay

A. Considerations for Kinase Activity Assay

The kinase that is added to the reaction can be obtained from different sources. The most straightforward option is to compare kinase activities present in cell extracts gained from overexpressing *versus* non-induced HEK293T Flp-In/tetR cells. Alternatively, a tagged kinase variant might be affinity-purified from cell extracts, or endogenous kinase could be immunoprecipitated with kinase-specific antibodies from tissue lysates. Using extracts from overexpressing cells reduces the number of handlings and simplifies analysis with the caveat that a less well-defined system is used. The presence of endogenous PI-specific kinases, phosphatases, and lipases may result in further modifications of the primary radioactive product. Furthermore cell extracts may contain additional modulators of the studied kinase and thereby narrow the window of maximal activation to be achieved by the addition of exogenous proteins or protein domains. Longer protein induction times will increase the amount of kinase within cell extracts, but increase the likelihood of toxic side effects.

The concentration and purity of the kinase under study can be increased by affinity purification. This also allows to directly assess kinase activity, minimizing effects of potential binding partners, while isolating the protein from its cellular environment. However, harsh elution conditions might reduce enzymatic activity. In the case of PIPKIγ, we faced an additional problem: Direct binding of a bivalent antibody directed against its C-terminal tail turned out to potently stimulate kinase activity, presumably by triggering homo-oligomerization (Krauss *et al.*, 2006). Antibodies targeting affinity tags may have similar effects. We therefore suggest to evaluate putative modulatory effects of the chosen antibody in preliminary experiments using cell lysates as a source.

B. Kinase Assay Using Lysates From Overexpressing Cells

We note that the assays described here were optimized for analysis of PIPKIγ activity. Hence, minor modifications and adaptations might be required for assaying other PI kinases.

Control non-overexpressing and kinase-overexpressing cell extracts have to be prepared. Expression of HA-tagged PI kinase in HEK293T Flp-In/tetR cells is induced with 1 μg/mL doxycycline (kinase) or with DMSO (control) for 16 h. Cells are harvested by trypsinization and washed once in PBS. In case of PIPKIγ expressed in HEK293T Flp-In/tetR cells, cell pellets can be frozen in liquid nitrogen and stored at −80 °C for at least 2−4 weeks. The cell pellet collected from a 15 cm dish is resuspended in 2.5 mL of isolation buffer (20 mM Hepes/KOH, pH 7.4, 100 mM KCl, 2 mM $MgCl_2$, supplemented with 1 mM PMSF and a mammalian protease inhibitor cocktail) and homogenized using a cell cracker (purchased from EMBL). Cell homogenates are centrifuged for 15 min at 20,000 x *g*, and the resulting supernatants are subsequently spun for 15 min at 150,000 x *g* to remove unbroken cells, precipitates, and cell debris. The final supernatant is harvested, supplemented with Triton X-100 to a final concentration of 0.1%, and used for further experiments. It contains about 2–3 mg/mL of total protein.

Note, that under these conditions cytosolic extracts are recovered. It should be considered, that several PI-metabolizing enzymes are modified by lipid anchors. Thus, isolation of such enzymes requires solubilization prior to the second centrifugation step. Additionally, PI4-kinases have been shown to become activated in presence of Triton-X-100 (Wong and Cantley, 1994), necessitating the use of alternative detergents such as CHAPS.

Within this setup it is possible to add activators directly to the reaction mixture. For example, we supplemented the extract with 2 μM His_6-tagged μ2 (amino acids 147–435, encoding the soluble PIPKI- and cargo-binding subdomain) purified from bacteria, alone or in combination with peptides (50 μM) comprising tyrosine-based sorting motifs or mutant peptides deficient in μ2 binding (Fig. 3). The addition of recombinant μ2 in complex with functional AP-2 μ binding peptides triggers a significant increase in PI kinase activity.

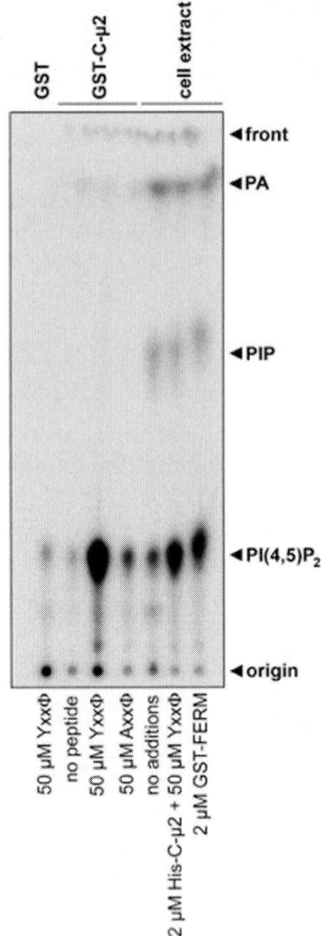

Fig. 3 Kinase assay read-out. Kinase assays were performed with either PIPK1γ purified from cell extracts using GST-μ2 (left side of the panel), or with HEK293T Flp-In/tetR PIPK1γ overexpressing cell extracts (right side of the panel). Cargo-derived μ2 binding peptide (YxxΦ) or control peptides (AxxΦ) were added to the reaction. As positive control we included the FERM domain of talin, which is a potent activator of PIPK1γ enzymatic activity (Di Paolo *et al.*, 2002). After the kinase reaction, lipids were extracted and separated by TLC chromatography, and then visualized by a radiation-sensitive screen. Mono-phosphorylated PIs (like PI(4)P) will run faster, compared to double-phosphorylated lipids (like PI(4,5)P_2). It is noteworthy that the use of complex protein mixtures gives rise to additional side products (PIP and phosphatidic acid (PA)), likely via the action of endogenous PI kinases, phosphatases, or lipases.

Unfortunately, our own efforts to identify key residues involved in complex formation between PIPKIγ and the cargo-binding subdomain of μ2 were unsuccessful. Such binding-deficient mutants of either binding partner would constitute ideal controls to evaluate the contribution of additional factors present in cell extracts and should be considered when studying PI-metabolizing enzymes.

As shown by others (Oude Weernink *et al.*, 2004), co-transfection of activating proteins may significantly increase PIPKI-activity present in cellular lysates. This method was successfully used to demonstrate the activating effects of Rac-, Rho- and Cdc42-GTPases on type I PIPK isozymes.

C. Kinase Assay Using Purified Enzyme

The most straightforward approach to characterize the activity of PI kinases is to purify the enzyme of interest from cell extracts using specific antibodies or affinity tags. The antibody can be either directed against an epitope-tag or the kinase itself. In case the chosen antibody displays no stimulatory effect on enzymatic activity, the immunoprecipitated material can be directly subjected to activity assays, in the presence or absence of the protein of interest. This approach was used to show that PI3KC2α, an enzyme generating 3-phosphorylated phosphoinositides, is activated by clathrin (Gaidarov *et al.*, 2001).

Qualitative results may also be gained from immunoprecipitiation experiments using a binding partner of the enzyme under investigation as bait. Talin has been immunoprecipitated from cells overexpressing two different splice variants of PIPKIγ, p87 and p90, which differ in a 28 amino acid splice insert required for its association with talin. Talin-associated PI(4)P 5-kinase activity was detectable only in p90-, but not in p87-overexpressing cells (Di Paolo *et al.*, 2002; Ling *et al.*, 2002).

In addition to antibody-based methods for purification of PI kinases, affinity purification involving known binding partners can be used. In the case of PIPKIγ, a GST-fusion of amino acids 147–435 of the μ2 subunit of the AP-2 complex has been successfully employed. First, GST-μ2 is isolated from 500 mL of overexpressing bacteria (induced for 3 h at 25 °C) using 400 μL glutathione-coupled beads in PBS containing 500 mM NaCl. The beads are extensively washed and finally resuspended in 500 μL PBS supplemented with 300 mM NaCl. In our hands, the protein amounts of GST-μ2 cannot be determined by conventional Bradford analysis. Therefore, aliquots containing different amounts of GST and of GST-μ2 are subsequently separated by SDS-gel-electrophoresis and visualized by Coomassie Brilliant Blue staining to monitor the purity of the protein and to determine its concentration.

Next, 50 μg of GST or GST-μ2 coupled to the GSH-beads are incubated for 2 h at 4 °C with 250 μg extract from HEK293T Flp-In/tetR cells overexpressing PI kinase (prepared as described above) in a final volume of 1 mL isolation buffer. The beads are washed three times in 700 μL isolation buffer and once in kinase buffer (25 mM Hepes/KOH, pH 7.2, 25 mM KCl, 2.5 mM magnesium acetate, 150 mM potassium glutamate, 10 μM $CaCl_2$, 0.1% Triton X-100) before resuspension in 1 mL of kinase buffer. Two hundred microliters (20% of the total material corresponding to 10 μg of GST-μ2) of the affinity-purified material are then transferred into a new reaction tube, centrifuged for 2 min at 1000 x *g*, and the supernatant is carefully removed. The residual beads are then supplemented with the reaction mix as indicated below.

A second pulldown assay performed in parallel serves to probe the retention of HA-PIPKIγ on GST-μ2, but not on GST. To this aim the resin is eluted with Laemmli sample buffer. Aliquots of these samples are then subjected to Western blot analysis using anti-HA antibodies. Besides GST, several control proteins may be included. Given that we observed the maximal stimulatory effect of GST-μ2 on PIPKIγ only upon addition of AP-2 μ-binding sorting signal peptides, we analyzed a GST-μ2 mutant (D176A/W421A) defective for binding to sorting signals (Krauss *et al.*, 2006). Such binding-defective mutants of the modulatory protein or of the PI-metabolizing enzyme might serve as important negative controls to detect potential activities of unknown kinases/phosphatases co-affinity-purified with the overexpressed enzyme.

D. Kinase Reaction

In principle, the substrate of the PI-metabolizing enzyme, PI(4)P in the case of PIPKI, is offered following solubilization in detergent-containing buffer or as a minor component of liposomes. Detergents may affect the activity of PI-metabolizing enzymes, as outlined above. Liposomes, on the other hand, more closely resemble the native state of a membrane, at the surface of which the PI is presented. However, their preparation is more time-consuming and demands for more material, as half of the substrate PI is sequestered within the inaccessible inner leaflet. Hence, apparent enzymatic activities are lower (Krauss *et al.*, 2003). As the presence of Triton-X-100 displayed no adverse effects on the activity of PIPKIγ, we used PI(4)P solubilized in kinase buffer as a substrate.

Per sample 10 μg of the isolated substrate PI of choice, stored as a solution in $CHCl_3$ under nitrogen at −80 °C, are dried and resuspended in 50 μL kinase buffer. The suspension is incubated on ice for 30 min, then briefly sonified for 2 min, and finally supplemented with 200 μM ATP and 10 μCi [γ-^{32}P] ATP. PI kinase is added from the appropriate source (see paragraphs above) and the mixture is incubated for 10–20 min at 37 °C.

E. Lipid Extraction

The kinase reaction is stopped by transferring samples on ice. The samples are immediately supplemented with 500 μL of a pre-cooled mixture of methanol:H_2O: HCl (20:20:1, v/v) and vigorously mixed for 2 min. After incubation at room temperature for 10 min, 550 μL of $CHCl_3$ are added, followed by vigorous shaking for 30 s. At this stage, the suspension should become cloudy due to the separation of aqueous and organic phases. Samples are then centrifuged for 5 min at 1200 × *g*.

The resulting bottom phase contains hydrophobic lipids dissolved in $CHCl_3$; proteins precipitated by the treatment are concentrated at the interphase. The upper phase is carefully taken off and transferred to a fresh Eppendorf cup; the interphase can be collected together with the aqueous phase at this point. The organic phase is transferred to a separate tube and re-extracted as described above. The organic phase collected from the second extraction step is combined with the first one, and the complete material is divided into two halves. Samples are thereafter supplemented with 500 μL methanol:H_2O:HCl (20:20:1, v/v), thoroughly mixed, incubated for 15 min at RT and centrifuged. The aqueous phase is removed by pipetting, and the organic phase is lyophilized. These tubes containing the reaction product can be stored at −20 °C for several days.

F. Thin-Layer Chromatography (TLC) Analysis

Samples are subjected to TLC. We use silica plates (10 x 20 cm; HPTLC Silica gel 60 F_{254}; Merck) pre-incubated for 30 min with 1% potassium oxalate/2 mM EDTA in methanol:H_2O (2:3, v/v). The plates are dried overnight and finally incubated at 150 °C for 30 min. The solvent consists of $CHCl_3$:acetone:methanol:acetic acid: H_2O (64:30:24:30:13, v/v) and is freshly prepared prior to use. A chromatography chamber is lined with Whatman paper, supplemented with this solvent and left for equilibration at RT for 1 h. Individual spots are carefully marked with a pencil, 2 cm above the outer edge of the TLC plate. About 12 individual samples can be analyzed on one plate. Dried samples are left in closed tubes at RT and are subsequently dissolved in 12 μL $CHCl_3$. Each sample is then spotted in 1.5 μL steps onto a premarked spot. To prevent expansion of the material on the plate, each spotted aliquot is dried under airflow before the next one is applied at the same place. Finally, the TLC plate is carefully placed in the chromatography chamber. Samples are separated during 3 h of incubation at RT. The plate is then placed under a hood and allowed to dry overnight. Finally, the plate is wrapped in transparent film and exposed to a detection screen in an autoradiography cassette for 24 h at −80 °C. Radioactive signals are detected by PhosphorImager analysis, which allows for a qualitative and quantitative detection of the radioactive products.

Alternatively, radioactive or non-radioactive products of a kinase reaction can also be separated by HPLC (Nasuhoglu *et al.*, 2002).

IV. Interpretation and Troubleshooting

The described technique allows for an efficient separation of differentially charged PIs. Highly phosphorylated PIs (e.g., PI(4,5)P_2) display a lower mobility as compared to PIs of a lower phosphorylation state (e.g., PI(4)P). A representative example is depicted in Fig. 3. Ideally, only one spot representing the reaction product should be detected on the plate. However, in particular if whole-cell lysates containing endogenous lipid-modifying enzymes (such as PI3-kinases, phospholipases,

$PI(4,5)P_2$-dependent 4-phosphatases etc.) have been used during the reaction, several by-products may be formed, including mono-phosphorylated PIs or lyso-lipids. These by-products may contain a radiolabeled phosphate, not necessarily attached to the 5-position. It may thus be advisable to adjust the assay conditions (incubation time, amount of cell lysate used as source for the PI-metabolizing enzyme) to minimize the complexity of the reaction products.

To detect modulatory effects of binding partners, it is important to determine the dynamic range of the kinase concentration and to adjust assay conditions for the linear range: Using too much enzyme may obscure stimulatory effects, whereas very low kinase concentrations may render inhibitory effects non-detectable. Furthermore, the specificity of a given kinase interactor should be controlled by (i) testing the concentration dependence of modulatory effects and (ii) by using appropriate mutants or other non-interacting proteins as controls.

V. Outlook

The sorting of membranes and proteins into transport vesicles requires the concerted action of coat proteins, PIs, and small GTPases. Within this context, PIs are synthesized *de novo* to initiate coat formation and rapidly degraded to facilitate fission and/or uncoating. We recently have gained first insights into how adaptors and small GTPases may control PI metabolism during AP-2/clathrin-dependent coat formation at the plasma membrane.

AP-2 associates with all three isoforms of type I PIPKs through interaction surfaces involving the kinase core domain and the cargo-binding domain of its μ2-subunit. Upon concomitant binding of AP-2 to tyrosine-based sorting signals, kinase activity is potently stimulated. Yet, not all cargo molecules internalized by CME encode such sorting motifs. How is PIP_2 synthesis coordinated with coat assembly in these cases? First answers to this question were gained for AP-2/clathrin-dependent endocytosis at neuronal synapses, where PIPKIγ-p90 represents the major PIP_2-synthesizing enzyme (Wenk *et al.*, 2001). PIPKIγ-p90 provides two additional, overlapping interaction surfaces for AP-2 within its 28 amino acid splice insert. Through this splice insert PIPKIγ-p90 associates with the AP-2β subunit, an interaction that is disrupted by the assembling clathrin coat (Thieman *et al.*, 2009). This splice insert also harbors a tyrosine-based sorting signal, which can effectively stimulate kinase activity and may thus boost PIP_2-synthesis in situations where non-conventional cargo is internalized in a clathrin/AP-2-dependent fashion (Kahlfeldt *et al.*, 2010). To what extent the tight coupling between AP-2 and PIPKIγ represents a specific adaptation of neurons established to maintain high capacity retrieval of SV membranes during sustained stimulation is unknown.

Further yet unresolved questions pertain to the coupling between PI synthesis and coat recruitment and assembly on other intracellular membranes. Membrane recruitment of several adaptors (AP-1, AP-3, GGAs) may depend on PI(4)P (Craige *et al.*,

2008; Demmel *et al.*, 2008; Wang *et al.*, 2003, 2007). However, these adaptors function at different organelles: For example, AP-1 and AP-3 display non-overlapping distributions on the TGN and on endosomes, respectively, and have been shown to regulate the formation of physically and functionally distinct carriers. GGA proteins are monomeric clathrin-binding adaptors with important roles in membrane traffic at the TGN and in the sorting of lysosomal enzymes. This raises an important question: Do all vesicle coats regulated by PI(4)P depend on the same PI4-kinase, or are different isoforms involved? Membrane translocation of all of these adaptors depends on the prior activation of the small GTPase Arf1. GTP-Arf1 in turn has been shown to associate with and activate type III PI4-kinase. Recent studies suggest that AP-3 forms a complex with PI4KIIα via an acidic cluster dileucine motif present in its N-terminal tail (Craige *et al.*, 2008). Its sister enzyme PI4KIIβ encodes a similar sequence that might confer specificity for interaction with other AP complexes. How such fine-tuning is achieved and to which extent different PI kinase isoforms may confer subpools of PIs dedicated to distinct membrane subdomains or transport routes remains to be elucidated.

Even less is known regarding the mechanisms that regulate the activity of PI phosphatases during fission or uncoating of vesicular carriers. The best-studied example is synaptojanin 1, an enzyme involved in the degradation of PIP_2 on AP-2/clathrin-coated vesicles, which facilitates uncoating (Cremona *et al.*, 1999). Recent studies have revealed that the 5-phosphatase activity of synaptojanin 1 is modulated by membrane curvature (Chang-Ileto *et al.*, 2011), thereby restricting synaptojanin-mediated PIP_2 hydrolysis to late stage intermediates undergoing membrane fission. Additional PI phosphatases have been implicated in CME and downstream endosomal sorting events. The accessory endocytic protein intersectin recruits the 5-phosphatase SHIP2, which preferentially acts on $PI(3,4,5)P_3$, to early-stage clathrin-coated pits, from which the enzyme is released prior to dynamin-mediated fission (Nakatsu *et al.*, 2010), suggesting a potential role for $PI(3,4,5)P_3$ in CME. The clathrin binding partner OCRL, a 5-phosphatase similar to synaptojanin, on the other hand, arrives at coated pits late and remains associated with endocytic structures during their maturation into peripheral early endosomes (Erdmann *et al.*, 2007). It thus appears that PI phosphatase recruitment to distinct subdomains and their activity is spatiotemporally controlled by interactions with coat components. The functional role of some of these enzymes at distinct stages of vesicle formation, however, remains largely elusive.

We also know little about how the dynamics of other clathrin-dependent and clathrin-independent carriers is orchestrated by PI-kinases and phosphatases. Activity assays such as the one described here may help to answer these questions.

Acknowledgments

Work in the authors' laboratories is funded by the German funding agency DFG (SFB740/ TP C8; HA 2686/ 2-1, 2-2).

References

Behnia, R., and Munro, S. (2005). Organelle identity and the signposts for membrane traffic. *Nature* **438**, 597–604.

Bonifacino, J. S., and Glick, B. S. (2004). The mechanisms of vesicle budding and fusion. *Cell* **116**, 153–166.

Chang-Ileto, B., Frere, S. G., Chan, R. B., Voronov, S. V., Roux, A., and Di Paolo, G. (2011). Synaptojanin 1-mediated PI(4,5)P2 hydrolysis is modulated by membrane curvature and facilitates membrane fission. *Dev. Cell* **20**, 206–218.

Craige, B., Salazar, G., and Faundez, V. (2008). Phosphatidylinositol-4-Kinase Type II alpha contains an AP-3–sorting motif and a kinase domain that are both required for endosome traffic. *Mol. Biol. Cell* **19**, 1415–1426.

Cremona, O., *et al.* (1999). Essential role of phosphoinositide metabolism in synaptic vesicle recycling. *Cell* **99**, 179–188.

De Matteis, M. A., and Luini, A. (2008). Exiting the Golgi complex. *Nat. Rev. Mol. Cell Biol.* **9**, 273–284.

Demmel, L., *et al.* (2008). The clathrin adaptor Gga2p is a phosphatidylinositol 4-phosphate effector at the Golgi exit. *Mol. Biol. Cell* **19**, 1991–2002.

Di Paolo, G., and De Camilli, P. (2006). Phosphoinositides in cell regulation and membrane dynamics. *Nature* **443**, 651–657.

Di Paolo, G., *et al.* (2002). Recruitment and regulation of phosphatidylinositol phosphate kinase type 1γ by the FERM domain of talin. *Nature* **420**, 85–89.

Erdmann, K. S., *et al.* (2007). A role of the Lowe syndrome protein OCRL in early steps of the endocytic pathway. *Dev. Cell* **13**, 377–390.

Gaidarov, I., and Keen, J. H. (1999). Phosphoinositide-AP-2 interactions required for targeting to plasma membrane clathrin-coated pits. *J. Cell Biol.* **146**, 755–764.

Gaidarov, I., Smith, M. E., Domin, J., and Keen, J. H. (2001). The class II phosphoinositide 3-kinase C2α is activated by clathrin and regulates clathrin-mediated membrane trafficking. *Mol. Cell* **7**, 443–449.

Haucke, V. (2005). Phosphoinositide regulation of clathrin-mediated endocytosis. *Biochem. Soc. Trans.* **33**, 1285–1289.

Jackson, L. P., Kelly, B. T., McCoy, A. J., Gaffry, T., James, L. C., Collins, B. M., Höning, S., Evans, P. R., and Owen, D. J. (2010). A large-scale conformational change couples membrane recruitment to cargo binding in the AP2 clathrin adaptor complex. *Cell* **141**, 1220–1229.

Kahlfeldt, N., Vahedi-Faridi, A., Koo, S. J., Schäfer, J. G., Krainer, G., Keller, S., Saenger, W., Krauss, M., and Haucke, V. (2010). Molecular basis for association of PIPKI γ-p90 with clathrin adaptor AP-2. *J. Biol. Chem.* **285**, 2734–2749.

Krauss, M., and Haucke, V. (2007). Phosphoinositide-metabolizing enzymes at the interface between membrane traffic and cell signalling. *EMBO Rep.* **8**, 241–246.

Krauss, M., Kinuta, M., Wenk, M. R., De Camilli, P., Takei, K., and Haucke, V. (2003). ARF6 stimulates clathrin/AP-2 recruitment to synaptic membranes by activating phosphatidylinositol phosphate kinase type Iγ. *J. Cell Biol.* **162**, 113–124.

Krauss, M., Kukhtina, V., Pechstein, A., and Haucke, V. (2006). Stimulation of phosphatidylinositol kinase type I-mediated phosphatidylinositol (4,5)-bisphosphate synthesis by AP-2 μ-cargo complexes. *Proc. Natl. Acad. Sci. U. S. A.* **103**, 11934–11939.

Ling, K., Doughman, R. L., Firestone, A. J., Bunce, M. W., and Anderson, R. A. (2002). Type Iγ phosphatidylinositol phosphate kinase targets and regulates focal adhesions. *Nature* **420**, 89–93.

McMahon, H. T., and Mills, I. G. (2004). COP and clathrin-coated vesicle budding: different pathways, common approaches. *Curr. Opin. Cell Biol.* **16**, 379–391.

Nakano-Kobayashi, A., *et al.* (2007). Role of activation of PIP5Kγ661 by AP-2 complex in synaptic vesicle endocytosis. *EMBO J.* **26**, 1105–1116.

Nakatsu, F., Perera, R. M., Lucast, L., Zoncu, R., Domin, J., Gertler, F. B., Toomre, D., and De Camilli, P. (2010). The inositol 5-phosphatase SHIP2 regulates endocytic clathrin-coated pit dynamics. *J. Cell Biol.* **190**, 307–315.

Nasuhoglu, C., Feng, S., Mao, J., Yamamoto, M., Yin, H. L., Earnest, S., Barylko, B., Albanesi, J. P., and Hilgemann, D. W. (2002). Nonradioactive analysis of phosphatidylinositides and other anionic phospholipids by anion-exchange high-performance liquid chromatography with suppressed conductivity detection. *Anal. Biochem.* **301**, 243–254.

Oude Weernink, P. A., Meletiadis, K., Hommeltenberg, S., Hinz, M., Ishihara, H., Schmidt, M., and Jakobs, K. H. (2004). Activation of type I phosphatidylinositol 4-phosphate 5-kinase isoforms by the Rho GTPases, RhoA, Rac1, and Cdc42. *J. Biol. Chem.* **279**, 7840–7849.

Rohde, G., Wenzel, D., and Haucke, V. (2002). A phosphatidylinositol (4,5)-bisphosphate binding site within μ2-adaptin regulates clathrin-mediated endocytosis. *J. Cell. Biol.* **158**, 209–214.

Thieman, J. R., Mishra, S. K., Ling, K., Doray, B., Anderson, R. A., and Traub, L. M. (2009). Clathrin regulates the association of PIPKIγ661 with the AP-2 adaptor β2 appendage. *J. Biol. Chem.* **284**, 13924–13939.

Varnai, P., and Balla, T. (2006). Live cell imaging of phosphoinositide dynamics with fluorescent protein domains. *Biochim. Biophys. Acta* **1761**, 957–967.

Varnai, P., Thyagarajan, B., Rohacs, T., and Balla, T. (2006). Rapidly inducible changes in phosphatidylinositol 4,5-bisphosphate levels influence multiple regulatory functions of the lipid in intact living cells. *J. Cell Biol.* **175**, 377–382.

Wang, J., Sun, H-Q., Macia, E., Kirchhausen, T., Watson, H., Bonifacino, J. S., and Yin, H. L. (2007). PI4P promotes the recruitment of the GGA adaptor proteins to the trans-Golgi network and regulates their recognition of the ubiquitin sorting signal. *Mol. Biol. Cell* **18**, 2646–2655.

Wang, Y. J., *et al.* (2003). Phosphatidylinositol 4 phosphate regulates targeting of clathrin adaptor AP-1 complexes to the Golgi. *Cell* **114**, 299–310.

Wenk, M. R. (2010). Lipidomics: New tools and applications. *Cell* **143**, 888–895.

Wenk, M. R., Pellegrini, L., Klenchin, V. A., Di Paolo, G., Chang, S., Daniell, L., Arioka, M., Martin, T. F., and De Camilli, P. (2001). PIP kinase Igamma is the major PI(4,5)P_2 synthesizing enzyme at the synapse. *Neuron* **32**, 79–88.

Wieffer, M., Maritzen, T., and Haucke, V. (2009). SnapShot: endocytic trafficking. *Cell* **137**, 382.e1-3.

Wong, K., and Cantley, C. L. (1994). Cloning and characterization of a human phosphatidylinositol 4-kinase. *J. Biol. Chem.* **269**, 28878–28884.

Zoncu, R., Perera, R. M., Sebastian, R., Nakatsu, F., Chen, H., Balla, T., Ayala, G., Toomre, D., and De Camilli, P. V. (2007). Loss of endocytic clathrin-coated pits upon acute depletion of phosphatidylinositol 4,5-bisphosphate. *Proc. Natl. Acad. Sci. U. S. A.* **104**, 3793–3798.

CHAPTER 12

Phosphoinositides at the Neuromuscular Junction of *Drosophila melanogaster*: A Genetic Approach

Jan R. Slabbaert[*,†,1], **Thang Manh Khuong**[*,†,1] **and Patrik Verstreken**[*,†]

[*]VIB Center for the Biology of Disease, Leuven, Belgium

[†]K.U. Leuven, Center for Human Genetics, Leuven, Belgium

Abstract

Phosphoinositides are critically important for numerous cellular signaling pathways such as membrane trafficking, cytoskeleton rearrangement, and ion channel regulation in eukaryotic organisms. The physiological relevance of phosphoinositide metabolism at the *Drosophila* neuromuscular junction (NMJ) has been illustrated using several mutants that lack crucial factors of phosphoinositide signaling. Although several decades of research in both *in vitro* and *in vivo* models have led to an understanding of the mechanisms of lipid–protein interactions and downstream

[1]Jan R. Slabbaert and Thang Manh Khuong contributed equally to this manuscript.

Copyright 2012, Elsevier Inc. All rights reserved.

0091-679X/10 $35.00
DOI 10.1016/B978-0-12-386487-1.00012-2

signaling, the details on how their temporal and spatial distribution is regulated at the sub-cellular level *in vivo* remains poorly understood. To obtain a better understanding of phosphoinositide signaling, detailed biochemical and cell biological approaches can best be combined with genetics. In this review, we present an overview of the methodologies available in the fruit fly *Drosophila melanogaster*, to genetically dissect the complex regulation of signaling pathways involving phosphoinositides.

I. Introduction

Inositol phospholipids regulate a large repertoire of cellular processes and therefore require specific mechanisms that control their local metabolism, dependent on the physiological context. Reversible phosphorylation of the phosphatidylinositol head group at positions D-3, D-4 and D-5 creates seven different phosphoinositides of which each has a unique subcellular distribution in subsets of membranes (De Matteis and Godi, 2004; Di Paolo and De Camilli, 2006). An important factor influencing this "compartmentalization" is the localization and activation of specific metabolizing enzymes often regulated by signaling pathways. Consequently, recruited phosphoinositide kinases and phosphatases can act as feedback regulators within the cell to alter local phosphoinositide content (reviewed in (Astle *et al.*, 2006; Balla and Balla, 2006; Doughman *et al.*, 2003; Foster *et al.*, 2003; Marone *et al.*, 2008; Ooms *et al.*, 2009; Shisheva, 2008)). Another mechanism that imposes specificity of phosphoinositide signaling pathways in the cell is that some phosphoinositides act as substrates for phospholipases thus generating a number of products acting as local second messengers (Berridge, 1983; Russell *et al.*, 1987). A final mode of specificity is co-incidence detection where specific recruitment to membranes is achieved by the simultaneous binding of proteins to the inositol head group and another protein, often a small GTPase (reviewed in (Carlton and Cullen, 2005)). Hence, the specific intracellular distribution and high turnover rate of phosphoinositides regulate various cellular processes such as signal transduction, cytoskeleton regulation, membrane trafficking, and cellular homeostasis.

Analyses of the phosphoinositide network require a multidisciplinary experimental approach that includes genetics, biochemistry and imaging. While *in vitro* methods based on metabolic labeling with [^{3}H]*myo*-inositol or [^{32}P]phosphate, mass measurements, and the use of pharmacological agents have been invaluable tools to characterize specific phosphoinositides (Arcaro and Wymann, 1993; Chilvers *et al.*, 1991; Palmer *et al.*, 1986), several modern methodologies have entered the field. For example, green fluorescent protein (GFP) linked to specific protein domains that possess natural phosphoinositide binding allow to follow the localization and turnover of these lipids in live cells (Gillooly *et al.*, 2000; Kontos *et al.*, 1998; Kutateladze *et al.*, 1999; Levine and Munro, 1998; Stauffer *et al.*, 1998; Varnai and Balla, 1998; Varnai *et al.*, 1999). Alternatively, the use of genetic gain- or loss of function analyses or acute manipulation of phosphoinositide levels revealed the phenotypic and cellular consequences of altering the phosphoinositide metabolism in their natural context (Belshaw *et al.*, 1996; Ozaki *et al.*, 2000; Suh *et al.*, 2006; Varnai *et al.*, 2006). These developments enable to combine the advantages of genetic studies with elegant tools to acutely manipulate phosphoinositides in combination with functional studies and therefore open avenues to comprehend new components of the phosphoinositide network.

The *Drosophila* larval neuromuscular junction (NMJ) harbors a unique set of characteristics that make it suitable to combine genetic, biochemical and imaging techniques to study the regulation of phosphoinositides. First, an extended set of tools to perform genetic manipulations is available to fruit fly researchers and allows in depth gain- or loss of function studies (discussed below (Venken and Bellen, 2007)). Second, individual synaptic boutons at the larval NMJ are large, up to several microns, and these structures are often located close to the muscle surface allowing *in vivo* imaging (Budnik *et al.*, 2006). Third, the morphology and physiology of the NMJ are well characterized, allowing for studies of synaptic function as well. These features can be combined with super resolution imaging technology and advanced electron microscopy techniques, enabling the visualization of phosphoinositide content, distribution and dynamics in an *in vivo* context. In this chapter we will first outline the toolbox available to study phosphoinositide-related signaling pathways in *Drosophila* followed by several illustrations of these tools at the larval NMJ.

II. Genetic Tools to Study Phosphoinositides in *Drosophila*

Phosphoinositides are involved in numerous membrane-related signaling events. $PI(4,5)P_2$ and $PI(3,4,5)P_3$ have been shown to be important for a wide range of cellular processes at the synapse through specific binding of their phosphorylated head groups to a variety of synaptic proteins (Wenk and De Camilli, 2004). Indeed, the regulated turn-over of $PI(4,5)P_2$ by $PI(4)P5K\gamma$ and Synaptojanin (Synj) (Fig. 1) coordinates the spatial and temporal (de)activation and recruitment of several regulatory proteins from the cytosol (Aikawa and Martin, 2003; Cremona *et al.*, 1999; Di Paolo *et al.*, 2004; Krauss *et al.*, 2003; Nakano-Kobayashi *et al.*, 2007; Verstreken *et al.*, 2003, 2009; Zheng and Bobich, 2004). While $PI(4,5)P_2$ is widely involved in endocytosis, it may also play a role during exocytosis in the "priming" of vesicles prior to fusion (Eberhard *et al.*, 1990; Hay and Martin, 1993). Its role involves interaction with synaptic vesicle proteins such as synaptotagmin (Bai *et al.*, 2004), plasma membrane proteins such as CAPS in dense core vesicle endocytosis (Grishanin *et al.*, 2004), or the production of the second messenger diacylglycerol (DAG), an important ligand for the priming factor Munc-13 (Lackner *et al.*, 1999; Rhee *et al.*, 2002). Synaptic development on the other hand requires dynamic remodeling of the actin filaments present in the nerve cells and $PI(4,5)P_2$ interacts with different actin binding proteins such as Wiscott-Aldrich syndrome protein (WSP) (Sechi and Wehland, 2000), profilin (Skare and Karlsson, 2002), cofilin (Ojala *et al.*, 2001), and gelsolin (Feng *et al.*, 2001). A more specific example includes the regulation of neurite elongation and branching by the Arf6/PI(4)P5K

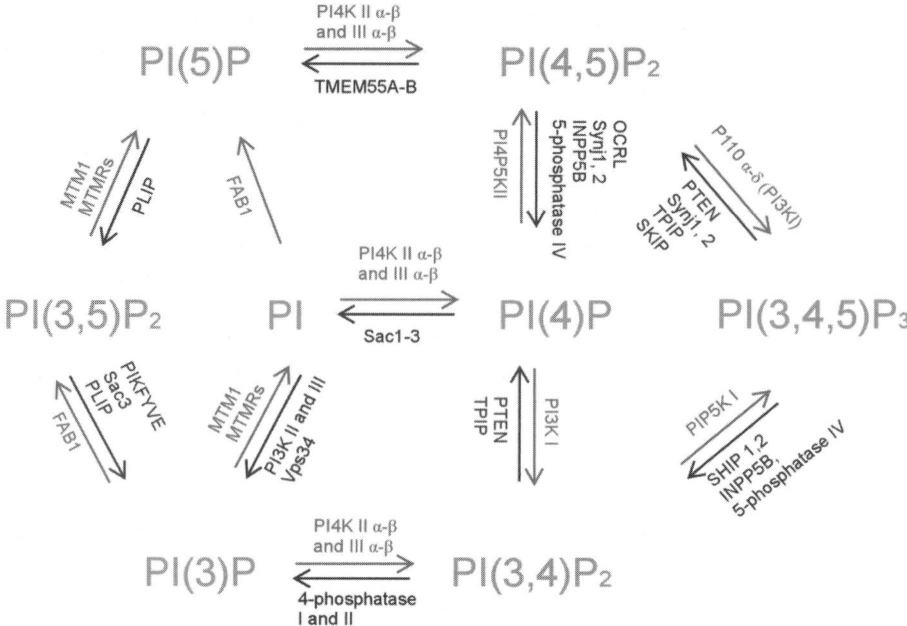

Fig. 1 Schematic representation of the phosphoinositide metabolism. The main pathways of phosphoinositide synthesis and degradation in mammalian cells are displayed in the diagram. Most of the indicated enzymes have clear homologs in flies, which can be found in Table I. (See color plate.)

pathway, possibly by the recruitment of the actin binding protein Mena to the growth cone by local accumulations of the PI(4)P5K product PI(4,5)P_2 (Hernandez-Deviez *et al.*, 2004). Regulation of synaptic morphology also involves the induction of dendritic arborization and spine formation through PI3K-Akt-mammalian target of rapamycin (mTOR) signaling pathway whose activation results in local accumulations of PI(3,4,5)P_3 (Jaworski *et al.*, 2005; Kumar *et al.*, 2005). The mechanisms of neuronal survival and proliferation requires signaling of other phosphoinositides including PI(3,4)P_2 and PI(3,4,5)P_3. However, their exact working mechanism in these processes is still being elucidated (Cantley, 2002; Jaworski *et al.*, 2005; Kumar *et al.*, 2005). Given the intricate involvement of phosphoinositide metabolism in synaptic function and development, it is critical to gain a complete understanding of the precise cellular mechanisms of action of each phosphoinositide species, to identify their metabolizing enzymes, and in particular, their precise sub-cellular localization. In the following sections, we focus on *Drosophila* and we will outline different tools that have been applied to study phosphoinositide function at the larval NMJ.

A. Targeting Phosphoinositide Metabolism

The completion of the *Drosophila* genome sequence has led to the identification of conserved genetic loci necessary for phosphoinositide metabolism and signaling

(Adams *et al.*, 2000; Lloyd *et al.*, 2000; Rubin *et al.*, 2000). A search of the vertebrate proteins implicated in phosphoinositide metabolism indicates fly homologues are very well conserved (Table I (Morrison *et al.*, 2000)) and our analyses identified five fly proteins that were not yet previously identified (Table I (Morrison *et al.*, 2000)). We discovered a fly homologue of all proteins searched, except for the SH2-domain–containing inositol 5′ phosphatase 1 and 2 (SHIP1 and 2). Also the *C. elegans* genome does not contain SHIP1 or 2 homologues, suggesting these proteins evolved later. Finally, while mammalian genomes often encode several isoforms of a protein, the fruit fly genome harbors only one gene, greatly facilitating genetic studies. Taken together, our searches suggest that phosphoinositide metabolism is very well conserved between flies and man, and given the limited number of isoforms or gene copies in fruit flies, *Drosophila* is ideally suited for genetic analyses of phosphoinositide metabolism.

Drosophila researchers have an extended set of tools at their disposal that allows them to genetically manipulate components of the phosphoinositide signaling cascade. First, an efficient transgenesis platform is in place, and nowadays, several companies are offering injection service. Transgenes are either inserted randomly in the genome, or within specific docking sites ensuring insertion within a specific genomic locus, holding the advantage that they enable identical expression conditions for different constructs used (Bischof *et al.*, 2007; Groth *et al.*, 2004; Venken *et al.*, 2006). Second, several genome-wide RNAi libraries exist and the transgenic hairpin expressing constructs are under the control of the UAS/GAL4 system (Brand and Perrimon, 1993) that enables temporally regulated tissue specific knock-down of gene expression (Dietzl *et al.*, 2007; Ni *et al.*, 2009). New RNAi collections are generated using specific genomic docking sites, enabling more reliable comparison of multiple RNAi lines. RNAi mediated knock down is fast, but sometimes may suffer from incomplete gene inactivation and off target effects. Therefore “classical mutations” are widely used. Genome wide disruption projects use transposable elements to disrupt gene function, and numerous genes harbor publicly available insertions (Bellen *et al.*, 2004; Bier *et al.*, 1989; Cooley *et al.*, 1988; Rubin and Spradling, 1982; Spradling *et al.*, 1999). While often the insertion itself disrupts gene function, in many cases transposon dysgenesis can be used to generate small local deletions, creating additional alleles (Rubin and Spradling, 1982). These, and numerous additional tools, including homologous recombination (Rong and Golic, 2000), recombination mediated cassette exchange (Choi *et al.*, 2009; Gao *et al.*, 2008), etc., allow to efficiently create loss of function alleles of phosphoinositide metabolizing enzymes. Rescue experiments of loss of function phenotypes are also straightforward using the classical Gal4/UAS cDNA over expression system (Brand and Perrimon, 1993), or with the development of numerous new tools in recent years (Venken *et al.*, 2006, 2008), using genomic rescue constructs. The main advantage of genomic constructs is that the gene of interest is expressed at near-endogenous levels without confounding effects of overexpression or ectopic expression of the gene of interest.

Similar to other small genetic model systems, *Drosophila* is well suited to conduct genetic screens, and several regulators of phosphoinositide metabolism proteins

Table I
Overview of the phosphoinositide kinases and phosphatases in *Drosophila* melanogaster.

PI kinase/phosphatase/ transfer proteins	Predominant substrate	Mammalian enzyme	Flybase ID	*Drosophila* gene	Reference in *Drosophila*
PI3K	PI(4,5)P_2	P110$\alpha-\delta$	FBgn0015279	CG4141/*Pi3K92E*	(MacDougall *et al.*, 1995)
	PI	PI(3)K-C2α,β,γ	FBgn0015278	CG11621/*Pi3K68D*	(MacDougall *et al.*, 1995; Molz *et al.*, 1996)
		Vps34	FBgn0015277	CG5373/*Pi3K59F*	(MacDougall *et al.*, 1995)
PI4K	PI	PI(4)K IIα	FBgn0037339	CG2929/*Pi4KIIalpha*	(Barylko *et al.*, 2002)
		PI(4)K IIβ	FBgn0037339	CG2929/*Pi4KIIalpha*	(Barylko *et al.*, 2002)
		PI(4)K IIIα	FBgn0040335	CG10260	
		PI(4)K IIIβ	FBgn0004373	CG7004/*fwd*	(Brill *et al.*, 2000)
PI3P5K	PI(3)P	PIKFYVE	FBgn0028741	CG6355/*fab1*	(Rusten *et al.*, 2006)
PI4P5K	PI(4)P	PIP5K1A or B	FBgn0039924	CG17471	
		PIP5K1A or B	FBgn0016984	CG9985/*sktl*	(Hassan *et al.*, 1998; Kania *et al.*, 1995; Knirr *et al.*, 1997)
		PIP5K1A or B	FBgn0034789	CG3682/*PIP5K59B*	
3-phosphatases	PI(3,4)P_2, PI(3,4,5)P_3	PTEN1,2, TPIP	FBgn0026379	CG5671/*Pten*	(Smith *et al.*, 1999)
	PI(5)P, PI(3,5)P_2	PLIP	FBgn0039111	CG10371/*plip*	(Pagliarini *et al.*, 2004)
	PI(3)P, PI(3,5)P_2	MTM1	FBgn0025742	CG9115/*mtm*	(Laporte *et al.*, 1998; Velichkova *et al.*)
	PI(3)P, PI(3,5)P_2	MTMRs	FBgn0025742	CG9115/*mtm*	(Laporte *et al.*, 1998; Velichkova *et al.*)
		MTMRs	FBgn0030735	CG3632	
		MTMRs	FBgn0028497	CG3530	
		MTMRs	FBgn0035945	CG5026	
4-phosphatases	PI(3,4)P_2	TypeI,II	FBgn0259166	CG42271	
	PI(4)P	Sac1-3[a]	FBgn0031611[a]	CG17840[a]	
		Sac1-3[b]	FBgn0035195[b]	CG9128/*sac1*[b]	(Wei *et al.*, 2003)
	PI(4,5)P_2	TMEM55A-B	FBgn0036058	CG6707	
5-phosphatases	PI(4,5)P_2, PI(3,4,5)P_3	Synj1,2	FBgn0034690	CG6562/*synj*	(Verstreken *et al.*, 2003)
		5-phosphatase IV[c]	Na[c]	Na[c]	
	PI(4,5)P_2	OCRL	FBgn0023508	CG3573/*ocrl*	(Ben El Kadhi *et al.*, 2011)
	PI(3,4,5)P_3	SKIP	FBgn0034179	CG6805	
		SKIP	FBgn0030761	CG9784	
		PIPP	FBgn0034179	CG6805	
		SHIP1,2[c]	Na[c]	Na[c]	
		72 kD 5-phosphatase	FBgn0036273	CG10426	
		5- phosphatase II/ INPP5B	FBgn0023508	CG3573	
Phosphatidylinositol transfer proteins		PITPNA-B	FBgn0260992	CG5269/vib	(Giansanti *et al.*, 2006)
		PITPNM1-2	FBgn0003218	CG11111/rdgB	(Hotta and Benzer, 1969; Vihtelic *et al.*, 1991)
		PITPNM3	FBgn0260992	CG5269/vib	(Giansanti *et al.*, 2006)

[a] Based on the yeast Fig4p protein sequence.
[b] Based on the yeast Sac1p protein sequence.
[c] No homolog in Drosophila melanogaster.

fwd, four wheel drive; INPP5B, inositol polyphosphate-5-phosphatase; 75kDa MTM, Myotubularin; MTMR, Myotubularin-related protein; Na, not available; OCRL, Oculocerebrorenal syndrome of Lowe; PIPP, Proline-rich inositol polyphosphate 5-phosphatase; PITP, Phosphatidylinositol transfer proteins; PLIP, Phosphoinositide lipid phosphatase; PTEN, Phosphatase and tensin homolog deleted on chromosome 10; rdgB, retinal degeneration B; SHIP, SH2-containing inositol phosphatase; SKIP, Skeletal muscle- and kidney-enriched inositol phosphatase; sktl, skittles; Synj, Synaptojanin; TPIP, TPTE and PTEN homologous inositol lipid phosphatase; TMEM, transmembrane protein; vib, vibrator.

Barylko, B., Wlodarski, P., Binns, D. D., Gerber, S. H., Earnest, S., Sudhof, T. C., Grichine, N., and Albanesi, J. P. (2002). Analysis of the catalytic domain of phosphatidylinositol 4-kinase type II. *J. Biol. Chem.* **277**, 44366–44375.

Ben El Kadhi, K., Roubinet, C., Solinet, S., Emery, G., Carreno, S., (2011). The inositol 5-phosphatase dOCRL controls PI(4,5)P_2 homeostasis and is necessary for cytokinesis. *Curr Biol.* 21, 1074–1079.

Brill, J. A., Hime, G. R., Scharer-Schuksz, M., and Fuller, M. T. (2000). A phospholipid kinase regulates actin organization and intercellular bridge formation during germline cytokinesis. *Development* **127**, 3855–3864.

Giansanti, M. G., Bonaccorsi, S., Kurek, R., Farkas, R. M., Dimitri, P., Fuller, M. T., and Gatti, M. (2006). The class I PITP giotto is required for Drosophila cytokinesis. *Curr. Biol.* **16**, 195–201.

Hotta, Y., and Benzer, S. (1969). Abnormal electroretinograms in visual mutants of Drosophila. *Nature* **222**, 354–356.

Kania, A., Salzberg, A., Bhat, M., D'Evelyn, D., He, Y., Kiss, I., and Bellen, H. J. (1995). P-element mutations affecting embryonic peripheral nervous system development in Drosophila melanogaster. *Genetics* **139**, 1663–1678.

Knirr, S., Breuer, S., Paululat, A., and Renkawitz-Pohl, R. (1997). Somatic mesoderm differentiation and the development of a subset of pericardial cells depend on the not enough muscles (nem) locus, which contains the inscuteable gene and the intron located gene, skittles. *Mech. Dev.* **67**, 69–81.

Laporte, J., Blondeau, F., Buj-Bello, A., Tentler, D., Kretz, C., Dahl, N., and Mandel, J. L. (1998). Characterization of the myotubularin dual specificity phosphatase gene family from yeast to human. *Hum. Mol. Genet.* **7**, 1703–1712.

MacDougall, L. K., Domin, J., and Waterfield, M. D. (1995). A family of phosphoinositide 3-kinases in Drosophila identifies a new mediator of signal transduction. *Curr. Biol.* **5**, 1404–1415.

Molz, L., Chen, Y. W., Hirano, M., and Williams, L. T. (1996). Cpk is a novel class of Drosophila PtdIns 3-kinase containing a C2 domain. *J. Biol. Chem.* **271**, 13892–13899.

Pagliarini, D. J., Worby, C. A., and Dixon, J. E. (2004). A PTEN-like phosphatase with a novel substrate specificity. *J. Biol. Chem.* **279**, 38590–38596.

Rusten, T. E., Rodahl, L. M., Pattni, K., Englund, C., Samakovlis, C., Dove, S., Brech, A., and Stenmark, H. (2006). Fab1 phosphatidylinositol 3-phosphate 5-kinase controls trafficking but not silencing of endocytosed receptors. *Mol. Biol. Cell* **17**, 3989–4001.

Smith, A., Alrubaie, S., Coehlo, C., Leevers, S. J., and Ashworth, A. (1999). Alternative splicing of the Drosophila PTEN gene. *Biochim. Biophys. Acta* **1447**, 313–317.

Velichkova, M., Juan, J., Kadandale, P., Jean, S., Ribeiro, I., Raman, V., Stefan, C., and Kiger, A. A. Drosophila Mtm and class II PI3K coregulate a PI(3)P pool with cortical and endolysosomal functions. *J. Cell Biol.* **190**, 407–425.

Vihtelic, T. S., Hyde, D. R., and O'Tousa, J. E. (1991). Isolation and characterization of the Drosophila retinal degeneration B (rdgB) gene. *Genetics* **127**, 761–768.

Wei, H. C., Shu, H., and Price, J. V. (2003). Functional genomic analysis of the 61D-61F region of the third chromosome of Drosophila melanogaster. *Genome* **46**, 1049–1055 have thus been identified. Classically, flies are mutagenized, either using chemical mutagens (EMS) or using transposons and progeny is screened for phenotypic abnormalities, but RNAi based screens are possible as well (Dietzl *et al.*, 2007; Lewis and Bacher, 1968; Ni *et al.*, 2009). Such efforts have led to the identification of for example, *skittles*, a 1-phosphatidylinositol-4-phosphate 5-kinase (PI(4)P5K), that modulates developmental signaling (Hassan *et al.*, 1998), as well as *tweek* and *synaptojanin* (*synj*), two genes that control synaptic vesicle endocytosis (Hassan *et al.*, 1998; Verstreken *et al.*, 2003, 2009). While genetic screens specifically designed to isolate phosphoinositide signaling molecules have yet to be performed, proteins involved have been found by screening for mutants in the pathways these phosphoinositide lipids are regulating, including nervous system development and synaptic transmission.

B. Visualization of Phosphoinositides at *Drosophila* Neuromuscular Junctions

Phosphatidylinositols are low abundance lipids, and the lack of high affinity antibodies hampers their localization in the cell. However, protein domains that recognize specific phosphatidylinositol species have been identified. Similar to other species, fusion proteins of such domains to GFP or Cherry expressed under control of the Gal4/UAS system have been used to visualize phosphoinositides in *Drosophila*. For example, at the NMJ, the PI(4,5)P_2 binding probe $PLC_{\delta 1}$-PH-GFP labels plasma membranes, the PI(3)P binding probe 2xFYVE(Hrs)-GFP marks endosomes and the GRP1-PH-GFP has been used to study PI(3,4,5)P_3 related signaling (Compagnon *et al.*, 2009; Khuong *et al.*, 2010; Verstreken *et al.*, 2009; Wucherpfennig *et al.*, 2003) (Fig. 2). Numerous additional probes have been described and used in various cell types. However, in fruit flies, only $PLC_{\delta 1}$-PH-GFP, $PLC_{\delta 1}$-PH-Cherry, 2xFYVE-GFP, 2xFYVE-Cherry, and GRP1-PH-GFP have been used (Britton *et al.*, 2002; Compagnon *et al.*, 2009; Khuong *et al.*, 2010; Wucherpfennig *et al.*, 2003.

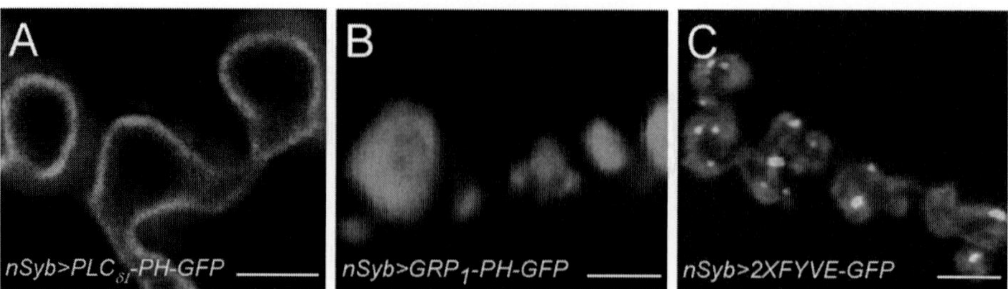

Fig. 2 NMJs of animals neuronally expressing phosphoinositide-binding protein domains. (A) NMJs expressing $PLC_{\delta 1}$-PH-GFP (green; nSyb-Gal4 > PLCδ1-PH-GFP) are labeled with anti-DLG, a mostly postsynaptic marker (blue). (B) NMJs expressing GRP1-PH-GFP (green; nSyb-Gal4 > GRP1-PH-GFP). (C) NMJs expressing 2xFYVE-GFP (green; nSyb-Gal4 > FYVE-GFP) are labeled with anti-Synaptotagmin, marking synaptic vesicles (red). Scale bar: 5 μm. (For interpretation of the references to color in this figure legend, the reader is referred to the web version of this book.)

Generating transgenic flies that allow tissue specific expression of additional probes will greatly facilitate *in vivo* phosphoinositide research.

PI(3)P is an endosomal marker that binds several different proteins including early endosomal autoantigen (EEA1), hepatocyte growth factor-regulated tyrosine kinase substrate (Hrs), and sorting nexins (SNX) (Vicinanza *et al.*, 2008). The FYVE finger domains of EEA1 and Hrs fused to fluorescent proteins have been widely used in yeast and mammalian cells to study vacuolar/lysosomal and endosomal sorting (Burd and Emr, 1998; Gillooly *et al.*, 2000; Katzmann *et al.*, 2003). In order to study the recycling pathways of synaptic vesicles, 2XFYVE(Hrs)-GFP has been used at the presynaptic terminals of the fly NMJ enabling to monitor endosomal compartments in vesicle trafficking at a live synapse (Uytterhoeven *et al.*, 2011; Wucherpfennig *et al.*, 2003). These studies showed that synaptic vesicles travel through endosomes and that these stations are important to sort dysfunctional synaptic vesicle components for degradation at the lysosome.

PI(4,5)P_2 is one of the best characterized phosphoinositide species and has also been studied in *Drosophila*. In the *Drosophila* germline, PH-PLC$_{\delta 1}$-GFP has been used to study the role of Rab5 and PI(4,5)P_2 during oogenesis. This study shows that in the germline cyst, PI(4,5)P_2 is necessary for endocytic vesicle formation and endocytosis of yolk protein and recruitment of Rab5 to the endosome. In motor neurons, PH-PLC$_{\delta 1}$-GFP or PH-PLC$_{\delta 1}$-Cherry decorate the plasma membrane at synaptic boutons (Fig. 2A) (Khuong *et al.*, 2010; Verstreken *et al.*, 2009). The probe labels PI(4,5)P_2 specifically, because fluorescence levels of the probe are increased in *synj* mutants (Verstreken *et al.*, 2009). Synj is an evolutionary conserved polyphosphoinositide phosphatase, and *synj1* knock-out mice show increased PI(4,5)P_2 levels as well (Cremona *et al.*, 1999). Interestingly, more recently, other mutants that alter synaptic PH-PLC$_{\delta 1}$-GFP levels have also been identified. For example Tweek, a large evolutionary conserved protein, was isolated in a genome wide screen for defects in synaptic transmission. While the exact role of Tweek in phosphoinositide metabolism is not known, PH-PLC$_{\delta 1}$-GFP levels are significantly reduced in the mutants, and stimulation of motor neurons shows an abnormal accumulation of PH-PLC$_{\delta 1}$-GFP in endocytic structures not observed in controls. In line with altered synaptic PI(4,5)P_2 handling, *tweek* mutants show synaptic vesicle endocytosis defects and concomitant deficits in NMJ morphology (Khuong *et al.*, 2010; Verstreken *et al.*, 2009). However, only recently introduced in *Drosophila*, PH- PLC$_{\delta 1}$-GFP has been used to study a wide range of membrane-related biological events including phagocytosis, regulation of cell shape and mobility and exocytosis. Thus, PH-PLC$_{\delta 1}$-GFP and related probes are great tools to assess phosphoinositide concentration and localization *in vivo*.

Receptor mediated production of PI(3,4,5)P_3 plays an important role in insulin stimulated glucose uptake, cell survival, chemotaxis and cell adhesion (Greenwood *et al.*, 2000; Halet *et al.*, 2008; Ikushima *et al.*, 2010). The PI(3,4,5)P_3 binding protein GRP1-PH-GFP has been used in several cell types including PC12 cells and adipocytes to monitor receptor mediated activation of the PI3K pathway. Due to low abundance of PI(3,4,5)P_3 in resting cells GRP1-PH-GFP is not bound to PI(3,4,5)P_3

and is localized mainly in the cytoplasm. However, upon activation of the class I PI3K pathway by addition of nerve and epidermal growth factor or insulin this probe localizes to the increased concentrations of PI(3,4,5)P_3 at the plasma membrane (Oatey *et al.*, 1999; Venkateswarlu *et al.*, 1998a, 1998b). In *Drosophila* S2 cells as well as in the epidermis and fat body of larvae, GRP1-PH-GFP localizes mainly in the cytoplasm and nucleus. Insulin-induced Class I PI3-K activity recruits the probe to the plasma membranes that can be abolished by the PI3K inhibitor Wortmannin, consistent with the results obtained in cells (Britton *et al.*, 2002). Similar to S2 cells, when expressed specifically in motor neurons, GRP1-PH-GFP localizes throughout the bouton lumen (TK & PV, unpublished data; Fig. 2B). Given the low levels of PI (3,4,5)P_3 in the cell, most of the probe is likely floating in the boutonic cytoplasm and specific patches of concentrated probe bound to PI(3,4,5)P_3 may not be easily revealed at the synapse.

Expression of specific protein domains fused to GFP may visualize the location and abundance of specific phosphoinositides; however, this manipulation also effectively reduces available binding sites for physiological relevant processes thus impeding with phosphoinositide signaling cascades (Field *et al.*, 2005; Holz *et al.*, 2000; Khuong *et al.*, 2010; Raucher *et al.*, 2000). While this dominant negative effect may be desired in some instances, in *Drosophila* several mechanisms to limit expression of probe in time exist and could be used to avoid long-term exposure to the phosphoinositide binding proteins. Some of these technologies include temperature-, hormone- or tetracycline-controlled expression of the probes in time. When combined with the UAS/Gal4 system, these tools allow researchers to define cell type and time controlled expression of GFP fused phosphoinositide binding proteins (Elliott and Brand, 2008).

III. Cellular Processes Regulated by Phosphoinositides at the Fly NMJ

The application of genetics linked with phenotypic analysis at *Drosophila* NMJs constitutes an ideal platform to investigate the cellular role of phosphoinositides in synaptic development and function. In the following section, we will illustrate how the NMJ has been used to elucidate an involvement of PI(4,5)P_2 in regulating NMJ morphology and synaptic vesicle cycling.

A. Phosphoinositides in the Regulation of NMJ Morphology

Phosphoinositides have been shown to play an important role in membrane shaping during physiological processes such as chemotaxis, cytokinesis, and morphogenesis (Saarikangas *et al.*, 2010). A well known example is the activation of neuronal Wiscott-Aldrich syndrome protein (WSP) by the cooperative binding to PI(4,5)P_2 and Cdc42 to mediate ARP2/3 dependent actin polymerization

(Saarikangas *et al.*, 2010). Recently, a role of PI(4,5)P_2 in restricting NMJ morphology has been found by using several methods to alter PI(4,5)P_2 synaptic levels. Over expression of $PLC_{\delta1}$-PH-GFP can shield PI(4,5)P_2 from its endogenous binding partners (Field *et al.*, 2005; Holz *et al.*, 2000; Raucher *et al.*, 2000) and results in extensive NMJ overgrowth (Khuong *et al.*, 2010). This defect is specific, because different genetic manipulations of PI(4,5)P_2 at the larval NMJ also indicate a PI(4,5)P_2 dependent restriction of NMJ growth. Indeed, over expression of Class I PI3K or expression of RNAi to biosynthetic enzymes such as PI4K, PI4P5K that produce PI(4)P and PI(4,5)P_2, respectively, show larger NMJs with more boutons (Howlett *et al.*, 2008; Khuong *et al.*, 2010; Knox *et al.*, 2007; Martin-Pena *et al.*, 2006). Furthermore, removing a single copy of the phosphoinositide phosphatase, *synj* restores NMJ overgrowth caused by the overexpression of PH-$PLC_{\delta1}$ (Khuong *et al.*, 2010). Thus, genetic manipulation of PI(4,5)P_2 levels at NMJs indicates these lipids limit NMJ growth.

The regulated turnover of phosphoinositides is of special importance to regulate specific downstream signaling cascades. These pathways can be studied by genetic targeting of phosphoinositide effectors. For example, overexpression of specific components of the classic insulin receptor signaling pathway, protein kinase B/Akt (PKB/AKT) and Glycogen synthase kinase 3 (GSK3) results in an increase in synapse number suggesting that these components play a regulatory role in synaptogenesis downstream of PI(3,4,5)P_3 (Martin-Pena *et al.*, 2006).

Alternatively phosphoinositide-binding domains in signaling proteins necessary to mediate downstream signaling can be mutated and be used in genetic epistasis experiments to further characterize the phosphoinositide network. For example, *wsp* mutants that specifically lack their PI(4,5)P_2 binding domain also show larger NMJ terminals, very similar to those observed by genetically reducing PI(4,5)P_2 and the defect in these *wsp* mutants cannot be exacerbated by additionally expressing PH-$PLC_{\delta1}$. Further experiments suggest that WSP achieves this function independently of its ability to bind Cdc42. These data indicate that at the NMJ, PI(4,5)P_2 restricts synaptic growth by mediating *wsp* activation (Khuong *et al.*, 2010; Tal *et al.*, 2002), in line with previous *in vitro* experiments where *wsp* is activated by PI(4,5)P_2 alone to mediate ARP2/3 dependent actin polymerization (Papayannopoulos *et al.*, 2005).

WSP activation independently from Cdc42 seems to be dependent on the fine regulation PI(4,5)P_2 levels. Biochemical data suggest that WSP can be regulated by subtle signal induced alterations of PI(4,5)P_2 distribution at the plasma membrane (Papayannopoulos *et al.*, 2005). In this context, Tweek, a novel protein recently identified in *Drosophila*, is implicated in regulating the PI(4,5)P_2 distribution at the plasma membrane (Verstreken *et al.*, 2009). *Tweek* mutants harbor lower availability of boutonic PI(4,5)P_2 and consequently excessive NMJ overgrowth. Genetic interaction experiments indicate that Tweek, PI(4,5)P_2, and WSP act in a common genetic pathway to control NMJ morphology (Khuong *et al.*, 2010). Thus, targeting different phosphoinositide network components, in combination with phosphoinositide imaging and genetic epistasis has allowed to dissect pathways of synaptic growth downstream of PI(4,5)P_2.

B. Phosphoinositides in Synaptic Transmission

Phosphoinositides play an important role in membrane trafficking and at the fly NMJ, research has mainly been focused on the role of PI(4,5)P_2. PI(4,5)P_2 is critical for clathrin-mediated endocytosis at NMJ synaptic boutons, and the lipid acts as a co-receptor for the recruitment of endocytic proteins such as α-adaptin, a member of the AP-2 adaptor complex and Like-AP180(LAP)/AP180, another clathrin adaptor protein (Balla *et al.*, 2009). *Drosophila* mutants that harbor reduced synaptic PI(4,5)P_2 (e.g., *tweek* mutants but also animals that over express PH-$PLC_{\delta 1}$ domains), show decreased levels and mislocalization of endocytic adaptor proteins, including α-adaptin, LAP/AP180 and StonedB, another adaptor like protein (Gonzalez-Gaitan and Jackle, 1997; Stimson *et al.*, 2001; Verstreken *et al.*, 2009; Zhang *et al.*, 1998). Increasing PI(4,5)P_2 in *tweek* mutants by removing a copy of *synj* can partially restore α-adaptin recruitment and synaptic vesicle recycling (Verstreken *et al.*, 2009). Abolishing *synj* function, resulting in increased boutonic PI(4,5)P_2, leads to an accumulation of protein-coated vesicles in electron microscopy (Dickman *et al.*, 2005; Gad *et al.*, 2000; Harris *et al.*, 2000; Stefan *et al.*, 2002; Van Epps *et al.*, 2004; Verstreken *et al.*, 2003). These data suggest that under the conditions tested, adaptor proteins show increased affinity for synaptic vesicle membranes (Cremona *et al.*, 1999; Verstreken *et al.*, 2009). These data corroborate findings in other systems and indicate that PI(4,5)P_2 recruits endocytic adaptors to the membrane to initiate synaptic vesicle formation.

Synj is critically important for synaptic vesicle recycling (Cremona *et al.*, 1999). At the fruit fly NMJ, loss of *synj* results in dramatic defects in synaptic vesicle endocytosis as evidenced by a defect to maintain neurotransmission during long periods of stimulation, a vast reduction in synaptic vesicle number as well as an accumulation of endocytic vesicles that harbor electron dense proteinaceous coats (Dickman *et al.*, 2005; Verstreken *et al.*, 2003). While in mouse a PI(4)P5K was shown to counteract Synj (Di Paolo *et al.*, 2004), such an enzyme has not been characterized at the *Drosophila* NMJ. Nonetheless, the data further support a phosphoinositide cycle where the phosphorylation of phosphoinositides recruits adaptors to initiate synaptic vesicle endocytosis. The dephosphorylation of phosphoinositides following vesicle formation then leads to shedding of the adaptor and clathrin coat liberating the newly formed vesicle into the cytoplasm. In this context, Endophilin (Endo) tightly binds Synj (Ringstad *et al.*, 1997, 1999) and association to Endo activates Synj enzymatic activity (Chang-Ileto *et al.*, 2011; Lee *et al.*, 2004). At the fly NMJ *endo* null mutants show phenotypes indistinguishable from *synj* mutants, and also *synj; endo* double mutants show a phenotype that is very similar to either single mutant (Verstreken *et al.*, 2003), observations also made at *C. elegans* NMJs (Schuske *et al.*, 2003). Similarly, *endo* mutants also show an accumulation of densely coated vesicles both at the NMJ as well as in mutant photoreceptor cells, and in *endo* null mutants, Synj is destabilized and mislocalized (Schuske *et al.*, 2003; Verstreken *et al.*, 2003). While these data suggest that Endo serves to recruit and activate Synj during vesicle formation, recent data using *C. elegans* suggests that the SH3 and proline rich domains that mediate Endo-Synj binding are dispensable for synaptic endocytosis, and further studies may thus be required (Bai *et al.*, 2010). Nonetheless, the phenotypic analysis of these mutants indicates a link to phosphoinositide metabolism and the regulation of synaptic vesicle recycling at the NMJ in *Drosophila*.

IV. Future Directions

Phosphoinositides play multiple roles in a variety of signaling pathways, and prolonged activation or inactivation of a PI-kinase or phosphatase, for example, using genetic ablation, may mask acute effects of the lipids on cellular physiology. Furthermore, proper activation of signaling may require acute activation and/or inactivation of PI-kinases or phosphatases at specific locations in the cell and these conditions are often difficult to recapitulate with systemic loss- or gain of function. While at the fly NMJ, only chronic manipulation of PI-kinase or phosphatase function has been employed, technologies designed to acutely alter local phosphoinositide levels are within reach.

An elegant strategy that addresses concerns of local and acute phosphoinositide production or break-down builds on the rapamycin induced heterodimerization of two proteins (FK506 and FRB), one linked to a targeting domain (e.g., a plasma membrane binding moiety, endosomal targeting sequence, etc.), the other linked to a specific PI-kinase or phosphatase domain. Addition of the membrane permeable rapamycin (or its inactive analogue iRap) results in the dimerization of both proteins and acute as well as local alteration of phosphoinositide species (Belshaw *et al.*, 1996). This strategy has been successfully used to manipulate $PI(4,5)P_2$ and $PI(3,4,5)P_3$ at the plasma membrane as well as PI(3)P at sorting endosomes and PI(4)P at the golgi complex of cells *in vitro* (Abe *et al.*, 2008; Fili *et al.*, 2006; Suh *et al.*, 2006; Szentpetery *et al.*, 2010; Varnai *et al.*, 2006; Zoncu *et al.*, 2007). Clearly this system holds great potential to bring complex cellular events regulated by phosphoinositides under molecular control, and would be an asset to study phosphoinositide signaling at the *Drosophila* NMJ.

A second approach encompasses the ability to acutely and locally inactivate protein function by virtue of fluorescein assisted light inactivation (FALI) in combination with the membrane permeable compound 4'5'-bis(1,3,2-dithioarsolan-2-yl)fluorescein (FlAsH) (Griffin *et al.*, 1998; Jacobson *et al.*, 2008). Transgenically expressed tetracysteine (4C) tagged proteins can be bound by the membrane permeable FlAsH. Excitation of FlAsH with $\sim$500 nm light results in the production of reactive oxygen species (ROS) within a radius of a few angstroms, inactivating the tagged protein within seconds (Beck *et al.*, 2002; Habets and Verstreken, 2010). This technology has been successfully used at the larval NMJ to study the role of synaptic proteins in vesicle fusion and recycling (Heerssen *et al.*, 2008; Kasprowicz *et al.*, 2008; Marek and Davis, 2002; Poskanzer *et al.*, 2003) but also holds potential to dissect the phosphoinositide network at the NMJ.

As indicated in Section IIB, the XFP probes used to measure phosphoinositide localization in cells may suffer from low signal to noise, hampering careful analysis of phosphoinositide dynamics at the NMJ. Indeed, the probes are usually over expressed and unbound fluorescent probe dwells in the cytoplasm, a particular problem for low-abundance phosphoinositides including PI(3,4,5)P_3. Although not yet applied at the fruit fly NMJ, FRET based probes hold the promise of increased signal to noise ratio. Here, the same phosphoinositide binding domain is fused to both CFP and YFP and co-expressed. Only when concentrated at phosphoinositide patches efficient CFP/YFP FRET will occur (Lankiewicz *et al.*, 1997; Pap *et al.*, 1993; Pollok and Heim, 1999; van der Wal *et al.*, 2001). Thus, FRET technology, and several related techniques (Ananthanarayanan *et al.*, 2005; Sato *et al.*, 2003), may allow increasing the resolution by which phosphoinositide dynamics can be visualized, particularly at the synaptic boutons of the *Drosophila* NMJ.

V. Conclusions

In this chapter we have highlighted the use of the *Drosophila* NMJ to gain further insight in the spatial and temporal regulation of phosphoinositides. The availability of a set of yet unused genetic tools in *Drosophila* together with the improvement of research tools such as the generation of more specific inhibitors, refined techniques for electron microscopy or live imaging should enable to answer numerous open questions that remain.

Extensive genetic modeling in combination with acute manipulation of phosphoinositide levels, novel imaging techniques, and functional assays may lead to the identification of novel pathways, components, and concepts of phosphoinositide signaling at the NMJ of *Drosophila*. This information will result in more specific insights in the fine regulation of phosphoinositides at synapses. Given the increasing implication of phosphoinositides in brain disorders such as Alzheimer's disease, syndrome of Down and bipolar disorder we will be able to apply this information to analyze candidate or disease related genes and screen for specific modulators of the phosphoinositide balance at the synapse. Further understanding of these diseases might provide valuable therapeutic targets and improved treatment of phosphoinositide related brain disorders.

Acknowledgments

We thank members of the Verstreken lab and Stein Aerts for constructive comments. Work in the Verstreken lab is supported by a Marie Curie Excellence grant (MEXT-CT-2006-042267); and ERC Starting Grant (260678), FWO grants (G074709, G094011N and G095511N), the Research Fund KU Leuven, a Methusalem grant of the Flemish Government and KULeuven, the Francqui Foundation, VIB, and an Institute for the Promotion of Innovation through Science and Technology in Flanders (IWT-Vlaanderen) O&O grant. JRS is supported by an Agency for Innovation by Science and Technology in Flanders (IWT) fellowship.

References

Abe, N., Inoue, T., Galvez, T., Klein, L., and Meyer, T. (2008). Dissecting the role of PtdIns(4,5)P2 in endocytosis and recycling of the transferrin receptor. *J. Cell Sci.* **121**, 1488–1494.

Adams, M. D., Celniker, S. E., Holt, R. A., Evans, C. A., Gocayne, J. D., Amanatides, P. G., Scherer, S. E., Li, P. W., Hoskins, R. A., Galle, R. F., George, R. A., Lewis, S. E., Richards, S., Ashburner, M., Henderson, S. N., Sutton, G. G., Wortman, J. R., Yandell, M. D., Zhang, Q., Chen, L. X., Brandon, R. C., Rogers, Y. H., Blazej, R. G., Champe, M., Pfeiffer, B. D., Wan, K. H., Doyle, C., Baxter, E. G., Helt, G., Nelson, C. R., Gabor, G. L., Abril, J. F., Agbayani, A., An, H. J., Andrews-Pfannkoch, C., Baldwin, D., Ballew, R. M., Basu, A., Baxendale, J., Bayraktaroglu, L., Beasley, E. M., Beeson, K. Y., Benos, P. V., Berman, B. P., Bhandari, D., Bolshakov, S., Borkova, D., Botchan, M. R., Bouck, J., Brokstein, P., Brottier, P., Burtis, K. C., Busam, D. A., Butler, H., Cadieu, E., Center, A., Chandra, I., Cherry, J. M., Cawley, S., Dahlke, C., Davenport, L. B., Davies, P., de Pablos, B., Delcher, A., Deng, Z., Mays, A. D., Dew, I., Dietz, S. M., Dodson, K., Doup, L. E., Downes, M., Dugan-Rocha, S., Dunkov, B. C., Dunn, P., Durbin, K. J., Evangelista, C. C., Ferraz, C., Ferriera, S., Fleischmann, W., Fosler, C., Gabrielian, A. E., Garg, N. S., Gelbart, W. M., Glasser, K., Glodek, A., Gong, F., Gorrell, J. H., Gu, Z., Guan, P., Harris, M., Harris, N. L., Harvey, D., Heiman, T. J., Hernandez, J. R., Houck, J., Hostin, D., Houston, K. A., Howland, T. J., Wei, M. H., Ibegwam, C., Jalali, M., Kalush, F., Karpen, G. H., Ke, Z., Kennison, J. A., Ketchum, K. A., Kimmel, B. E., Kodira, C. D., Kraft, C., Kravitz, S., Kulp, D., Lai, Z., Lasko, P., Lei, Y., Levitsky, A. A., Li, J., Li, Z., Liang, Y., Lin, X., Liu, X., Mattei, B., McIntosh, T. C., McLeod, M. P., McPherson, D., Merkulov, G., Milshina, N. V., Mobarry, C., Morris, J., Moshrefi, A., Mount, S. M., Moy, M., Murphy, B., Murphy, L., Muzny, D. M., Nelson, D. L., Nelson, D. R., Nelson, K. A., Nixon, K., Nusskern, D. R., Pacleb, J. M., Palazzolo, M., Pittman, G. S., Pan, S., Pollard, J., Puri, V., Reese, M. G., Reinert, K., Remington, K., Saunders, R. D., Scheeler, F., Shen, H., Shue, B. C., Siden-Kiamos, I., Simpson, M., Skupski, M. P., Smith, T., Spier, E., Spradling, A. C., Stapleton, M., Strong, R., Sun, E., Svirskas, R., Tector, C., Turner, R., Venter, E., Wang, A. H., Wang, X., Wang, Z. Y., Wassarman, D. A., Weinstock, G. M., Weissenbach, J., Williams, S. M., WoodageT., Worley, K. C., Wu, D., Yang, S., Yao, Q. A., Ye, J., Yeh, R. F., Zaveri, J. S., Zhan, M., Zhang, G., Zhao, Q., Zheng, L., Zheng, X. H., Zhong, F. N., Zhong, W., Zhou, X., Zhu, S., Zhu, X., Smith, H. O., Gibbs, R. A., Myers, E. W., Rubin, G. M., and Venter, J. C. (2000). The genome sequence of Drosophila melanogaster. *Science* **287**, 2185–2195.

Aikawa, Y., and Martin, T. F. (2003). ARF6 regulates a plasma membrane pool of phosphatidylinositol (4,5)bisphosphate required for regulated exocytosis. *J. Cell Biol.* **162**, 647–659.

Ananthanarayanan, B., Ni, Q., and Zhang, J. (2005). Signal propagation from membrane messengers to nuclear effectors revealed by reporters of phosphoinositide dynamics and Akt activity. *Proc. Natl. Acad. Sci. U. S. A.* **102**, 15081–15086.

Arcaro, A., and Wymann, M. P. (1993). Wortmannin is a potent phosphatidylinositol 3-kinase inhibitor: the role of phosphatidylinositol 3,4,5-trisphosphate in neutrophil responses. *Biochem. J.* **296**(Pt 2), 297–301.

Astle, M. V., Seaton, G., Davies, E. M., Fedele, C. G., Rahman, P., Arsala, L., and Mitchell, C. A. (2006). Regulation of phosphoinositide signaling by the inositol polyphosphate 5-phosphatases. *IUBMB Life* **58**, 451–456.

Bai, J., Hu, Z., Dittman, J. S., Pym, E. C., and Kaplan, J. M. (2010). Endophilin functions as a membrane-bending molecule and is delivered to endocytic zones by exocytosis. *Cell* **143**, 430–441.

Bai, J., Tucker, W. C., and Chapman, E. R. (2004). PIP2 increases the speed of response of synaptotagmin and steers its membrane-penetration activity toward the plasma membrane. *Nat. Struct. Mol. Biol.* **11**, 36–44.

Balla, A., and Balla, T. (2006). Phosphatidylinositol 4-kinases: old enzymes with emerging functions. *Trends Cell Biol.* **16**, 351–361.

Balla, T., Szentpetery, Z., and Kim, Y. J. (2009). Phosphoinositide signaling: new tools and insights. *Physiology (Bethesda)* **24**, 231–244.

Beck, S., Sakurai, T., Eustace, B. K., Beste, G., Schier, R., Rudert, F., and Jay, D. G. (2002). Fluorophore-assisted light inactivation: a high-throughput tool for direct target validation of proteins. *Proteomics* **2**, 247–255.

Bellen, H. J., Levis, R. W., Liao, G., He, Y., Carlson, J. W., Tsang, G., Evans-Holm, M., Hiesinger, P. R., Schulze, K. L., Rubin, G. M., Hoskins, R. A., and Spradling, A. C. (2004). The BDGP gene disruption project: single transposon insertions associated with 40% of Drosophila genes. *Genetics* **167**, 761–781.

Belshaw, P. J., Ho, S. N., Crabtree, G. R., and Schreiber, S. L. (1996). Controlling protein association and subcellular localization with a synthetic ligand that induces heterodimerization of proteins. *Proc. Natl. Acad. Sci. U. S. A.* **93**, 4604–4607.

Berridge, M. J. (1983). Rapid accumulation of inositol trisphosphate reveals that agonists hydrolyse polyphosphoinositides instead of phosphatidylinositol. *Biochem J.* **212**, 849–858.

Bier, E., Vaessin, H., Shepherd, S., Lee, K., McCall, K., Barbel, S., Ackerman, L., Carretto, R., Uemura, T., and Grell, E., *et al.* (1989). Searching for pattern and mutation in the Drosophila genome with a P-lacZ vector. *Genes Dev.* **3**, 1273–1287.

Bischof, J., Maeda, R. K., Hediger, M., Karch, F., and Basler, K. (2007). An optimized transgenesis system for Drosophila using germ-line-specific phiC31 integrases. *Proc. Natl. Acad. Sci. U. S. A.* **104**, 3312–3317.

Brand, A. H., and Perrimon, N. (1993). Targeted gene expression as a means of altering cell fates and generating dominant phenotypes. *Development* **118**, 401–415.

Britton, J. S., Lockwood, W. K., Li, L., Cohen, S. M., and Edgar, B. A. (2002). Drosophila's insulin/PI3-kinase pathway coordinates cellular metabolism with nutritional conditions. *Dev. Cell* **2**, 239–249.

Budnik, V., Gorczyca, M., and Prokop, A. (2006). Selected methods for the anatomical study of Drosophila embryonic and larval neuromuscular junctions. *Int. Rev. Neurobiol.* **75**, 323–365.

Burd, C. G., and Emr, S. D. (1998). Phosphatidylinositol(3)-phosphate signaling mediated by specific binding to RING FYVE domains. *Mol. Cell* **2**, 157–162.

Cantley, L. C. (2002). The phosphoinositide 3-kinase pathway. *Science* **296**, 1655–1657.

Carlton, J. G., and Cullen, P. J. (2005). Coincidence detection in phosphoinositide signaling. *Trends Cell Biol.* **15**, 540–547.

Chang-Ileto, B., Frere, S. G., Chan, R. B., Voronov, S. V., Roux, A., and Di Paolo, G. (2011). Synaptojanin 1-mediated PI(4,5)P2 hydrolysis is modulated by membrane curvature and facilitates membrane fission. *Dev. Cell* **20**, 206–218.

Chilvers, E. R., Batty, I. H., Challiss, R. A., Barnes, P. J., and Nahorski, S. R. (1991). Determination of mass changes in phosphatidylinositol 4,5-bisphosphate and evidence for agonist-stimulated metabolism of inositol 1,4,5-trisphosphate in airway smooth muscle. *Biochem J.* **275**(Pt 2), 373–379.

Choi, C. M., Vilain, S., Langen, M., Van Kelst, S., De Geest, N., Yan, J., Verstreken, P., and Hassan, B. A. (2009). Conditional mutagenesis in Drosophila. *Science* **324**, 54.

Compagnon, J., Gervais, L., Roman, M. S., Chamot-Boeuf, S., and Guichet, A. (2009). Interplay between Rab5 and PtdIns(4,5)P2 controls early endocytosis in the Drosophila germline. *J. Cell Sci.* **122**, 25–35.

Cooley, L., Kelley, R., and Spradling, A. (1988). Insertional mutagenesis of the Drosophila genome with single P elements. *Science* **239**, 1121–1128.

Cremona, O., Di Paolo, G., Wenk, M. R., Luthi, A., Kim, W. T., Takei, K., Daniell, L., Nemoto, Y., Shears, S. B., Flavell, R. A., McCormick, D. A., and De Camilli, P. (1999). Essential role of phosphoinositide metabolism in synaptic vesicle recycling. *Cell* **99**, 179–188.

De Matteis, M. A., and Godi, A. (2004). PI-loting membrane traffic. *Nat. Cell Biol.* **6**, 487–492.

Di Paolo, G., and De Camilli, P. (2006). Phosphoinositides in cell regulation and membrane dynamics. *Nature* **443**, 651–657.

Di Paolo, G., Moskowitz, H. S., Gipson, K., Wenk, M. R., Voronov, S., Obayashi, M., Flavell, R., Fitzsimonds, R. M., Ryan, T. A., and De Camilli, P. (2004). Impaired PtdIns(4,5)P2 synthesis in nerve terminals produces defects in synaptic vesicle trafficking. *Nature* **431**, 415–422.

Dickman, D. K., Horne, J. A., Meinertzhagen, I. A., and Schwarz, T. L. (2005). A slowed classical pathway rather than kiss-and-run mediates endocytosis at synapses lacking synaptojanin and endophilin. *Cell* **123**, 521–533.

Dietzl, G., Chen, D., Schnorrer, F., Su, K. C., Barinova, Y., Fellner, M., Gasser, B., Kinsey, K., Oppel, S., Scheiblauer, S., Couto, A., Marra, V., Keleman, K., and Dickson, B. J. (2007). A genome-wide transgenic RNAi library for conditional gene inactivation in Drosophila. *Nature* **448**, 151–156.

Doughman, R. L., Firestone, A. J., and Anderson, R. A. (2003). Phosphatidylinositol phosphate kinases put PI4,5P(2) in its place. *J. Membr. Biol.* **194**, 77–89.

Eberhard, D. A., Cooper, C. L., Low, M. G., and Holz, R. W. (1990). Evidence that the inositol phospholipids are necessary for exocytosis. Loss of inositol phospholipids and inhibition of secretion in permeabilized cells caused by a bacterial phospholipase C and removal of ATP. *Biochem. J.* **268**, 15–25.

Elliott, D. A., and Brand, A. H. (2008). The GAL4 system : a versatile system for the expression of genes. *Methods Mol. Biol.* **420**, 79–95.

Feng, L., Mejillano, M., Yin, H. L., Chen, J., and Prestwich, G. D. (2001). Full-contact domain labeling: identification of a novel phosphoinositide binding site on gelsolin that requires the complete protein. *Biochemistry* **40**, 904–913.

Field, S. J., Madson, N., Kerr, M. L., Galbraith, K. A., Kennedy, C. E., Tahiliani, M., Wilkins, A., and Cantley, L. C. (2005). PtdIns(4,5)P2 functions at the cleavage furrow during cytokinesis. *Curr. Biol.* **15**, 1407–1412.

Fili, N., Calleja, V., Woscholski, R., Parker, P. J., and Larijani, B. (2006). Compartmental signal modulation: Endosomal phosphatidylinositol 3-phosphate controls endosome morphology and selective cargo sorting. *Proc. Natl. Acad. Sci. U. S. A.* **103**, 15473–15478.

Foster, F. M., Traer, C. J., Abraham, S. M., and Fry, M. J. (2003). The phosphoinositide (PI) 3-kinase family. *J. Cell Sci.* **116**, 3037–3040.

Gad, H., Ringstad, N., Low, P., Kjaerulff, O., Gustafsson, J., Wenk, M., Di Paolo, G., Nemoto, Y., Crun, J., Ellisman, M. H., De Camilli, P., Shupliakov, O., and Brodin, L. (2000). Fission and uncoating of synaptic clathrin-coated vesicles are perturbed by disruption of interactions with the SH3 domain of endophilin. *Neuron* **27**, 301–312.

Gao, G., McMahon, C., Chen, J., and Rong, Y. S. (2008). A powerful method combining homologous recombination and site-specific recombination for targeted mutagenesis in Drosophila. *Proc. Natl. Acad. Sci. U. S. A.* **105**, 13999–14004.

Gillooly, D. J., Morrow, I. C., Lindsay, M., Gould, R., Bryant, N. J., Gaullier, J. M., Parton, R. G., and Stenmark, H. (2000). Localization of phosphatidylinositol 3-phosphate in yeast and mammalian cells. *EMBO J.* **19**, 4577–4588.

Gonzalez-Gaitan, M., and Jackle, H. (1997). Role of Drosophila alpha-adaptin in presynaptic vesicle recycling. *Cell* **88**, 767–776.

Greenwood, J. A., Theibert, A. B., Prestwich, G. D., and Murphy-Ullrich, J. E. (2000). Restructuring of focal adhesion plaques by PI 3-kinase. Regulation by PtdIns (3,4,5)-p(3) binding to alpha-actinin. *J. Cell Biol.* **150**, 627–642.

Griffin, B. A., Adams, S. R., and Tsien, R. Y. (1998). Specific covalent labeling of recombinant protein molecules inside live cells. *Science* **281**, 269–272.

Grishanin, R. N., Kowalchyk, J. A., Klenchin, V. A., Ann, K., Earles, C. A., Chapman, E. R., Gerona, R. R., and Martin, T. F. (2004). CAPS acts at a prefusion step in dense-core vesicle exocytosis as a PIP2 binding protein. *Neuron* **43**, 551–562.

Groth, A. C., Fish, M., Nusse, R., and Calos, M. P. (2004). Construction of transgenic Drosophila by using the site-specific integrase from phage phiC31. *Genetics* **166**, 1775–1782.

Habets, R. L. P., and Verstreken, P. (2010). Acute Inactivation of Proteins Using FlAsH-FALI. *In* "Drosophila Neurobiology: A Laboratory Manual," (B. Zhang, M. R. Freeman, S. Waddell, eds.), pp. 393–409. John Inglis.

Halet, G., Viard, P., and Carroll, J. (2008). Constitutive PtdIns(3,4,5)P3 synthesis promotes the development and survival of early mammalian embryos. *Development* **135**, 425–429.

Harris, T. W., Hartwieg, E., Horvitz, H. R., and Jorgensen, E. M. (2000). Mutations in synaptojanin disrupt synaptic vesicle recycling. *J. Cell Biol.* **150**, 589–600.

Hassan, B. A., Prokopenko, S. N., Breuer, S., Zhang, B., Paululat, A., and Bellen, H. J. (1998). skittles, a Drosophila phosphatidylinositol 4-phosphate 5-kinase, is required for cell viability, germline development and bristle morphology, but not for neurotransmitter release. *Genetics* **150**, 1527–1537.

Hay, J. C., and Martin, T. F. (1993). Phosphatidylinositol transfer protein required for ATP-dependent priming of Ca(2+)-activated secretion. *Nature* **366**, 572–575.

Heerssen, H., Fetter, R. D., and Davis, G. W. (2008). Clathrin dependence of synaptic-vesicle formation at the Drosophila neuromuscular junction. *Curr. Biol.* **18**, 401–409.

Hernandez-Deviez, D. J., Roth, M. G., Casanova, J. E., and Wilson, J. M. (2004). ARNO and ARF6 regulate axonal elongation and branching through downstream activation of phosphatidylinositol 4-phosphate 5-kinase alpha. *Mol. Biol. Cell* **15**, 111–120.

Holz, R. W., Hlubek, M. D., Sorensen, S. D., Fisher, S. K., Balla, T., Ozaki, S., Prestwich, G. D., Stuenkel, E. L., and Bittner, M. A. (2000). A pleckstrin homology domain specific for phosphatidylinositol 4, 5-bisphosphate (PtdIns-4,5-P2) and fused to green fluorescent protein identifies plasma membrane PtdIns-4,5-P2 as being important in exocytosis. *J. Biol. Chem.* **275**, 17878–17885.

Howlett, E., Lin, C. C., Lavery, W., and Stern, M. (2008). A PI3-kinase-mediated negative feedback regulates neuronal excitability. *PLoS Genet.* **4**, e1000277.

Ikushima, M., Ishii, M., Ohishi, M., Yamamoto, K., Ogihara, T., Rakugi, H., and Kurachi, Y. (2010). ANG II inhibits insulin-mediated production of PI 3,4,5-trisphosphates via a Ca2+-dependent but PKC-independent pathway in the cardiomyocytes. *Am. J. Physiol. Heart Circ. Physiol.* **299**, H680–H689.

Jacobson, K., Rajfur, Z., Vitriol, E., and Hahn, K. (2008). Chromophore-assisted laser inactivation in cell biology. *Trends Cell Biol.* **18**, 443–450.

Jaworski, J., Spangler, S., Seeburg, D. P., Hoogenraad, C. C., and Sheng, M. (2005). Control of dendritic arborization by the phosphoinositide-3'-kinase-Akt-mammalian target of rapamycin pathway. *J. Neurosci.* **25**, 11300–11312.

Kasprowicz, J., Kuenen, S., Miskiewicz, K., Habets, R. L., Smitz, L., and Verstreken, P. (2008). Inactivation of clathrin heavy chain inhibits synaptic recycling but allows bulk membrane uptake. *J. Cell Biol.* **182**, 1007–1016.

Katzmann, D. J., Stefan, C. J., Babst, M., and Emr, S. D. (2003). Vps27 recruits ESCRT machinery to endosomes during MVB sorting. *J. Cell Biol.* **162**, 413–423.

Khuong, T. M., Habets, R. L., Slabbaert, J. R., and Verstreken, P. (2010). WASP is activated by phosphatidylinositol-4,5-bisphosphate to restrict synapse growth in a pathway parallel to bone morphogenetic protein signaling. *Proc. Natl. Acad. Sci. U. S. A.* **107**, 17379–17384.

Knox, S., Ge, H., Dimitroff, B. D., Ren, Y., Howe, K. A., Arsham, A. M., Easterday, M. C., Neufeld, T. P., O'Connor, M. B., and Selleck, S. B. (2007). Mechanisms of TSC-mediated control of synapse assembly and axon guidance. *PLoS One* **2**, e375.

Kontos, C. D., Stauffer, T. P., Yang, W. P., York, J. D., Huang, L., Blanar, M. A., Meyer, T., and Peters, K. G. (1998). Tyrosine 1101 of Tie2 is the major site of association of p85 and is required for activation of phosphatidylinositol 3-kinase and Akt. *Mol. Cell Biol.* **18**, 4131–4140.

Krauss, M., Kinuta, M., Wenk, M. R., De Camilli, P., Takei, K., and Haucke, V. (2003). ARF6 stimulates clathrin/AP-2 recruitment to synaptic membranes by activating phosphatidylinositol phosphate kinase type Igamma. *J. Cell Biol.* **162**, 113–124.

Kumar, V., Zhang, M. X., Swank, M. W., Kunz, J., and Wu, G. Y. (2005). Regulation of dendritic morphogenesis by Ras-PI3K-Akt-mTOR and Ras-MAPK signaling pathways. *J. Neurosci.* **25**, 11288–11299.

Kutateladze, T. G., Ogburn, K. D., Watson, W. T., de Beer, T., Emr, S. D., Burd, C. G., and Overduin, M. (1999). Phosphatidylinositol 3-phosphate recognition by the FYVE domain. *Mol. Cell* **3**, 805–811.

Lackner, M. R., Nurrish, S. J., and Kaplan, J. M. (1999). Facilitation of synaptic transmission by EGL-30 Gqalpha and EGL-8 PLCbeta: DAG binding to UNC-13 is required to stimulate acetylcholine release. *Neuron* **24**, 335–346.

Lankiewicz, L., Malicka, J., and Wiczk, W. (1997). Fluorescence resonance energy transfer in studies of inter-chromophoric distances in biomolecules. *Acta Biochim. Pol.* **44**, 477–489.

Lee, S. Y., Wenk, M. R., Kim, Y., Nairn, A. C., and De Camilli, P. (2004). Regulation of synaptojanin 1 by cyclin-dependent kinase 5 at synapses. *Proc. Natl. Acad. Sci. U. S. A.* **101**, 546–551.

Levine, T. P., and Munro, S. (1998). The pleckstrin homology domain of oxysterol-binding protein recognises a determinant specific to Golgi membranes. *Curr. Biol.* **8**, 729–739.

Lewis, E. B., and Bacher, F. (1968). Method of feeding ethyl methane sulfonate (EMS) to Drosophila males. *Drosoph. Inf. Serv.* **43**, 1993.

Lloyd, T. E., Verstreken, P., Ostrin, E. J., Phillippi, A., Lichtarge, O., and Bellen, H. J. (2000). A genome-wide search for synaptic vesicle cycle proteins in Drosophila. *Neuron* **26**, 45–50.

Marek, K. W., and Davis, G. W. (2002). Transgenically encoded protein photoinactivation (FlAsH-FALI): acute inactivation of synaptotagmin I. *Neuron* **36**, 805–813.

Marone, R., Cmiljanovic, V., Giese, B., and Wymann, M. P. (2008). Targeting phosphoinositide 3-kinase: moving towards therapy. *Biochim. Biophys. Acta* **1784**, 159–185.

Martin-Pena, A., Acebes, A., Rodriguez, J. R., Sorribes, A., de Polavieja, G. G., Fernandez-Funez, P., and Ferrus, A. (2006). Age-independent synaptogenesis by phosphoinositide 3 kinase. *J. Neurosci.* **26**, 10199–10208.

Morrison, D. K., Murakami, M. S., and Cleghon, V. (2000). Protein kinases and phosphatases in the Drosophila genome. *J. Cell Biol.* **150**, F57–F62.

Nakano-Kobayashi, A., Yamazaki, M., Unoki, T., Hongu, T., Murata, C., Taguchi, R., Katada, T., Frohman, M. A., Yokozeki, T., and Kanaho, Y. (2007). Role of activation of PIP5Kgamma661 by AP-2 complex in synaptic vesicle endocytosis. *EMBO J.* **26**, 1105–1116.

Ni, J. Q., Liu, L. P., Binari, R., Hardy, R., Shim, H. S., Cavallaro, A., Booker, M., Pfeiffer, B. D., Markstein, M., Wang, H., Villalta, C., Laverty, T. R., Perkins, L. A., and Perrimon, N. (2009). A Drosophila resource of transgenic RNAi lines for neurogenetics. *Genetics* **182**, 1089–1100.

Oatey, P. B., Venkateswarlu, K., Williams, A. G., Fletcher, L. M., Foulstone, E. J., Cullen, P. J., and Tavare, J. M. (1999). Confocal imaging of the subcellular distribution of phosphatidylinositol 3,4,5-trisphosphate in insulin- and PDGF-stimulated 3T3-L1 adipocytes. *Biochem. J.* **344**(Pt 2), 511–518.

Ojala, P. J., Paavilainen, V., and Lappalainen, P. (2001). Identification of yeast cofilin residues specific for actin monomer and PIP2 binding. *Biochemistry* **40**, 15562–15569.

Ooms, L. M., Horan, K. A., Rahman, P., Seaton, G., Gurung, R., Kethesparan, D. S., and Mitchell, C. A. (2009). The role of the inositol polyphosphate 5-phosphatases in cellular function and human disease. *Biochem. J.* **419**, 29–49.

Ozaki, S., DeWald, D. B., Shope, J. C., Chen, J., and Prestwich, G. D. (2000). Intracellular delivery of phosphoinositides and inositol phosphates using polyamine carriers. *Proc. Natl. Acad. Sci. U. S. A.* **97**, 11286–11291.

Palmer, S., Hawkins, P. T., Michell, R. H., and Kirk, C. J. (1986). The labelling of polyphosphoinositides with [32P]Pi and the accumulation of inositol phosphates in vasopressin-stimulated hepatocytes. *Biochem. J.* **238**, 491–499.

Pap, E. H., Bastiaens, P. I., Borst, J. W., van den Berg, P. A., van Hoek, A., Snoek, G. T., Wirtz, K. W., and Visser, A. J. (1993). Quantitation of the interaction of protein kinase C with diacylglycerol and phosphoinositides by time-resolved detection of resonance energy transfer. *Biochemistry* **32**, 13310–13317.

Papayannopoulos, V., Co, C., Prehoda, K. E., Snapper, S., Taunton, J., and Lim, W. A. (2005). A polybasic motif allows N-WASP to act as a sensor of PIP(2) density. *Mol. Cell* **17**, 181–191.

Pollok, B. A., and Heim, R. (1999). Using GFP in FRET-based applications. *Trends Cell Biol.* **9**, 57–60.

Poskanzer, K. E., Marek, K. W., Sweeney, S. T., and Davis, G. W. (2003). Synaptotagmin I is necessary for compensatory synaptic vesicle endocytosis in vivo. *Nature* **426**, 559–563.

Raucher, D., Stauffer, T., Chen, W., Shen, K., Guo, S., York, J. D., Sheetz, M. P., and Meyer, T. (2000). Phosphatidylinositol 4,5-bisphosphate functions as a second messenger that regulates cytoskeleton-plasma membrane adhesion. *Cell* **100**, 221–228.

Rhee, J. S., Betz, A., Pyott, S., Reim, K., Varoqueaux, F., Augustin, I., Hesse, D., Sudhof, T. C., Takahashi, M., Rosenmund, C., and Brose, N. (2002). Beta phorbol ester- and diacylglycerol-induced augmentation of transmitter release is mediated by Munc13s and not by PKCs. *Cell* **108**, 121–133.

Ringstad, N., Gad, H., Low, P., Di Paolo, G., Brodin, L., Shupliakov, O., and De Camilli, P. (1999). Endophilin/SH3p4 is required for the transition from early to late stages in clathrin-mediated synaptic vesicle endocytosis. *Neuron* **24**, 143–154.

Ringstad, N., Nemoto, Y., and De Camilli, P. (1997). The SH3p4/Sh3p8/SH3p13 protein family: binding partners for synaptojanin and dynamin via a Grb2-like Src homology 3 domain. *Proc. Natl. Acad. Sci. U. S. A.* **94**, 8569–8574.

Rong, Y. S., and Golic, K. G. (2000). Gene targeting by homologous recombination in Drosophila. *Science* **288**, 2013–2018.

Rubin, G. M., and Spradling, A. C. (1982). Genetic transformation of Drosophila with transposable element vectors. *Science* **218**, 348–353.

Rubin, G. M., Yandell, M. D., Wortman, J. R., Gabor Miklos, G. L., Nelson, C. R., Hariharan, I. K., Fortini, M. E., Li, P. W., Apweiler, R., Fleischmann, W., Cherry, J. M., Henikoff, S., Skupski, M. P., Misra, S., Ashburner, M., Birney, E., Boguski, M. S., Brody, T., Brokstein, P., Celniker, S. E., Chervitz, S. A., Coates, D., Cravchik, A., Gabrielian, A., Galle, R. F., Gelbart, W. M., George, R. A., Goldstein, L. S., Gong, F., Guan, P., Harris, N. L., Hay, B. A., Hoskins, R. A., Li, J., Li, Z., Hynes, R. O., Jones, S. J., Kuehl, P. M., Lemaitre, B., Littleton, J. T., Morrison, D. K., Mungall, C., O'Farrell, P. H., Pickeral, O. K., Shue, C., Vosshall, L. B., Zhang, J., Zhao, Q., Zheng, X. H., and Lewis, S. (2000). Comparative genomics of the eukaryotes. *Science* **287**, 2204–2215.

Russell, J. K., Torres, B. A., and Johnson, H. M. (1987). Phospholipase A2 treatment of lymphocytes provides helper signal for interferon-gamma induction. Evidence for second messenger role of endogenous arachidonic acid. *J. Immunol.* **139**, 3442–3446.

Saarikangas, J., Zhao, H., and Lappalainen, P. (2010). Regulation of the actin cytoskeleton-plasma membrane interplay by phosphoinositides. *Physiol. Rev.* **90**, 259–289.

Sato, M., Ueda, Y., Takagi, T., and Umezawa, Y. (2003). Production of PtdInsP3 at endomembranes is triggered by receptor endocytosis. *Nat. Cell Biol.* **5**, 1016–1022.

Schuske, K. R., Richmond, J. E., Matthies, D. S., Davis, W. S., Runz, S., Rube, D. A., van der Bliek, A. M., and Jorgensen, E. M. (2003). Endophilin is required for synaptic vesicle endocytosis by localizing synaptojanin. *Neuron* **40**, 749–762.

Sechi, A. S., and Wehland, J. (2000). The actin cytoskeleton and plasma membrane connection: PtdIns (4,5)P(2) influences cytoskeletal protein activity at the plasma membrane. *J. Cell Sci.* **113**(Pt 21), 3685–3695.

Shisheva, A. (2008). PIKfyve: Partners, significance, debates and paradoxes. *Cell Biol. Int.* **32**, 591–604.

Skare, P., and Karlsson, R. (2002). Evidence for two interaction regions for phosphatidylinositol(4,5)-bisphosphate on mammalian profilin I. *FEBS Lett.* **522**, 119–124.

Spradling, A. C., Stern, D., Beaton, A., Rhem, E. J., Laverty, T., Mozden, N., Misra, S., and Rubin, G. M. (1999). The Berkeley Drosophila Genome Project gene disruption project: Single P-element insertions mutating 25% of vital Drosophila genes. *Genetics* **153**, 135–177.

Stauffer, T. P., Ahn, S., and Meyer, T. (1998). Receptor-induced transient reduction in plasma membrane PtdIns(4,5)P2 concentration monitored in living cells. *Curr. Biol.* **8**, 343–346.

Stefan, C. J., Audhya, A., and Emr, S. D. (2002). The yeast synaptojanin-like proteins control the cellular distribution of phosphatidylinositol (4,5)-bisphosphate. *Mol. Biol. Cell* **13**, 542–557.

Stimson, D. T., Estes, P. S., Rao, S., Krishnan, K. S., Kelly, L. E., and Ramaswami, M. (2001). Drosophila stoned proteins regulate the rate and fidelity of synaptic vesicle internalization. *J. Neurosci.* **21**, 3034–3044.

Suh, B. C., Inoue, T., Meyer, T., and Hille, B. (2006). Rapid chemically induced changes of PtdIns(4,5)P2 gate KCNQ ion channels. *Science* **314**, 1454–1457.

Szentpetery, Z., Varnai, P., and Balla, T. (2010). Acute manipulation of Golgi phosphoinositides to assess their importance in cellular trafficking and signaling. *Proc. Natl. Acad. Sci. U. S. A.* **107**, 8225–8230.

Tal, T., Vaizel-Ohayon, D., and Schejter, E. D. (2002). Conserved interactions with cytoskeletal but not signaling elements are an essential aspect of Drosophila WASp function. *Dev. Biol.* **243**, 260–271.

Uytterhoeven, V., Kuenen, S., Kasprowicz, J., Miskiewicz, K., and Verstreken, P. (2011). Loss of skywalker reveals synaptic endosomes as sorting stations for synaptic vesicle proteins. *Cell* **145**, 117–132.

van der Wal, J., Habets, R., Varnai, P., Balla, T., and Jalink, K. (2001). Monitoring agonist-induced phospholipase C activation in live cells by fluorescence resonance energy transfer. *J. Biol. Chem.* **276**, 15337–15344.

Van Epps, H. A., Hayashi, M., Lucast, L., Stearns, G. W., Hurley, J. B., De Camilli, P., and an Brockerhoff, S. E. (2004). The zebrafish nrc mutant reveals a role for the polyphosphoinositide phosphatase synaptojanin 1 in cone photoreceptor ribbon anchoring. *J. Neurosci.* **24**, 8641–8650.

Varnai, P., and Balla, T. (1998). Visualization of phosphoinositides that bind pleckstrin homology domains: calcium- and agonist-induced dynamic changes and relationship to myo-[3H]inositol-labeled phosphoinositide pools. *J. Cell Biol.* **143**, 501–510.

Varnai, P., Rother, K. I., and Balla, T. (1999). Phosphatidylinositol 3-kinase-dependent membrane association of the Bruton's tyrosine kinase pleckstrin homology domain visualized in single living cells. *J. Biol. Chem.* **274**, 10983–10989.

Varnai, P., Thyagarajan, B., Rohacs, T., and Balla, T. (2006). Rapidly inducible changes in phosphatidylinositol 4,5-bisphosphate levels influence multiple regulatory functions of the lipid in intact living cells. *J. Cell Biol.* **175**, 377–382.

Venkateswarlu, K., Gunn-Moore, F., Oatey, P. B., Tavare, J. M., and Cullen, P. J. (1998a). Nerve growth factor- and epidermal growth factor-stimulated translocation of the ADP-ribosylation factor-exchange factor GRP1 to the plasma membrane of PC12 cells requires activation of phosphatidylinositol 3-kinase and the GRP1 pleckstrin homology domain. *Biochem. J.* **335**(Pt 1), 139–146.

Venkateswarlu, K., Oatey, P. B., Tavare, J. M., and Cullen, P. J. (1998b). Insulin-dependent translocation of ARNO to the plasma membrane of adipocytes requires phosphatidylinositol 3-kinase. *Curr. Biol.* **8**, 463–466.

Venken, K. J., and Bellen, H. J. (2007). Transgenesis upgrades for Drosophila melanogaster. *Development* **134**, 3571–3584.

Venken, K. J., He, Y., Hoskins, R. A., and Bellen, H. J. (2006). P[acman]: a BAC transgenic platform for targeted insertion of large DNA fragments in D. melanogaster. *Science* **314**, 1747–1751.

Venken, K. J., Kasprowicz, J., Kuenen, S., Yan, J., Hassan, B. A., and Verstreken, P. (2008). Recombineering-mediated tagging of Drosophila genomic constructs for in vivo localization and acute protein inactivation. *Nucleic Acids Res.* **36**, e114.

Verstreken, P., Koh, T. W., Schulze, K. L., Zhai, R. G., Hiesinger, P. R., Zhou, Y., Mehta, S. Q., Cao, Y., Roos, J., and Bellen, H. J. (2003). Synaptojanin is recruited by endophilin to promote synaptic vesicle uncoating. *Neuron* **40**, 733–748.

Verstreken, P., Ohyama, T., Haueter, C., Habets, R. L., Lin, Y. Q., Swan, L. E., Ly, C. V., Venken, K. J., De Camilli, P., and Bellen, H. J. (2009). Tweek, an evolutionarily conserved protein, is required for synaptic vesicle recycling. *Neuron* **63**, 203–215.

Vicinanza, M., D'Angelo, G., Di Campli, A., and De Matteis, M. A. (2008). Function and dysfunction of the PI system in membrane trafficking. *EMBO J.* **27**, 2457–2470.

Wenk, M. R., and De Camilli, P. (2004). Protein-lipid interactions and phosphoinositide metabolism in membrane traffic: insights from vesicle recycling in nerve terminals. *Proc. Natl. Acad. Sci. U. S. A.* **101**, 8262–8269.

Wucherpfennig, T., Wilsch-Brauninger, M., and Gonzalez-Gaitan, M. (2003). Role of Drosophila Rab5 during endosomal trafficking at the synapse and evoked neurotransmitter release. *J. Cell Biol.* **161**, 609–624.

Zhang, B., Koh, Y. H., Beckstead, R. B., Budnik, V., Ganetzky, B., and Bellen, H. J. (1998). Synaptic vesicle size and number are regulated by a clathrin adaptor protein required for endocytosis. *Neuron* **21**, 1465–1475.

Zheng, Q., and Bobich, J. A. (2004). ADP-ribosylation factor6 regulates both [3H]-noradrenaline and [14C]-glutamate exocytosis through phosphatidylinositol 4,5-bisphosphate. *Neurochem. Int.* **45**, 633–640.

Zoncu, R., Perera, R. M., Sebastian, R., Nakatsu, F., Chen, H., Balla, T., Ayala, G., Toomre, D., and De Camilli, P. V. (2007). Loss of endocytic clathrin-coated pits upon acute depletion of phosphatidylinositol 4,5-bisphosphate. *Proc. Natl. Acad. Sci. U. S. A.* **104**, 3793–3798.

CHAPTER 13

Devising Powerful Genetics, Biochemical and Structural Tools in the Functional Analysis of Phosphatidylinositol Transfer Proteins (PITPs) Across Diverse Species

James M. Davison, Vytas A. Bankaitis and Ratna Ghosh

Department of Cell & Developmental Biology, Lineberger Comprehensive Cancer Center, School of Medicine, University of North Carolina at Chapel Hill, Chapel Hill, North Carolina, USA

METHODS IN CELL BIOLOGY, VOL 108
Copyright 2012, Elsevier Inc. All rights reserved.

0091-679X/10 $35.00
DOI 10.1016/B978-0-12-386487-1.00013-4

Abstract

The minor cellular lipid phosphoinositides represents key regulators of diverse intracellular processes such as signal transduction at membrane-cytosol interface, regulation of membrane trafficking, cytoskeleton organization, nuclear events and the permeability, and transport functions of the membrane. The heterogeneous subcellular localization of phosphoinositides and their multiple and co-operative membrane-protein recognition mechanisms contribute to a "coincidence detection code" for the membrane–cytosol interactions in eukaryotic signaling networks. Such a "coincidence detection code" relies on the fine coordination of the broader lipid metabolism and organization, and their coupling to dedicated physiological processes. The phosphatidylinositol transfer proteins (PITPs) play a key regulatory role, essentially as "coincidence detectors" or "nanoreactors" in this "signal detection code" that spatially and temporally coordinate the diverse aspects of lipid metabolome with phosphoinositide signaling to effect various cellular functions. The integral role of PITPs in the highly conserved eukaryotic signal transduction strategy is amply demonstrated by the mammalian diseases associated with the derangements in the function of these proteins, to stress response and developmental regulation in plants, to fungal dimorphism and pathogenicity, to membrane trafficking in yeast and higher eukaryotes. The study of PITPs is fundamental to understanding of how the phosphoinositide signal transduction network is regulated and integrated to the larger lipid metabolome in diverse cellular processes. To comprehend how the PITPs integrate phosphoinositide signaling to broader lipid metabolome in diverse cellular processes, it is necessary to devise methods that can correlate the biochemical properties of these non-enzymatic proteins to biologically relevant functional insights. In this chapter, we present combinatorial approaches that primarily employ genetics and structural tools to assess the functional role of PITPs in yeast, plant and mammalian systems. An elaborate discussion on the various genetic models devised for interpreting the functional role of PITPs in relation to their operational assays has been included. We also describe the structural and biophysical methods that have advanced our understanding of how these proteins operate as "nanoreactor" molecules.

I. Introduction

The phosphatidylinositol transfer proteins (PITPs) are an ancient and unique protein superfamily that is ubiquitously present in eukaryotes including fungi, plants and mammals. The PITPs regulate key interfaces between lipid signaling, lipid metabolism, and membrane trafficking/signaling in eukaryotes. The key regulatory roles of PITPs in eukaryotic signal transduction control is evinced by the PITP mutations that cause cancer, neurodegenerative disorders and lipodystrophies in mammals, metabolic and developmental derangements in yeast, failure in embryogenesis and photoreceptor biogenesis in zebrafish, defects in peripheral nervous system in flies and stress response and developmental anomalies in plants. The PITP protein superfamily are highly conserved and bifurcates into two distinct evolutionary groups based on their primary sequence homology and structural folds, namely the Sec14-like PITPs and the steroidogenic acute response protein related START-like PITPs. The fungal and plant PITPs are Sec14-like and are highly similar to each other while the metazoan PITPs from mammals, flies, fish, and worms display high primary sequence identity and comprises the START-like PITP group. Metazoans also possess Sec14-like proteins that share homology with plant and fungal Sec14-like proteins. The two PITP groups do not have any sequence similarity, though they exhibit nearly identical *in vitro* biochemical activities.

The PITPs are non-enzymatic proteins that bind PtdIns (phosphatidylinositol) and PtdCho (phosphatidylcholine) in mutually exclusive way and were initially characterized by their ability to mediate the energy independent transfer of PtdIns or PtdCho monomers between membrane bilayers *in vitro*. While the classical PITPs are defined as proteins that could transfer PtdIns and PtdCho between membranes *in vitro*, this definition no longer holds true as many PITPs bind and transfer PtdIns but not PtdCho. The non-enzymatic nature of the PITPs makes their operational assays (by which they are characterized) extremely difficult to interpret, leading to conceptual difficulties in relating their biochemical functions with obvious functional insights. Primarily this is the reason why studies on understanding of how these proteins function as molecules lacked pace. However, a large advance on the PITP biology came through the biochemical, genetic, and structural tools developed in the last several years. These studies have shown that PITPs are not mere PtdIns carriers as earlier thought, but they execute a particularly interesting and novel mode of PIP signal transduction control by regulating the PtdIns accessibility to interfacial PtdIns-kinase enzymes.

An understanding of the PITP function has revealed novel mechanisms of diversification for PIP signaling. Current studies indicates that the PIP regulatory outcome is neither exclusively determined by the chemical nature of the PIP, nor by identity of the PtdIns kinase that produces it, rather PITPs specify biological outcomes for PtdIns kinase activities and in the process effect PIP signaling. The concept of regulatory roles of PITPs are primarily derived from studies on the prototypical Sec14 protein which has proven tractable to comprehensive employ of structural, biochemical and genetic approaches directed toward functional elucidation of mechanisms. The collective data demonstrate that stimulated PIP synthesis lies at the core of how PITPs regulate signaling *in vivo*, but the mechanism by which PITPs stimulate PtdIns kinases remains a central question. Historically, PITPs were thought to stimulate PIP synthesis by facilitating PtdIns transport from PtdIns-rich endoplasmic reticulum (ER) to sites of PtdIns kinase action that is, *trans*-Golgi network (TGN) or plasma membrane (PM) based on the *in vitro* transfer assays. However, the Sec14 studies have revealed very distinct mechanisms that do not involve inter-membrane phospholipid (PL) transfer. The findings show that Sec14 co-ordinates PtdCho and PIP metabolism in the TGN/endosomal system in a transfer-independent fashion, so that key components of the membrane trafficking machinery are active. Novel conclusions on the PITP/PtdIns kinase relationship are forthcoming from these studies that showed that functionally specified PIP pools are defined by the identity of PITP that couples to the lipid kinase. The collective data posit that Sec14 functions as "nanoreactor" which effect primed presentation of PtdIns to the biologically inadequate PtdIns kinase. This PITP/PtdIns kinase relationship assigns instructive roles for PITPs in specifying outcomes for PtdIns kinase action, and holds large implications on how the eukaryotic signaling landscape is regulated. These emerging concepts recommend PITPs as discerning entry points for viewing highly specific aspects of PIP signaling. Key to our understanding of the mechanisms of diversification for PIP signaling pathways and their link to broader lipid metabolism rely on how the PITPs functions as single molecules and how they are integrated into the landscape of eukaryotic signal transduction.

II. Rationale

The PITP proteins were first characterized based on the biochemical PtdIns and PtdCho transfer assays between membranes *in vitro* (Helmkamp *et al.*, 1974; van Paridon *et al.*, 1987; Venuti and Helmkamp, 1988). These radiolabeled assays though operationally defined these proteins for the first time, they were not able to correlate the *in vitro* lipid transfer activity to obvious functional insights. While much of the early thinking was driven by this biochemical activity, it is now clear that these proteins play important complex regulatory roles, of which transport function is just one. Advances in the PITP field came through incisive molecular studies employing comprehensive use of biochemical, genetic, and structural approaches directed toward elucidation of functional mechanisms.

Genetic approaches have significantly advanced our understanding on these proteins. Yeast genetics has provided important insights on the function of these proteins in the complex regulation PtdCho, PtdIns, phosphatidic acid (PA), diacylglycerol (DAG), and phosphoinositide metabolism thereby providing appropriate lipid environment for membrane trafficking through TGN. Crucial insights on how Sec14 translates its PtdIns/PtdCho transfer activity to membrane trafficking was gained from detailed analyses of "bypass mutants" that relieve cells of the Sec14 requirement for Golgi function and cell viability in yeast (Cleves *et al.*, 1989, 1991a; Fang *et al.*, 1996; Li *et al.*, 2002). The rationale behind the analyses of these "bypass mutations" was that they generated physiological conditions that mimic the consequences of Sec14 function *in vivo* and therefore their detailed studies were expected to identify effectors that respond to Sec14 activity. Similarly, other genetic screens identified "loss of function mutations" that exacerbated defects associated with reduced Sec14 function, are thought to identify potential regulators of Sec14 and its effectors (Xie *et al.*, 1998; Yanagisawa *et al.*, 2002). Analyses of the various Sec14 bypass mutations led to a common proposition, according to which the physiological function of the Sec14 in vesicle trafficking is the coordination of

PtdCho and PtdIns metabolism so as to generate a lipid environment in the *trans*-Golgi membranes that is conducive to optimal activity of core components of Golgi TGN vesicle biogenesis/budding machinery (Cleves *et al.*, 1991b). Based on genetic studies, the current model of Sec14 pathway proposes Sec14's regulatory role at the interface of lipid metabolism and lipid signaling in Golgi secretory function. In this pathway, Sec14 functions to generate a low PtdCho and high DAG lipid environment that is conducive to the optimal activation of specific ARFGAP effector proteins which recruits vesicle budding machinery for vesicle formation and budding at Golgi membranes. The ideas drawn from the yeast genetic studies have provided important insights on the mechanisms of Sec14 function and are being extended broadly across the Sec14-superfamily.

Although the genetic studies established the regulatory role of these proteins at the interface of complex lipid metabolism and lipid signaling, the fundamental nature of this regulation could only be realized through structural approaches. These approaches employ structural/mutational studies and biophysical methods to probe the structural dynamics of the Sec14-like protein molecules. In this regard, significant advance has been made toward the structural and mutational studies of yeast Sec14. The Sec14 crystal structure has provided invaluable insights on how members of this protein family bind PL substrate and how it catalyzes PL exchange reaction. The 2.5 Å crystal structure of detergent-bound Sec14 described a novel protein fold that serves as a model for the entire Sec14-like PITP proteins in fungi, plants, and mammals (Sha *et al.*, 1998). Further, the successful crystallization of PL bound forms of Sfh1, the protein most similar to Sec14, advanced our understanding of how members of Sec14 family bind monomeric PLs and how this unique binding mechanism is translated to its function in regulating phosphoinositide homeostasis (Schaaf *et al.*, 2008). The structural studies correlate to and explain the molecular basis of the genetic studies. Available evidence proposes that a heterotypic PtdIns/PtdCho exchange reaction uniquely prime Sec14 for productive presentation of PtdIns to PtdIns-kinases, which are inefficient interfacial enzymes otherwise. In other words, Sec14 employs its heterotypic PtdIns- or PtdCho exchange activities to bind or sense local PtdCho and simultaneously prime a PtdIns presentation unit or "nanoreactor" that stimulates PtdIns-4-OH kinases (Ile *et al.*, 2006; Schaaf *et al.*, 2008). Another important outcome of the structural approach is the discovery of the well conserved crystal structure based PtdIns- and PtdCho- binding bar codes in Sec14 and Sec14-like proteins (Bankaitis *et al.*, 2010). These bar codes predict most Sec14-like proteins bind PtdIns and that most will fail to bind PtdCho. This is particularly relevant to human disease as several naturally occurring mutations in human PITPs directly involve residues of the predicted PtdIns binding bar codes of the corresponding Sec14-like proteins/domains.

The emerging evidence derived from the combinatorial approaches support "nanoreactor" mechanistic model versus the historically accepted "transfer" models for Sec14-like PITP functions and possibly also for the START-like PITPs. Unlike transfer model the "nanoreactor" model do not describe PITPs as *trans*-organelle lipid carriers, rather they define PITPs as PtdIns scaffolding molecules with intrinsic regulatory potential. In this chapter, we describe the *in vitro* approaches which includes radiolabeling transfer assays and permeabilized cell based reconstitution assays, *in vivo* approaches which employ genetic knockdown and complementation experiments, radiolabeling and fluorescent assays, and the structural approaches which involves site directed mutagenesis and biophysical tools for studying the PITP proteins in yeast, plant, and mammalian systems. In each section, we will discuss the principle, the procedure, and the advantages and limitations of the methods in the context of the model system employed. Such combinatorial approaches are providing important insights on how PITPs operate at the molecular level and have mostly advanced our perception on diverse biological functions of PITPs.

III. *In Vitro* Approaches

A. *In Vitro* Biochemical Assays

PITPs have historically been defined by *in vitro* transfer assays, whose translation to biological function is problematic – as discussed in this section. The transfer assays were used to describe protein-mediated phospholipid transport between membranes in 1968 and 1969 (Akiyama and Sakagami, 1969; McMurray and Dawson, 1969; Wirtz and Zilversmit, 1968). These assays involved labeling of microsomal lipids with ^{32}P and ^{14}C and tracking the mobilization of radioactivity from the "donor" microsomes to "acceptor" mitochondria. From the specific perspective of PITPs, initial characterizations identified both phosphatidylinositol (PtdIns) and phosphatidylcholine (PtdCho) as transfer substrates with PtdIns-transfer being the most robust. This assay was further refined in the mid-1970s to include liposomes as the acceptor-membrane fraction and, during this time, the proteins catalyzing the exchange reaction were first purified and characterized (Helmkamp *et al.*, 1974; Kamp *et al.*, 1975). These simple transfer assays are utilized to this day, and represent convenient biochemical readouts for PITP activity.

The operational ease and convenience of transfer assays notwithstanding, issues of how transfer activities are to be functionally interpreted immediately present themselves. A genuine carrier activity is one such interpretation and this interpretation is generally accepted as a principle of mechanism for PITPs. However, this is a rather uncritical interpretation. At its core, the *in vitro* transfer reaction is an equilibration assay where the PITP "transports" signal down an effective concentration gradient. When viewed from this perspective, PITPs are technically phospholipid exchange proteins. As discussed in the following sections, one can imagine diverse ways such exchange reactions could relate to biological function of the PITPs. For simple purposes of measurement, however, transfer assays provide high throughput and cost effective ways to characterize transfer protein's lipid interactions. Herein, we describe basic transfer assays. Later, we also discuss other coupled activity assays referred as "*in vitro* functional assays" used to assess PITP activity in purportedly genuine physiological contexts. We close the section with a discussion of the limitations of these *in vitro* approaches from mechanistic points of view.

1. Radiolabeled Transfer Assays for PITPs

Principle

The radiolabeled transfer assays are biochemical measurements of the transfer of a radiolabeled phospholipid from a donor to an acceptor membrane (Helmkamp *et al.*, 1974; Phillips *et al.*, 1999; Schaaf *et al.*, 2008). These membrane fractions are distinguished by their sedimentation properties and are separated by centrifugation following incubation with a transfer protein. The gain in radioactivity of the accepting membrane fraction is measured after centrifugation. The detected radioactivity reports the transfer protein i.e., PITP-dependent transfer events (Fig. 1). Other protocols measure the loss of radioactivity from the "donor" membrane.

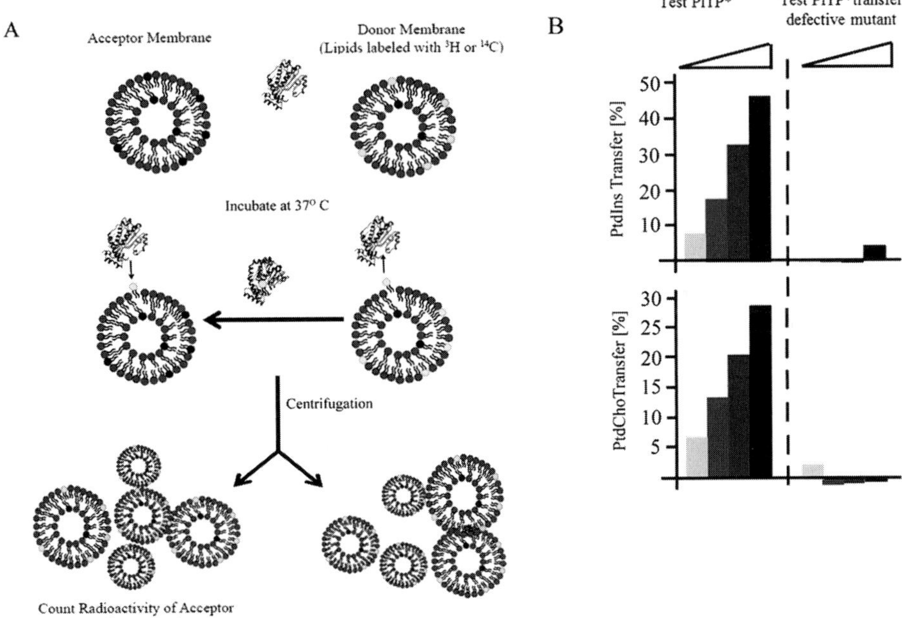

Fig. 1 Radiolabeled transfer Assays. (A) Flow chart of transfer assay. The assay consists of acceptor and donor membranes and a transfer protein. These components are incubated at 37 °C for 30 minutes and then separated by centrifugation. The radiolabeled PLs are shown in light shades of grey. Radioactivity of acceptor membrane fraction is measured on a scintillation counter. **(B) Graphical representation of data for radiolabeled transfer assays.** The assays are performed on a wild type test PITP and a PITP* transfer defective mutant, which is defective for both PtdIns and PtdCho binding. Purified proteins or cytosol fractions are used in these assays. PtdIns transfer assays and PtdCho transfer assays are performed with increasing concentrations of cytosol or protein as indicated. The different concentrations of the protein/cytosol are shown in distinct shades of the bars. Data is presented as a percent transfer.

PtdIns Transfer Assay

This assay is designed to monitor PtdIns transfer. The ratio of microsomal lipid to liposomes and actual amount of PtdIns between the two membranes favors transfer from microsomes (donating membrane) to liposomes (accepting membrane).

Materials

(i) 98% PtdCho/2% PtdIns lipid cocktail [Advanti Lipids (PtdIns: 84044C, and PtdCho: 840054C)].
This cocktail is dried down under liquid N_2 gas and then reconstituted in 10 mL of chloroform:methanol (1:1). Store at −20 °C in an air tight glass tube until use.
(ii) Rat liver microsmes (RLM) with labeled [^{3}H] PtdIns.
(iii) 0.2 M Sodium acetate pH 5.0.
(iv) Cytosol or recombinant proteins as a source of PITP.
(v) SET buffer: 0.25 M sucrose, 1 mM EDTA, and 10 mM Tris-HCl, pH 7.4
Low Phosphate (LP) assay buffer: 300 mM NaCl, 25 mM Na_2HPO_4, pH 7.5
High Phosphate (HP) assay buffer: 300 mM NaCl, 50 mM Na_2HPO_4, pH 7.5

Transfer activity can vary depending on the assay buffer. The low phosphate lysis buffer is suitable for Sec14 transfer assays, while the SET buffer works well for mammalian PITP transfer assays. This protocol uses LP assay buffer, but other buffers may be more suitable depending on the transfer protein.

Preparation of PtdIns labeled microsomes

All steps must be completed at 4 °C

1. Extract livers from six rats that are fasted overnight. Homogenize as a 20–30% solution in cold SET buffer, first in a blender followed by a douncer.
2. Separate rat liver microsomes (RLMs) from cellular debris by centrifugation at 1000 g for 10 min. Transfer the supernatant to a fresh tube and centrifuge at 20,000 *g* for 20 min. Transfer the supernatant and give a final spin at 35,000 rpm for 90 min to pellet down the microsomal fraction. Decant supernatant and resuspend the microsomal pellet in 80 mL of cold 20 mM Tris-HCl, pH 7.4.
3. Label inositol phospholipids by incubating crude RLMs in presence of 10 mM $MnCl_2$ with 1mCi of [^{3}H]-inositol for 2 h at 37 °C. During this incubation, endogenous enzymes will incorporate radiolabeled PtdIns in microsomes.
4. Centrifuge the labeled RLMs for 60 min at 35,000 rpm and then resuspend the microsomal pellet in cold 10 mM Tris-HCl with 2 mM inositol, pH8.6. Again, centrifuge the labeled RLMs for 60 min at 35,000 rpm and resuspend the RLM pellet in cold 1 mM Tris-HCl with 2 mM inositol, pH 8.6. Finally, centrifuge the RLMs for 60 min at 35,000 rpm and resuspend the pellet in cold SET buffer. Count samples on scintillation counter under the [^{3}H] channel. Freeze in 0.8 mL aliquots at −20 °C with target radioactivity of ∼1000 cpm/μL.

Procedure for PtdIns transfer assay

The assay procedure is for analysis of 50 samples, including background transfer controls and total radioactivity controls.

1. Prepare liposomes by pipetting 12.5 mL of PtdIns/PtdCho cocktail into a chloroform-resistant tube. Dry down the lipids under liquid nitrogen gas and resuspend in 1 ml of LP assay buffer. Sonicate the lipid suspension for 1 min on low amplitude and then place it on ice for one minute. Repeat the sonication procedure until the solution becomes clear. The sonication generates PtdIns/PtdCho liposomes that serve as acceptor membranes.
2. Prepare the donor membrane by adding 200 μL of 0.2 M sodium acetate (pH 5.0) to 1 mL of [^{3}H] labeled inositol RLMs. The microsomes are centrifuged at 13,000 rpm for 10 min and resuspend the pellet in 1 mL of LP buffer. Dounce homogenize this solution and add it to the PtdIns/PtdCho liposomes prepared earlier. Raise the final volume of the microsome-liposome mixture to 12.5 mL.
3. Aliquot the purified protein or cytosolic fractions into eppendroff tubes and raise the final volume of all the tubes to 500 μL with LP buffer.
4. Add 500 μL of the RLM-liposome mix to each sample tube and vortex. For measuring background transfer activity, include control that includes only LP assay buffer and no protein/cytosol.
5. Incubate the assay mixture at 37 °C for 30 min. Add 200 μL of 0.2 M sodium acetate to each sample to stop the transfer. Centrifuge the samples at 13,000 rpm for 10 min.
6. To measure transfer activity, add 1 mL of supernatant from each sample to scintillation vials containing 4.5 mL of scintillation cocktail. Include controls for background transfer and total radioactivity. Count the radioactivity of each sample on a scintillation counter using the [^{3}H] setting.

PtdCho/Sphingomyelin Transfer Assay

This assay is designed to monitor PtdCho or sphingomyelin transfer. In this assay, the radiolabeled [^{14}C] PtdCho/ [^{3}H] PtdCho or [^{14}C] sphingomyelin (SM) liposomes act as donating membrane fraction, are incubated with bovine heart mitochondria (BHM) which serve as the accepting membrane fraction.

Materials

(i) Radiolabeled PtdCho or sphingomyelin [American Radiolabeled Chemicals, Inc: [^{14}C] PtdCho (0376), [^{3}H] PtdCho (0284) and [^{14}C]Sphingomyelin (0772)].
(ii) Cold PtdCho or radiolabeled Sphingomyelin [Avanti Polar Lipids PtdCho (840054C) and Sphingomyelin (860061)].
(iii) Bovine heart mitochondria (BHM)
(iv) 14.3% sucrose

(v) 10% SDS
(vi) Cytosol or recombinant proteins as source of PITP
(vii) Composition of SET buffer, Low phosphate (LP) and High phosphate (HP) assay buffers are prepared as described above.

For SET buffer for BHM preparation: See BHM preparation below for specific SET recipes.

Preparation of bovine heart mitochondria

1. Trim off fat and chop muscle tissue from bovine heart (1.5–2.5 kg) into 0.5–1 cm^3 pieces. Wash 2–3 times with SET buffer (0.25 M Sucrose, 20 mM EDTA, 5 mM Tris-HCl) to rinse away blood. Homogenize the heart in a blender and make a 30% (w/v) solution in 20 mM EDTA SET buffer. Adjust pH of homogenate to pH 7.4 using NaOH.
2. Isolate the BHM by centrifuging the homogenate for 20 min at 1200 *g* to pellet down nuclei and any unbroken cells. Pour supernatant through four layers of cheese cloth and adjust pH to 7.4 using NaOH. Next, centrifuge the supernatant for 20 min at 16,000 *g* to isolate mitochondria. Decant supernatant and grey gelatinous layer.
3. Resuspend the pellet in 4–5 mL 10 mM EDTA SET buffer and then dounce homogenize the resuspended pellet. Raise volume to 300 mL with 10 mM EDTA SET buffer and then adjust the pH to 7.4. Centrifuge the BHM for 20 min at 17,000 *g*. Decant supernatant, and resuspend pellet in 50 mL of 10 mM EDTA. Store BHM in 0.8 mL aliquots at −20 °C.

Procedure for PtdCho / sphingomyelin transfer assay

Assay procedure is for analysis of 50 samples, including background transfer controls and total radioactivity controls. PtdCho and Sphingomyelin transfer assays follow the same protocol. The only deviation between the protocols is the lipid type used in the liposomes (donating membrane fraction). This protocol uses LP Lysis buffer, but other buffers may be more suitable depending on the transfer protein.

1. Prepare individual 100 mg/mL stock solutions each of non-radioactive sphingomyelin and [^{14}C] labeled sphingomyelin for sphingomyelin transfer assay. Similarly, prepare individual 100 mg/mL stock solutions each of non-radioactive PtdCho and [^{14}C] or [^{3}H] labeled PtdCho for PtdCho transfer assays.
2. Prepare donor membranes by pipetting 20 μL of unlabeled lipid and 5 μL of radiolabeled lipid into a chloroform-resistant tube. Dry down the lipids with liquid nitrogen gas and then resuspend in 1 mL of LP lysis buffer. Sonicate the lipid suspension for 1 min and keep the suspension on ice for 1 min. Repeat this sonication procedure until solution is clear. The sonication generates liposomes and these serve as the donor membrane fraction.
3. To prepare the acceptor membrane fraction, dounce homogenize 12.5 mL of BHM in 1 mL of LP lysis buffer. Add the homogenized BHM to the liposomes and raise final volume to 2 mL.

4. Aliquot purified protein or cytosolic fractions into sample tubes and raise the final volumes to 500 μL with LP assay buffer. Add 200 μL of the BHM-liposome mix to each sample tube and vortex. Incubate samples at 37 °C for 30 min.
5. Stop transfer assays by incubating the samples on ice. Using an 18-gauge needle, add 500 μL of cold 14.3% sucrose at the bottom of the eppendroff tubes for each sample. Do not mix the sucrose and the incubated sample. Pellet down the mitochondria through the sucrose cushion at 12,000–14,000 rpm for 10 min.
6. Decant supernatant without disturbing the mitochondrial pellet. Add 100 μL of 10% SDS to each pellet and resuspend by pipetting. If the pellet does not dissolve, boil the samples for 5 min.
7. Transfer the contents of each sample tube into scintillation vials containing 4.5 mL of scintillation solution. Include controls for background transfer and total radioactivity. If the samples were boiled, wash the reaction with 100 μL of methanol before transferring to scintillation vial. Count the radioactivity of each sample on scintillation counter using the [^{14}C] or [^{3}H] setting depending on the label.

Analysis of Ligand Binding Affinity of PITPs by Surface Plasmon Resonance (SPR)

It is highly desirable to have available methods for measuring PITP activity that are not constructed directly on the transfer assay platform. Since PITPs are not enzymes, classic kinetic biochemistry approaches cannot be confidently applied to these proteins. However, the transfer assays can now be coupled to surface plasmon resonance (SPR) techniques to measure binding kinetics for these proteins. SPR is becoming a popular method to determine phospholipid binding specificities for proteins of interest. Influences of other membrane parameters, such as curvature and charge, on protein activities and protein-membrane interactions are also amenable to analysis by this technique (Cho *et al.*, 2001). A recent example of how SPR has been used to measure membrane association and dissociation kinetics for a member of the Sec14 superfamily, involves SPR analysis for the ligand binding affinity for Sec14-like protein "Clavesin 1." This protein preferentially binds PtdIns(3,5)P_2 with a K_d of 11.3 μM, a binding specificity and affinity consistent with the role of Clavesin 1 as a regulator of neuronal lysosome morphology (Katoh *et al.*, 2009).

Principle

SPR is an optical technique that utilizes properties of a metal/dielectric interface phenomenon called "surface plasma" to measure protein binding affinities for its ligands. Upon excitation by an electron or light beam, electromagnetic waves, defined as "surface plasma" propagate along the "excited" surface. Surface plasma transfers resonance energy to an evanescent wave at a specific incident angle, termed the "surface plasmon resonance." The "surface plasma," and the 'SPR,' is sensitive to any change in the refractive index along the surface. Therefore, any change in the refraction index of the medium changes the SPR (Fig. 2).

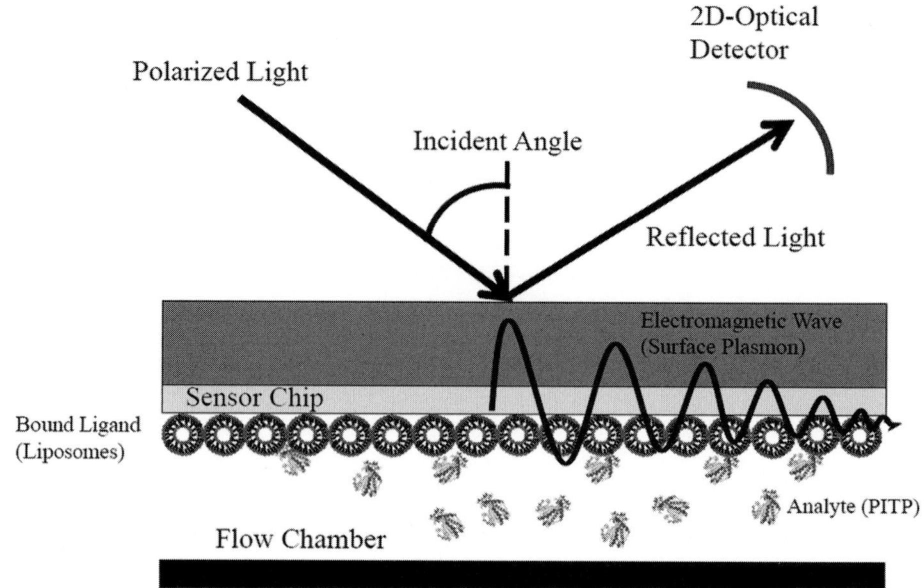

Fig. 2 **Surface Plasmon Resonance detection of PITP ligand binding affinities of PtdIns and PIPs.** Diagram representing the optical principle of SPR in detection of ligand binding affinities for any protein. Polarized light excites a metal plate and is reflected at a particular incident angle. At a certain incident angle, termed the 'surface plasmon resonance', electromagnetic waves propagate along the excited metal. Any change in the refractive index between the gold plate and dielectric media directly affects the propagating electromagnetic wave and is measured by the change in the incident angle termed the 'surface plasmon resonance'. Using microfluidic techniques, PITPs are allowed to bind a ligand covalently attached to the metal sensor chip. This binding modifies the refractive index of the metal/dielectric interface thereby adjusting the incident angle of the reflected light. The change in incident angle is a measurement of PITP's binding affinity to the immobilized lipids. (For color version of this figure, the reader is referred to the web version of this book.)

Procedure

In a SPR assay that interrogates protein binding to a lipid surface, a phospholipid bilayer or liposome is displayed onto the sensor surface. The query protein (PITP) is passed under continuous flow over the immobilized lipid surface. As the protein binds ligand and accumulates onto the sensor surface, it alters the refractive index of the electromagnetic wave and consequently, the SPR. This change in SPR is measured in real time, and the result is plotted as response, or "resonance units" versus time. The SPR signal reports protein mass at the resonant interface. By collecting SPR data at various protein input concentrations, protein–lipid k_a (association) and k_d (dissociation) constants are derived.

B. *In Vitro* Functional Assays

The "resolution and reconstitution" approach is a powerful method for dissecting enzymatic networks in a manner that provides detailed mechanistic insights into how components of a complex biochemical pathway function. This approach has been applied to a variety of signal transduction pathways. In this regard, permeabilized cell based systems are established for study of PtdIns signaling cascades, and these are shown to be PITPα–dependent *ex vivo* (Cunningham *et al.*, 1995, 1996; Hay and Martin, 1993; Hay *et al.*, 1995; Kauffmann-Zeh *et al.*, 1995). The advantage of such systems is that these more closely mimic physiological settings because the functional readouts play out in the context of a relatively undisturbed cellular endomembrane system. Moreover, these systems permit the coupling of PITP activity to other activities so that functional measurements can be made without reliance on their lipid transfer activities. The two PITP-dependent permeabilized cell models we will focus on here are reconstitutions of EGFR signaling and regulated noradrenaline secretion.

Other measurements are more minimalistic than the permeabilized cell assays and seek to incorporate information regarding PITP function from *in vivo* systems into development of physiologically relevant *in vitro* assays – particularly from the standpoint of potentiation of phosphoinositide synthesis by PtdIns kinases (Hubner *et al.*, 1998; Schaaf *et al.*, 2008). Compared to the reconstitution approach, coupled PtdIns kinase assays provide a more reductionist approach in addressing a specific aspect of phosphoinositide signaling. Below, we summarize technical aspects of these assays, and discuss what we learn from them.

1. *In vitro* Permeabilized Cell-based Reconstitution Assays

The *in vitro* biochemical reconstitution assays show that secretory granule priming for regulated exocytosis and phospholipase C signaling in permeabilized HL60 cells depends on PITPs (Hay and Martin, 1993; Cunningham *et al.*, 1996). Priming represents PITP-stimulated synthesis of PIP_2, which facilitates the Ca^{2+}-dependent fusion of secretory granules to PM. Similarly, PLC signaling depends on rate of inositol triphosphate production that is dictated by PITP through stimulation of PIP_2 synthesis. Essentially the primed exocytosis and the PLCγ signaling assays score PtdIns-4-P or PIP_2 synthesis. These two reconstitution assays are routinely used for biochemical analyses of PITPs and are described in the following sections.

Reconstitution of Ca^{2+}-Activated Noradrenaline Secretion as Mediated by PITPs

Principle

This cytosol-dependent assay was developed in the laboratory of Tom Martin and measures the secretion of noradrenaline from permeabilized PC12 cells, a neuronal precursor cell line. Permeabilization is achieved by mechanical means using shear forces. Exocytosis of radiolabeled noradrenaline is primed by three cytosolic proteins – one of which is the START-like PITPα, one is a PtdIns-4-phosphate 5-OH kinase, and the last component is a secretory granule tethering factor termed CAPS. This set of discoveries was enormously important because it represented the first reconstitution of a regulated secretory event and it established an important role for

$PtdIns(4,5)P_2$ in exocytosis. For the purpose of this review, it also implicated a role for PITPα in neurotransmission – a conclusion that will be further discussed below. This protocol is derived from Hay and Martin, 1993 and Hay *et al.,* 1995.

Materials

(i) [^{3}H] nor-adrenaline
(ii) PC12 cells
(iii) Stainless steel ball homogenizer for cell cracking
(iv) Pre-incubation buffer: 10 mM EGTA
(v) Priming buffer: 20 mM Hepes, 120 mM Potassium glutamate, 20 mM Potassium acetate, 2 mM EGTA, and 0.1% bovine serum albumin (BSA), pH 7.2
(vi) Mg-ATP
(vii) Rat brain cytosol
(viii) Recombinant or purified PITP

Procedure

1. Permeabilize [^{3}H]noradrenaline labeled PC12 cells by cell cracking technique (Martin, 1989). Briefly force the cell suspensions through the narrow hole between a bored hole and a tungsten carbide ball. Once passed through, cells remain structurally intact in appearance, but are rendered permeable to high molecular weight probes.
2. Pre-incubate permeabilized cells with buffer containing 10 mM EGTA, and wash extensively to remove soluble cellular components.
3. Add PITP of interest to washed permeabilized cells to prime Ca^{2+}-activated secretion in priming buffer and 2 mM Mg-ATP and incubate at 30 °C for 30 min. Terminate priming by placing samples on ice. Centrifuge at 800 g and resuspend the pellet in fresh priming buffer.
4. Add ~0.5 mg/mL rat brain cytosol and 1 μM Ca^{2+} and incubate at 30 °C for 3 min to trigger secretion of [^{3}H]noradrenaline.
5. Centrifuge at 800 g and transfer the supernatant. Measure the radioactivity of the supernatant containing secreted [^{3}H]noradrenaline on a scintillation counter on [^{3}H] settings. The amount of Ca^{2+}-dependent secretion of [^{3}H] noradrenaline reflects the priming activity of the PITP.

Reconstitution of PLCγ Signaling as Mediated by PITPs

Principle

Permeabilized cell-based assays that reconstitute cytosol-dependent growth factor receptor signaling by activating PLCγ, and measure release of soluble inositol-phosphates as read-out, have been reported. Cellular PtdIns pools are metabolically radiolabeled to steady state with [^{3}H]-inositol. The cells are permeabilized with streptolysin O, cytosolic proteins leak out, and fractionated cytosol is used to reconstitute the complete reaction and to resolve the cytosolic factors required for reconstitution of optimal activity. Again, PITPα reconstitutes as such a factor, and the conclusion of these studies is that PITPα is obligatorily involved in stimulating epidermal growth factor (EGFR) signaling. It is proposed to do so by potentiating PtdIns(4,5)P_2 production via PtdIns transfer from the ER to the PM. Subsequent PLC-mediated hydrolysis of PtdIns(4,5)P_2 results in enhanced release of soluble inositol phosphates. The basic protocol for this assay is assembled from various sources (Cockcroft *et al.*, 1987; Cunningham *et al.*, 1995, 1996; Kauffmann-Zeh *et al.*, 1995).

Materials

(i) RBL-2H3 or HL60 cells
(ii) Purified PITP and PLCγ1
(iii) [^{3}H]Inositol, GTP[γ-s] and DNP-HAS (2,4-Dinitrophenyl hapten conjugated to Human Serum Albumin)
(iv) Dowex anion exchange resin
(v) Buffers:

Priming media: RPMI 1640 media supplemented with1 μg/mL IgE anti-DNP and 12.5% fetal calf serum.

Pipes buffer: 20 mM Pipes, 137 mM NaCl, 2.7 mM KCl, 1 mg/mL glucose, 1 mg/mL BSA, pH 6.8.

Ca^{2+} buffer: 100 nM Ca^{2+} buffered in 3 mM EGTA, 1 mM Mg-ATP

Assay buffer: 1 μM Ca^{2+} buffered in 3 mM EGTA, 2 mM Mg-ATP, 2 mM $MgCl_2$, 10 mM LiCl.

Procedure

1. Radiolabel RBL-2H3 or HL60 cells by addition of 1 μCi/mL [^{3}H]Inositol to growth medium and culture for 48 h and then resuspend and incubate in priming media for 90 min at 37 °C.
2. Wash cells and resuspend in Pipes buffer and permeabilize by addition of streptomycin O (0.6 units/mL final concentration) in Ca^{2+} buffer for 10 min. Centrifuge permeabilized cells and resuspend in Pipes buffer.
3. Aliquot cells into tubes containing assay buffer, GTP[γ-s] (10 μM final), DNP-HAS (40 ng/mL final conc.), PITP, and PLCγ1 in a final assay volume of 100 μL. Incubate the samples for 20 min at 37 °C.
4. Stop assay reaction by adding equal volume of chloroform:methanol:water (0.5:1:0.4).

5. To extract inositol phosphates (IPs), generate a two-phase system by adding 1 mL chloroform:water (1:1). Separate the aqueous phase containing the IPs from free inositol and glyercophophoinositols by passage through Dowex anion exchange resin. Remove free inositol from the sample by washing with water. Wash with of 60 mM sodium tetraborate/5 mM sodium formate to remove glycerophospholipids.
6. Elute individual IPs from the column by increasing concentrations of formic acid given below:

   (i) Inositol monophosphates (IP_1): 0.2 M formic acid
   (ii) Inositol bisphophates(IP_2): 0.4 M formic acid
   (iii) Inositol trisphosphate(IP_3): 1 M formic acid

   For most experiments, bulk measurement of IPs is sufficient to measure signaling activation, thus, it is only necessary to elute with 1 M ammonium formate/0.1 M formic acid. However, for the PITP-PLCγ activation assay, it is recommended to measure IP_3 since PIP_2 is the preferred substrate for PLCγ.
7. Alternatively, separate IPs and count radioactivity immediately by HPLC connected to a scintillation counter. Radioactivity of IPs is measured via scintillation counter on [^{3}H] settings.

2. PtdIns 4-OH IIIβ Kinase Assay

Principle

Assays have been devised in attempts to reconstitute PITP-mediated stimulation of specific PtdIns kinase activities in defined *in vitro* systems. These assays employ recombinant PtdIns-kinase subunits, liposomes, and the PITP of choice. Kinase activity is measured in a standard manner that is, by monitoring incorporation of radiolabeled ^{32}P into phosphoinositide (Fig. 3). The assay we best understand uses PtdIns 4-OH kinase IIIβ as query enzyme and it is the one we feature here.

Materials

(i) Kinase Buffer: 370 mM NaCl, 35–42 mM $MgCl_2$, 1.7 mM EGTA, and 50 mM Tris-HCl (pH 7.5).
(ii) Uni-lamellar liposomes (Avanti PtdCho and PtdIns).
Dry down lipids under liquid N_2, reconstitute in water and vortex. Prepare uni-lamellar liposomes following the Avanti extruder protocol.
(iii) 10 mg/mL stock of Bovine Serum Albumin (BSA)
(iv) 10% Triton X-100
(v) Purified PITP
(vi) Purified PtdIns 4-OH IIIβ Kinase

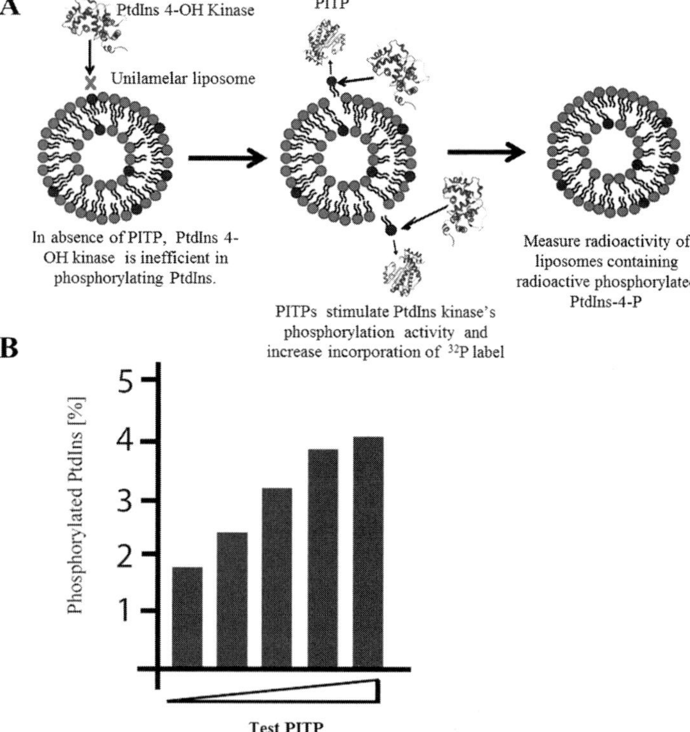

Fig. 3 PtdIns 4-OH IIIβ Kinase activity assay. (A) Flow chart of PtdIns 4-OH kinase assay. The assay consists of a unilamilar liposome, ATPγ[^{32}P], PtdIns 4-OH IIIβ Kinase and a test PITP, incubated at 37 °C. In absence of PITP, PtdIns 4-OH Kinase IIIβ cannot efficiently bind and phosphorylate its natural substrate, PtdIns. However, in the presence of a PITP, the kinase phosphorylates the lipid with higher efficiency, suggesting PITPs aids in accessibility of PtdIns substrate to the kinase. The radioactive incorporation of ^{32}P in PtdIns is shown in light shades of grey. The radioactive incorporation of ^{32}P in PtdIns is depicted in red. **(B) Graphical representation of the PtdIns 4-OH kinase assay data.** This assay is done with purified test PITP as shown. The assay is performed with increasing concentrations of the proteins and data portrayed as a percent total phosphorylatablePtdIns. (See color plate.)

(vii) ATP Mix: 2 mM cold ATP and 1 μCi/μL [γ-^{32}P] ATP in Kinase Buffer
(viii) Extraction Buffers: EBA: 33.4% Methanol, 66.6% Chloroform, 30 mM HCl

EBB: 0.6 M HCl
EBC: 50% Methanol, 47% 0.6 M HCl, 3% Chloroform

Procedure

The assay is modified protocols of PtdIns-kinase assays (Hubner *et al.*, 1998; Jones *et al.*; 1998; Schaaf *et al.*, 2008)

1. Prepare the reaction mix containing uni-lamellar liposomes, BSA (0.1 mg/mL final), 10% Triton X-100 (0.4% final), 0–2.5 μg of PITP, 0–2.5 μg of PtdIns-4-P kinase and [γ-^{32}P] ATP mix (2.5 μCi/reaction). Make up the volume with kinase buffer to a reaction volume of 60 μL. Incubate the assay at 30 °C for 30 min.
2. Stop the reaction by adding 1 mL of EBA. Generate a two-phase system by adding 200 μL of EBB to each sample, vortex for 5 s. Centrifuge for 5 min at 13,000 rpm. Transfer the organic phase to fresh tube. Add 500 μL of EBC, vortex for 5 s and centrifuge for 5 min at 13,000 rpm. Transfer the organic phase into a scintillation vial containing 4.5 mL of scintillation cocktail. Count radioactivity in a scintillation counter using [^{32}P] settings.

C. Credibility of *In Vitro* Assays: "*In Vitro* Veritas"?

The acid test of any *in vitro* biochemical assay, or any *in vitro* reconstitution system, is that it describes a functional insight that is experimentally validated in a genuine physiological context. It is in the context of PITP biology that the assays described above have failed. We focus our discussion on the Sec14 and Sec14-like PITPs because these are particularly well-understood lipid transfer proteins. We start with the transfer assay – a system that simply measures a "downhill" transport of PtdIns. An *in vivo* rationale for such an equilibrating activity is that ER membranes are PtdIns rich and the PM (where significant phosphoinositide signaling occurs) is PtdIns poor. This concept is old, and relied on what was known about the intracellular locations of PtdIns biosynthesis at that time (Michel, 1975). However, this concept has never been adequately tested in the mammalian system where it was born.

We argue that several aspects of the available data regarding PITP function gleaned from *in vivo* experiments as briefly summarized here, are difficult to reconcile with PITPs being mere "PtdIns carrier" proteins. First, Sec14 levels in cells are in large excess above the threshold required for viability (Salama *et al.*, 1990). If Sec14 is a PtdIns-transfer protein, then a low threshold for PtdIns transfer must be sufficient for yeast cell viability. Low threshold arguments are frustrating models because these can be reduced to the point that these are not subject to satisfactory experimental test. Second, a strong prediction of transfer models is that supply of membranes with PtdIns by alternative mechanisms should render PITP-dependent PtdIns-transport mechanisms irrelevant. However, elevating PtdIns levels in yeast cell membranes to 40 mol% did not circumvent the Sec14 requirement in protein transport from TGN which argued against the PtdIns transfer function for Sec14 (Jones *et al.*, 1998). Third, genetic manipulation of the yeast such that PtdIns is the major glycerophospholipid in membranes (a situation that should solve all PtdIns supply demands) failed to relieve cells of the essential Sec14 requirement (Cleves *et al.*, 1991a, 1991b). Perhaps most strikingly, genetic ablation for a specific pathway for PtdCho biosynthesis, or in specific pathways for PIP degradation, rendered Sec14 requirement for yeast membrane trafficking competence and cell viability. These data indicated that Sec14 regulates lipid metabolism, and not PtdIns transport (Cleves *et al.*, 1991a, 1991b). Finally, transfer models demand a specific vectorality of PtdIns transfer – a property determined by the relative PITP affinity for PtdIns vs PtdCho. Yet, specific reductions in PtdIns binding affinity of mutant Sec1 $^{K66,\ 239A}$ rescued the *sec14*^{ts} mutation comparable to the wild type Sec14 protein (Phillips *et al.*, 1999). Can we envision PITP mechanisms that reconcile *in vitro* transfer activities with these confounding biological data? As discussed later (Section V) in this chapter, we suggest Sec14-like PITPs, at least, are sensors that use their phospholipid-exchange activities to regulate phosphoinositide synthesis. In these scenarios the exchange reaction is not used for PtdIns-transport purposes, but to potentiate PtdIns-kinase activities by presenting substrate to the enzyme – a substrate that the PtdIns-kinase cannot efficiently access on its own.

As about the reconstitution systems discussed above, the cracked cell system of Martin described the first reconstitution of a regulated secretory event and established an important role for PtdIns(4,5)P_2 in exocytosis. The involvements of PtdIns-4-phosphate 5-OH kinase and the CAPS protein in this process have been amply validated by *in vivo* studies (Hay *et al.*, 1995; Klenchin and Martin, 2000; Loyet *et al.*, 1998; Walent *et al.*, 1992; Wiedemann *et al.*, 1996). Further, these PtdIns signaling cascades in nor-adrenaline secretion and EGFR signaling were shown to reconstitute only upon introduction of PITPα (Hay and Martin, 1993; Hay *et al.*, 1995; Kauffmann-Zeh *et al.*, 1995). The hypothesis derived from these results was that the PITPα is an obligate potentiater of PIP_2 synthesis, the major signaling lipid in these cascades. However, knockout mouse studies indicate no obvious role for PITPα in neurotransmission (Alb *et al.*, 2003). Similarly, animal studies clearly demonstrated that PITPα is not obligatorily required for EGFR signaling – the mouse phenotypes show no indications of any obvious deficit in EGFR signaling (Alb *et al.*, 2003). The dissonance between *in vitro* and *in vivo* systems with regard to physiological roles for PITPα function is observed in other contexts as well. In permeabilized cell systems, PITPα stimulates secretory vesicle and immature granule budding from hepatocyte and neuroendocrine TGN (Jones *et al.*, 1998; Ohashi *et al.*, 1995), the priming and fusion of secretory granules in mast cells (Pinxteren *et al.*, 2001), and PM receptor/G-protein coupled PIP hydrolysis by PLC (Cunningham *et al.*, 1995, 1996). Yet, none of these functional couplings are validated in mouse models (Alb *et al.*, 2002, 2003).

Why this dissonance arises between the *in vitro* and *in vivo* systems with regard to PITPs function *in vivo*? One possibility is that these biological activities are all PITP-dependent, but that PITPα is simply the one that reconstitutes – even though it is not the operative PITP *in vivo*. The second possibility is that the reconstituted systems are all phosphoinositide-dependent (this is virtually certain), but that cell permeabilization and/or membrane isolation processes activates lipases and phosphatases that degrade phosphoinositides. Thus, reconstituted systems operate from artifactual phosphoinositide-deficits, and any mechanism that reconstitutes phosphoinositide levels (i.e., PITPs) will score as activator in such assays. Consistent with such a view, the structurally unrelated yeast Sec14 substitutes for PITPα in the priming and fusion of secretory granules in mast cells and neuroendocrine cells, and budding of TGN-derived vesicles (Ohashi *et al.,* 1995; Pinxteren *et al.*, 2001). In addition, PITPα, PITPβ, and yeast Sec14 function with equal efficacy in reconstituted *trans*-membrane signaling reactions (Cunningham *et al.*, 1996). The PITP promiscuity in such reconstitutions argues against PITPs entering into dedicated physical interactions with PIP metabolic machines in the *in vitro* systems. This lack of PITP specificity observed *in vitro* is at odds with *in vivo* data that clearly demonstrate a tight physiological coupling to specific biological functions. It is abundantly clear that physiologically relevant assays for PITP function are essential for advancing the field. In recent years, a combinatorial approach primarily employing genetic models and structural tools are proving instrumental in elucidating the *in vivo* functions of PITPs. These approaches will be discussed in Section IV and V of this chapter.

Where the *in vivo* and *in vitro* data do agree is that PITPs stimulate phosphoinositide production, although how these proteins do so (i.e., "transfer" versus "nanoractor" models) is a central question in the field now. We close this section with some comments regarding coupled PITP/PtdIns kinase assays that seek to examine the functional relationship between these two proteins. The specific case upon which we focus involves the 3-4-fold stimulation of recombinant PtdIns 4-OH kinase activity by Sec14 *in vitro*. Examination of the data reveals that PtdIns 4-OH kinase, even in the presence of Sec14, only phosphorylates some 5% of the available PtdIns. This translates to production of only ~0.372 μg (420 pmols) of product in the presence of 2.5 μg (25.7 pmols) of PtdIns 4-OH kinase and 2.5 μg of Sec14 (i.e., *ca.* 16 catalytic cycles per enzyme per 30 min incubation (~0.5 cycles per kinase molecule per min) (Schaaf *et al.*, 2008). This is clearly an extremely poor catalytic efficiency, and it turns out to represent the sum of PtdIns-kinase-mediated phosphorylation of liposomal and Sec14-bound PtdIns. Catalytic efficiency increases another 20-fold (even in the absence of Sec14; *ca.* 11 catalytic cycles per kinase molecule per min) when PtdIns is presented in Triton X-100 micelles, demonstrating the kinase is active but cannot efficiently access liposomal PtdIns. These data make the point that it is important to quantify catalytic efficiencies in these assays as observed PITP-dependent "stimulation" of kinase activity are misleading (Schaaf *et al.*, 2008).

IV. *In Vivo* Approaches

A. Genetic Models

The emerging view is that PITPs are not just passive mediators of lipid transport but function in complex ways by modulating phospholipid metabolic pathways and impacting many cellular processes including lipid mediated signaling pathways and membrane trafficking. This concept on the Sec14-like and START-like PITP functions came from yeast, mice, and *Drosophila* and plant genetic studies. In fact, yeast genetics proved particularly expedient in deciphering the underlying core biochemical mechanism of PITPs that is, their regulatory role in lipid metabolism and membrane trafficking, and is increasingly used as a routine system in elucidating the functional role of novel PITPs.

1. Yeast Genetic System

Yeast Sec14 Complementation Assays

Principle

The yeast genetic system is a powerful tool for incisive dissection of mechanisms of PITP's physiological function, and the first clear insights on the physiological function of PITPs came from this system. The molecular identity for *SEC14*, the essential involvement of its product for membrane trafficking through the distal stages of the secretory pathway and yeast viability was established by characterization of *sec14* null and Sec14 temperature sensitive mutant *sec14-1*^{ts} strains (Table I) (Bankaitis *et al.*, 1989, 1990). The null mutants were lethal and completed one or two rounds of cell divison before cessation of growth indicating *SEC14* codes for an essential function in yeast vegetative growth (Bankaitis *et al.*, 1989). The *sec14-1*^{ts} referred as CTY1-1A was isolated and characterized as the conversion of residue 266

Table I
Yeast strains

Strain	Genotype	References
Sec14 proficient strain		
CTY182	*MATa ura3-52 Δhis3-200 lys2-801 SEC14*	Bankaitis *et al.*, 1989
***sec14* deletion/ mutant**		
CTY558	*MATα ade2 ade3 leu2 Δhis3-200 ura3-52 sec14Δ1::HIS3/pCTY11*	Bankaitis *et al.*, 1989, 1990
CTY1-1A	*MATa ura3-52 Δhis3-200 lys2-801 sec14-1*^{ts}	Bankaitis *et al.*, 1989, 1990
CTY2-1C	*MATα ade2-101 sec14-1*^{ts}	Cleves *et al.*, 1989
Sec14 bypass mutants		
***cki* mutant**		
CTY303	*MATaura3-52 lys2-801 Δhis3-200 Δsec14, cki1::HIS3*	Li *et al.*, 2000[a]
***sac1* mutant**		
CTY100	*MATa ura3-52 Δhis3-200 lys2-801 sec14-1*^{ts}*sac1-26*^{cs}	Cleves *et al.*, 1991a
***cct* mutant**		
CTY102	*MATa, ura3-52 Δhis3-200 lys2-801 sec14-1*^{ts}*cct-2*	Fang *et al.*, 1996
***kes1* mutant**		
CTY159	*MATa ura3-52 Δhis3-200 lys2-801 sec14-1*^{ts} *kes1-1*	Li *et al.*, 2000[a]

[a] Li, X., Routt, S., Xie, Z., Cui, X., Fang, M., Kearns, M. A., Bard, M., Kirsch, D., Bankaitis, V. A. (2000). *Mol. Biol. Cell* **11**, 1989–2005.

of the Sec14 primary translation product from Gly to Asp (Cleves *et al.*, 1989; Bankaitis *et al.*, 1990; Novick *et al.*, 1980). The *sec14-1*^{ts} missense mutation profoundly decreases Sec14 function and leads to Golgi secretory blockage under restrictive temperature condition (at 37 °C) as judged by recessive Ts-lethal phenotype. The conditional lethal phenotype of *sec14-1*^{ts} provides a selection for Ts$^+$ revertants and a model complementation system for Sec14-like activity.

The detailed analyses of yeast "bypass Sec14" mutants that no longer require Sec14 for Golgi function or cell viability provided valuable insights into how Sec14 translates its PtdIns/PtdCho transfer activity to membrane trafficking. There are seven non-essential genes, ablation of any one of which is sufficient for "bypass *sec14*" (Table I). The rationale behind the analyses of these "bypass mutations" is that these generated physiological conditions that mimic the consequences of Sec14 function *in vivo* and therefore their analyses identify effectors that respond to Sec14 activity. Five of the bypass Sec14 mutations represent structural genes for PtdCho biosynthesis namely, *cki1*(choline kinase), *pct* (phosphocholine transferase), *cct* (choline cytidyltransferase), *bsd1* and *bsd2* (Cleves *et al.*, 1991a, Mcgee *et al.*, 1994). A second class of Sec14 bypass mutant is *sac1*, which is a gene that encodes integral membrane protein of ER and Golgi, whose dysfunction results in a deranged inositol phospholipid metabolism that correlated precisely with bypass Sec14 condition as well as inositol auxotrophy (Cleves *et al.*, 1989; Rivas *et al.*, 1999; Whitters *et al.*, 1993).The final gene identified as bypass Sec14 function is *kes1* that exhibits lipid binding properties relevant to *in vivo* function. The Kes1 is amongst one of the seven members of the yeast oxysterol binding protein (OSBP) family and is unique antagonist of Sec14-dependent Golgi function. The Kes1 is a peripheral Golgi protein whose localization on Golgi membranes is crucial for its function. Deficiency of Kes1 results in bypass Sec14 while its overexpression reverses the bypass Sec14 phenotypes of known bypass Sec14mutations (Fang *et al.*, 1996; Li *et al.*, 2002). Analyses of the various Sec14 bypass mutations led to a common proposition, according to which the physiological function of the Sec14 in vesicle trafficking is the coordination of PtdCho and PtdIns metabolism so as to generate a lipid environment in the *trans*-Golgi membranes that is conducive to optimal activity of core components of Golgi TGN vesicle biogenesis/budding machinery (Cleves *et al.*, 1991b). Essentially, these bypass *sec14* mutant strains (Table I) represents Sec14 devoid (*sec14* null or conditional *sec14-1*^{ts} mutant) yeast strains, which provides excellent *in vivo* model system for assessing the PITP specific activity (i.e., PtdIns/PtdCho transfer or PIP stimulation) of any test PITPs.

Plasmid Shuffle Assay for Yeast sec14Δ Complementation

Principle

Plasmid shuffle/colony sectoring experiment is employed to test the ability of the cloned prospective PITPs to complement or suppress the lethality associated with the inheritance of *sec14*Δ null alleles by haploid *S. cerevisiae* strains. The *sec14*Δ null strain, CTY558 (Table I) carries plasmid CTY11, a YEp (*LEU2,ADE3*) vector, where the *sec14* gene is under the control of an attenuated *sec14* promoter and drives the synthesis of Sec14 in yeast at a rate similar to that normally sustained by the genomic *SEC14* locus. Strain CTY558 is absolutely dependent on pCTY11 for viability, as this plasmid complements the haploid lethal *sec14 Δ1::HIS3* allele. The CTY558 forms uniformly red colonies on all media, a characteristic phenotype of *ade2* strains of *S. cerevisiae*, whereas *ade2, ade3* double mutants are white. Thus loss of pCTY11 from CTY558 can be scored visually by appearance of white colonies in background of otherwise red colonies (Fig. 4A). The test Sec14-like protein when complements Sec14 is able to substitute pCTY11 in sustaining the viability of the *sec14 Δ1:: HIS3* strain, resulting in appearance of visual white colonies.

Procedure

1. Transform *sec14*Δ null strain CTY558 with the plasmid YEp having test Sec14-like PITP with *URA* selection marker.
2. Select for Ura^+ Leu^+ transformants demanding the presence of both pCTY11 (with *LEU* marker) and the test construct plasmids (have a *URA* marker) in the transformants in YNB (yeast nitrogen base) plates supplemented with glucose and adenine. All the colonies will be *ade2* mutants and hence red in color.
3. Plate transformants in serial dilutions on YPD (yeast extract peptone dextrose) plates that relieve the strains of any kind of auxotrophy. The CTY558/YEp *sec14*-like test

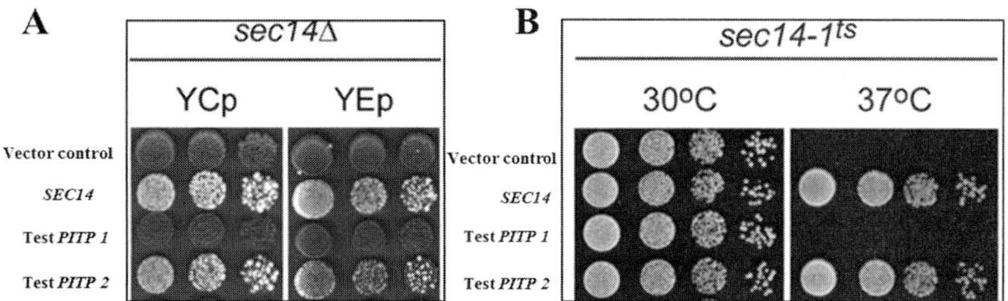

Fig. 4 Yeast complementation assays. (A) Plasmid shuffle assay for yeast *sec14*Δ complementation. An *ade2ade3 sec14*Δ/YEp (*SEC14,LEU2, ADE3*) yeast strain (CTY558) is transformed with the indicated YCp (URA3) or high copy YEp (URA3) plasmids comprising either the *SEC14* (positive control) or the test *PITP* cDNAs. The empty vector serves as a negative control. The transformed colonies are dilution spotted on to YPD plates, where all nutritional selection is removed. Functionality of the test construct products (in this case test *PITP2*) is manifested as appearance of white segregant colonies that acquire leucine auxotrophy, signifying loss of parental YEp (*SEC14, LEU2, ADE3*) plasmid. Retention of the parental plasmid (i.e non-functionality of test PITPs) is scored by the red colonies (colonies seen as grey shades in the figure), as is shown for test *PITP1*. **(B) Yeast dilution assays for yeast *sec14-1^{ts}* complementation.** The *sec14-1^{ts}* yeast strain transformed with YCp plasmids carrying the designated cDNAs are spotted in 10-fold dilution series on to YPD plates and incubated both at 30 °C and 37 °C temperatures. Rescue of growth defect at 37 °C reports functionality of test constructs. YCp (*URA3*) and YCp (*SEC14,URA3*) serve as negative and positive controls respectively. As shown in figure, test *PITP1* does not complement, whereas test *PITP2* complements *sec14-1^{ts}* defects. (For interpretation of the references to color in this figure legend, the reader is referred to the web version of this book.)

construct transformants will yield white colonies at a very high frequency, only if the test Sec14-like gene complements yeast *sec14*. If the test *sec14*-like gene do not complement, all the colonies will be red at this stage.

4. Re-streak the white colonies on Leu^- and Ura^- plates to check the loss of pCTY11 over the YEp test construct plasmid. The white segregants will be Leu^- and Ura^+, signifying the loss of pCTY11 and retention of the test construct plasmid. This will indicate that the test Sec14-like gene is able to complement/substitute for yeast *sec14*.

Yeast Dilution Assays for Yeast sec14-1^{ts} Complementation

Principle

The conditional temperature sensitive lethal phenotype of *sec14-1^{ts}* (CTY1-1A) (Table I) provides a selection for Ts^+ revertants and a model complementation system for Sec14-like activity. The recessive nature of the *sec14-1^{ts}* allows identification of *sec14* homologs by complementation. Pseudo-reversion analysis of sec14-1^{ts} is employed to identify gene products that complements or modulates Sec14 activity or is responsive to it. CTY1-1A strain is transformed with yeast centromeric plasmid carrying the test cDNA gene and the *URA3* gene. The desired transformants are obtained by simultaneous selection for uracil prototrophy and temperature resistance. Purified transformants are screened for their growth and ability to form single colonies at the permissive and non-permissive temperature at 25 and 37 °C respectively by serial dilution on selection plates (Fig. 4B). In case of cDNA library screening, the cDNA expression library is transformed into CTY1-1A and Ura^+ transformants are selected at 25 °C. A total of approximately 20,000 Ura^+ transformants are screened for growth at restrictive temperature by replica plating onto uracil deficient minimal media. The Ts^+ colonies obtained are then analyzed by two criteria to demonstrate that the Ts^+ phenotype is due to plasmid linked trait. First, isolation and characterization of spontaneous segregates that lose plasmid under non-selective conditions reveals that the plasmid cured derivatives fail to grow at 37 °C. Second, the plasmids recovered from Ts^+ Ura^+ transformants by transformation into *E. coli* are subsequently transformed into CTY1-1A, which reveals a complete coincidence of inheritance of both Ura^+ and Ts^+ in the transformants. The restriction analysis of *sec14-1^{ts}* complementing cDNA provides the nucleotide sequence identity.

Procedure

1. Transform the CTY1-1A strain with the yeast centromeric plasmid with *URA* marker, containing the test cDNA or with empty vector as control.
2. Select the CTY1-1A transformants initially at permissive temperature at 28 °C in selective medium lacking uracil.

3. Evaluate complementation of the temperature sensitive phenotype of the *sec14-1* allele after 4 days of growth on selective medium without uracil at 37 °C.
4. Evaluate eight independent CTY1-1A transformants containing each plasmid by serial dilution on replica plates. Grow each replica plate at permissive and restrictive temperature at 28 and 37 °C respectively. The transformants growing at both the temperatures are the ones that complement *sec14-1*^{ts} (CTY1-1A).

Invertase Secretion Assays for Screening the Secretory Efficiency in Yeast sec14^{ts} Complementation

Principle

The *sec14*^{ts} is classified as conditionally defective in Golgi transport processes, based upon the observation that such yeast exhibit a marked exaggeration of Golgi related structures (berkeley bodies) under non-permissive conditions and that intracellular invertase pool that accumulates under such conditions consists of enzyme that exhibits fully matured outer glycosyl chains (Esmon *et al.*, 1981; Novick *et al.*, 1980). The complementation of *sec14*^{ts} strains by the test constructs is confirmed by measuring the secretory competence. Analysis of secretory competence of *sec14*^{ts} involves the measurement of intracellular invertase accumulation after a long exposure of cells to restrictive temperature. The secretion competence is measured by the secretion index that relates extracellular secreted invertase to total cellular invertase as described by Salama *et al.* (1990) (Fig. 5A). The secretion indices of wild type and *sec14*^{ts} strains represents measure of normal secretory proficiency and the magnitude of the *sec14*^{ts} Golgi secretory block respectively.

Procedure

1. Grow the CTY1-1A transformants with the test cDNA construct or the empty vector control to mid logarithmic phase in 2% glucose YP (yeast extract peptone) medium at 25 °C with shaking. In 2% glucose concentration in the media, the secretory invertase synthesis is repressed.
2. Pellet cells, wash with two volumes of water, resuspend in prewarmed in 0.1% glucose YP medium. Incubate at 30 °C for 30 min to allow secretory invertase synthesis. Subsequently, the cultures are shifted to 37 °C for 2 h.
3. After 2 h, adjust the samples to 10 mM NaN_3, wash twice with ice-cold 10 mM NaN_3 and resuspend in 0.5 mL of the same.
4. Split each sample in two equal aliquots. For each of the non-permeabilized sample, adjust volume to 0.5 ml with 250 μL of 10 mM NaN_3. Adjust the samples to be permeabilized similarly with 10 mM NaN_3 and 0.2% TritonX-100 and subject each to one cycle of freeze thaw. Use these samples to measure external and total invertase activities respectively using the assay of Goldstein and Lampen (1979). Calculate the invertase secretion index as secreted invertase/ total invertase. Invertase units are described as nanomoles of glucose produced per minute at 30 °C.

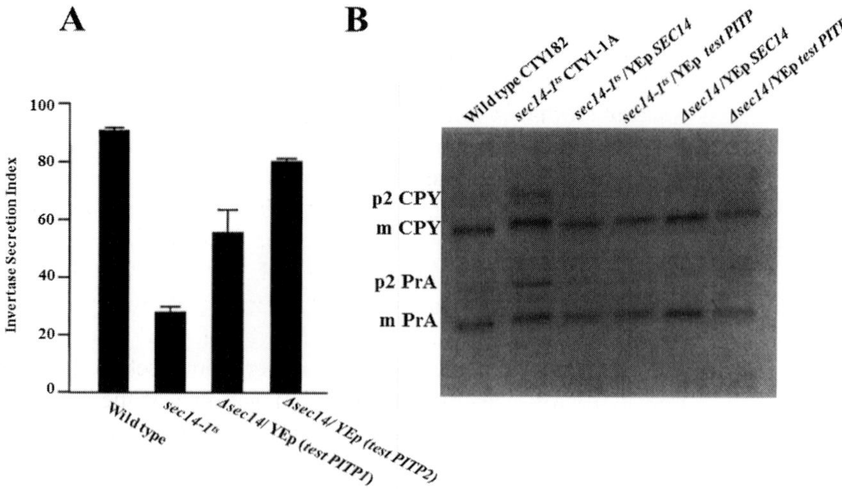

Fig. 5 Measurement of yeast secretory competence by invertase assays and kinetics of vacuolar protein biogenesis. (A)Graphical representation of data for invertase secretion assay. Efficiency of invertase secretion at 37 °C for Δ*sec14* strains carrying indicated YEp plasmid clones of test PITPs referred as *PITP1* and *PITP2*. The secretion index relates extracellular secreted invertase to total cellular invertase. The secretion indices of wild type and *sec14-1^{ts}* strains represent measures of normal secretory proficiency and the magnitude of the *sec14-1^{ts}* Golgi secretory block respectively. The secretory competence of yeast Δ*sec14* strains affected with Golgi secretory defects are significantly restored nearly to wild type levels on expression of *PITP1* and *PITP2* respectively. These PITPs are representatives Sec14-like PITPs. **(B) Diagrammatic representation of data for analysis of vacuolar protein biogenesis by immunoprecipitation of CPY and PrA.** Designated yeast strains (indicated above corresponding lanes) are grown in 2% glucose minimal media at 30 °C and subsequently shifted to 37 °C for 30mins, pulse radiolabeled with ^{35}S-aminoacids. CPY and PrA cross reactive materials are evaluated by immunoprecipitation, SDS-PAGE and autoradiography. As seen, wild type strain CTY182 exhibits essentially all of its radiolabeled CPY and PrA as the corresponding mature (mCPY or mPrA) vacuolar species. The *sec14-1^{ts}*(CTY1-1A) exhibit defects in CPY and PrA biogenesis at 37 °C and is indicated by recovery of some of the total radiolabeled material in CPY and PrA precursor (p2CPY and pPrA) fractions respectively. Expression of *SEC14* and test *PITP* (which complements yeast *sec14*) restored wild type kinetics for biogenesis of CPY and PrA to the vacuole in *sec14-1^{ts}* mutants at 37 °C and also in normally inviable *sec14*Δ mutants.

Kinetics of Vacuolar Protein Biogenesis in Screening of Yeast sec14^{ts} Complementation

Principle

As a screen for the *sec14^{ts}* complementation, kinetics of the biogenesis of two vacuolar proteins is routinely employed. The biogenesis of these two proteins is indicative of the Golgi apparatus derived vesicular trafficking function. The two proteinases of the vacuolar lumen CPY (carboxypeptidase Y) and PrA (proteinase A) exhibit distinct precursor forms (i.e., a core glycosylated p1 form representative of material in transit through the ER and Golgi apparatus and p2 species that has acquired additional glycosyl modifications in Golgi apparatus) and mature forms that permit monitoring of the progress of radiolabeled proteinase through the secretory pathway to vacuole (Stevens *et al.*, 1982; Woolford *et al.*, 1986). In particular, as the proteolytic maturation that is associated with activation of these p2 zymogen forms is thought to occur in vacuole, the maturation event provides a convenient indicator for the arrival of these proteinases at the vacuole and hence a measure of Golgi vesicular trafficking efficiency. In this experiment, appropriate yeast strains are grown at restrictive temperature and subsequently pulse radiolabeled with ^{35}S-aminoacids for 10 min and subjected to 20 min chase with excess unlabeled methionine and cysteine. CPY and PrA cross-reactive materials are extracted from total proteins by immuno-precipitation and the products are evaluated by SDS–PAGE and autoradiography as described by Bankaitis *et al.* (1989). Under this experimental set up, the wild type CTY182 exhibit essentially all of its radiolabeled CPY and PrA as the corresponding mature vacuolar species (Fig. 5B) and only trace quantities of precursor forms are detected (<5% of total signal). The isogenic *sec14^{ts}* mutant exhibit defects in CPYand PrA biogenesis at 37 °C as indicated by the recovery of some 40 and 30% of total radiolabeled material in CPY and PrA precursor fractions respectively. Any test cDNA that complements *sec14* is supposed to restore the wild type kinetics for biogenesis of CPYA and PrA to vacuole in *sec14^{ts}* mutants and the normally in viable *sec14Δ* null mutant incubated at 37 °C.

Procedure

1. Grow the CTY1-1A transformants with the test cDNA construct or the empty vector control to an early mid-logarithmic growth stage in glucose 2% minimal media at 30 °C with shaking. Shift the cultures to 37 °C for 30 min.
2. Radiolabel cultures (cells with an OD_{600} of 0.5) with 150 μCi of [^{35}S]-cysteine and [^{35}S]-methionine for 10 min.
3. Initiate a 20 min chase by addition of unlabeled methionine and cysteine to a final concentration of 1%.
4. After 20 min add trichloroacetic acid (TCA) to each culture to a final concentration of 5%. Solubilize the total protein precipitates and immunoprecipitate using CPY and PrA rabbit polyclonal antibodies. Evaluate the cross-reactive immunoprecipitates by SDS–PAGE and autoradiography.

PIP Quantification Assays for PIP Profile in Screening of Yeast sec14 Complementation

Principle

The *in vitro* reconstitution and *in vivo* assays have demonstrated that PITPs stimulate PtdIns-4-P and PIP_2 synthesis (Hay and Martin, 1993; Phillips *et al.*, 2006). Current evidence proposes that Sec14 employs its heterotypic PtdIns or PtdCho exchange activities to bind or sense local PtdCho and simultaneously prime a PtdIns presentation unit or nanoreactor that stimulates PtdIns-4-OH kinases

(Schaaf *et al.*, 2008). Therefore another routine criterion to test yeast *sec14* complementation implicates PIP quantification assays to investigate the PIP profiles. For this, the yeast *sec14* bypass mutants are readily employed as host systems for assessing the PIP stimulating activity of any test PITP that complements yeast *sec14*. Each test protein is expressed into *sec14Δ cki* (CTY303) yeast strain. This "bypass *sec14*" strain exhibits basal PIP levels, due to its Sec14 deficiency and provides a context for assessing the ability of a protein to modulate PIP metabolism *in vivo* (Philips *et al.*, 1999; Routt *et al.*, 2005). Further, to assess the Sec14-mediated activation of yeast PtdIns-4-OH kinase in its native environment, the Sac1 PIP phosphatase-mediated degradation of PtdIns-4-P generated by the Stt4 PtdIns-4-P kinase is exploited. The *sec14-1^{ts}sac1-26* (CTY100) yeast mutant strain is used for this assay. This strain sustains cell viability under Sec14-deficient conditions and impedes degradation of PtdIns-4-P. Due to Sac deficiency, this strain accumulates PtdIns-4-P to levels 10-fold in excess of wild type PtdIns-4-P and PtdIns-3-P levels, thereby simplifying specific identification of PtdIns-4-P and analysis of its production rates. The 10-fold accumulation of PtdIns-4-P in Sac1-deficient yeast reports Stt4 activity. For PIP analysis, phosphoinositides are metabolically radiolabeled over 10 cell doublings to steady state in appropriate minimal media containing [^{3}H] *myo*-inositol. The cells are pelleted, washed and lysed in 4.5% perchloric acid with bead beating. The total cellular lipids are extracted from the membranes in cell suspension, the polar phospholipids are extracted in acidified extraction solvent, resolved by TLC, identified by co-chromatography with radiolabeled standards, and Stt4-dpendent PtdIns-4-P accumulation is quantified by autoradiography (Fig. 6A). For quantitative analysis of total phosphoinositides, the total bulk lipids are deacylated by base hydrolysis (Stolz *et al.*, 1998). The deacylated products (glycerophosphoinositides) are resolved and quantified using anion exchange chromatography using Whatmann Partisphere SAX HPLC system (Fig. 6B).

Analysis and Quantification of Stt4-Dependent Accumulation of PtdIns-4-P by TLC

Procedure

1. Transform the *sec14-1^{ts}sac1-26* yeast mutant (CTY100) with the test constructs or the empty control vector. Grow the transformants to late-logarithmic phase in radiolabeled minimal media with [^{3}H]-myoinositol (20 μCi/mL) to steady state at 30 °C. Initiate labeling of the cultures with cell concentrations at A_{600} of 0.1–0.2 and grow with shaking until late-log phase (A_{600} = 0.6–1.0).
2. Shift late-logarithmic phase cultures to 37 °C for 2 h. Kill the cells by adding TCA to 5% final concentration and incubate the cultures on ice for 15 min.
3. Pellet the cells and wash twice with 0.5 mL of cold water. Resuspend the pellet in 0.5 mL of 4.5% perchloric acid, add $^1/_2$ volume of glass beads, and vortex the cell suspension for 1 min for 10 min with incubation on ice between rounds of vortexing.

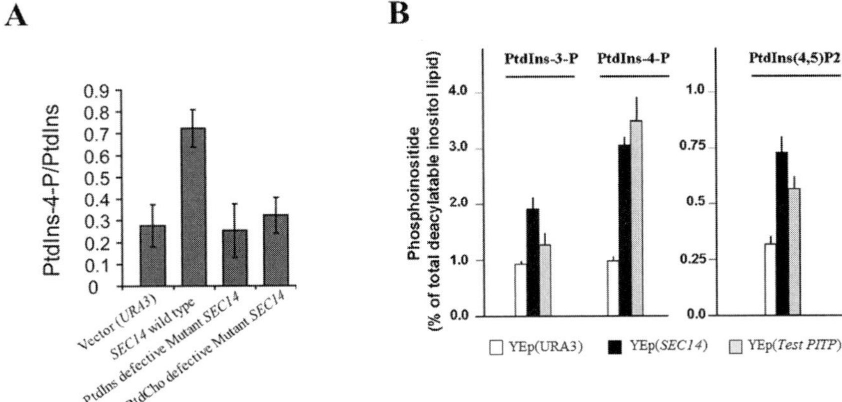

Fig. 6 PIP analysis and quantification. (A) Analysis of Stt4-dependent accumulation of PtdIns-4-P by TLC. A *sec14-1*^{ts}*sac1-26* yeast mutant (CTY100) and isogenic derivatives expressing the indicated Sec14 or Sec14 mutant proteins (which are PtdIns or PtdCho transfer defective) are radiolabeled to steady state with [^{3}H]-inositol and shifted to 37 °C for 2hrs. PtdIns-4-P and other Ins-phospholipids extracted, resolved by TLC and identified by co-chromatography with radiolabeled standards. The Stt4 dependent PtdIns-4-P accumulation is detected and quantified by quantification of the autoradiograhs of the TLC plates. The *URA3* and *SEC14* derivatives represent negative and positive controls respectively. **(B) Analysis of phosphoinositides by HPLC.** Isogenic derivatives of *sec14Δ* yeast strain CTY303 carrying designated YEp plasmids were radiolabeled to steady state with [^{3}H]-inositol. PIPs extracted deacylated and quantified by HPLC. PtdIns-3-P, PtdIns-4-P and PIP_2 levels are as indicated. The YEp (*URA3*) and YEp(*SEC14*) derivatives serve as negative and positive controls respectively while YEp (*Test PITP*) represents the construct for a representative PITP which complements yeast *SEC14*.

4. Transfer the cell suspension to new microfuge tube and pellet membranes 14,000 rpm for 10 min. Wash the pellet in 0.5 mL of 100 mM EDTA (pH 7.4).
5. Resuspend the pellet in 0.5 mL of $CHCl_3$:CH_3OH:1N HCl (4:2:0.015) and shake at 800 rpm in eppendroff thermomixer for 10 min at RT.
6. Convert in a two-phase system by adding 0.1 mL of 0.6 M HCl, vortex vigorously for 5 min in cold room. Centrifuge at 14,000 rpm for 5 min and transfer organic bottom layer to new centrifuge tube. Clean the organic phase by adding 0.25 mL of CH_3OH:HCl:$CHCl_3$ (2:2:0.1), vortex vigorously and transfer organic bottom layer to new centrifuge tube.
7. Dry the organic phase containing polar phospholipids in speed vac and resuspend in 30 μL of $CHCl_3$. Spot them onto TLC plate (Partisil LK6DF Silica Gel40 Å, from Whatmann). Pre-treat the plates with 1% potassium oxalate, 1 mM EDTA in CH_3OH/H_2O (2:3) and bake at 100 °C for 1 h prior to run. Run the TLC in ammonia solvent system [$CHCl_3$:CH_3OH:H_2O:NH_3 (2.4:2:0.35:0.25)]. The various polar PLs are identified on TLC by co-chromatographing standard radiolabels.
8. Expose the TLC plates to X-ray films in Kodak low energy intensifying screen and visualize the radiolabeled polar phospholipids and quantify them by phosphorimaging.

Analysis and Quantification of Phosphoinositides by HPLC

Procedure

1. Grow derivatives of the *cki sec14*Δ yeast mutant (CTY303) or *sec14-1^{ts}sac1-26* yeast mutant (CTY100) carrying the test constructs or the empty control vector to late-logarithmic phase in radiolabeled minimal media with [^{3}H]-*myo*-inositol (20 μCi/mL) to steady state for 18 h at 30 °C.
2. Kill the cells by adding TCA to 5% final concentration and incubate the cultures on ice for 15 min. Pellet the cells and wash twice with 0.5 mL of cold water. Resuspend the pellet in 0.5 mL of 4.5% perchloric acid, add $^1/_2$ volume of glass beads and vortex the cell suspension for 1 min for 10 min with incubation on ice between rounds of vortexing.
3. Transfer the cell suspension to new microfuge tube and pellet membranes 14,000 rpm for 10 min. Wash the pellet in 0.5 mL of 100 mM EDTA (pH 7.4).
4. Extract the lipids twice in 0.25 mL of ethanol:water:diethylether:pyridine (15:15:5:1) at 60 °C for 2 h. Spin at max speed and transfer the supernatant in fresh centrifuge tube. Dry the supernatant fluids in speed vac.
5. Add 0.3 mL of deacylation buffer containing methanol:water:1-butanol:methylamine (3:2:0.0.1:3). Resuspend the pellet and incubate at 53 °C for 2 h. Stop deacylation by adding 0.15 mL of ice-cold 1-propanol and speed vac the deacylated lipids to dryness.
6. Resuspend the pellet in 0.3 ml of solvent butanol:petroleum ether:ethylformate (20:4:1), add 0.3 mL of water. Vortex to mix and spin at 13,000 rpm for 2 min. Take out the organic upper phase and discard. Add 0.3 mL of solvent, vortex and spin to separate the aqueous phase. Transfer the bottom aqueous phase containing the extracted glycerophosphoinositides to a fresh centrifuge tube. Dry the extracted lipid in speed vac to complete dryness.
7. Dissolve the dried samples in suitable volume of 10 mM ammonium phosphate pH 3.5. Inject the deacylated products to a 4.6 x 125-mm Patrisphere SAX-10 column (Whatman), and elute with linear gradient of 10–340 mM ammonium phosphate over 15 min, 340–1.02 M ammonium phosphate over 7.5 min, and isocratic 1.02 M ammonium phosphate for 5 min on a Beckman System Gold HPLC system. Measure radioactivity using the BetaRAM™ in-line detector (INUS, Tamps, FL).

Yeast sec14-1^{ts} in Identification of sec14 Bypass Genes

The conditional lethal phenotype of *sec14-1^{ts}* provides a powerful selection system for Ts$^+$ revertants. The *sec14-1^{ts}* is a readily revertible allele (~10^{-6} Ts$^+$ revertants per cell per generation). Isolation and characterization of 107 spontaneous Ts$^+$ revertants from *sec14-1^{ts}* haploid strains – CTY1-1A and CTY2-1C led to the identification of *sec14* bypass genes. Standard dominance tests and meiotic segregation analyses showed that 26 of the mutants exhibited a recessive suppression of *sec14-1*^{ts} (termed as bypass suppressor recessive, *bsr*) while 81 exhibited dominant suppressor phenotype (termed as bypass suppressor dominant, *bsd*). The 107 suppressors identified six suppressor genes, of which four (*sac11, bsr2*(*cct*), *bsr3* (*kes1*), and *bsr4* (*cki*)) were defined by recessive mutations, whereas two (*bsd1* and *bsd2*) were defined by dominant mutations. Recent studies have identified *bsd1* as *kes1* mutation and *bsd2* as *sac1* mutation. The ability of *bsr* and *bsd* mutations to restore wild type growth properties to yeast strains carrying *sec14* disruption alleles suggested that the bypass suppression was efficient. The identification of the *sec14* bypass genes provided the first link between biosynthetic protein transport, intracellular phospholipid signaling, and phospholipid biosynthesis via CDP-choline pathway. The recognition of the role of Sec14 in controlling the secretory competence of yeast Golgi membranes through modulation of their phospholipid content underscored the importance of genetic approaches in elucidation of the role of PITPs in inter-compartmental protein transport.

2. Metazoan PITP Model Systems

In vivo studies demonstrate that PITPs control the interface between membrane trafficking and lipid metabolic pathways in yeast. By contrast, the physiological functions for metazoan PITPs, which are structurally unrelated to yeast PITPs, are not understood in cellular or organismal levels. Biochemical assays show their function as soluble factors that stimulate various reconstitutions of PIP dependent functions in permeabilized mammalian cells. These functions include regulated and constitutive membrane trafficking and phospholipase C dependent signaling through G-protein coupled receptors discussed above (Cunnigham *et al.*, 1996; Hay and Martin, 1993; Ohashi *et al.*, 1995). However, the predictions derived from the *in vitro* systems consistently failed *in vivo* tests and therefore they complicated the interpretation of the PITP's physiological role. This dissention is mainly due to PIP "run down" artifacts. Cell permeabilization evokes PIP degradation and imposes PIP-deficits on reconstituted PIP-dependent systems. These conditions impose lack of PITP specificity in these assays. Contrary to these assays, genetic studies are showing exquisite functional specificities of PITPs. These genetic models are forthcoming and have provided initial clues on metazoan PITP's function in receptor signaling and membrane trafficking.

Mouse PITP Model

The gene ablation approaches showed that PITPβ plays an essential housekeeping function, as its deficiency results in catastrophic failure early in murine embryonic development (Alb *et al.*, 2002). Initial clues concerning the cellular role of PITPβ came from studies where both the splice variants of PITPβ were depleted by RNAi in HeLa cells (Carvou *et al.*, 2010). Analyses of the membrane transport in these cells identified a specific defect in retrograde traffic from Golgi to the ER mediated by COPI- coated vesicles. The retrograde transport defect observed in PITPβ-knockdown cells could be rescued when wild type PITPβ was re-expressed. On the other hand, reduction of PITPα to 18% of wild type levels leads to vibrator neurodegenerative disorder (Hamilton *et al.*, 1997). Further, the data culled from the PITPα null mouse model provided crucial insights on the relationship between PITPα and signaling receptors (Alb *et al.*, 2003). The mouse model is robust and PITP$\alpha^{0/0}$ neonates are born alive, but present fully penetrant neurodegeneration, pancreatic compromise and chylomicron retention disease. The neurodegenerative disease is marked by defective myelination of motor neurons, aponecrosis and CNS infiltration by inflammatory cells.

The PITP$\alpha^{0/0}$ inflammatory cells are highly active, which is in contrast to the *in vitro* reconstitution studies that assigned essential roles for PITPα in potentiating compound exocytic responses to inflammatory cells. The PITP$\alpha^{0/0}$ syndromes are not influenced by genetic background and hence they simplify genetic interaction studies in authentic *in vivo* contexts. The double mutant analysis for surveying the functional interactions of PITPα with signaling receptors is feasible in this mouse model. Besides, an allelic series of hypomorphic mice with graded levels of PITPα activity is developed which provides an excellent system that affords flexibility with regard to model mouse or cell-based system for assessing PITPα/signaling receptor relationships (Alb *et al.*, 2007). These mouse lines are robust source for embryonic fibroblasts (MEFs), cerebellar granule cells, cortical neurons, dorsal root ganglia neurons that are readily maintained *ex vivo*, thus providing appropriate systems for study. Also, the PITP$\alpha^{0/0}$ ES cells exhibit robust tumorigenicity and enhanced EGFR signaling. These murine or the cell models are employed to address the question of how PITPs functionally interface with signaling receptors. Models for dissecting the relationships between PITP and EGFR/DCC signaling include primary and immortalized *pitp*$\alpha^{0/0}$ MEFs and cortical neurons. Thus mouse PITP model system will provide mechanistic insights on how the PITPs regulate signaling receptors and will help understand the neuro-degenerative and chylomicron retention disease that afflicts PITPα-deficient mice.

Drosophila PITP Model

Inactivation of the membrane bound PITP of *Drosophila*-RdgB results in an inherited form of light enhanced retinal degeneration (Milligan *et al.*, 1997). The *rdgB* mutant flies exhibit the morphological hallmarks of photoreceptor cell degeneration several days after ecloison, and abnormal light response shortly after fly's initial exposure to light (Harris and Stark, 1977; Stark *et al.*, 1983). Genetic epistasis analyses suggested that RdgB functions downstream of both rhodopsin and phospholipase C in the visual transduction cascade. The studies demonstrated that the complete repertoire of RdgB functions essential for normal photo-transduction resides in its PITP domain. Also, genetic interaction studies showed that PITPα deficiency potentiates EGFR signaling and defects in peripheral nervous system in *Drosophila* (unpublished data). These *in vivo* analyses provided the first clues on the function of any membrane bound PITP class and represents a model system for the study of PITPs, in particular for membrane bound class II PITPs. Besides, the single class I PITP protein in *Drosophila* has provided the first genetic evidence of the concerted function of class I PITPs with the PtdIns-4- kinaseβ (Giansanti *et al.*, 2006).

Zebrafish PITP Model

Zebrafish (*Danio rerio*) is emerging as a popular model vertebrate system, which resembles mammals in its development. In this system gene knockdown by morpholino oligonucleotides is easily accomplished. Zebrafish express mammalian set of metazoan PITPs plus a unique PITPβ like isoform termed PITPng. Since mice ablated in PITPβ are lethal and does not lead to term, zebrafish serves as an excellent vertebrate model system for characterizing the function of PITPβ. Zebrafish PITPβ expression is specifically robust in double cone cells of zebrafish retina. Morpholino-mediated protein knockdown experiments demonstrated that dysfunction of zebrafish PITPβ function compromises biogenesis and or maintenance of double cone photoreceptor cell outer segments in developing retina. In contrast to PITPβ morphants, knockdown of PITPα expression results in early developmental failure (Ile *et al.*, 2010). In another report, morpholino mediated knock-down of PITPα expression in zebrafish embryos demonstrated dose dependent defects in motor neuron axons and reduced number of spinal cord neurons (Xie *et al.*, 2005). Collective data suggest that while zebrafish PITPα has a crucial role in early embryonic development likely in neuronal development, PITPβ perhaps supports a high capacity membrane trafficking pathway required for biogenesis/maintenance in double cone photoreceptor cells.

3. Plant Sec14-like Multi-Domain PITP Model System

The Sec14-like modular protein family is large in eukaryotes and constitutes an intriguing class which includes mammalian Sec14 like proteins like RasGAP neurofibromin, specific RhoGAPs, PTP/MEG2, plant Sec14-nodulin and plant and mammalian Sec14-GOLD proteins. These modular arrangements serve as platforms for interesting biological functions with broad implications in eukaryotic cell biology. However, these modular Sec14-like PITP class remain largely understudied. Lack of functional analyses is largely the result of multi-domain Sec14 expression being limited to higher eukaryotes. In this regard, plant model system *Arabidopsis thaliana* is proving valuable in studying the function of modular Sec14 proteins. Multi-domain Sec14 proteins are highly represented in *Arabidopsis* genome which encodes 13 Sec14-nodulin and 6 Sec14-GOLD proteins. Multiple T-DNA insertion lines for each of these genes are available in TAIR (The *Arabidopsis* Information Resource) database.

Characterization of one of the Sec14-nodulin gene *AtSfh1* from the T-DNA insertion lines have demonstrated its role in developmentally regulated polarized membrane trafficking program in root hair biogenesis (Vincent *et al.*, 2005). Loss of *AtSFH1* in the T-DNA insertion lines results in derangement of the tip directed organization of PIP_2 pools in *AtSfh1-/-* root hairs which is accompanied by specific disruption of the tip oriented actin microfilament, microtubule network, and Ca^{2+} gradient. These multiple defects are manifested primarily from collapse of AtSfh1 driven PIP_2-dependent polarized membrane trafficking pathway. Specific ablation of either PtdIns or PtdCho-binding activity of Sec14 or only the nodulin domain results in functional inactivation of the protein (unpublished data). The *Arabidopsis* data on Sec14 modular protein is particularly forthcoming in supporting the nanoreactor concept for modular Sec14-proteins. Present evidence proposes AtSfh1 as a "membrane bound PITP nanoreactor" in which the Sec14 domain of AtSfh1 is the "PtdIns presentor" for PtdIns 4-OH kinase that ultimately promotes synthesis of PIP_2. The nodulin domain defines the spatial territory of AtSfh1, as it sequesters PIP_2 by high affinity electrostatic interactions thereby imposing a non-random tip oriented distribution of specific PIP_2 pools in the root hair PM. In another study, a role of the Sec14-GOLD protein PATL1 was shown in the formation and maturation of the cell plate during late telophase stage of cytokinesis (Peterman *et al.*, 2004). *Arabidopsis* provided the first clues on function of any modular Sec14 protein. The flexibility of the *Arabidopsis* model system affords unique opportunities to study the fascinating biology of Sec14-modular proteins in eukaryotes.

B. Fluoroscent-based Localization Assays

Fluorescence based approaches employ the use of fluorescent tags for the subcellular localization/co-localization of PITPs and cellular phosphoinositide pools. Subcellular localization of PITPs utilizes immunofluroscence or expression of recombinant fluorescent tags fused to PITPs *in vivo*. The localization of PIP pools is achieved with the use of fluorescent tag fused to specific PIP-recognition domains.

1. Fluorescence-based Localization Studies of PITPs

The subcellular localization of PITPs is demonstrated using immunofluorescence or recombinant expression of fluorescent tag fused PITPs. In immunofluorescence, the primary antibody bound to PITP is detected by the secondary antibody that is bound to a fluorescent probe. Commonly fluroscein isothiocyanate (FITC, fluoresces green under UV light) and rhodamine (fluoresces red) are used as fluorescent probes. Immuno-fluorescence can be detected by either a UV standard fluorescence microscope or confocal microscopy. The recombinant fluorescent-tagged proteins can be expressed *in vivo* using GFP, RFP, YFP, or CFP protein expression vectors.

When the cell expresses the protein, it contains the fluorescent marker that allows location of the protein to be detected by fluorescence microscopy.

The fluorescent fused subcellular localization markers (Table II) are often used along with fluorescent-tagged PITPs to image the co-localization of PITPs for specific organelle distribution. The co-localization is further confirmed by using PITP antibody and sub-cellular marker antibodies in western blotting to identify the presence of PITPs in specific cell fractions by sub-cellular fractionation

Table II
Subcellular localization markers and antibodies

Subcellular Marker/Antibodies	Subcellular region localized	References
Fluorescent tagged markers		
GFP-NLS of SV40 large antigen (in triplicate)	Nucleus (with heavily stained nucleolus)	Kalderon *et al.*, 1984[a]; Lanford *et al.*, 1986[b]
GFP-N-terminal human β 1,4-galactosyltransferase	Golgi	Watzele *et al.*, 1990[c]
GFP-mitochondial targeting sequence cytochrome c oxidase subunit VIII	Mitochondria	Rizzuto *et al.*, 1989[d]
GFP-N-terminal calreticulin ER targeting signal + C-terminal KDEL-ER retention signal	Endoplasmic Reticulum (ER)	Fliegel *et al.*, 1989[e]
GFP-C-terminal peroxisomal targeting signal 1 tripeptide (SKL)	Peroxisome	Weimer *et al.*, 1997[f]
Vacuole syntaxin-GFP	Vacuole	Uemura *et al.*, 2002[g]
GFP- mouse Talin	Actin	Kost *et al.*, 1998[h]
Histone 3-GFP	Chromosome	Fang and Spector, 2005[i]
Subcellular marker antibodies		
Anti-ERp61 antibody	Recognizes ER Chaperone protein	Urade *et al.*, 1997[j]
Anti-actin antibody	Recognizes actin filaments	Lessard, 1988[k]
Anti-PMP70 antibody	Recognizes a peroxisomal membrane protein	Imanaka *et al.*, 2000[l]
Anti-GM130 antibody	Recognizes a 130 kDa Golgi matrix protein	Nakamura *et al.*, 1997[m]

[a] Kalderon, D., Roberts, B. L., Richardson, W. D., and Smith, A. E. (1984). *Cell* **39**, 499–509.
[b] Lanford, R. E., Kanda, P., and Kennedy, R.C. (1986). *Cell* **46**:575–582.
[c] Watzele, G., and Berger, E. G. (1990). *Nucleic Acids. Res.***18(23)**, 7174.
[d] Rizzuto, R., Nakase, H., Darras, B., Francke, U., Fabrizi, G. M., Mengel, T., Walsh, F., Kadenbach, B., DiMauro, S., Schon, E. A. (1989). *J. Biol. Chem.* **264(18)**, 10595-10600.
[e] Fliegel, L., Burns, K., MacLennan, D. H., Reithmeier, R. A., and Michalak, M. (1989). *J. Biol. Chem.***264**, 21522-21528.
[f] Wiemer, E. A., Wenzel, T., Deernick, T. J., Ellisman, M. H., and Subramani, S. (1997). *J. Cell Biol.* **136**, 71-80.
[g] Uemura, T., Yoshimura, S. H., Takeyasu, K., and Sato, M. H. (2002). *Genes Cells***7**, 743–753.
[h] Kost, B., Spielhofer, P. and Chua, N. H. (1998). *Plant J.***16**, 393–401.
[i] Fang, Y. and Spector, D. L. (2005). *Mol. Biol.Cell* **16**, 5710–5718.
[j] Urade, R., Oda, T., Ito, H., Moriyama, T., Utsumi, S., and Kito, M. (1997). *J. Biochem. (Tokyo)* **122(4)**, 834-842.
[k] Lessard, J. (1988). *Cell Motil.Cytoskeleton* **10(3)**, 349-362.
[l] Imanaka, T., Aihara, K., Suzuki, Y., Yokota, S., and Osumi, T. (2000). *Cell Biochem. Biophys.***32**, 131-138.
[m] Nakamura, N., Lowe, M., Levine, T., Rabouille, C., and Warren, G. (1997). *Cell* **89(3)**, 445-455.

experiments. The subcellular marker antibodies (Table II) are usually used in conjunction with immunofluorescence/immunohistochemistry or western blot techniques to identify the localization of the protein of interest. The subcellular markers help to conclusively identify the organelle or region of the cell where PITPs are localized. Apart from the subcellular markers, several fluorescent dyes like Nile Red (for lipid particles), DAPI (for nuclei), or vital marker fluorescent dyes like FM4-64 (endocytic marker) are used to monitor organelle organization and dynamics.

By indirect immunofluorescence using antibodies, PITPα was detected in the nucleus and cytoplasm of Swiss mouse 3T3 fibroblasts and PITPβ associated with perinuclear Golgi system (De vries *et al.*, 1995). In another study, PITPα and PITPβ covalently labeled with the sulfo-indocyanine fluorescent dyes Cy3 and Cy5 was localized to nucleus/cytoplasm and to perinuclear membrane structures respectively (De vries *et al.*, 1996). Later, a functional PITPβ-GFP protein expressed in MEFs and COS7 cells were localized to Golgi membranes (Philips *et al.*, 1996). Double-labeled immunofluorescence using PITP antibody in combination with compatible antibodies raised against markers for specific Golgi compartments (TGN38), showed PITPβ localization to TGN. Similarly, co-localization of yeast Sec14 with the yeast Golgi marker Kex2 was demonstrated in double label immunofluorescence experiments (Cleves *et al.*, 1991a). In another study, through expression of the C-terminal yEGFP fusions, the Sec14 homolog proteins Sfh2, Sfh4, and Sfh5 were localized to cytosol and microsomes (Schnabl *et al.*, 2003).

2. Localization of PIP Pools with Fluoroscent Tagged Protein Domains

The PITPs role in PIP homeostasis was established by the PITP/PtdIns-kinase relationship. The PITP/PtdIns-kinase relationship assigns instructive roles for PITPs in specifying the outcomes of PtdIns-kinase action. In this the PITP acts as a nanoreactor that traps PtdIns in a transition state during heterotypic PtdCho/PtdIns exchange reactions and these trapped PtdIns become accessible to PtdIns kinase, which are otherwise biologically inadequate interfacial enzymes (Schaaf *et al.*, 2008). Therefore investigation of the PIP distribution sets the stage for interpreting the biology of PITPs, and links the homeostasis of specific PIP pools to PITP regulation. The functions of PITPs are inferred in the context of PITP localization, its co-localization with specific PIP pools or PtdIns kinases, and the consequences of PITP deletion on PIP homeostasis. The PIP distribution map is particularly required for establishment of the physical relationship between PITP and PIP pools, and this allows recognition of derangements in PIP pools upon functional ablation of a PITP.

Available technique in live cell applications for detection of PtdIns pools at single cell level involves fluorescent biosensor-driven PIP imaging strategies. These biosensors rely on the natural protein domains that recognize specific cellular phosphoinositides. These modules are fused to fluorescent proteins, which allow detection of dynamic changes in phosphoinositides (Balla and Varnai, 2009). These include PH domain to detect PtdIns (4,5)P_2, PtdIns (3,4,5)P_3, PtdIns-4-P, and FYVE domain to detect PtdIns-3-P (Table III). In spite the limitations of this method, it provided enormous information on PIP distribution and dynamics in a great number of systems and has become a standard method in cell biology. These PIP binding modules is used as probes for sensitive detection of their target PIPs in enzyme assays, in cell and tissue extracts and in ultrastructural studies of lipid localization using quantitative immuno-electron microscopy. The advantage of using these domains is that they display high degree of selectivity for specific phosphoinositide species appropriate for detection of lipids at the levels that occur in cells. However, the PIP recognition domains have some limitations. Firstly, the overexpressed fluorescent-tagged PIP binding domains may interfere with cellular processes, perhaps by sequestering lipid pools for example, high expression of the PLCδ PH-GFP causes morphological changes that include rounding of the cells, loss of attachment, and development of intracellular vesicles. These toxic effects are due to the inhibition of the connections between the cytoskeleton and PM by PLCδ PH-GFP (Raucher *et al.*, 2000). Secondly, the interaction of the PIP binding domains with membrane lipids may be inhibited or competed by certain metabolites within the cytosol (e.g., inhibition of PLCδ PH-GFP labeling by inositol trisphosphate

Table III
Fluorescent fused PIP recognition modules used in live cell imaging

PIP recognition domain	Live cell localization	References
PtdIns (4,5)P_2		
PLCδ_1-PH	Plasma membrane	Stauffer *et al.*, 1998[a]; Varnai and Balla, 1998[b]
Tubby domain	Plasma membrane plus cleavage furrow	Santagata *et al.*, 2001[c]
PtdIns (3,4,5)P_3		
GRP1-PH	Plasma membrane	Venkateswarlu *et al.*, 1998a[d]
ARNO-PH	Plasma membrane	Venkateswarlu *et al.*, 1998b[e]
Cytohesin-1-PH	Plasma membrane	Nagel *et al.*, 1998[f]
Btk-PH	Plasma membrane	Varnai *et al.*, 1999[g]
PtdIns (3,4,5)P_3/ PtdIns (3,4)P_2		
Akt-PH	Plasma membrane	Watton and Downward, 1999[h]
PDK1	Plasma membrane	Komander *et al.*, 2004[i]
PtdIns (3,4)P_2		
TAPP1-PH	Plasma membrane	Kimber *et al.*, 2002[j]
PtdIns(3,5)P_2		
Ent3p-ENTH	Yeast pre-vacuole	Friant *et al.*, 2003[k]
Svp1p	Yeast vacuole	Dove *et al.*, 2004[l]
PtdIns-3-P		
FYVE (Hrs,EEA1)	Early endosome, yeast vacuole	Gillooly *et al.*, 2000[m]
P40phox-PX	Early endosome	Ellson *et al.*, 2001[n]

(Continued)

Table III *(Continued)*

PIP recognition domain	Live cell localization	References
PtdIns-4-P		
OSH 2-2xPH	Plasma membrane	Roy and Levine, 2004[o]
OSBP-PH	Golgi and plasma membrane	Levine and Munro, 1998[p]
FAPP1-PH	Golgi and plasma membrane	Godi *et al.*, 2004[q]
PtdIns-5-P		
3xPHD(ING2)	Nucleus and plasma membrane	Gozani *et al.*, 2003[r]

[a] Stauffer, T.P., Ahn, S., and Meyer, T. (1998). *Curr. Biol.* **8**:343-346

[b] V'arnai, P., and Balla, T. (1998). *J. Cell Biol.***143**, 501-510.

[c] Santagata, S., Boggon, T. J., Baird, C. L., Gomez, C. A., Zhao, J., Shan, W. S., Myszka, D. G., and Shapiro, L. (2001). *Science* **292**, 2041-2050.

[d] Venkateswarlu, K., Gunn-Moore, F., Oatey, P. B., Tavare, J. M., and Cullen, P. J. (1998a). *Biochem. J.***335**, 139-146.

[e] Venkateswarlu, K., Oatey, P. B., Tavare, J. M., and Cullen, P. J. (1998b). *Curr. Biol.* **8**, 463-466.

[f] Nagel, W., Schilcher, P., Zeitlmann, L., and Kolanus, W. (1998). *Mol. Biol. Cell* **9**, 1981-1994.

[g] V'arnai, P., Rother, K. I., and Balla, T. (1999). *J. Biol. Chem.* **274**, 10983-10989.

[h] Watton, J., and Downward, J. (1999). *Curr. Biol.* **9**, 433-436.

[i] Komander, D., Fairservice, A., Deak, M., Kular, G. S., Prescott, A. R., Peter Downes, C., Safrany, S. T., Alessi, D. R., and van Aalten, D. M. (2004). *EMBO J.* **23**, 3918-3928.

[j] Kimber, W. A., Trinkle-Mulcahy, L., Cheung, P. C., Deak, M., Marsden, L. J., Kieloch, A., Watt, S., Javier, R. T., Gray, A., Downes, C. P., Lucocq, J. M., and Alessi, D. R. (2002). *Biochem. J.* **361**, 525-536.

[k] Friant, S., Pecheur, E. I., Eugster, A., Michel, F., Lefkir, Y., Nourrisson, D., and Letourneur, F. (2003). *Dev. Cell* **5**, 499-511.

[l] Dove, S. K., Piper, R. C., McEwen, R. K., Yu, J. W., King, M. C., Hughes, D. C., Thuring, J., Holmes, A. B., Cooke, F. T., Michell, R. H., Parker, P. J., and Lemmon, M. A. (2004). *EMBO J.* **23**, 1922-1933.

[m] Gillooly, D. J., Morrow, I. C., Lindsay, M., Gould, R., Bryant, N. J., Gaullier, L. M., Parton, G. P., and Stenmark, H. (2000). *EMBO J.* **19**, 4577-4588.

[n] Ellson, C. D., Gobert-Gosse, S., Anderson, K. E., Davidson, K., Erdjument-Bromage, H., Tempst, P., Thuring, J. W., Cooper, M. A., Lim, Z. Y., Holmes, A. B., Gaffney, P. R. J., Coadwell, J., Chilvers, E. R., Hawkins, P. T., and Stephens, L. R. (2001). *Nat. Cell Biol.***3**, 679-682.

[o] Roy, A. and Levine, T. P. (2004). *J. Biol. Chem.* **279**:44683-44689.

[p] Levine, T. P. and Munro, S. (1998). *Curr. Biol.***8**,729-739.

[q] Godi, A., Di Campi, A., Konstantakopoulos, A., Di Tullio, G., Alessi, D. R., Kular, G. S., Daniele, T., Marra, P., Lucocq, J. M., and De Matteis, M. A. (2004). *Nat. Cell Biol.* **6**, 393-404.

[r] Gozani, O., Karuman, P., Jones, D. R., Ivanov, D., Cha, J., Logovskoy, A. A., Baird, C. L., Zhu, H., Field, S. J., Lessnick, S. L., Villasenov, J., Mehrotra, B., Chen, J., Rao, V. R., Brugge, J. S., Ferguson, C. G., Payrastre, B., Myszka, D. G., Cantley, L. C., Wagner, G., Divecha, N., Prestwich, G. D., and Yuan, J. (2003). *Cell* **114**, 99-111.

generated by PLC mediated signaling). These methods have been described in detail elsewhere (Downes *et al.*, 2003, 2005).

The limitations of employing protein domain fluorescent protein chimeras in live cells were overcome by developing alternative method in which the lipids are detected post-fixation without interfering with the biological process. Post-fixation detection of lipids has been achieved with anti-phosphoinositide antibodies or recombinant inositol binding domains that are fused to both GST and GFP tags in immuno-cytochemical or electron microscopy applications. The principle is to expose ultrathin sections of tissue to the recombinant GST and GFP fused PIP domain probes and to detect them with commercially available antibodies to GST and through fluorescence. The advantage of this technique is that these detect lipids in intact cells without any distortion caused by the detection process and many cells could be analyzed with no time constraints. These fixed cell EM studies revealed the presence of lipids in compartments where *in vivo* PH domain imaging failed for example, PtdIns (4,5)P_2 in Golgi was detected with GST-fused PLCδ1-PH-GFP post fixation (Watt *et al.*, 2002). However, post-fixation has its own drawbacks. First, the fixation process has a major influence on what phosphoinositide pools are visible to the antibodies or the GST-fused inositide binding module. Second, the sensitivity of these methods is hard to evaluate and leads to variability in results.

A quantitative assessment of membrane localization of fluorescent PIP probes is achieved with the use of FRET (Fluorescence Resonance Energy Transfer) technology. The FRET principle was used in several studies to obtain a signal that reflects true binding of PIP-binding probes to membranes. The method is based on co-expression of CFP- and YFP-tagged versions of the same PIP probes. These fluorophores are the most widely used pairs for FRET studies. The principle is when the donor (CFP) and the acceptor (YFP) are within FRET distance ($<$8 nm) there will be energy transfer from CFP emission (475 nm) to YFP emission (525 nm) when using CFP excitation (430 nm). The principle is applicable to lipid binding domains when expressed both as CFP and YFP fusion protein. The two fluorophores will show efficient energy transfer when their attached PIP binding domains are bound to PIPs at the membrane. Like intermolecular FRET, intramolecular FRET is also exploited for PIP detection on membranes. In this, both the fluorophores are attached to same PIP-recognition probes. Lipid binding induces molecular rearrangement that changes the distance of the two fluorophores and hence FRET signal. Such probe based on Grp1-PH domain was targeted to different membranes for PtdIns (3,4,5)P_3 detection (Sato *et al.*, 2003).

The PIP localization methodology is successfully used in establishment of physical relationship between PITP and PIP pools. The derangement in PIP pools upon functional ablation of a Sec14-like PITP AtSfh1 was demonstrated in *Arabidopsis*. Genetic ablation of *AtSFH1* showed that it is an essential component in regulation of the polarized membrane trafficking program in root hair morphogenesis (Vincent *et al.*, 2005). Loss of *AtSFH1* resulted in derangement of the tip directed organization of PIP_2 pools in *AtSfh1-/-* root hairs and this was demonstrated using PLCδ1-PH-GFP localization (Fig. 7). The PIP_2 derangement is accompanied by specific disruption of the tip oriented actin microfilament, microtubule network, and Ca^{2+} gradient. These multiple defects are manifested primarily from collapse of AtSfh1 driven PIP_2-dependent polarized membrane trafficking pathway. Like this study, the PIP localization approach will prove instrumental in guiding interpretations of genetic and biochemical analyses of other PITPs. The PITP/PtdIns kinase relationship is established in yeast studies that suggest functionally specified PIP pools are defined by the identity of the PITP which couples to the PtdIns kinase (Bankaitis *et al.*, 2010; Routt *et al.*, 2005;

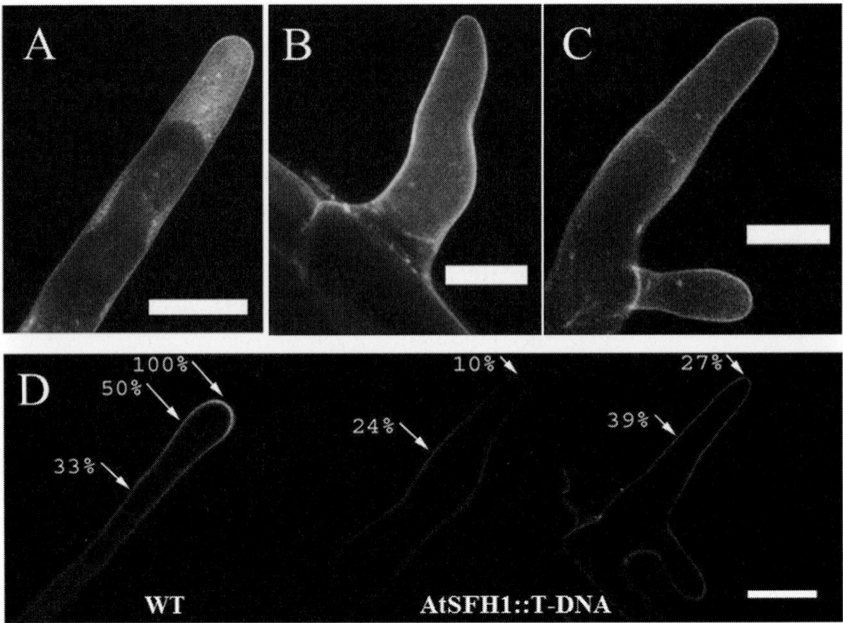

Fig. 7 Detection of deranged tip directed PIP_2 pools in Sec14-like PITP, *AtSFH1* ablated *Arabidopsis* mutant plants by fluorescence imaging. Localization of PIP_2 pools using PH-PLCδ1-YFP expression in 3d-old seedlings. **(A) Wild type plants (B) *AtSFH1::T-DNA* single root hairs. (C) *AtSFH1::T-DNA* double root hairs.** The tip directed PIP_2 localization as seen in wild type root hairs, is disrupted in single and double root hairs of mutant plants. **(D) Ratiometric imaging of PH-PLCδ1-YFP/FM1-43 fluorescence.** The 3 day old wild type seedlings expressing PH-PLCδ1-YFP were labeled with FM1-43 under conditions that selectively label plasma membrane. The fluorescence ratio is measured at points indicated by arrows. PHPLCδ1-YFP/FM1-43 fluorescence in wild type root hair apex is set as 100% and all other ratio indicated is normalized to this value. Scale bar- 20 μm (Adapted from Vincent *et al*, 2005. Originally published in Journal of Cell Biology. doi: 10.1083). (For interpretation of the references to color in this figure legend, the reader is referred to the web version of this book.)

Schaaf *et al.*, 2008). Further, the biochemical studies suggest that the START like PITPs resemble Sec14-like PITPs in their mechanism of action as PtdIns kinase presenting units (Ile *et al.*, 2010). Thus the PIP localization assays will prove crucial in unraveling the role of PITPs as discriminating portals for highly specific aspects of PIP signaling.

V. Structural Approach

The *in vivo* genetic approaches described above provided important insights on the function of these proteins in the complex regulation of lipid metabolism, membrane trafficking, and PIP signaling. However, an understanding of the fundamental nature of this regulation and the molecular mechanism by which the Sec14 superfamily of proteins execute their critical biological function require detailed structural and mechanistic analyses. In this regard, the structural and biophysical studies of Sec14 proved instrumental in deriving the molecular basis for Sec14-like function, which perhaps also applies for START-like PITP function. The various structural approaches employed for studying the PITPs are outlined in the following sections. We will mainly focus our discussion on Sec14-like PITPs, as the structural studies on these proteins were most significant in elucidating the operation of PITPs at the molecular level. In future, these approaches will hold greatest implications on the less known metazoan Sec14-like and START-like PITPs. At the end of this section we will briefly comment on the current structural studies on START-like PITPs, which show similar PtdIns binding and transfer activity as Sec14-like proteins, yet structurally distinct from them.

A. Insights Gained From Crystal Structures of Sec14-like PITPs

The Sec14 crystal structure provided valuable insights on how members of this protein family bind phospholipid (PL) substrate and how it catalyzes PL exchange reaction (Ryan *et al.*, 2007; Sha *et al.*, 1998; Schaaf *et al.*, 2008). Other available crystal structures of Sec14-like proteins include ligand bound form and apo-versions of α-tocopherol transfer protein (α-TTP) (Min *et al.*, 2003), the mammalian Sec14-GOLD protein Sec14L2 (Stocker and Baumann, 2003) and detergent-bound and PL bound forms of the neurofibromin Sec14-like domain (Welti *et al.*, 2007). All these structures show that Sec14 domains adopt similar folds termed the "Sec14-fold," an approximately 280-residue two lobed globular structure (Fig. 8A). The crystal structure of apo-Sec14 comprises 12 α-helices, six β strands, and eight 3_{10}-helices that cooperate to form a large hydrophobic pocket, which accommodates a single PL monomer. An unusual hydrophobic helix gating module (A10/T4 helix) composed of both α-helical and 3_{10}-helical elements gates the hydrophobic PL binding pocket. Additional structural and mechanistic insights on Sec14 function were gained with the high-resolution crystal structure of yeast Sec14 homolog Sfh1 in complex with biologically relevant PLs (Schaaf *et al.*, 2008).

The most striking difference between the apo-Sec14 and holo-Sfh1 structure is the repositioning of the helix gating module. The gating module is flipped open in Sec14 domain apo-structure and is closed in holo-structures. The apo-structure most likely describes a transitional conformation attained by the protein docked on the surface of the target membranes upon ejection of the bound PL and prior to reloading of another phospholipid molecule. Further, comparison of the various holo-Sfh1 revealed distinct binding strategies for PtdIns and amino-PL headgroups (i.e., PtdCho or phosphatidylethanolamine (PE)) for Sec14 and Sec14-like proteins. In contrast to amino-PL headgroup which are sequestered deep into Sfh1 cavity stabilized by tyrosine mediated cation-π interactions, the inositol headgroup is configured near the protein surface by extensive hydrogen bonding and electrostatic interactions. All the holo-Sfh1 configurations were consistent with the structural features of closed forms of

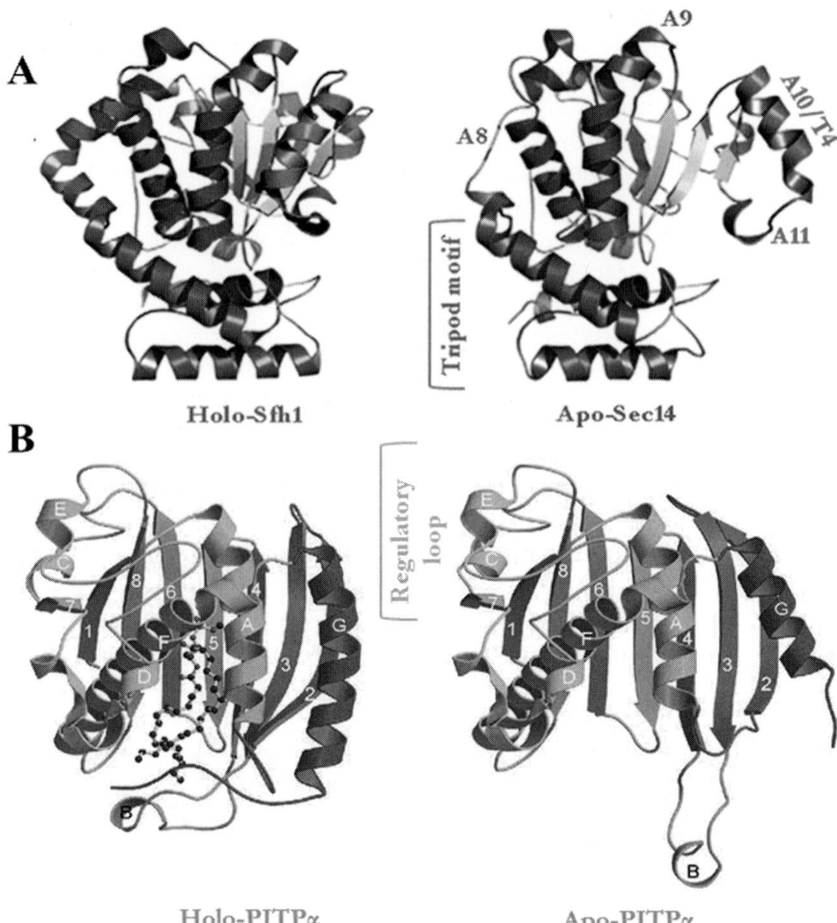

Fig. 8 Crystal structure of Sec14-like and START-like proteins. (A) Crystal structure of holo-Sfh1 (left) and apo-Sec14 (right) of yeast. The Sec14 fold consists of twelve α-helices, six β-strands and eight 3_{10} helices. The hydrophobic pocket of the protein is formed of six β-strands which form the floor and three α-helices A8, A9 and A11 which forms the walls of the pocket. The pocket is gated by A10/T4 helix gating module. The N-terminal A1-A4 α-helices fold into a tripod like motif (indicated in bracket) and consist of the bulk of 129 aminoacids required for Sec14 targeting to Golgi membranes. **(B) Crystal structure of closed (left) and open (right) conformation of PITPα from rat and mouse respectively.** The structure contains four functional regions i.e a lipid binding core (alpha-helix-A and F and beta sheet-1-6, 8) with flexible part of β-strand (beta-sheets 2-4), a regulatory loop (alpha helix-C-E and beta sheet-7), the C-terminal region (alpha helix-G) and the lipid exchange loop (alpha helix-B). The lipid molecule in the closed form is PtdCho. The eight β sheets are numbered from 1-8. The α-helices are represented as A-G. (See color plate.)

other Sec14-like proteins. Another distinctive feature derived from the holo-Sfh1 crystal structure is their ability to accommodate within their PL binding pocket, PL molecules with different volumes, without significant effect on shape of the protein surface. This is achieved in part by a water flux mechanism, where the unoccupied PtdCho and PtdIns headgroup binding site is loaded with ordered water molecules in the respective Sec14-PtdIns and Sec14-PtdCho complexes. Such flux of water molecules into and out of the hydrophobic pocket is proposed to play a key role in overcoming the differences in Sec14 relative binding affinities for PtdIns and PtdCho making the heterotypic exchange reactions feasible.

B. Site Directed Mutational Studies Based on Ligand Bound Sec14-like PITP Crystals

The structure-guided generation of mutant Sec14-like proteins with headgroup specific PL binding/transfer defects was particularly useful in providing important clues on the individual contribution of each PL binding/transfer activity of Sec14 to biological function. Functional analyses of the Sec14 PtdIns and PtdCho binding mutations, $Sec14^{R65,T236D}$ and $Sec14^{S173,T175I}$ showed that both PtdIns and PtdCho binding activities of Sec14 are required in *cis* to potentiate Sec14-mediated PIP synthesis by activation of PtdIns-4-OH kinases *in vivo*. This revealed the role of heterotypic exchange reactions of Sec14 in stimulation of PtdIns-4-OH kinase activity, in which accessible "PtdCho" are required to drive the exchange reactions. These results assigned a dual role of Sec14, one as a PtdCho sensor and other as a PtdIns-presenting nanoreactor, which transmits PtdCho metabolic information to PIP synthesis.

These structural studies provided a conceptual framework for how Sec14-like PITPs couple metabolic cues to the action of interfacial lipid modifying enzymes like PtdIns-4-OH kinases, and this perhaps represent a common theme in signal transduction regulation. Available evidence proposes that a heterotypic PtdIns/PtdCho exchange reaction uniquely prime Sec14 for productive presentation of PtdIns to PtdIns-kinases, which are inefficient interfacial enzymes otherwise. These structural studies support a "nanoreactor" mechanistic model versus the historically accepted transfer models for Sec14-like PITP functions (Bankaitis *et al.*, 2010; Ile *et al.*, 2006). The nanoreactor concept proposes that Sec14 and other Sec14-like PITPs act as "primed lipid biosensors or nanoreactors" that couple binding of lipids other than PtdIns (sensor function) to a PtdIns-presentation activity that potentiates the PtdIns-kinase activity by making PtdIns a better substrate for the enzyme.

The crystal structure based predictive bioinformatics approaches permitted the assignment of primary sequence bar codes for PtdIns and PtdCho binding for the Sec14 superfamily members. The primary sequence comparisons clearly identify "PtdIns binding bar code" across this superfamily, however they do not show extensive conservation of the "PtdCho binding bar codes." Indeed, several Sec14-like protein members lacking the PtdCho binding bar code bind hydrophobic ligands with diverse chemical properties (Meier *et al.*, 2003; Welti *et al.*, 2007). The "PtdIns binding barcode" is particularly relevant to human disease as several naturally occurring mutations in human proteins directly involve residues of the predicted PtdIns binding bar codes of the corresponding Sec14-like proteins/domains. Besides, the fact that the exchange of two disparate ligands is required for Sec14-mediated stimulation of PIP synthesis suggests that a two ligand nanoreactor mechanism for PtdIns kinase activation might be broadly used by the Sec14 superfamily of proteins to integrate the metabolism of wide range of lipids or lipophilic molecules (serving as Sec14-protein ligands) to phosphoinositide synthesis thereby coordinating diverse aspects of lipid metabolome with common phosphoinositide signaling pathways in specific physiological processes. At present neither PtdIns binding, nor the physiological relevance of such binding for most of the metazoan Sec14-like PITPs is demonstrated. Future structural work on these proteins will address these questions.

C. Conformational Dynamics of Sec14

Molecular dynamics simulations (MDS) of Sec14 provided the first detailed insight into the conformational transitions of a PITP engaged in phospholipid exchange (Ryan *et al.*, 2007). The simulations showed that an unusual hydrophobic helix gating module (A10/T4 helix) gates the hydrophobic PL binding pocket of the protein. This helix represents the most hydrophobic patch on the surface of the open Sec14 conformer. On these unusual structural accounts, it is proposed that A10/T4 exhibits enough conformational flexibility to penetrate the membrane bilayer during PL transfer reaction. Further, the simulations predicted that these helical gate dynamics are controlled by the compact gating module that regulates an extensive H-bond network through which conformational information is transduced to the helical gate upon membrane binding. These molecular dynamic predictions were confirmed by site directed mutational analyses, and were consistent with Sec14 apo- and holo-structures. Moreover, these conformational dynamics analyses identified common conformational substructures that are affected by mutations associated with inherited human diseases of the Sec14 superfamily (Ryan *et al.*, 2007).

D. Biophysical Analyses of Ligand Bound Sec14-like PITP by Electron Paramagnetic Resonance (EPR) Spectroscopy

EPR is used to study protein–lipid macromolecular environments like the organization/stoichiometry of membrane proteins (Marsh, 1997; Páli *et al.*, 2006) and molecular dynamics of protein-lipid exchange (Horváth *et al.*, 1994). This technique is useful, among other things, for measuring the electrostatic properties of phospholipids incorporated in the lipid binding pocket of a protein. Using nitroxide spin labeling of the phospholipid ligand, the dynamics, electrostatic state, and mobility of the phospholipid can be measured as it resides within a protein. Essentially, the data gathered from EPR elucidate the microenvironment of the binding pocket and solvent accessibility of the bound phospholipid.

EPR was used in measuring the microenvironment of PtdCho molecules bound to Sec14 (Smirnova *et al.*, 2006). In this study, carbons along the *sn-2* chain of PtdCho were spin labeled with nitroxide groups and upon excitation, the electron spin of the free electron pairs of the nitroxide groups were measured. Depending on the environment of nitroxide group, the paramagnetic relaxation was enhanced or quenched and read out as "signal." These experiments demonstrated that carbons C_5, C_7, and C_{12} of the *sn-2* chain of the bound PtdCho are highly aprotic, with the nitroxide group existing almost entirely in the non-hydrogen-bonded form, indicating they are strongly shielded from hydrophilic solute molecules (Smirnova *et al.*, 2006). These experiments demonstrated that Sec14 has tight interactions with the PtdCho headgroup and proximal region of the bound phospholipid acyl chain, but its interactions with the distal regions of the acyl chain were less robust.

High-field multi-frequency EPR experiments on Sec14 clearly demonstrated the hydrophobicity of the Sec14 PL binding pocket (Smirnova *et al.*, 2007). EPR measurements showed that the hydrophobicity parameters of the PL binding pocket is such that it offers an environment similar to that provided by membrane leaflet. Thus, incorporation of a PL from membrane into the Sec14 interior, and *vice versa*, is primarily driven by simple partitioning of a PL between two chemically equivalent environments.

E. Functional Engineering of a Sec14-like PITP Through Directed Evolution Screens

A novel directed evolutionary approach was exploited to resurrect activity to a functionally dead Sec14-like protein, Sfh1, the closest Sec14 homolog (Schaaf *et al.*, 2011). In the directed evolution screen, missense mutations are incorporated into the *SFH1* gene by error-prone PCR and *in vivo* gap repair, and reconstituted plasmids driving the expression of mutagenized genes are introduced into *sec14-1^{ts}ura3-52* yeast strain. The transformants are co-selected for growth at 37 °C and uracil prototrophy and subjected to detailed analyses including biochemical transfer and *in vivo* genetic assays. Unexpectedly, the single substitutions that revive Sec14-like activity in Sfh1 alter residues conserved between Sec14 and Sfh1. The activated Sfh1* must execute heterotypic PtdIns/PtdCho-exchange reactions to potentiate PtdIns 4-OH kinase activity *in vivo*. Biochemical, biophysical, structural, and computational approaches demonstrate that the activation mechanism reconfigured atomic interactions between amino acid side chains and internal water in an unusual hydrophilic microenvironment within the hydrophobic Sfh1 phospholipid binding cavity. These altered dynamics reconstituted a functional "gating module" that controls phospholipid cycling into and out of Sfh1* hydrophobic pocket.

This directed evolution approach revealed an unexpectedly flexible functional engineering of "Sec14-like PITPs" that was invisible to the standard bioinformatics, crystallographic, and rational mutagenesis approaches. The study also held out the prospect that, plasticity in conformational coupling could be employed for the purpose of bypassing the normal requirements for specific structural elements in

Sec14 or a model Sec14-like protein. Such engineering strategy is promising in relation to human diseases related to mutations in Sec14-like proteins. A number of inherited human disease mutations in proteins of the Sec14 superfamily compromise the G-module, the structural unit that transmits conformational information to the helical gate (Ryan *et al.*, 2007). With this approach, it may prove feasible to reactivate the mutant proteins with small molecules that reprogram the conformational transitions of the Sec14-like protein.

F. START-like PITPs: Similar or Dissimilar to Sec14-like PITPs?

The "START-like PITPs" do not show any sequence or structural similarity to Sec14 or Sec14-like PITPs. However, both the classes of PITPs exhibit similar *in vitro* activity that is, they both transfer PtdIns and PtdCho. Besides, expression of classI START-like PITPs in yeast complements Sec14 defects (Skinner *et al.*, 1993). Similarly, the Sec14 is able to replace PITPα's role in stimulation of PLC signaling in permeabilized mammalian cells (Cunnigham *et al.*, 1996). Though these studies suggest some conservation of the mechanistic modes between the two types of PITPs, it is still an open question whether these converge to common functional mechanisms (i.e., a case of convergent evolution) or their shared biochemical activities are coincidental without mechanistic similarity.

The distinct phenotypes of START-like PITP knock-down both in genetic model systems as in mammals and flies and in cultured cell lines (as discussed in section IV) clearly indicate that these PITPs play specific roles in different biological functions like neuronal development, membrane trafficking, cell signaling, and cytokinesis. Therefore not only ablation of any single PITP leads to a number of different cell specific defects, but also different orthologs of START-like PITPs have distinct phenotypes at the organismal level. The most challenging question is what is the underlying mechanism of action of the START-like PITPs in these diverse contexts? On the basis of cell line based *in vitro* reconstitution systems described in previous sections, the START-like PITPs are proposed to potentiate PtdIns-kinase activities. The stimulation of PtdIns-kinases by PITPs leads to increased PIP_2 production which in turn activates PIP_2 dependent signaling events like PLC signaling and Ca^{2+}-activated secretory granule exocytosis (Cockcroft, 1998; Cunningham *et al.*, 1995, 1996; Hay *et al.*, 1995). Also, the PITPα and PITPβ enhance the activity of PtdIns-3-kinase and PtdIns-4-kinase *in vitro* and when added to permeabilized cells (Fensome *et al.*, 1996; Kular *et al.*, 1997, 2002; Panaretou *et al.*, 1997; Way *et al.*, 2000). Besides, genetic evidence from *Drosophila* show that its single classI PITP functions in concert with PtdIns-4-kinaseβ and the *C. elegans* show that a class I PITP functions together with Vps34, a classIII PtdIns-3-kinase (Giansanti *et al.*, 2007; Lee *et al.*, 2008). Thus, like Sec14-like PITPs, START-like PITPs are also involved in stimulation of PIP synthesis. Interestingly, vertebrate START-like PITPs stimulate PtdIns-kinase activities under conditions of PtdIns surfeit – suggesting these too function as nanoreactors for PtdIns-presentation (Ile *et al.*, 2010).

The available crystal structures of START-like PITP include apo-PITPα, PtdIns, and PtdCho bound forms of PITPα and PtdCho bound form of PITPβ (Schouten *et al.*, 2002; Tilley *et al.*, 2004; Vordtriede *et al.*, 2005). They all exhibit a "START-fold" that is structurally distinct from the "Sec14 fold" (Fig. 8B). The START-fold constitutes a hydrophobic pocket comprised of eight β-strands and two α-helices that can accommodate a single phospholipid molecule, either a PtdIns or a PtdCho. Like Sec14-like proteins, the PL binding pocket is also guarded by a "lid" that comprises a C-terminal α-helix (the G-helix) and an 11-aminoacid extension. In the lipid free apo-structure, the G-helix is dislodged while in the holo-structures it is apposed to the PL cavity. The PL is buried deep into the protein with the polar head group interacting with aminoacid residues located toward the distal end of the cavity. As in Sec14-like proteins, the PL has access to the membrane interface only when the lid is dislodged in the protein's "open" state. In this state the protein exposes the hydrophobic residues that allow it to remain associated with the membrane. The protein is thus in "closed" state for transport through aqueous compartments while it is in "open state" for lipid exchange. The structural studies also revealed a loop with adjacent Trp residues in PITPα. Some reports, based on *in vitro* analyses, suggest that this motif is important for membrane docking and lipid transfer activity of the protein (Shadan *et al.*, 2008; Tilley *et al.*, 2004). Other *in vitro* and *in vivo* studies demonstrate this Trp–Trp motif, while required for stable association of PITPβ with Golgi membranes *in vivo*, is neither important for the types of transient membrane interactions that accompany lipid exchange reactions, nor for biological function in vertebrate context (Ile *et al.*, 2010; Philips *et al.*, 2006).

Comparison of the structures of PITPα complexed with either PtdCho or PtdIns reveals that unlike Sec14, both the lipids attain similar poses within the lipid binding cavity and the backbones of both structures are able to be superimposed. The only notable difference is that the inositol headgroup makes more hydrogen bonds with specific aminoacid residues than the choline moiety and this explains its higher affinity for PtdIns than PtdCho (Tilley *et al.*, 2004; Vordtriede *et al.*, 2005; Yoder *et al.*, 2001). In the holo-structures the PL is buried deep into the cavity with the head group region posed toward the distal end of the PL binding cavity. Though the PL binding strategy in the holo-forms does not favor a Sec14-like nanoreactor/PtdIns presentation mode, the apo-PITPα conformer displays an open channel that provides access to headgroup binding region. Thus this channel could provide a path via which a lipid kinase could potentially access a PITP bound PtdIns headgroup. Therefore, a nanoreactor/PtdIns presentation mechanism for START-like PITPs that involves PITP-PtdIns kinase interactions during the interfacial lipid exchange reaction remains plausible.

The mutational studies of theSTART-like PITPs are important for solving the individual roles of PtdIns and PtdCho binding/transfer in its biological functions. The first PtdIns transfer defective rat PITPα was generated by random mutagenesis. This study characterized Thr59 and Glu248 as amino acid residues specifically required for PtdIns binding/transfer activity (Alb *et al.*, 1995). Further, it was established that the PtdIns binding is an essential functional property of PITPα *in vivo*, as mice expressing the PtdIns defective PITPα as the sole source of PITPα exhibited phenotypes that recapitulated those of authentic PITPα nullizygotes (Alb *et al.*, 2007). Later, structure based mutational approaches generated both PtdIns deficient and PtdCho deficient mutant proteins respectively (Carvou *et al.*, 2010; Tilley *et al.*, 2004).These mutant proteins were assessed for their biological function in different permeabilized cell reconstitution assays. PITPα deficient in PtdIns binding/transfer was not able to restore the PIP_2 dependent PLC signaling in cytosol depleted HL60 cells. Similarly, the PtdIns or PtdCho binding/transfer deficient PITPβ proteins were unable to rescue the PtdIns-4-P dependent COP1-mediated retrograde transport defect in PITPβ knock-down cells. The non-functionality of the PtdCho binding mutant suggests that PITPβ does more than just potentiating the PtdIns-4-P synthesis. It is possible that PtdCho bound PITPβ might regulate other activities that are required for the protein to execute its function at the Golgi. Whether both PtdIns and PtdCho bound activities are required to reside in the same protein was not examined. It is important to note that the mutant proteins in this study were tested for their function in permeabilized reconstitution assays. For reasons described in previous sections, these reconstitution assays are neither reliable, nor physiologically faithful, measures of the *in vivo* function of the protein in question. Therefore, *in vivo* genetic assays are essential to validate the significance of the specific mutations on the activity of these proteins. Future studies toward this goal will determine whether a heterotypic exchange reaction similar to Sec14-like proteins works for the START-like PITPs too. Whether all functional activities of START-like PITPs depend on both PtdIns and PtdCho binding/transfer will need to be tested in authentic physiological contexts.

VI. Conclusions and Summary

The PITP protein family has come a long way since their discovery as "lipid transfer" proteins. Recent progress, from various lines of research, are providing unprecedented insights into their complex regulatory roles at the interface of broader lipid metabolism, PIP signal transduction, and membrane trafficking. This advance in the PITP field is the outcome of a combinatorial approach employing biochemical, biophysical, structural, and computational methods in the context of appropriate genetic model systems. These recent developments contribute to a physical appreciation of how these proteins integrate lipid metabolism, membrane trafficking, and signaling. These studies propose "nanoreactor mechanistic model" for Sec14 and Sec14-like PITPs according to which these proteins act as "primed lipid biosensors or nanoreactors" that couple binding of lipids other than PtdIns (sensor function) to a PtdIns-presentation activity in a heterotypic PtdIns/PtdCho exchange reaction, that potentiates the PtdIns-kinase activity by making PtdIns a better substrate for the enzyme. In this way, they couple metabolic cues to the action of interfacial lipid modifying enzymes in regulation of PIP signaling or membrane trafficking. However, it is evident that the mechanism of these proteins will vary individually. For example, the metazoan specific START-like PITPs are structurally unrelated to the Sec14-like PITPs, yet they bind PtdIns and PtdCho in the same binding site. The functional distinction of these PITPs as "nanoreactors for PIP signaling" versus "lipid carriers" can be experimentally examined with the biochemical methods and structural approaches in yeast genetic system described in this chapter. Finally, the facile model systems devised for each metazoan START-like PITP class will greatly assist in identifying their functional role.

Undoubtedly the structural approach is emerging as a powerful tool in deciphering the molecular basis of the PITP function. Knowledge of the PITP structure at the membrane bilayer under conditions when PITP is "active" will provide a valuable insight into its action. At present, our knowledge of when and where PITPs are active in living cells is limited. The engineering of reliable conformational biosensors is required to address these problems. Also, our understanding of the spatial and temporal regulation of lipid metabolic flux is inadequate. Given the role of PITPs in regulation of lipid metabolism, navigating the technical obstacles associated with acquisition of such information is a prerequisite in deciphering the functional mechanism of PITPs. Hence, future development of the novel and improved technologies to study membrane lipid/protein dynamics in living cells is instrumental in unraveling the biological functions of this family of enigmatic proteins.

References

Alb, J. G. Jr., Phillips, S. E., Rostand, K., Cui, X., Pinsteren, J., Cotlin, L., Manning, T., Guo, S., York, J. D., Sontheimer, H., Collawn, J. F., and Bankaitis, V. A. (2002). Genetic ablation of phosphatidylinositol transferprotein function in murine embryonic stem cells. *Mol. Biol. Cell* **13**, 739–754.

Alb, J. G. Jr., Cortese, J. D., Phillips, S. E., Albin, R. L., Nagy, T. R., Hamilton, B. A., and Bankaitis, V. A. (2003). Mice lacking phosphatidylinositol transfer protein-alpha exhibit spinocerebellar degeneration, intestinal and hepatic steatosis, and hypoglycemia. *J. Biol. Chem.* **278**, 33501–33518.

Alb, J. G. Jr., Phillips, S. E., Wilfley, L. R., Philpot, B. D., and Bankaitis, V. A. (2007). Thepathologies associated with functional titration of phosphatidylinositol transfer protein alpha activity in mice. *J. Lipid Res.* **48**, 1857–1872.

Akiyama, M., and Sakagami, T. (1969). Exchange of mitochondrial lecithin and cephalin with those in rat liver microsomes. *Biochim. Biophys. Aeta* **187**, 105–112.

Balla, T., and Varnai, P. (2009). Visualization of cellular phosphoinositidepools with GFP-fused protein-domains. *Curr. Protoc. Cell Biol.* **24**, Unit 24.4.

Bankaitis, V. A., Malehorn, D. E., Emr, S. D., and Greene, R. (1989). The *Saccharomyces cerevisiae* SEC14 gene encodes a cytosolic factor that is required for transport of secretory proteins from the yeast Golgi complex. *J. Cell Biol.* **108**, 1271–1281.

Bankaitis, V. A., Aitken, J. R., Cleves, A. E., and Dowhan, W. (1990). An essential role for a phospholipid transfer protein in yeast Golgi function. *Nature* **347**, 561–562.

Bankaitis, V. A., Mousley, C. J., and Schaaf, G. (2010). The Sec14 superfamily and mechanisms for crosstalk between lipid metabolism and lipid signaling. *Trends Biochem. Sci.* **35**, 150–160.

Carvou, N., Holic, R., Li, M., Futter, C., Skippen, A., and Cockcroft, S. (2010). Phosphatidylinositol- and phosphatidylcholine-transfer activity of PITPbeta is essential for COPI-mediated retrograde transport from the Golgi to the endoplasmic reticulum. *J. Cell Sci.* **123**, 1262–1273.

Cho, W., Bittova, L., and Stahelin, R. V. (2001). Membrane binding assays for peripheral proteins. *Anal. Biochem.* **296**, 153–161.

Cleves, A. E., Novick, P. J., and Bankaitis, V. A. (1989). Mutations in the *SAC1* gene suppress defects in yeast Golgi and yeast actin function. *J. Cell Biol.* **109**, 2939–2950.

Cleves, A. E., Mcgee, T. P., Whitters, E. A., Champion, K. M., Aitken, J. R., Dowhan, W., Goebl, M., and Bankaitis, V. A. (1991a). Mutations in the CDP-choline pathway for phospholipid biosynthesis bypass the requirement for an essential phospholipid transfer protein. *Cell* **64**, 789–800.

Cleves, A., Mcgee, T., and Bankaitis, V. A. (1991b). Phospholipid transfer proteins: a biological debut. *Trends Cell Biol.* **1**, 30–34.

Cockcroft, S., Howell, T. W., and Gomperts, B. D. (1987). Two G-proteins act in series to control stimulus-secretion coupling in mast cells: Use of neomycin to distinguish between G-protesin controlling polyphosphinositide phosphodiesterase and exocytosis. *J. Cell. Biol.* **105**, 2745–2750.

Cockcroft, S. (1998). Phosphatidylinositol transfer proteins: a requirement in signal transduction and vesicle traffic. *Bioessays* **20**, 423–432.

Cunningham, B. C., Thomas, G. M. H., Ball, A., Hiles, L., and Cockcroft, S. (1995). Phosphatidylinositol transfer protein dictates the rate of inositol trisphosphate production by promoting synthesis of PIP_2. *Curr. Biol.* **5**, 775–783.

Cunningham, E., Tan, S. K., Swigart, P., Hsuan, J., Bankaitis, V. A., and Cockcroft, S. (1996). The yeast and mammalian isoforms of phosphatidylinositol transfer protein can all restore phospholipase C-mediated inositol lipid signaling in cytosol-depleted RBL-2H3 and HL-60 cells. *Proc. Natl. Acad. Sci. U. S. A.* **93**, 6589–6593.

De Vries, K. J., Heinrichs, A. A., Cunningham, E., Brunink, F., Westerman, J., Somerharju, P. J., Cockcroft, S., Wirtz, K. W., and Snoek, G. T. (1995). An isoform of the phosphatidylinositol-transfer protein transfers sphingomyelin and is associated with the Golgi system. *Biochem. J.* **310**, 643–649.

De Vries, K. J., Westerman, J., Bastiaens, P. I., Jovin, T. M., Wirtz, K. W., and Snoek, G. T. (1996). Fluorescently labeled phosphatidylinositol transfer protein isoforms (alpha and beta), microinjected into fetal bovine heart endothelial cells, are targeted to distinct intracellular sites. *Exp. Cell Res.* **227**, 33–39.

Downes, C. P., Gray, A., Watt, S. A., and Lucocq, J. M. (2003). Advances in procedures for the detection and localization of inositol phospholipid signals in cells, tissues, and enzyme assays. *Methods Enzymol* **366**, 64–84.

Downes, C. P., Gray, A., and Lucocq, J. M. (2005). Probing phosphoinositide functions in signaling and membrane trafficking. *Trends Cell Biol.* **15**, 259–268.

Esmon, B., Novick, P., and Schekman, R. (1981). Compartmentalized assembly of oligosaccharides on exported glycoproteins in yeast. *Cell* **25**, 451–460.

Fang, M., Kearns, B. G., Gedvilaite, A., Kagiwada, S., Kearns, M., Fung, M. K., and Bankaitis, V. A. (1996). Kes1p shares homology with human oxysterol binding protein and participates in a novel regulatory pathway for yeast Golgi-derived transport vesicle biogenesis. *EMBO J.* **15**, 6447–6459.

Fensome, A., Cunningham, E., Prosser, S., Tan, S. K., Swigart, P., Thomas, G., Hsuan, J., and Cockcroft, S. (1996). ARF and PITP restore GTP gamma S-stimulated protein secretion from cytosol-depleted HL60 cells by promoting PIP_2 synthesis. *Curr. Biol.* **6**, 730–738.

Giansanti, M. G., Bonaccorsi, S., Kurek, R., Farkas, R. M., Dimitri, P., Fuller, M. T., and Gatti, M. (2006). The class I PITP giotto is required for *Drosophila* cytokinesis. *Curr. Biol.* **16**, 195–201.

Giansanti, M. G., Belloni, G., and Gatti, M. (2007). Rab11 is required for membrane trafficking and actomyosin ring constriction in meiotic cytokinesis of *Drosophila* males. *Mol. Biol. Cell* **18**, 5034–5047.

Goldstein, A., and Lampen, J. O. (1979). β-D-fructofuranoside fructohydrolase from yeast. *Methods Enzymol.* **42**, 504–511.

Hamilton, B. A., Smith, D. J., Mueller, K. L., Kerrebrock, A. W., Bronson, R. T., van Berkel, V., Daly, M. J., Kruglyak, L., Reeve, M. P., Nemhauser, J. L., Hawkins, T. L., Rubin, E. M., and Lander, E. S. (1997). The vibrator mutation causes neurodegeneration via reduced expression of PITP alpha: positional complementation cloning and extragenic suppression. *Neuron* **18**, 711–722.

Harris, W. A., and Stark, W. S. (1977). Hereditary retinal degeneration in *Drosophila melanogaster*: a mutant defect associated with the phototransductionprocess. *J. Gen. Physiol.* **69**, 261–291.

Hay, J. C., and Martin, T. F. J. (1993). Phosphatidylinositol transfer protein is required for ATP-dependent priming of Ca^{2+}-activated secretion. *Nature* **366**, 572–575.

Hay, J. C., Fisette, P. L., Jenkins, G. H., Fukami, K., Takenawa, T., Anderson, R. A., and Martin, T. F. J. (1995). ATP-dependent phosphorylation required for Ca^{2+}-activated secretion. *Nature* **374**, 173–177.

Helmkamp, G. M., Harvey, M. S., Wirtz, K. W., and Van Deenen, L. L. (1974). Phospholipid exchange between membranes. Purification of bovine brain proteins that preferentially catalyze the transfer of phosphatidylinositol. *J. Biol. Chem.* **25**, 6382–6389.

Horváth, L. I., Brophy, P. J., and Marsh, D. (1994). Microwave frequency dependence of ESR spectra from spin labels undergoing two-site exchange in myelin proteolipid membranes. *J. Magn. Reson.* **105**, 120–128.

Hubner, S., Couvillon, A. D., Kas, J. A., Bankaitis, V. A., Vegners, R., Carpenter, C. L., and Janmey, P. A. (1998). Enhancement of phosphoinositide 3-kinase (PI 3-kinase) activity by membrane curvature and inositol-phospholipid-binding peptides. *Eur. J. Biochem.* **258**, 846–853.

Ile, K. E., Schaaf, G., and Bankaitis, V. A. (2006). Phosphatidylinositol transfer proteins and cellular nanoreactors for lipid signaling. *Nat. Chem. Biol.* **2**, 576–583.

Ile, K. E., Kassen, S., Cao, C., Vihtehlic, T., Shah, S. D., Mousley, C. J., Alb, J. G. Jr., Huijbregts, R. P., Stearns, G. W., Brockerhoff, S. E., Hyde, D. R., and Bankaitis, V. A. (2010). Zebrafish class 1 phosphatidylinositol transfer proteins: PITPbeta and double cone cell outer segment integrity in retina. *Traffic* **11**, 1151–1167.

Jones, S. M., Alb, J. G., Phillips, S. E., Bankaitis, V. A., and Howell, K. E. (1998). A phosphatidylinositol-3-kinase and phosphatidylinositol transfer protein cooperate to drive phosphatidylinositol-3-phosphate-dependent formation of constitutive transport vesicles from the trans-Golgi Network. *J. Biol. Chem.* **273**, 10349–10354.

Kamp, H. H., Wirtz, K. W., and Van Deenen, L. L. (1975). Protein-mediated transfer of phosphatidulcholine between membranes and liposomes. *Biochem. Soc. Trans.* **16**, 601–605.

Katoh, Y., Ritter, B., Gaffry, R., Blondeau, F., Honing, S., and Mcpherson, P. S. (2009). The clavesin family, neuron-specific lipid- and clathrin-binding Sec14 proteins regulating lysosomal morphology. *J. Biol. Chem.* **284**, 27646–27654.

Kauffmann-Zeh, A., Thomas, G. M., Ball, A., Prosser, S., Cunningham, E., Cockcroft, S., and Hsuan, J. J. (1995). Requirement for phosphatidylinositol transfer protein in epidermal growth factor signaling. *Science* **26**, 1188–1190.

Klenchin, V. A., and Martin, T. F. (2000). Priming in exocytosis: Attaining fusion-competence after vesicle docking. *Biochimie* **82**, 399–407.

Kular, G., Loubtchenkov, M., Swigart, P., Whatmore, J., Ball, A., Cockcroft, S., and Wetzker, R. (1997). Co-operation of phosphatidylinositol transfer protein with phosphoinositide 3-kinase gamma in the formylmethionyl-leucylphenylalanine-dependent production of phosphatidylinositol 3,4,5-trisphosphate in human neutrophils. *Biochem. J.* **325**, 299–301.

Kular, G. S., Chaudhary, A., Prestwich, G., Swigart, P., Wetzker, R., and Cockcroft, S. (2002). Co-operation of phosphatidylinositol transfer protein with phosphoinositide 3-kinase gamma *in vitro*. *Adv. Enzyme Regul.* **42**, 53–61.

Lee, I., Lehner, B., Crombie, C., Wong, W., Fraser, A. G., and Marcotte, E. M. (2008). A single gene network accurately predicts phenotypic effects of gene perturbation in *Caenorhabditis elegans*. *Nat. Genet.* **40**, 181–188.

Li, X., Rivas, M. P., Fang, M., Marchena, J., Mehrotra, B., Chaudhary, A., Feng, L., Prestwich, G. D., and Bankaitis, V. A. (2002). Analysis of oxysterol binding protein homologue Kes1p function in regulation of Sec14p-dependent protein transport from the yeast Golgi complex. *J. Cell Biol.* **157**, 63–77.

Loyet, K. M., Kowalchyk, J. A., Chaudhary, A., Chen, J., Prestwich, G. D., and Martin, T. F. J. (1998). Specific binding of phosphatidylinositol 4,5-bisphosphate to calcium-dependent activator protein for secretion (CAPS), a potential phosphoinositide effector protein for regulated exocytosis. *J. Biol. Chem.* **273**, 8337–8343.

Martin, T. F. J. (1989). Cell cracking: Permiabilizing cells to macromelcular probes. *Meth. Enzymol.* **168**, 225–233.

Marsh, D. (1997). Stoichiometry of lipid–protein interaction and integral membrane protein structure. *Eur. Biophys. J.* **26**, 203–208.

Mcgee, T. P., Skinner, H. B., Whitters, E. A., Henry, S. A., and Bankaitis, V. A. (1994). A phosphatidylinositol transfer protein controls the phosphatidylcholine content of yeast Golgi membranes. *J. Cell Biol.* **124**, 273–287.

McMurray, W. C., and Dawson, R. M. C. (1969). Phospholipid exchange reactions within the liver cell. *Biochem. J.* **12**, 91–108.

Meier, R., Tomizaki, T., Schulze-Briese, C., Baumann, U., and Stocker, A. (2003). The molecular basis of vitamin E retention: structure of human alpha-tocopherol transfer protein. *J. Mol. Biol.* **331**, 725–734.

Michel, G., Rodolakis, A., and Starka, J. (1975). Synthesis and turnover of phospholipids in *Escherichia coli* cells with inhibited DNA synthesis. *Ann. Microbiol.* **126**, 237–285.

Milligan, S. C., Alb, J. G. Jr., Elagina, R. B., Bankaitis, V. A., and Hyde, D. R. (1997). The phosphatidylinositol transfer protein domain of *Drosophila* retinal degeneration B protein is essential for photoreceptor cell survival and recovery from light stimulation. *J. Cell Biol.* **139**, 351–363.

Min, K. C., Kovall, R. A., and Hendrickson, W. A. (2003). Crystal structure of human alpha-tocopherol transfer protein bound to its ligand: implications for ataxia with vitamin E deficiency. *Proc. Natl. Acad. Sci. U. S. A.* **100**, 14713–14718.

Novick, P., Field, C., and Schekman, R. (1980). Identification of 23 complementation groups required for post-translational events in the yeast secretory pathway. *Cell* **21**, 205–215.

Ohashi, M., de Vries, K. J., Frank, R., Snoek, G., Bankaitis, V. A., Wirtz, K., and Huttner, W. B. (1995). A role for phosphatidylinositol transfer protein in secretory vesicle formation. *Nature* **377**, 544–547.

Páli, T., Bashtovyy, D., and Marsh, D. (2006). Stoichiometry of lipid interaction with transmembrane proteins, deduced from the 3-D structures. *Prot. Sci.* **15**, 1153–1161.

Panaretou, C., Domin, J., Cockcroft, S., and Waterfield, M. D. (1997). Characterization of p150, an adaptor protein for the human phosphatidylinositol (PtdIns) 3-kinase. *Substrate presentation by phosphatidylinositol transfer protein to the p150. Ptdins 3-kinase complex. J. Biol. Chem.* **272**, 2477–2485.

Peterman, T. K., Ohol, Y. M., Mcreynolds, L. J., and Luna, E. J. (2004). Patellin1, a novel Sec14-like protein, localizes to the cell plate and binds phosphoinositides. *Plant Physiol.* **136**, 3080–3094.

Phillips, S. E., Sha, B., Topalof, L., Xie, Z., Alb, J. G., Klenchin, V. A., Swigart, P., Cockcroft, S., Martin, T. F., Luo, M., and Bankaitis, V. A. (1999). Yeast Sec14p deficient in phosphatidylinositol transfer activity is functional in vivo. *Mol. Cell* **4**, 187–197.

Phillips, S. E., Ile, K. E., Boukhelifa, M., Huijbregts, R. P., and Bankaitis, V. A. (2006). Specific and non-specific membrane-binding determinants cooperate in targeting phosphatidylinositol transfer protein beta-isoform to the mammalian trans-Golgi network. *Mol. Biol. Cell* **17**, 2498–2512.

Pinxteren, J. A., Gomperts, B. D., Rogers, D., Phillips, S. E., Tatham, P. E., and Thomas, G. M. (2001). Phosphatidylinositol transfer proteins and protein kinase C make separate but non-interacting contributions to the phosphorylation state necessary for secretory competence in rat mast cells. *Biochem. J.* **356**, 287–296.

Raucher, D., Stauffer, T., Chen, W., Shen, K., Guo, S., York, J. D., Sheetz, M. P., and Meyer, T. (2000). Phosphatidylinositol 4,5-bisphosphate functions as a second messenger that regulates cytoskeleton-plasma membrane adhesion. *Cell* **100**, 221–228.

Rivas, M. P., Kearns, B. G., Xie, Z., Guo, S., Sekar, M. C., Hosaka, K., Kagiwada, S., York, J. D., and Bankaitis, V. A. (1999). Pleiotropic alterations in lipid metabolism in yeast *sac1* mutants: relationship to "bypass Sec14p" and inositol auxotrophy. *Mol. Biol. Cell* **10**, 2235–2250.

Routt, S. M., Ryan, M. M., Tyeryar, K., Rizzieri, K. E., Mousley, C., Roumanie, O., Brennwald, P. J., and Bankaitis, V. A. (2005). Nonclassical PITPs activate PLD via the Stt4p PtdIns-4-kinase and modulate function of late stages of exocytosis in vegetative yeast. *Traffic* **6**, 1157–1172.

Ryan, M. M., Temple, B. R. S., Phillips, S. E., and Bankaitis, V. A. (2007). Conformational dynamics of the major yeast phosphatidylinositol transfer protein Sec14p: insight into the mechanisms of phospholipid exchange and diseases of Sec14p-like protein deficiencies. *Mol. Biol. Cell* **18**, 1928–1942.

Salama, S. R., Cleves, A. E., Malehorn, D. E., Whitters, E. A., and Bankaitis, V. A. (1990). Cloning and characterization of *Kluyveromyces lactis* SEC14, a gene whose product stimulates Golgi secretory function in Saccharomyces cerevisiae. *J. Bacteriol.* **172**, 4510–4521.

Sato, M., Ueda, Y., Takagi, T., and Umezawa, Y. (2003). Production of PtdInsP3 at endomembranes is triggered by receptor endocytosis. *Nat. Cell Biol.* **5**, 1016–1022.

Schaaf, G., Ortlund, E. A., Tyeryar, K. R., Mousley, C. J., Ile, K. E., Garrett, T. A., Ren, J., Woolls, M. J., Raetz, C. R., Redinbo, M. R., and Bankaitis, V. A. (2008). Functional anatomy of phospholipid binding and regulation of phosphoinositide homeostasis by proteins of the sec14 superfamily. *Mol. Cell* **29**, 191–206.

Schaaf, G., Dynowski, M., Mousley, C. J., Shah, S. D., Yahn, P., Winklbauer, E. M., de Campos, M. K., Trettin, K., Quinones, M. C., Smirnova, T. I., Yanagisawa, L. L., Ortlund, E. A., and Bankaitis, V. A. (2011). Resurrection of a functional phosphatidylinositol transfer protein from a pseudo-Sec14 scaffold by directed evolution. *Mol. Biol. Cell* **22**, 892–905.

Schnabl, M., Oskolkova, O. V., Holic, R., Brezna, B., Pichler, H., Zagorsek, M., Kohlwein, S. D., Paltauf, F., Daum, G., and Griac, P. (2003). Subcellular localization of yeast Sec14 homologues and their involvement in regulation of phospholipid turnover. *Eur. J. Biochem.* **270**, 3133–3145.

Schouten, A., Agianian, B., Westerman, J., Kroon, J., Wirtz, K. W., and Gros, P. (2002). Structure of apo-phosphatidylinositol transfer protein alpha provides insight into membrane association. *EMBO J.* **21**, 2117–2121.

Sha, B., Phillips, S. E., Bankaitis, V. A., and Luo, M. (1998). Crystal structure of the *Saccharomyces cerevisiae* phosphatidylinositol -transfer protein. *Nature* **391**, 506–510.

Shadan, S., Holic, R., Carvou, N., Ee, P., Li, M., Murray-Rust, J., and Cockcroft, S. (2008). Dynamics of lipid transfer by phosphatidylinositol transfer proteins in cells. *Traffic* **9**, 1743–1756.

Skinner, H. B., Alb, J. G., Whitters, E. A., Helmkamp, G. M., and Bankaitis, V. A. (1993). Phospholipid transfer activity is relevant to but not sufficient for the essential function of yeast Sec14 gene product. *EMBO J.* **12**, 4775–4784.

Smirnova, T. I., Chadwick, T. G., MacArthur, R., Poluektov, O., Song, L., Ryan, M. M., Schaaf, G., and Bankaitis, V. A. (2006). The chemistry of phospholipid binding by the *Saccoromyces cerevisiae* phosphatidylinositol transfer protein Sec14p as determined by EPR spectroscopy. *J. Biol. Chem.* **281**, 34897–34908.

Smirnova, T. I., Chadwick, T. G., Voinov, M. A., Poluektov, O., van Tol, J., Ozarowski, A., Schaaf, G., Ryan, M. M., and Bankaitis, V. A. (2007). Local polarity and hydrogen bonding inside the Sec14p phospholipid-binding cavity: high-field multi-frequency electron paramagnetic resonance studies. *Biophys. J.* **92**, 3686–3695.

Stark, W. S., Chen, D. -M., Johnson, M. A., and Frayer, K. L. (1983). The *rdgB* gene of *Drosophila*: retinal degeneration in different alleles and inhibition by *norpA*. *J. Insect Physiol.* **29**, 123–131.

Stevens, T., Esmon, B., and Schekman, R. (1982). Early stages in the yeast secretory pathway are required for transport of carboxypeptidase Y to the vacuole. *Cell* **30**, 439–448.

Stocker, A., and Baumann, U. (2003). Supernatant protein factor in complex with RRR-alpha-tocopherylquinone: a link between oxidized Vitamin E and cholesterol biosynthesis. *J. Mol. Biol.* **332**, 759–765.

Stolz, L. E., Kuo, W. J., Longchamps, J., Sekhon, M. K., and York, J. D. (1998). Inp51, a yeast inositol phosphate 5-phosphatase required for phosphatidylinositol 4,5-bisphoaphate homeostasis and whose absence confers a cold-resistant phenotype. *J. Biol. Chem.* **273**, 11852–11861.

Tilley, S. J., Skippen, A., Murray-Rust, J., Swigart, P. M., Stewart, A., Morgan, C. P., Cockcroft, S., and McDonald, N. Q. (2004). Structure-function analysis of human phosphatidylinositol transfer protein alpha bound to phosphatidylinositol. *Structure* **12**, 317–326.

van Paridon, P. A., Visser, A. J., and Wirtz, K. W. (1987). Binding of phospholipidsto the phosphatidylinositol transfer protein from bovine brain as studied by steady-state and time-resolved fluorescence spectroscopy. *Biochim. Biophys. Acta* **898**, 172–180.

Venuti, S. E., and Helmkamp, G. M. J. (1988). Tissue distribution, purification and characterization of rat phosphatidylinositol transfer protein. *Biochim. Biophys. Acta* **946**, 119–128.

Vincent, P., Chua, M., Nogue, F., Fairbrother, A., Mekeel, H., Xu, Y., Allen, N., Bibikova, T. N., Gilroy, S., and Bankaitis, V. A. (2005). A Sec14p-nodulin domain phosphatidylinositol transfer protein polarizes membrane growth of *Arabidopsis thaliana* root hairs. *J. Cell Biol.* **168**, 801–812.

Vordtriede, P. B., Doan, C. N., Tremblay, J. M., Helmkamp, G. M. Jr., and Yoder, M. D. (2005). Structure of PITPbeta in complex with phosphatidylcholine: comparison of structure and lipid transfer to other PITP isoforms. *Biochemistry* **44**, 14760–14771.

Walent, J. H., Porter, B. W., and Martin, T. F. (1992). A novel 145 kd brain cytosolic protein reconstitutes Ca($^{2+}$)- regulated secretion in permeable neuroendocrine cells. *Cell* **70**, 765–775.

Watt, S. A., Kular, G., Fleming, I. N., Downes, C. P., and Lucocq, J. M. (2002). Subcellular localization of phosphatidylinositol 4,5-bisphosphate using the pleckstrin homology domain of phospholipase C delta1. *Biochem. J.* **363**, 657–666.

Way, G., O'luanaigh, N., and Cockcroft, S. (2000). Activation of exocytosis by cross-linking of the IgE receptor is dependent on ADPribosylation factor 1-regulated phospholipase D in RBL-2H3 mast cells: evidence that the mechanism of activation is via regulation of phosphatidylinositol 4,5-bisphosphate synthesis. *Biochem. J.* **346**, 63–70.

Welti, S., Fraterman, S., D'angelo, I., Wilm, M., and Scheffzek, K. (2007). The sec14 homology module of neurofibromin binds cellular glycerophospholipids: mass spectrometry and structure of a lipid complex. *J. Mol. Biol.* **366**, 551–562.

Whitters, E. A., Cleves, A. E., Mcgee, T. P., Skinner, H. B., and Bankaitis, V. A. (1993). SAC1p is an integral membrane protein that influences the cellular requirement for phospholipid transfer protein function and inositol in yeast. *J. Cell Biol.* **122**, 79–94.

Wiedemann, C., Schafer, T., and Burger, M. M. (1996). Chromaffin granule-associated phosphatidulinositol 4-kinase activity is required for stimulated secretion. *EMBO J.* **1**, 2094–2101.

Wirtz, K. W. A., and Zilversmit, D. B. (1968). Exchange of phospholipids between liver mitochondria and microsomes in vitro*. *J. Biol. Chem.* **243**, 3596–3602.

Woolford, C. A., Daniels, L. B., Park, F. J., Jones, E. W., Van Arsdell, J. N., and Innis, M. A. (1986). The PEP4 gene encodes an aspartyl protease implicated in the posttranslational regulation of *Saccharomyces cerevisiae* vacuolar hydrolases. *Mol. Cell. Biol.* **6**, 2500–2510.

Xie, Y., Ding, Y. Q., Hong, Y., Feng, Z., Navarre, S., Xi, C. X., Zhu, X. J., Wang, C. L., Ackerman, S. L., Kozlowski, D., Mei, L., and Xiong, W. C. (2005). Phosphatidylinositol transfer protein-alpha in netrin-1-induced PLC signalling and neurite outgrowth. *Nat. Cell Biol.* **7**, 1124–1132.

Xie, Z., Fang, M., Rivas, M. P., Faulkner, A. J., Sternweis, P. C., Engebrecht, J. A., and Bankaitis, V. A. (1998). Phospholipase D activity is required for suppression of yeast phosphatidylinositol transfer protein defects. *Proc. Natl. Acad. Sci. U. S. A.* **95**, 12346–12351.

Yanagisawa, L. L., Marchena, J., Xie, Z., Li, X., Poon, P. P., Singer, R. A., Johnston, G. C., Randazzo, P. A., and Bankaitis, V. A. (2002). Activity of specific lipid-regulated ADP ribosylation factor-GTPase-activating proteins is required for Sec14p-dependent Golgi secretory function in yeast. *Mol. Biol. Cell* **13**, 2193–2206.

Yoder, M. D., Thomas, L. M., Tremblay, J. M., Oliver, R. L., Yarbrough, L. R., and Helmkamp, G. M. (2001). Structure of a multifunctional protein. Mammalian phosphatidylinositol transfer protein complexed with phosphatidylcholine. *J. Biol. Chem.* **276**, 9246–9252.

CHAPTER 14

Genome-Wide Screens for Gene Products Regulating Lipid Droplet Dynamics

Weihua Fei and Hongyuan Yang

School of Biotechnology and Biomolecular Sciences, the University of New South Wales, Sydney, Australia

Abstract

Lipid droplets (LDs) are emerging as dynamic cellular organelles that play a key role in lipid and membrane homeostasis. Abnormal lipid droplet dynamics are associated with the pathophysiology of many metabolic diseases, such as obesity, diabetes, atherosclerosis, fatty liver, and even cancer. Understanding the molecular mechanisms governing the dynamics of LDs, namely, their biogenesis, growth, maintenance, and degradation, will not only shed light on the cellular functions of LDs, but also provide additional clues to treatment of metabolic diseases. Genome-wide screen is a powerful approach to identify genetic factors that regulate lipid droplet dynamics. Here, we summarize recent genome-wide studies using yeast and *Drosophila* cells to understand the cellular dynamics of LDs. The results suggest that

Copyright 2012, Elsevier Inc. All rights reserved.

0091-679X/10 $35.00
DOI 10.1016/B978-0-12-386487-1.00010-9

the genome-wide screens should be carried out in multiple organisms or cells, and using different nutritional conditions.

I. Introduction

A. Lipid Droplets: Specialized Organelles to Store Neutral Lipids

Lipid droplets (LDs) are found in almost all eukaryotic cells and also in some prokaryotes. LDs consist of a hydrophobic core of neutral lipids, mainly triacylglycerol (TAG) and/or sterol ester (SE), which is delimited by a monolayer of polar lipids with attached or embedded proteins. It is believed that LDs are synthesized from the endoplasmic reticulum (ER) in all eukaryotic cells, including yeast, though the exact mechanism remains an enigma (Murphy and Vance, 1999; Martin and Parton, 2006; Ploegh, 2007; Robenek *et al.*, 2006; Walther and Farese, 2009a; Ohsaki *et al.*, 2009). So far only the synthesis of neutral lipids has been found to be essential in the formation of LDs. The yeast mutant strain *dga1*Δ*lro1*Δ*are1*Δ*are2*Δ, which is deficient in the formation of TAG and SE, does not produce LDs (Sandager *et al.*, 2002; Oelkers *et al.*, 2002). Interestingly, the adipocytes differentiated from DGAT1$^{-/-}$ DGAT2$^{-/-}$ mouse embryonic fibroblast cells also lack LDs (Harris *et al.*, 2011).

The structure of yeast LDs is similar to that of their mammalian and plant counterparts. In addition, molecular mechanisms that govern the dynamics of LDs, including their formation, growth, maintenance, and degradation, appear to be conserved from yeast to human (Walther and Farese, 2009b). Therefore, the study of LDs using yeast as a model may contribute to a better understanding of the cell biology and physiology of human LDs.

B. Growth Stages of Yeast Cells and Metabolism of Storage Neutral Lipids

It has long been recognized that under laboratory culture conditions exponentially growing yeast cells maintain low levels of storage lipids, both TAG and SE, but their synthesis increases dramatically upon entry into late exponential or early stationary phase (Bailey and Parks, 1975; Taylor and Parks, 1979). The rise of cellular TAG and SE content in stationary phase is partly, if not all, attributable to conversion from pre-synthesized phospholipids and free sterols, respectively (Taylor and Parks, 1979; Taylor and Parks, 1978). When cell growth resumes in fresh media, TAG is then utilized for synthesis of membrane phospholipids (Taylor and Parks, 1979), while SE serves as a supply of free sterols (Taylor and Parks, 1978). In line with this, Tgl4, a yeast TAG lipase, was recently found to be phosphorylated and activated by Cdk1 to provide precursors for membrane phospholipid synthesis when stationary phase cells were inoculated into fresh media (Kurat *et al.*, 2009). Given that the requirement for membrane lipid synthesis oscillates as the nutrient availability of yeast cells fluctuates, storage lipids appear to function as a temporary reserve for future membrane biosynthesis when conditions become favorable. The yeast *dga1Δlro1Δare1Δare2Δ* mutant underwent massive proliferation of membranes and cell death when challenged with exogenous unsaturated fatty acids, reflecting the importance of the formation of neutral lipids and LDs as a buffering system for membrane synthesis (Petschnigg *et al.*, 2009; Garbarino *et al.*, 2009).

C. Crosstalk Between the Metabolic Pathways of Storage Lipids and Membrane Lipids in Yeast

The metabolism of storage lipids is inextricably interwoven with that of membrane lipids. First, TAG and phospholipid synthesis shares a common intermediate, phosphatidic acid (PA) (Fig. 1A). In yeast, through CDP-DAG, PA can be converted into any major phsopholipid species including phosphatidylserine (PS) and its derivatives phosphatidylethanolamine (PE) and phosphatidylcholine (PC), phosphatidylinositol (PI), and cardiolipin. It should be noted that synthesis of PS from CDP-DAG is missing from higher eukaryotic cells including mammals (Dowhan, 1997). PA can also be hydrolyzed by PA phosphatases to diacylglycerol (DAG), which is used either for synthesis of PC and PE via the Kennedy pathway or for synthesis of TAG. It appears that PA is preferentially channeled into phospholipids by yeast cells in exponential phase but into TAG in stationary phase. Consistent with this notion, PA phosphatase activity was found to be markedly elevated in stationary phase (Hosaka and Yamashita, 1984).

Moreover, membrane phospholipids undergo rapid turnover. In addition to remodeling of acyl chains by deacylation–reacylation at the sn-1 or sn-2 position, phospholipids can be deacylated at both sn-1 and sn-2 positions by phospholipase B enzymes (Plb1, Plb2, Plb3, and Nte1), generating glycerophosphodiesters (Merkel *et al.*, 1999; Fernandez-Murray and McMaster, 2007). Hydrolysis of glycerophosphodiesters by phosphodiesterase such as Gde1 produces glycerol-3-phosphate, which can be further converted to PA by stepwise acylations (Fig. 1A). Phospholipids can also be hydrolyzed by phospholipase D enzyme, Pld1/Spo14, resulting in the formation of PA (Rose *et al.*, 1995). As a result, phospholipid turnover provides substrates for TAG synthesis.

On the other hand, accumulated cellular TAG, when hydrolyzed by lipases (Tgl3, Tgl4, and Tgl5), generates DAG. As mentioned above, DAG can be used for PC and PE synthesis via the Kennedy pathway if choline and ethanolamine is available. DAG can also be converted by the DAG kinase (Dgk1) into PA, while PA can be synthesized into all major phospholipids through CDP-DAG. Evidence has been shown very recently that Dgk1 is required to convert TAG-derived DAG to PA for the synthesis of membrane phospholipids Fakas *et al.*, 2011.

Yeast sterols comprise ergosterol and its upstream precursors including, but not limited to, those shown in Fig. 1B. Ergosterol is the predominant sterol in plasma membranes and secretory vesicles, the fractions with the highest sterol levels; yet its precursors are present at higher percentages in other subcellular membranes which exhibit lower sterol contents (Zinser *et al.*, 1993). Both ergosterol and

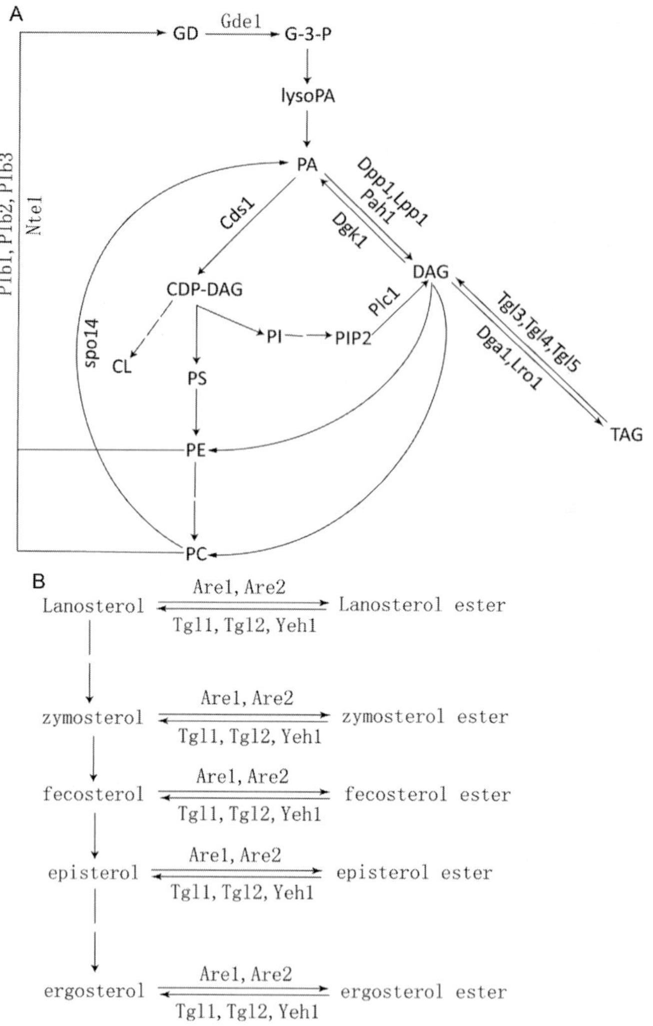

Fig. 1 Metabolic crosstalk between storage lipids and membrane lipids in *Saccharomyces cerevisiae*. (A) between TAG and phospholipids; (B) between sterols and sterol esters. TAG, triacylglycerol, DAG, diacylglycerol, PA, phosphatidic acid, CDP-DAG, cytidine diphosphate diacylglycerol, PS, phosphatidylserine, PE, phosphatidylethanolamine, PC, phosphatidylcholine, PI, phosphatidylinositol, PIP2, phosphatidylinositol bisphosphate, CL, cardiolipin, GD, glycerophosphodiester, G-3-P, glycerol-3-phosphate, lysoPA, 1-acylglycerol-3-phosphate.

precursors can be esterified to neutral SEs. The percentage of esterified precursors among all SEs gradually rises in stationary phase, accompanied by a decrease in the percentage of ergosterol esters (Bailey and Parks, 1975). Esterified ergosterol precursors cannot be directly synthesized into ergosterol (Bailey and Parks, 1975); instead they have to be hydrolyzed in order to be converted to the final product, ergosterol (Wagner *et al.*, 2009).

D. Compartmentalization of the Synthesis and Mobilization of Neutral Lipids

The yeast acyl-CoA:sterol acyltransferases (Are1 and Are2) catalyze the esterification of sterols (Yang *et al.*, 1996; Yu *et al.*, 1996). The yeast acyl-CoA: diacylglycerol acyltransferase (DGAT), Dga1, and phospholipid:diacylglycerol acyltransferase (PDAT), Lro1, catalyze the esterification of DAG by using acyl-CoA and phospholipid as the acyl donor, respectively (Oelkers *et al.*, 2002; Dahlqvist *et al.*, 2000; Sorger and Daum, 2002; Oelkers *et al.*, 2000). Are1 and Are2 catalyze the minute formation of TAG when Dga1 and Lro1 are absent (Sandager *et al.*, 2002). Dga1 localizes both to the ER membrane and lipid droplet surface, and the majority of DGAT activity is associated with the lipid droplet fraction (Sorger and Daum, 2002). In contrast, Lro1, Are1, and Are2 localize to the ER membrane only (Zweytick *et al.*, 2000; Sorger and Daum, 2003). Therefore, SE synthesis takes place at the ER, and TAG formation at both the ER and LDs.

The budding yeast has three TAG lipases (Tgl3, Tgl4, and Tgl5) and three SE hydrolases (Tgl1, Yeh1, and Yeh2). All three TAG lipases and two main SE hydrolases (Tgl1 and Yeh1) localize to LDs (Athenstaedt and Daum, 2003; Athenstaedt and Daum, 2005; Jandrositz *et al.*, 2005; Koffel *et al.*, 2005), except that Yeh2 resides in the plasma membrane (Koffel *et al.*, 2005; Mullner *et al.*, 2005). As a result, the breakdown of TAG and hydrolysis of most SE occurs at the LDs to generate DAG, free sterols, and fatty acids.

E. Unanswered Questions Surrounding the Dynamics of Lipid Droplets

In addition to how LDs are formed at the ER, many other questions remain unanswered regarding the dynamics of LDs. For instance, how is the synthesis of neutral lipids and LDs controlled at the cellular and molecular level? Do LDs undergo fission or fusion? How do LDs mature? What factors determine the spatial relationship between LDs and the ER, mitochondria, peroxisomes, plasma membrane, and vacuoles? How do cells sense the need of membrane lipid synthesis and hence supply substrates from breaking down TAG and SE? To answer some of these questions, genome-wide screens in genetically amenable systems such as yeast may represent a viable and strategic approach.

II. Genome-Wide Screen of Yeast Deletion Mutants for Changes in the Dynamics of Lipid Droplets

Three genome-wide studies were recently carried out to identify genetic factors that affect LD dynamics in yeast (Fei *et al.*, 2008; Mullner *et al.*, 2005; Szymanski *et al.*, 2007; Fei *et al.*, 2011a). The first two studies used rich YPD media to culture yeast cells; while our group focused on the changes in the number and size of LDs, Dr. Goodman and colleagues examined more parameters. Aiming to identify additional mutants that synthesize supersized LDs (which will be discussed later), our second screen used defined minimal media (synthetic glucose media) instead of YPD media (Fei *et al.*, 2011a). Below is the screening protocol used in our screens.

A. Screening Protocol

1. Materials

Various yeast strain collections are available from Thermo Scientific.
96-well plates and 96-pin replicators
96-well plate shaker and centrifuge
Yeast culture media as required (e.g., YPD rich media, synthetic dextrose media, synthetic glycerol media. . .)
Nile red solubilized in acetone (1 mg/mL)
Fluorescence microscope

2. Screening Procedure

(1) Dispense 150 μL of YPD media into each well of 96-well plates.
(2) Inoculate yeast cells from stock plates into 96-well plates with the aid of 96-pin replicators.
(3) Incubate cell culture on a 96-well plate shaker with vigorous shaking for 24 h.
(4) Prepare 96-well target plates by dispensing 150 μL of selected media as required by experiments into each well.
(5) Refresh yeast cells pre-cultured in YPD media into selected media with the aid of 96-pin replicators.
(6) Incubate cell culture on a 96-well plate shaker with vigorous shaking until desired growth phase as determined by OD_{600}.
(7) Pellet cells in a plate centrifuge.
(8) Remove culture media.
(9) Resuspend cells in ~50 μL of cold PBS supplemented with 20 μg/mL of Nile red and keep plates on ice.
(10) Load 3 μL of each sample unto slides and cover with coverslips.
(11) Examine the number, shape, size, or distribution of cytoplasmic LDs under fluorescence microscope using filters similar to Leica I3 (BP 450–490, LP 515) which preferentially detects LDs.
(12) Record the strains that display LDs distinct from those of wildtype strain in their number, shape, size, or distribution.
(13) Inoculate the recorded strains in 2 mL of media of interest in 10 mL culture tubes and incubate until desired growth phase. Confirm the LD phenotype in these strains.

B. Screening Results

1. Results of Screens Performed Using Rich YPD Media

'In our first screen, we found that most wildtype yeast cells and the majority of deletion mutants grown overnight in YPD media to early stationary phase accumulated 3–7 LDs per cell. We stringently classified the mutants that accumulated less than three LDs in the majority of cells as *fld* (few LDs) strains and the mutants that accumulated more than seven LDs as *mld* (many LDs) strains

(Fei *et al.*, 2008). A total of 17 *fld* mutants were identified, among which the strain deleted for *YLR404w/FLD1* synthesized supersized LDs with a diameter of over 1 μm (Fig. 2). 116 *mld* strains were also identified; among them 14 severe *mld* strains synthesized more than 11 LDs in the majority of cells (Fig. 2).

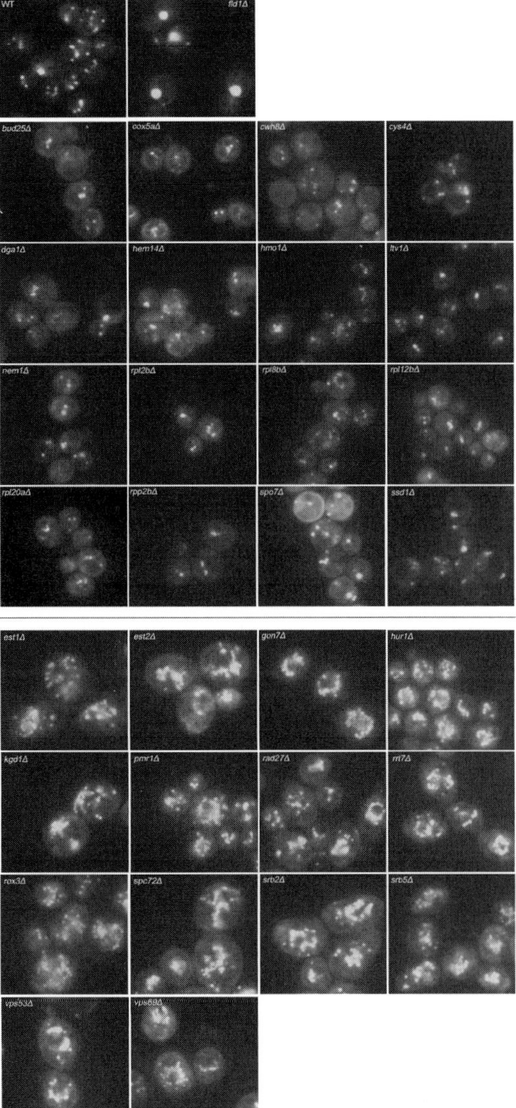

Fig. 2 *fld* (few lipid droplets) and severe *mld* (many lipid droplets) yeast mutants identified in Fei *et al.* (2008). Cells were cultured in rich YPD media until early stationary phase, stained with Nile red, and observed under the fluorescence microscope. Top panel, wildtype and *fld* mutants; bottom panel, severe *mld* mutants. (For color version of this figure, the reader is referred to the web version of this book.)

Aiming to identify ER factors involved in LD assembly, Dr. Goodman's group also cultured yeast strains in YPD media and screened for changes in LD number, size, distribution, resolution from the cytoplasm, or staining intensity both in mid-log and stationary phase. Fifty-nine strains were found to display different phenotypes from wild type (Szymanski *et al.*, 2007).

Both screens noted significant variability in LDs among cells of any particular strain, even wild type. Therefore, the screening results largely depend on the selection criteria that are arbitrarily determined. This might explain why there is only an overlap of 16 genes between the two screens (Fig. 3A). In addition, as a stringent cut-off was adopted in our screen, only *VMA6*, *VMA8*, *VMA13*, and *VMA21* were recorded as *mld* genes (Fei *et al.*, 2008). However, thin layer chromatography (TLC) analysis revealed that most viable *v*acuolar *m*embrane *A*TPase (*vma*) mutants exhibited a strong increase in cellular SE level (Fig. 3B). Accordingly, it may be necessary to develop reliable computational analysis software to reduce "false negatives."

Importantly, *YLR404w/FLD1* was identified in both screens (Fei *et al.*, 2008; Szymanski *et al.*, 2007). When cultured in rich YPD media, ~70% of *fld1*Δ cells

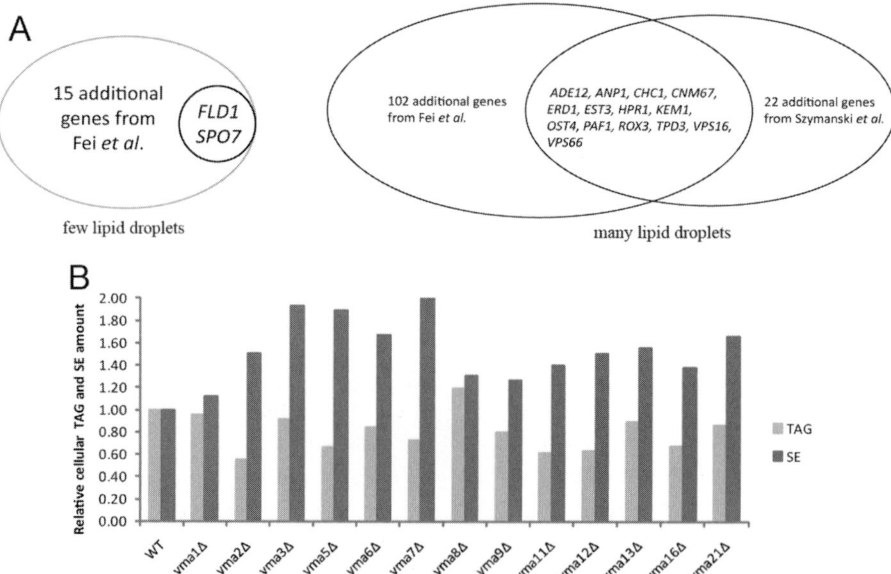

Fig. 3 (A). Overlap among the genes identified in Fei *et al.* and Szymanski *et al.* (Fei *et al.*, 2008; Szymanski *et al.*, 2007). (B). Quantitative analysis of cellular TAG and SE in wildtype and *vma* (vacuolar membrane ATPase) mutants. Cells were cultured in rich YPD media until early stationary phase. Lipids were extracted and separated via thin layer chromatography. Bands corresponding to TAG and SE were scanned and densitometric units compared. Among the 13 *vma* mutants, only *vma6*Δ, *vma8*Δ, *vma13*Δ, and *vma21*Δ were recorded as *mld* mutants due to high stringency (see text for details).

contained clustered LDs of complex morphology, which appeared like grapes, whereas ~20% of cells displayed supersized LDs, whose volume could be up to 50 times that of wild type LDs (Fei *et al.*, 2008; Szymanski *et al.*, 2007). Intriguingly, when cultured in synthetic glucose media, supersized LDs could be observed in ~70% of *fld1*Δ cells, accompanied by a decrease in the formation of LD cluster in ~20% of cells (Fei *et al.*, 2008). The formation of supersized LDs is most likely a result of fusion of smaller droplets. Indeed, fusion of LDs could be easily observed under the microscope in *fld1*Δ cells, especially when cells were cultured in synthetic glucose media (Fei *et al.*, 2008). In addition to aberrant LD morphology, the deletion of *FLD1* also affected the acyl chain compositions of phospholipids and TAG: a shift from long-chain (18:1) to medium/short-chain (16:0, 14:0, and 12:0) fatty acid incorporation into all major phospholipids and TAG was noticed (Fei *et al.*, 2008).

Yeast Fld1 is homologous to human seipin encoded by BSCL2 gene (Fei *et al.*, 2011b; Tian *et al.*, 2011; Cui *et al.*, 2011). Homozygous non-sense mutations or missense mutation (A212P) of BSCL2 is implicated in congenital generalized lipodystrophy (Magre *et al.*, 2001), whereas heterozygous mutation (N88S or S90L) is associated with motor neuron disease (Windpassinger *et al.*, 2004). Intriguingly, like wild type seipin, N88S or S90L could complement the yeast *fld1*Δ cells; in contrast, A212P lost this function (Fei *et al.*, 2008; Szymanski *et al.*, 2007).

In addition to identifying *FLD1* as a determinant of the size and morphology of LDs, our screen also revealed that conditions of ER stress induce neutral lipid synthesis and LD formation (Fei *et al.*, 2009). In agreement with this finding, yeast cells channel substrates for phospholipid synthesis into synthesis of neutral lipids to form LDs when ER-to-Golgi trafficking is blocked (Gaspar *et al.*, 2008). This may also explain why many mutants defective in vesicular transport were found to accumulate more LDs (Fei *et al.*, 2008).

2. Identification of Yeast Strains that form Supersized LDs Using Defined Minimal Media Besides *fld1*Δ

The increase in the percentage of supersized LD-forming *fld1*Δ cells from ~20% in rich YPD media to ~70% in synthetic glucose media led us to speculate that we might uncover more yeast mutants which synthesized supersized LDs if cells were cultured in synthetic glucose media. We reasoned that various nutrients present in YPD media might suppress the formation of supersized LDs in these mutants. We therefore re-screened the entire collection of viable yeast deletion mutants (~4800) grown on minimal media for supersized LDs. We also included the collection of mutants where all essential genes are controlled by the $TetO_7$-promoter, which can be switched off efficiently (Mnaimneh *et al.*, 2004). Remarkably, in addition to *FLD1*, we identified nine other genes whose deletions or repression resulted in the biogenesis of supersized LDs (Fei *et al.*, 2011c). They include *CHO2* and *OPI3* which encode enzymes catalyzing the methytransferases to convert PE to PC, *CDS1* encoding CDP-DAG synthase, *INO2* and *INO4* which encode transcriptional factors positively regulating phospholipid synthesis, *CKB1* and *CKB2*

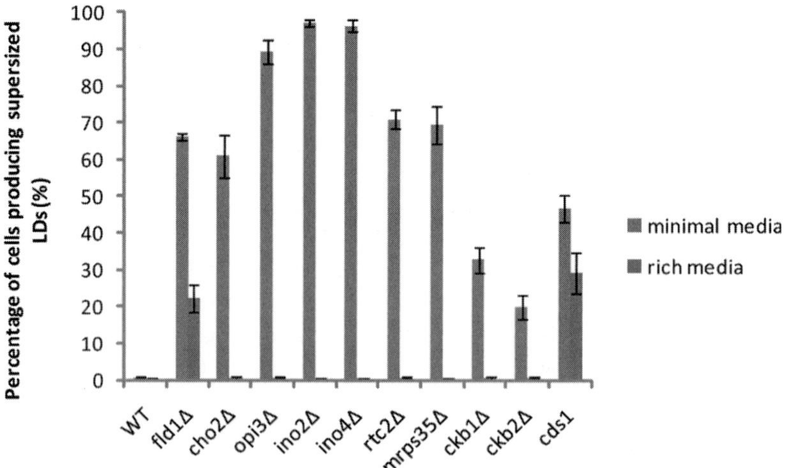

Fig. 4 Percentage of cells that produced supersized LDs when strains were cultured in rich (YPD) media and minimal (synthetic glucose) media. Cells were cultured to early stationary phase, stained with Nile red, and observed under the fluorescence microscope. Two hundred cells were examined for each sample and the percentage of supersized LD-forming cells was calculated. Experiments were done in triplicates and data presented as Mean ± SD. For details please refer to (Fei *et al.*, 2011a). (For color version of this figure, the reader is referred to the web version of this book.)

encoding regulatory subunits of casein kinase 2, as well as *RTC2* and *MRPS35* with unknown functions. Supporting our speculation, when these mutants were cultured in rich media, none of them produced supersized LDs except *cds1* in which the percentage of cells forming supersized LDs also decreased (Fig. 4). Moreover, addition of inositol or choline to the minimal media dramatically inhibited the formation of supersized LDs in all or most strains. We also discovered by lipidomic analysis that a common feature of these mutants including *fld1*Δ is an elevated intracellular PA level (Fei *et al.*, 2011a). Reducing PA in these mutants strongly inhibited the biogenesis of supersized LDs. In addition, PA facilitated the coalescence of artificial LDs. Our data suggest that metabolism of phospholipids, PA in particular, has a great impact on the size of LDs.

III. Additional Insights From Genome-Wide Studies in *Drosophila* Cells

Recently two independent genome-wide RNA interference (RNAi) screens were carried out in *Drosophila* cells to identify key genes involved in the regulation of LD dynamics. Guo *et al.* used S2 cells and examined alterations in droplet number, size and dispersion, whereas Beller *et al.* used Kc167 cells and examined differences in cellular lipid storage by computational analysis of ratio of LD-to-nucleus cross-sectional area (Guo *et al.*, 2008; Beller *et al.*, 2008). The phenotype of LDs in

S2 cells differs significantly from that of Kc167 cells. Upon oleate treatment, S2 cells showed clustering of many small LDs resembling grapes, which were often near the nucleus, while Kc167 cells produced one or several large LDs plus some small ones. Both screens identified that components of Arf1-COPI machinery are required for efficient lipolysis. However, Guo *et al.* additionally found that knockdown of these genes resulted in the dispersion of LDs in S2 cells (Guo *et al.*, 2008). Moreover, knockdown of Cct1, Cct2, CG2201 (which is predicted to have choline kinase activity), SREBP, SCAP, or FAS in S2 cells induced LDs to fuse, resulting in the formation of large LDs (Guo *et al.*, 2008). It is not surprising that these six genes were not identified in the screen by Beller *et al.*, given that large LDs were routinely observed in Kc167 cells. This indicates that it is critical to choose multiple organisms or cell lines to carry out the screens, given the dynamic nature of LDs.

IV. Concluding Remarks

The importance of LDs has been increasingly recognized, yet the molecular mechanisms underlying their biogenesis, growth, and catabolism remain largely unknown (Farese and Walther, 2009). Genetic screen was fundamental in many research areas such as vesicular trafficking and cell cycle control. Undoubtedly this approach will again be highly valuable to our understanding of the cell biology of LDs.

So far, genome-wide screens in yeast and *Drosophila* cells have successfully identified several key genes involved in the regulation of lipid droplet dynamics (Fei *et al.*, 2008; Szymanski *et al.*, 2007; Fei *et al.*, 2011a; Guo *et al.*, 2008; Beller *et al.*, 2008). Further studies of these genes and protein products may point towards new directions in lipid droplet research. Moreover, the gene lists generated in these screens could be further examined, which may identify new players important for lipid droplet dynamics.

Given the heterogeneity of LDs in different cell types as shown in the *Drosophila* S2 and Kc167 cells, it could be extremely important to choose the "right" cell type. For instance, if our aim is to study the fusion/fission of LDs and to identify factors that either promote the fusion of LDs or block their fission, we might choose a cell type that normally synthesizes only one or few giant LDs (e.g., white adipocytes). Genome-wide knockout or RNAi screen may be performed using this cell type to search for cells which produce many smaller LDs instead of one or few giant LDs. Conversely, if our aim is to identify factors which prevent the fusion of LDs, we may choose a cell type which normally synthesizes only many small LDs (e.g., brown adipocytes) and search for cells which form giant LDs instead of wildtype small LDs after genetic manipulations. Similarly, if our aim is to identify factors contributing to the clustering or dispersion of LDs, a wildtype cell type with an opposite phenotype should be used. From these studies we may understand how a white adipocyte produces a unilocular giant LD while a brown adipocyte produces many small LDs and pinpoint how the size and distribution of LDs is controlled in other cells/tissues.

Not only the "right" cell type is critical, selection of culture media can also be important. Based on our finding that the size of LDs in 10 yeast mutants cultured in YPD media differed significantly to those in synthetic glucose media, we may have to choose the 'right' media to study LD dynamics. Ideally, both rich and minimal media (if available for the cell type of interest) should be used and results compared.

In summary, the dynamic nature of LDs requires "dynamic" experimental procedures/approaches. Although a number of genetic screens have already been carried out, better-designed screens, especially in higher organisms, will undoubtedly yield additional insights into the regulation of the dynamics of LDs.

Acknowledgments

This work is supported by a research grant from the Australian Research Council (DP0984902). H. Yang is a Future Fellow of the Australian Research Council.

References

Murphy, D. J., and Vance, J. (1999). Mechanisms of lipid-body formation. *Trends Biochem. Sci.* **24**, 109–115.

Martin, S., and Parton, R. G. (2006). Lipid droplets: a unified view of a dynamic organelle. *Nat. Rev. Mol. Cell Biol.* **7**, 373–378.

Ploegh, H. L. (2007). A lipid-based model for the creation of an escape hatch from the endoplasmic reticulum. *Nature* **448**, 435–438.

Robenek, H., Hofnagel, O., Buers, I., Robenek, M. J., Troyer, D., and Severs, N. J. (2006). Adipophilin-enriched domains in the ER membrane are sites of lipid droplet biogenesis. *J. Cell Sci.* **119**, 4215–4224.

Walther, T. C., and Farese Jr., R. V. (2009a). The life of lipid droplets. *Biochim. Biophys. Acta* **1791**, 459–466.

Ohsaki, Y., Cheng, J., Suzuki, M., Shinohara, Y., Fujita, A., and Fujimoto, T. (2009). Biogenesis of cytoplasmic lipid droplets: from the lipid ester globule in the membrane to the visible structure. *Biochim. Biophys. Acta* **1791**, 399–407.

Sandager, L., Gustavsson, M. H., Stahl, U., Dahlqvist, A., Wiberg, E., Banas, A., Lenman, M., Ronne, H., and Stymne, S. (2002). Storage lipid synthesis is non-essential in yeast. *J. Biol. Chem.* **277**, 6478–6482.

Oelkers, P., Cromley, D., Padamsee, M., Billheimer, J. T., and Sturley, S. L. (2002). The DGA1 gene determines a second triglyceride synthetic pathway in yeast. *J. Biol. Chem.* **277**, 8877–8881.

Harris, C. A., Haas, J. T., Streeper, R. S., Stone, S. J., Kumari, M., Yang, K., Han, X., Brownell, N., Gross, R. W., Zechner, R., and Farese, R. V. (2011). DGAT enzymes are required for triacylglycerol synthesis and lipid droplets in adipocytes. *J. Lipid Res.* **52**, 657–667.

Walther, T. C., and Farese Jr., R. V. (2009b). The life of lipid droplets. *Biochim. Biophys. Acta* **1791**, 459–466.

Bailey, R. B., and Parks, L. W. (1975). Yeast sterol esters and their relationship to the growth of yeast. *J. Bacteriol.* **124**, 606–612.

Taylor, F. R., and Parks, L. W. (1979). Triaglycerol metabolism in Saccharomyces cerevisiae. Relation to phospholipid synthesis. *Biochim. Biophys. Acta* **575**, 204–214.

Taylor, F. R., and Parks, L. W. (1978). Metabolic interconversion of free sterols and steryl esters in Saccharomyces cerevisiae. *J. Bacteriol.* **136**, 531–537.

Kurat, C. F., Wolinski, H., Petschnigg, J., Kaluarachchi, S., Andrews, B., Natter, K., and Kohlwein, S. D. (2009). Cdk1/Cdc28-dependent activation of the major triacylglycerol lipase Tgl4 in yeast links lipolysis to cell-cycle progression. *Mol. Cell* **33**, 53–63.

Petschnigg, J., Wolinski, H., Kolb, D., Zellnig, G., Kurat, C. F., Natter, K., and Kohlwein, S. D. (2009). Good fat, essential cellular requirements for triacylglycerol synthesis to maintain membrane homeostasis in yeast. *J. Biol. Chem.* **284**, 30981–30993.

Garbarino, J., Padamsee, M., Wilcox, L., Oelkers, P. M., D'Ambrosio, D., Ruggles, K. V., Ramsey, N., Jabado, O., Turkish, A., and Sturley, S. L. (2009). Sterol and diacylglycerol acyltransferase deficiency triggers fatty acid-mediated cell death. *J. Biol. Chem.* **284**, 30994–31005.

Dowhan, W. (1997). Molecular basis for membrane phospholipid diversity:why are there so many lipids? *Ann. Rev. Biochem.* **66**, 199–232.

Hosaka, K., and Yamashita, S. (1984). Regulatory role of phosphatidate phosphatase in triacylglycerol synthesis of Saccharomyces cerevisiae. *Biochim. Biophys. Acta* **796**, 110–117.

Merkel, O., Fido, M., Mayr, J. A., Pruger, H., Raab, F., Zandonella, G., Kohlwein, S. D., and Paltauf, F. (1999). Characterization and function in vivo of two novel phospholipases B/lysophospholipases from Saccharomyces cerevisiae. *J. Biol. Chem.* **274**, 28121–28127.

Fernandez-Murray, J. P., and McMaster, C. R. (2007). Phosphatidylcholine synthesis and its catabolism by yeast neuropathy target esterase 1. *Biochim. Biophys. Acta* **1771**, 331–336.

Rose, K., Rudge, S. A., Frohman, M. A., Morris, A. J., and Engebrecht, J. (1995). Phospholipase D signaling is essential for meiosis. *Proc. Natl. Acad. Sci. U. S. A.* **92**, 12151–12155.

Fakas, S., Konstantinou, C., and Carman, G. M. (2011). DGK1-encoded diacylglycerol kinase activity is required for phospholipid synthesis during growth resumption from stationary phase in Saccharomyces cerevisiae. *J. Biol. Chem.* **286**, 1464–1474.

Zinser, E., Paltauf, F., and Daum, G. (1993). Sterol composition of yeast organelle membranes and subcellular distribution of enzymes involved in sterol metabolism. *J. Bacteriol.* **175**, 2853–2858.

Wagner, A., Grillitsch, K., Leitner, E., and Daum, G. (2009). Mobilization of steryl esters from lipid particles of the yeast Saccharomyces cerevisiae. *Biochim. Biophys. Acta* **1791**, 118–124.

Yang, H., Bard, M., Bruner, D. A., Gleeson, A., Deckelbaum, R. J., Aljinovic, G., Pohl, T. M., Rothstein, R., and Sturley, S. L. (1996). Sterol esterification in yeast: a two-gene process. *Science* **272**, 1353–1356.

Yu, C., Kennedy, N. J., Chang, C. C., and Rothblatt, J. A. (1996). Molecular cloning and characterization of two isoforms of Saccharomyces cerevisiae acyl-CoA:sterol acyltransferase. *J. Biol. Chem.* **271**, 24157–24163.

Dahlqvist, A., Stahl, U., Lenman, M., Banas, A., Lee, M., Sandager, L., Ronne, H., and Stymne, S. (2000). Phospholipid:diacylglycerol acyltransferase: an enzyme that catalyzes the acyl-CoA-independent formation of triacylglycerol in yeast and plants. *Proc. Natl. Acad. Sci. U. S. A.* **97**, 6487–6492.

Sorger, D., and Daum, G. (2002). Synthesis of triacylglycerols by the acyl-coenzyme A:diacyl-glycerol acyltransferase Dga1p in lipid particles of the yeast Saccharomyces cerevisiae. *J. Bacteriol.* **184**, 519–524.

Oelkers, P., Tinkelenberg, A., Erdeniz, N., Cromley, D., Billheimer, J. T., and Sturley, S. L. (2000). A lecithin cholesterol acyltransferase-like gene mediates diacylglycerol esterification in yeast. *J. Biol. Chem.* **275**, 15609–15612.

Zweytick, D., Leitner, E., Kohlwein, S. D., Yu, C., Rothblatt, J., and Daum, G. (2000). Contribution of Are1p and Are2p to steryl ester synthesis in the yeast Saccharomyces cerevisiae. *Eur. J. Biochem.* **267**, 1075–1082.

Sorger, D., and Daum, G. (2003). Triacylglycerol biosynthesis in yeast. *Appl Microbiol. Biotechnol.* **61**, 289–299.

Athenstaedt, K., and Daum, G. (2003). YMR313c/TGL3 encodes a novel triacylglycerol lipase located in lipid particles of Saccharomyces cerevisiae. *J. Biol. Chem.* **278**, 23317–23323.

Athenstaedt, K., and Daum, G. (2005). Tgl4p and Tgl5p, two triacylglycerol lipases of the yeast Saccharomyces cerevisiae are localized to lipid particles. *J. Biol. Chem.* **280**, 37301–37309.

Jandrositz, A., Petschnigg, J., Zimmermann, R., Natter, K., Scholze, H., Hermetter, A., Kohlwein, S. D., and Leber, R. (2005). The lipid droplet enzyme Tgl1p hydrolyzes both steryl esters and triglycerides in the yeast Saccharomyces cerevisiae. *Biochim. Biophys. Acta* **1735**, 50–58.

Koffel, R., Tiwari, R., Falquet, L., and Schneiter, R. (2005). The Saccharomyces cerevisiae YLL012/YEH1, YLR020/YEH2, and TGL1 genes encode a novel family of membrane-anchored lipases that are required for steryl ester hydrolysis. *Mol. Cell Biol.* **25**, 1655–1668.

Mullner, H., Deutsch, G., Leitner, E., Ingolic, E., and Daum, G. (2005). YEH2/YLR020c encodes a novel steryl ester hydrolase of the yeast Saccharomyces cerevisiae. *J. Biol. Chem.* **280**, 13321–13328.

Fei, W., Shui, G., Gaeta, B., Du, X., Kuerschner, L., Li, P., Brown, A. J., Wenk, M. R., Parton, R. G., and Yang, H. (2008). Fld1p, a functional homologue of human seipin, regulates the size of lipid droplets in yeast. *J. Cell Biol.* **180**, 473–482.

Szymanski, K. M., Binns, D., Bartz, R., Grishin, N. V., Li, W. P., Agarwal, A. K., Garg, A., Anderson, R. G., and Goodman, J. M. (2007). The lipodystrophy protein seipin is found at endoplasmic reticulum lipid droplet junctions and is important for droplet morphology. *Proc. Natl. Acad. Sci. U. S. A.* **104**, 20890–20895.

Fei, W., Shui, G., Zhang, Y., Krahmer, N., Ferguson, C., Kapterian, T. S., Lin, R. C., Dawes, I. W., Brown, A. J., Li, P., Huang, X., Parton, R. G., Wenk, M. R., Walther, T. C., and Yang, H. (2011a). A role for phosphatidic acid in the formation of "supersized" lipid droplets. *PLoS Genet.* **7:** e1002201.

Fei, W., Du, X., and Yang, H. (2011b). Seipin, adipogenesis and lipid droplets. *Trends Endocrinol. Metab.* **22**, 204–210.

Tian, Y., Bi, J., Shui, G., Liu, Z., Xiang, Y., Liu, Y., Wenk, M. R., Yang, H., and Huang, X. (2011). Tissue-Autonomous Function of Drosophila Seipin in Preventing Ectopic Lipid Droplet Formation. *PLoS Genet.* **7**, e1001364.

Cui, X., Wang, Y., Tang, Y., Liu, Y., Zhao, L., Deng, J., Xu, G., Peng, X., Ju, S., Liu, G., and Yang, H. (2011). Seipin ablation in mice results in severe generalized lipodystrophy. *Human Mol. Genet.* **20**, 3022–3030.

Magre, J., Delepine, M., Khallouf, E., Gedde-Dahl Jr., T., Van Maldergem, L., Sobel, E., Papp, J., Meier, M., Megarbane, A., Bachy, A., Verloes, A., d'Abronzo, F. H., Seemanova, E., Assan, R., Baudic, N., Bourut, C., Czernichow, P., Huet, F., Grigorescu, F., de Kerdanet, M., Lacombe, D., Labrune, P., Lanza, M., Loret, H., Matsuda, F., Navarro, J., Nivelon-Chevalier, A., Polak, M., Robert, J. J., Tric, P., Tubiana-Rufi, N., Vigouroux, C., Weissenbach, J., Savasta, S., Maassen, J. A., Trygstad, O., Bogalho, P., Freitas, P., Medina, J. L., Bonnicci, F., Joffe, B. I., Loyson, G., Panz, V. R., Raal, F. J., O'Rahilly, S., Stephenson, T., Kahn, C. R., Lathrop, M., and Capeau, J. (2001). Identification of the gene altered in Berardinelli-Seip congenital lipodystrophy on chromosome 11q13. *Nat. Genet.* **28**, 365–370.

Windpassinger, C., Auer-Grumbach, M., Irobi, J., Patel, H., Petek, E., Horl, G., Malli, R., Reed, J. A., Dierick, I., Verpoorten, N., Warner, T. T., Proukakis, C., Van den Bergh, P., Verellen, C., Van Maldergem, L., Merlini, L., De Jonghe, P., Timmerman, V., Crosby, A. H., and Wagner, K. (2004). Heterozygous missense mutations in BSCL2 are associated with distal hereditary motor neuropathy and Silver syndrome. *Nat. Genet.* **36**, 271–276.

Fei, W., Wang, H., Fu, X., Bielby, C., and Yang, H. (2009). Conditions of endoplasmic reticulum stress stimulate lipid droplet formation in Saccharomyces cerevisiae. *Biochem. J.* **424**, 61–67.

Gaspar, M. L., Jesch, S. A., Viswanatha, R., Antosh, A. L., Brown, W. J., Kohlwein, S. D., and Henry, S. A. (2008). A block in endoplasmic reticulum-to-Golgi trafficking inhibits phospholipid synthesis and induces neutral lipid accumulation. *J. Biol. Chem.* **283**, 25735–25751.

Mnaimneh, S., Davierwala, A. P., Haynes, J., Moffat, J., Peng, W. T., Zhang, W., Yang, X., Pootoolal, J., Chua, G., Lopez, A., Trochesset, M., Morse, D., Krogan, N. J., Hiley, S. L., Li, Z., Morris, Q., Grigull, J., Mitsakakis, N., Roberts, C. J., Greenblatt, J. F., Boone, C., Kaiser, C. A., Andrews, B. J., and Hughes, T. R. (2004). Exploration of essential gene functions via titratable promoter alleles. *Cell* **118**, 31–44.

Fei, W., Shui, G., Zhang, Y., Krahmer, N., Ferguson, C., Kapterian, T. S., Lin, R. C., Dawes, I. W., Brown, A. J., Li, P., Huang, X., Parton, R. G., Wenk, M. R., Walther, T. C., and Yang, H. (2011c). A role for phosphatidic Acid in the formation of "supersized" lipid droplets. *PLoS Genet.* **7**, e1002201.

Guo, Y., Walther, T. C., Rao, M., Stuurman, N., Goshima, G., Terayama, K., Wong, J. S., Vale, R. D., Walter, P., and Farese, R. V. (2008). Functional genomic screen reveals genes involved in lipid-droplet formation and utilization. *Nature* **453**, 657–661.

Beller, M., Sztalryd, C., Southall, N., Bell, M., Jackle, H., Auld, D. S., and Oliver, B. (2008). COPI complex is a regulator of lipid homeostasis. *PLoS Biol.* **6**, e292.

Farese Jr., R. V., and Walther, T. C. (2009). Lipid droplets finally get a little R-E-S-P-E-C-T. *Cell* **139**, 855–860.

PART III

Imaging

CHAPTER 15

The Three Dimensionality of Cell Membranes: Lamellar to Cubic Membrane Transition as Investigated by Electron Microscopy

Ketpin Chong and Yuru Deng

Cubic Membrane Laboratory, Department of Physiology, Yong Loo Lin School of Medicine, National University of Singapore, Singapore

METHODS IN CELL BIOLOGY, VOL 108
Copyright 2012, Elsevier Inc. All rights reserved.

0091-679X/10 $35.00
DOI 10.1016/B978-0-12-386487-1.00015-8

Abstract

Biological membranes are generally perceived as phospholipid bilayer structures that delineate in a lamellar form the cell surface and intracellular organelles. However, much more complex and highly convoluted membrane organizations are ubiquitously present in many cell types under certain types of stress, states of disease, or in the course of viral infections. Their occurrence under pathological conditions make such three-dimensionally (3D) folded and highly ordered membranes attractive biomarkers. They have also stimulated great biomedical interest in understanding the molecular basis of their formation. Currently, the analysis of such membrane arrangements, which include tubulo-reticular structures (TRS) or cubic membranes of various subtypes, is restricted to electron microscopic methods, including tomography. Preservation of membrane structures during sample preparation is the key to understand their true 3D nature. This chapter discusses methods for appropriate sample preparations to successfully examine and analyze well-preserved highly ordered membranes by electron microscopy. Processing methods and analysis conditions for green algae (*Zygnema sp.*) and amoeba (*Chaos carolinense*), mammalian cells in culture and primary tissue cells are described. We also discuss methods to identify cubic membranes by transmission electron microscopy (TEM) with the aid of a direct template matching method and by computer simulation. A 3D analysis of cubic cell membrane topology by electron tomography is described as well as scanning electron microscopy (SEM) to investigate surface contours of isolated mitochondria with cubic membrane arrangement.

I. Introduction

Our understanding of cellular ultrastructures is mainly based on the phospholipid bilayer membrane arrangements in two-dimensional (2D) projected transmission electron microscopic (TEM) micrographs. However, for 50 decades, cell membranes with crystalline arrangements, such as "cubic membranes" have been frequently reported in TEM studies (Almsherqi *et al.*, 2006, 2009; Landh, 1995). They often appeared under stress or disease conditions including myopathy, autoimmune disease and in virus infected cells (Almsherqi *et al.*, 2009; Goldsmith *et al.*, 2004; Grimley and Schaff, 1976; Tinari *et al.*, 1996). However, such complex yet highly organized membrane structures are frequently reported in the literature without understanding of their genuine organization in three dimensions (Landh, 1995).

Two main factors render the recognition and interpretation of these highly organized yet convoluted membrane structures problematic. First, the output of TEM is the 2D projection of a three-dimensional (3D) membrane arrangement. As such, the global cell topology delineated by bilayer membranes and their continuity in 3D space remains open to interpretation, in particular, if the lattice size of ordered membrane structures is in the same range as the thin sections typically employed in TEM studies (50–90 nm). Second, TEM sample preparation requires multiple steps, typically initiated by various types of fixation (chemical fixation, cryo-fixation) that may induce alterations of the ultrastructure and therefore lead to misinterpretation of the genuine membrane arrangements.

Given the development of computer and TEM technologies, we are now able to use mathematically well-defined surface modeling approaches to precisely recognize and describe cubic membrane arrangements. In this chapter, we will mainly focus on the effects of chemical fixation on highly ordered membrane organizations in their native state. Since such techniques are widely established and available to cell biologists, we would like to point out in this chapter critical steps in sample preparation and data interpretation to yield reliable results from chemical fixation approaches. It should be noted, however, that numerous alternative techniques exist, in particular cryo-fixation, which will be briefly discussed and that avoid the introduction of potential artifacts by chemical fixation. Access to such microscopic technologies, however, may be limited to many cell biologists. The aim of this chapter is, therefore, to facilitate the preservation and identification of cubic membrane arrangements by conventional TEM techniques. Recognizing highly ordered non-lamellar membrane arrangements (especially cubic membranes) in their genuine structure(s) serves as a first step to unfold the biological functions of highly folded and convoluted membrane arrangements.

A. Cell Membrane Organizations

Cells and cellular organelles are delineated by membranes that consist of a bilayer of phospholipids, sphingolipids, glycolipids, and imbedded sterols, as well as numerous integral and peripheral proteins that fulfill specific functions in the membrane. These membranes segregate cell internal constituents from the external environment and take part in the regulation and maintenance of cellular homeostasis and inter-organelle interaction. Membranes also determine the size, shape, and origin of subcellular organelles. The fundamental shapes of biological membranes are usually perceived as single or stacked flat sheets that may appear fenestrated and as spheres (vesicles) or tubes. This view is mainly derived by TEM image analysis of ultra-thin section (70–90 nm) of a specimen. However, this is rather a generalization of membrane morphology and it is well known that a number of membrane structures exist in cells with highly complex 3D periodic arrangements (Almsherqi *et al.*, 2009; Landh, 1995). These three-dimensionally arranged membranes appear to be ubiquitous in all cell types, both under physiological or pathological conditions. Although they may appear more often in certain cell types, these membranes are not limited to particular categories of cells. Furthermore, complex membrane organizations apparently evolve from almost any cytomembrane, that is, the plasma membrane, endoplasmic reticulum (ER), nuclear envelope, inner mitochondrial membrane, and the Golgi apparatus (Almsherqi *et al.*, 2009).

Ever since the invention of the electron microscope and the development of techniques for the processing of biological samples, highly ordered intracellular membrane organizations that yield complex patterns in TEM images have been the subject of investigation and controversy. Categories of such patterned membrane configurations that display symmetry include one-dimensional (1D) periodic lamellar shape, 2D periodic hexagonal packing of tubes, and 3D periodic organizations such as cubic membranes (Almsherqi *et al.*, 2009). Cubic membranes represent 3D periodic organizations that are described by mathematically well-defined triply periodic minimal or level surfaces (Almsherqi *et al.*, 2006; Landh, 1995).

The description of 3D membrane arrangements is based on the interpretation of 2D TEM images and, therefore, prone to misconception. As a consequence, highly organized (cubic) membrane morphologies appear under numerous nicknames in the literature (Almsherqi *et al.*, 2009). Even today, identifying the topology of 3D periodic cubic membranes in living systems is extremely limited since no appropriate analysis tools are available to depict such structures with lattice sizes that typically occur in the 50–1000 nm range. Thus, electron microscopy and tomography are currently the only techniques to reveal non-lamellar membrane configurations in cells.

B. Transmission Electron Microscopy and Membrane-bound Cell Ultrastructure

Electron microscopes exist in two major forms: the Transmission Electron Microscope (TEM) and the Scanning Electron Microscope (SEM). TEM generates a 2D projected image of a thin section of a 3D object (typically in the range of 50–90 nm thick). SEM, on the other hand, allows visualization of the surface contour of a specimen, giving a 3D impression of the object, but without any structural details of its interior.

One of the first discoveries using TEM was the confirmation of cell membranes as phospholipid bilayers (Robertson, 1959, 1960). Moreover, conventional TEM studies provided the framework that membranes constitute the architecture of cells and intracellular organelles. The profile of lipid bilayer typically appears as a "trilaminar membrane" with the polar region of the head group of the phospholipid and majority of the non-polar tail group stained, whereas the center of the bilayer appears translucent. Hence, a typical image of a phospholipid bilayer in high resolution (x 400 k) is its appearance as a "railroad track." At lower magnification, membranes typically appear as a single electron-dense line.

A fundamental biological maxim denotes the majority of cellular membranes as lamellar, sheet-like structures. However, membranes frequently appear in a variety of arrangements such as spheres, tubes, and highly curved vesicles. It is also known that pure lipid mixtures in aqueous buffers *in vitro* tend to form non-lamellar structures, dependent on the lipid composition (Bouligand, 1991; Larsson, 1989; Lewis *et al.*, 1997). Curvature of membranes may be induced by certain classes of phospholipids, but also by specific proteins and protein–protein interactions, examples of which are the COP coat proteins involved in vesicular trafficking and the reticulons that regulate the tubular shape of ER membranes (Lee *et al.*, 2005). Cell membrane remodeling in response to environmental stimuli, such as temperature, pH or ion concentration, is often associated with alterations of the membrane curvature induced by changes in membrane protein and lipid composition

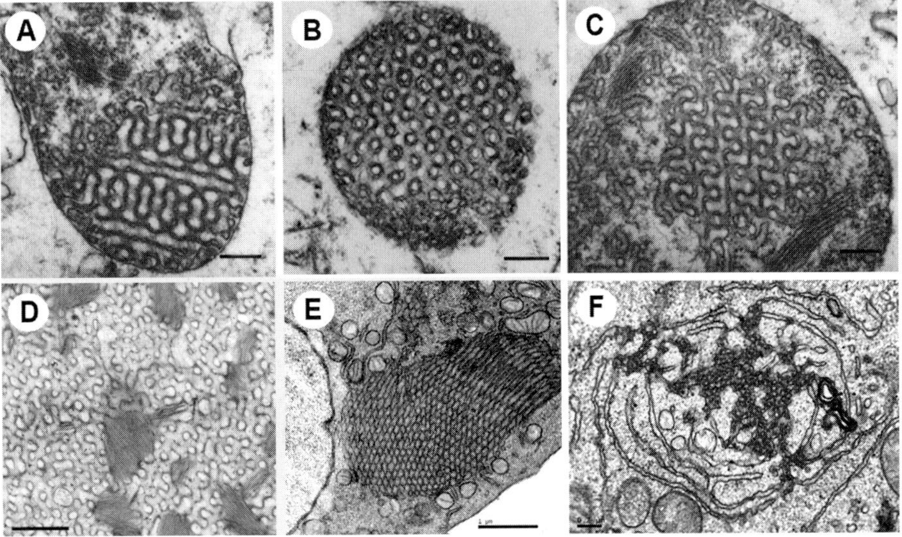

Fig. 1 TEM micrographs show a few examples of well-preserved highly ordered membrane organizations in different cell types under various stress conditions. (A–C) Different facets of cubic membrane morphology found in mitochondrial inner membranes of 7-days starved amoeba (*Chaos carolinense*). The electron-dense lines represent phospholipid bilayer-based biological membranes. Note the parallel arrangement of two bilayer membranes separating matrix from inter-membrane space. Bars = 250 nm each. (D) Chloroplast membrane morphology in the green alga *Zygnema sp.* (LB 923). TEM micrograph shows thylakoid membrane transition from lamellar to cubic arrangement in *Zygnema* at the end of log phase of cell growth. Sample was fixed using the method illustrated in **Fig. 2A**. Bar = 1 μm. (E) Crystalloid endoplasmic reticulum (ER) membranes appear in the compactin-resistant UT-1 cells grown in the presence of 40 μM compactin for 72 h. Bar = 1 μm. (F) Typical tubulo-reticular membrane structures (TRS) observed in human B-lymphocytes treated with 10,000 IU interferon-α for 72 h. Bar = 200 nm.

(Almsherqi *et al.*, 2006; Brown *et al.*, 2003; de Kruijff, 1997; Kooijman *et al.*, 2005). The fact that such membrane alterations typically occur only locally and transiently creates an additional challenge as to their identification and characterization.

C. Cubic Membranes: Facts or Artifacts

The broad range of interpretations for 3D membrane structures depicted in TEM micrographs has resulted in a significant controversy whether these are indeed fact or artifact (Landh, 1995). To date, cubic membranes are the best characterized highly ordered 3D membrane arrangements in biological systems. Their spatial arrangement can be mathematically described as triply periodic minimal surfaces; thus, comparing TEM images to computer simulated 2D projections allows the unequivocal identification in these images ("Direct Template Matching;" Cubic Membrane Simulation Projection (QMSP) program, developed by M. Mieczkowski) (Deng and Mieczkowski, 1998). Experimentally, electron tomography is required to confirm any cubic membrane arrangement predicted from the direct template matching technique (Fig. 5C).

Indeed, cubic membranes have been identified in thousands of TEM micrographs in published articles over the past decades by using this technique. Since buffers and fixatives used in chemical fixation of specimen greatly differ in these articles, it strongly suggests that the appearances of cubic membranes were not artifacts as a result of a certain protocol (Almsherqi *et al.*, 2009; Landh, 1995). In fact, these publications supported the fact that primary fixation might be the key to preserve the cubic membrane structures. Indeed, the initial steps of collecting and processing biological samples appear to play a key role in the preservation of highly organized membrane structures in biological system, with minimal risk of artifacts.

II. Cell Models to Study Non-lamellar Membrane Organizations

Highly ordered yet convoluted membranes with unusual configurations depicted in TEM micrographs have been observed in various cell types and in virtually any membrane-bound cell organelles, as previously summarized (Almsherqi *et al.*, 2009). Although any biological membrane could adopt non-lamellar morphology, membrane transformation is often associated with cellular stress condition. Thus, several factors (e.g., cell type, the nature and duration of the stress condition) need to be considered in choosing the appropriate cell model to study highly organized membrane structures.

A. Selecting Cell Types

Here, we describe several cell systems that have been under investigation in our laboratory for many years that yield 3D membrane arrangements (cubic membranes, tubulo-reticular structures (TRS), crystalloid ER) under defined experimental conditions, as references. Application of the methods outlined below, however, is certainly not restricted to these cell types.

Unicellular organisms: Amoeba (*Chaos carolinense*) and green algae (*Zygnema sp.*).

Mammalian cell lines: UT-1 (mutant Chinese hamster ovary (CHO) cells that are resistant to compactin) and human B-lymphocytes.

Cubic membranes are observed in the mitochondrial inner membranes of amoeba *Chaos carolinense* upon starvation (Fig. 1A–C), and thylakoid membranes of the filamentous green alga *Zygnema* at the end of logarithmic growth (Fig. 1D). Other types of highly ordered non-lamellar membrane organizations include induced crystalloid endoplasmic reticulum (cER) with hexagonal arrangement and TRS observed in compactin-treated UT-1 cells (Fig. 1E) and interferon-α-treated human B-lymphocytes (Fig. 1F), respectively.

These cell types have been chosen because they contain highly ordered membrane arrangements that can be easily induced by well-defined stress conditions such as starvation or drug treatment. Focusing on the best-characterized examples of (induced) cubic membranes in this chapter, the subsequent image analyses beyond TEM sample preparation are mainly derived from the unicellular amoeba *Chaos carolinense*.

B. Induction of Non-Lamellar Membrane Arrangements

Here we summarize the conditions used for each cell type to induce membrane rearrangements, as shown in Fig. 1.

1. Amoeba (*Chaos carolinense*)

Amoeba *Chaos* cells were originally obtained from Carolina Biological Supply Co. and adapted to the mass culture (Tan *et al.*, 2005). Amoebae are fed with *Paramecium multimicronucleatum* (axenic culture was kindly provided by Dr. Richard Allen, University of Hawaii). *Chaos* mitochondrial inner membrane rearranges to cubic membrane organization under starvation condition (food withdrawal). This membrane morphological transition is reversible upon re-feeding (Deng and Mieczkowski, 1998). Cubic membranes in the mitochondria of starved amoeba *Chaos* are to date the best-characterized highly organized membrane structure.

2. Green Algae (*Zygnema sp.*)

Fresh water filamentous green algae (*Zygnema sp.* strain LB923) obtained from University of Texas at Austin (UTEX) were grown in bold 1NV medium (Starr and Zeikus, 1993) at 22 °C under 16/8 h light/dark cycle at a light intensity of 3200 lux. Well-defined lamellar to cubic membrane transitions were observed in the chloroplasts of these algae at the end of logarithmic growth (Deng, 1998).

3. Mammalian Cell: UT-1 Cell Line

UT-1 cells are mutant CHO cells that are adapted to grow in the presence of compactin, a competitive inhibitor of HMG-CoA reductase. Cells were purchased from European Collection of Cell Cultures (ECACC) and grown in a monolayer at 37 °C in 5% carbon dioxide in a medium cocktail, as previously described (Chin *et al.*, 1982). Crystalloid ER membranes appear in UT-1 cells treated with 40 μM compactin for 72 h.

4. Mammalian Cells: Human B-lymphocytes

Human B-lymphocytes (male) (Coriell Cell Repositories, GM03798), immortalized by Epstein-Barr virus (provided by Dr. Prakash Hande, National University of Singapore), were cultured at 37 °C in 5% carbon dioxide in RPMI medium 1640

(Invitrogen), complete with 2 mM L-glutamine and 15% fetal bovine serum. Tubulo-reticular membrane structures are observed in these cells after incubation with 10,000 IU Interferon-α 2a (PromoKine) for 72 h.

III. Understanding Highly Ordered Membrane Arrangements through Transmission Electron Microscopy and Computer Simulation

A. Transmission Electron Microscopy (TEM) Sample Preparation – Maintaining Highly Ordered Membrane Structures

The highly ordered membrane structures can be induced in a large variety of cells. Here, a few examples of the cell types were chosen to highlight certain issues in preserving these induced membranes within the cells. We would like to emphasize that the following methods are not restricted to the cell models mentioned in this chapter. They are also applicable to other biological specimens including unicellular organisms, cultured cells and tissues, but also to isolated subcellular membrane fractions.

1. Preparation Steps for the Fixation of Unicellular Organisms, Cultured Cells, Tissues, and Isolated Mitochondria

Primary fixation plays a significant role in preventing artifacts in TEM micrographs. Fixing the samples is straightforward but certain details are often overlooked. Here, we implement a two-step method – collecting and fixing the sample – to effectively preserve highly ordered membrane structures.

Washed filaments of *Zygnema* are poured into the opening of a small container and the filaments are collected with a gauze filter (Fig. 2A). Filaments in this container are then submerged into 2% glutaraldehyde (GA): 2% formaldehyde for fast fixation. Although GA is the preferred primary fixative in our protocols, formaldehyde used in this case enhanced cell wall penetration for improved fixation of *Zygnema* cells (Bozzola and Russell, 1999).

Since amoeba *Chaos* cells are large in size, they can be simply transferred into a container with 2.5% GA without introducing any culture medium using a modified glass pipette (Asahi Techno Glass, USA) with long and narrow tip of approximately 1 mm diameter (Fig. 2B).

Isolated cubic mitochondria (obtained by subcellular fractionation) from starved amoeba *Chaos* also require removal of liquid surrounding the organelle pellet. This can be accomplished with a fine-tip plastic pipette (Copan Diagnostics Inc.) and a soft tissue paper (Kimtech Kimwipes). 2.5% GA is quickly poured over the mitochondria pellet and the sample is fixed overnight for subsequent processing steps.

Fixation of cells in culture may also follow the same protocol. Suspension cells such as human B-lymphocytes should first be centrifuged and collected as a pellet. Subsequent fixation is similar to that of the isolated amoeba mitochondria. Adherent

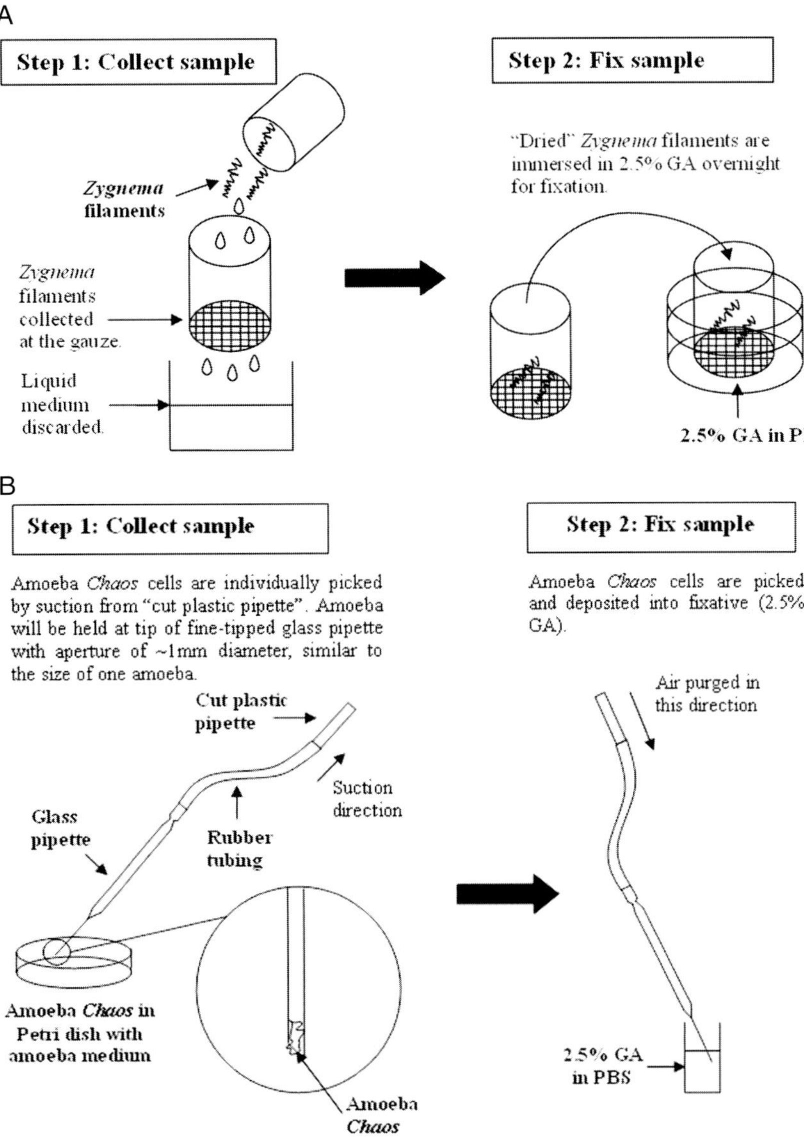

Fig. 2 Two-step process for collecting and primary fixation of samples from various cells and tissues to effectively preserve highly ordered membrane structures for TEM studies. (A) Unicellular organisms, such as *Zygnema* suspended in wash buffer are filtered through a small cylindrical container; cells collected at the gauze mesh are subsequently immersed in fixative. (B) Amoeba *Chaos* cells are transferred one at a time into primary fixative without introducing growth medium by using a fine-tip glass pipette. (C) Tissue sample from the heart of mongrel dog is excised at a proper size of about 0.5 mm in diameter using a biopsy needle (Temno) and directly injected into the fixative. It is important that the sample is free of blood and tissue fluid for a proper primary fixation.

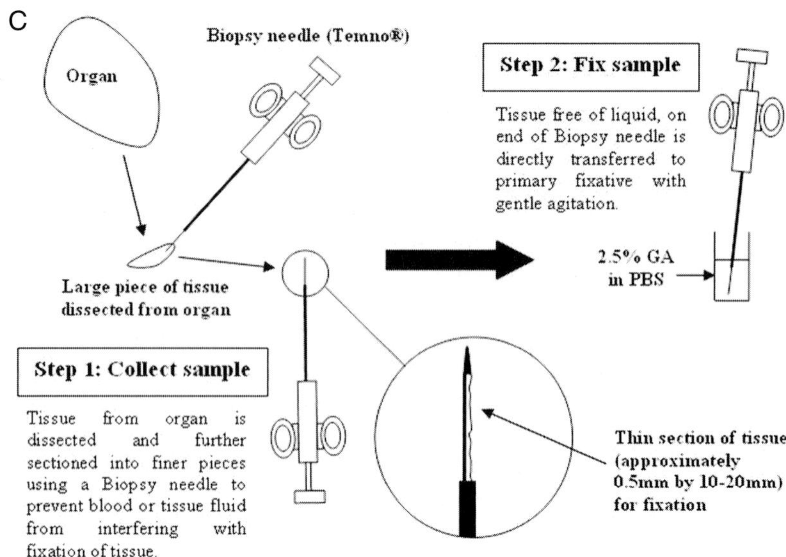

Fig. 2 *(Cont.)*

UT-1 cells, in contrast, are directly fixed in the culture flask. Culture medium is poured out of the container and the monolayer of cells washed once with phosphate buffered saline (PBS). PBS is discarded and any excess liquid in the flask is siphoned off using a fine-tip glass pipette. 2.5% GA in PBS buffer is then added to the culture flask to cover the entire layer of cells as quickly as possible, and cells are fixed overnight.

For tissue samples, it is important to remove a thin section of tissue that is also free of blood and tissue fluid, typically by biopsy. In our example, we prepared heart tissue of a mongrel dog (Fig. 2C). Dipping the biopsy needle carrying the tissue directly into primary fixative (2.5% GA) eliminates any excess fluid that may cover the specimen prior to its fixation.

Standard scheme for TEM sample preparation (Bozzola and Russell, 1999)

a. **Primary fixation**: All samples used in our experiments are primarily fixed with 2–2.5% GA of EM grade (Agar Scientific) or 2% GA: 2% formaldehyde (1:1) at 4 °C for 4 h to overnight followed by washing 3–4 times using cacodylate buffer (Electron Microscopy Sciences) or PBS (Sigma–Aldrich).
b. **Secondary fixation**: For post fixation samples are incubated with 1% osmium tetroxide (OsO_4) (Ted Pella, Inc.) for 1 h at room temperature (RT).
c. **Dehydration**: The sample is subjected to sequential dehydration by immersion in a graded series of ethanol/acetone (Fisher Scientific) with increasing strength of dehydration (25, 50, 75, and 100%) at 10 min each. After dehydration, the sample is exposed to 100% acetone.

d. **Infiltration**: Epoxy resin, Araldite (Ted Pella, Inc.) is introduced into the dehydrated specimen gradually to replace the acetone. Araldite:acetone in the ratio of 1:1 is added to the sample at RT for 30 min and replaced with 6:1 Araldite: acetone and incubated overnight in vacuum. Pure Araldite is subsequently introduced at RT and replaced with new Araldite at 40 °C for 30 min, 45 °C for 1 h and finally 50 °C for 1 h in an oven. During infiltration, the vial containing the specimen is sealed to prevent accumulation of moisture. Specimen is incubated with gentle agitation.
e. **Embedding and Curing:** At the final step of infiltration, pure resin specimen is transferred into a mold for embedment. The mold is then placed in an oven with increasing temperature (40 °C for 30 min, 45 °C for 1 h, 50 °C for 1 h and 60 °C for 24 h). Curing is the final step, which takes 12 h to a few days for the epoxy Araldite mixture to polymerize, ultimately forming a solid block.
f. **Ultramicrotomy and Staining:** The plastic block is sectioned into slices of approximately 50–90 nm thickness using an ultramicrotome (Leica) and stained in 2% uranyl acetate (Electron Microscopy Sciences) followed by Reynold's lead citrate. Specimen are viewed and examined under the TEM (JEM1010, JEOL Ltd.).

2. Significance of Primary Fixation for TEM Results

Unicellular Organisms

Excellent preservation of the thylakoid membranes from *Zygnema* cells in TEM studies have been obtained in our studies (Fig. 1D) by following a modification of fixation and staining protocols of McLean and Pessoney (1970). We removed the growth medium/buffer from *Zygnema* filaments by filtering out the wash buffer as shown in Fig. 2A. The thylakoid membranes of *Zygnema* at the end of the logarithmic phase of growth display lamellar-cubic transition (Fig. 1D).

Conversely, amoeba *Chaos* cells that were simply detached from the culture dish using a fine-tip plastic pipette and transferred to 2.5% GA for fixation shows the downside of fixing specimen with too much additional liquid. The introduction of some amoeba medium when fixing the cells caused the mitochondrial outer membrane to swell and subsequent disintegration of inner membranes (Fig. 3A).

Figure 3B demonstrates comparatively better preserved mitochondrial inner membranes. However, in this example certain sections of the mitochondrial inner membranes appear to harbor cubic morphology with continuous membranes while others appear to be discontinuous (refer to section IIIC for further discussion). This observation, however, is not unusual since cubic membrane regions may develop only locally in mitochondria and do not necessarily occupy the entire organelle (Landh, 1995). This phenomenon was also described and confirmed by Deng and Mieczkowski (1998) through TEM serial sectioning and EM tomography techniques.

On another note, starved amoeba *Chaos* mitochondria shown in Fig. 1A–C displayed the different facets of cubic membranes when the mitochondria were sliced at

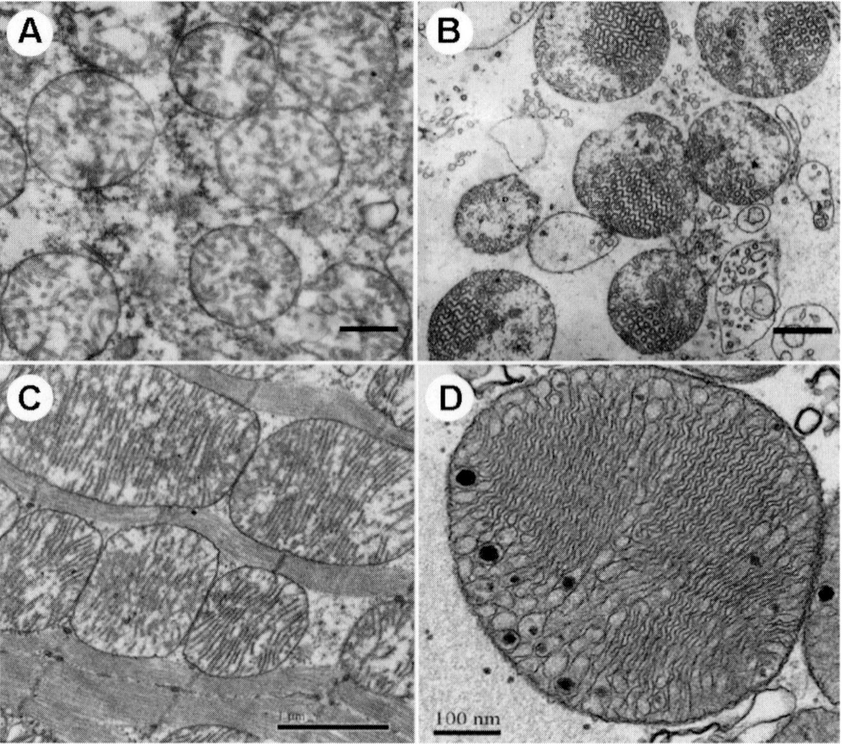

Fig. 3 Effects of sample preparation on preservation of 3D membrane structures. (A) Amoeba *Chaos* cells were primarily fixed in the presence of excess medium. Mitochondria show disintegrated tubular cristae, which is an artifact due to the poor sample fixation. Bar = 1 μm. (B) Amoeba *Chaos* cells were fixed using the method described in **Fig. 2B**. Mitochondria with highly ordered inner membrane arrangements have been well preserved by this primary fixation method. Bar = 1 μm. (C) Mitochondria of mongrel dog heart tissue cells (from a biopsy) appear swollen with destroyed cristae due to poor primary fixation. Bar = 1 μm. (D) Inner mitochondrial membranes reveal a well-preserved "zig-zag" pattern with proper primary fixation. Bar = 100 nm.

different directions. This illustrates an important notion that the membrane arrangements that do exist as a complex 3D configurations may yield multiple 2D "faces" in TEM projections, dependent on the section width and direction of (random!) cutting.

Cultured Cells and Mammalian Tissue

Fixation of cultured cells (typically mammalian cells) is fairly straightforward. Similar to filaments of *Zygnema*, fixation of cultured cells in accordance to our protocol (refer to section I) typically preserves non-lamellar membrane structures, for example cER, which consists of parallel stacks of large tubules in a hexagonal arrangement in UT-1 cells and TRS in B-lymphocytes (refer to Fig. 1E, F).

On the other hand, tissue samples taken from animal models are often contaminated with blood and tissue fluid. When tissues were sectioned from the heart of Mongrel dog by using a scalpel and fixed directly in 2.5% GA, the TEM micrograph illustrates mitochondria with apparently disintegrated cristae (Fig. 3C). Although the tissue was quickly cut into smaller pieces to enhance fixative penetration, the specimen was covered by traces of blood and tissue fluid when it was dipped into the fixative. As emphasized above, any form of excess fluid surrounding the specimen prior to fixation interferes with the penetration rate of the fixative, promoting cell disruption, and cell death. Swollen mitochondria with disrupted membranes are indicative of induced cell death during the fixation step (Fernandez-Gomez *et al.*, 2005; Pfeiffer *et al.*, 1995), as illustrated in Fig. 3C.

Strikingly, taking a biopsy (Fig. 2C) greatly improved the preservation of mitochondrial cristae in the heart tissue of Mongrel dog during fixation: the sample appears clean and clear, with mitochondria containing remarkably defined "zig-zag" cristae (Fig. 3D).

B. Methods to Identify Cubic Membranes

As outlined above, cubic membranes are characterized by their well-defined periodic structure in three dimensions. Since the extension to which cubic membrane organizations exist in different cell types, it is often difficult to judge the intactness of membrane components without the complete understanding of 3D membrane configuration. This is especially true when the cubic membrane is sectioned in different directions, which indeed yields different appearances of the same structure.

1. Computer Simulation of TEM Micrographs and Direct Template Matching

In an effort to comprehend cubic membranes in 3D from 2D TEM micrographs, computer simulations of 2D projections of these membrane structures can be generated based on the mathematically well-defined 3D models (Fig. 4) (Deng and Mieczkowski, 1998; von Schnering and Nesper, 1991). Using QMSP, a library of 2D projections of cubic membranes can be computed (Fig. 4A) that are based on the type of cubic surfaces, variable viewing direction, specimen thickness, and the number of parallel membranes (Almsherqi *et al.*, 2006; Deng and Mieczkowski, 1998; Landh, 1995). Three fundamental types of cubic membranes can be identified, based on double diamond (D), gyroid (G), and primitive (P) surfaces (Landh, 1995).

By applying the computer simulation program (QMSP), the following parameters are specified (Deng and Mieczkowski, 1998):

a. The type of periodic cubic surfaces (D, G, or P).
b. Projection direction [h k l] specified by coordinate triples.
c. The surface parameter, that is, the number of membranes for each surface desired.
d. The front and back microtome plane to specify the position of the simulated microtome planes and thickness of the section.

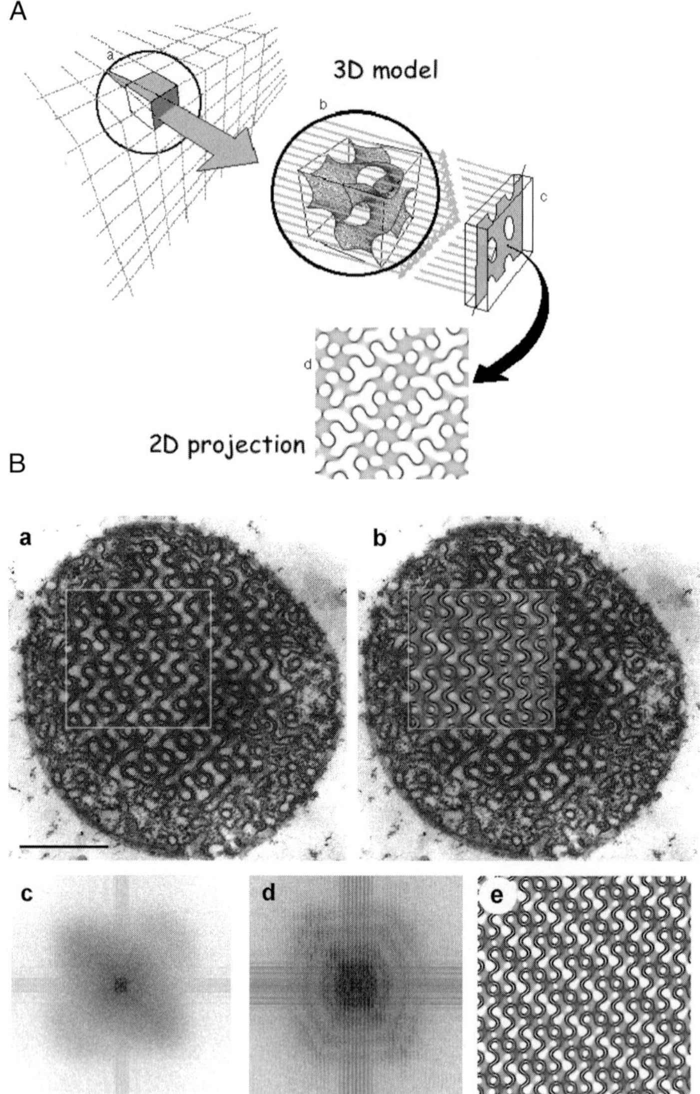

Fig. 4 (A) Computer simulation of TEM images. (a) An example of a 3D object which allows the penetration of the electron beam. (b) Cubic membrane surface in the 3D model of one unit cell. (c) Projection plane of the electron beam and recorded image. (d) 2D projection map generated by QMSP that is used for matching the corresponding patterned membrane structures in the TEM micrograph (After Almsherqi *et al.* (2005), with permission). (B) TEM micrograph of a starved amoeba *Chaos* mitochondrion matched to the projection of the double diamond (D)-based cubic morphology. (a) Area of interest (AOI) is boxed on TEM micrograph. Bar = 500 nm. (b) A theoretical 2D projection map (template) is matched to AOI from (a) by overlay. (c) The Fourier power spectrum on AOI in (a) and match in (b), respectively. (e) The template [viewed along lattice direction [5 3 1], with 0.2 unit cell thickness of the thin section and surface potential ±0.4].

These computer-generated 2D projection maps are subsequently used to match the TEM micrograph (Direct Template Matching method (Deng and Mieczkowski, 1998; Landh, 1995)). Figure 4 illustrates the principle of computer simulation of TEM by using the starved amoeba *Chaos* mitochondria as an example to demonstrate how cubic membrane arrangements are unequivocally identified and analyzed.

Image analysis is typically performed using public domain image processing program package such as NIH Image (Image-J) or Adobe Photoshop®. Briefly, the area of interest (AOI) on TEM micrographs containing cubic membranes is selected and compared by trial and error to the library of projection maps (Fig. 4B). Once the fine details of simulated projection match closely to those of the TEM micrographs, the selected theoretical 2D projection is overlaid onto the subdomain of the TEM micrograph, scaled, rotated, and translated until similar areas of electron density are brought into coincidence. The projection direction and 2D order of the subdomain of identical AOI from both TEM micrograph and projection is initially assessed by its power spectrum followed by Fourier transformation (Brenner and Rader, 1976). The entire process is repeated until an acceptable match is found. Resolution of the match can be further evaluated using a cross-correlation function (Unwin and Henderson, 1975) in Fourier space between the subdomain and the refined template.

The successful correlation of a micrograph to a theoretical reference structure further supports the interpretation that the cubic membrane structure observed is authentic.

2. Electron Tomography – 3D Reconstruction of Cubic Membranes

Although the cubic structure is mathematically well defined, computer simulated projection does not provide direct information of the actual 3D membrane topology, specifically the connectivity of internal compartments within these cubic mitochondria (Deng *et al.*, 1999). Thus, electron tomography (ET) is performed to further characterize the 3D structure of cubic membranes in the mitochondria of amoeba *Chaos*. ET reconstruction of membranes and characterization of their periodicity shows that the inner membranes of *Chaos* mitochondria fold into an arrangement with cubic symmetry (Fig. 5) (Deng *et al.*, 1999). The applied sample preparation techniques for ET were identical as for TEM except that the sections were cut at greater thickness (0.2–2 μm) compared to a TEM thin section (50–90 nm).

A standard protocol for electron tomography and surface rendering (refer to Deng *et al.*, 1999) is as follows:

a. Sections of fixed, plastic embedded amoeba sample preparations are cut at approximate 250-nm thickness and post stained with uranyl acetate and lead citrate.
b. Gold particles of 10-nm thickness are then coated on one surface of the section as reference points for image alignment.

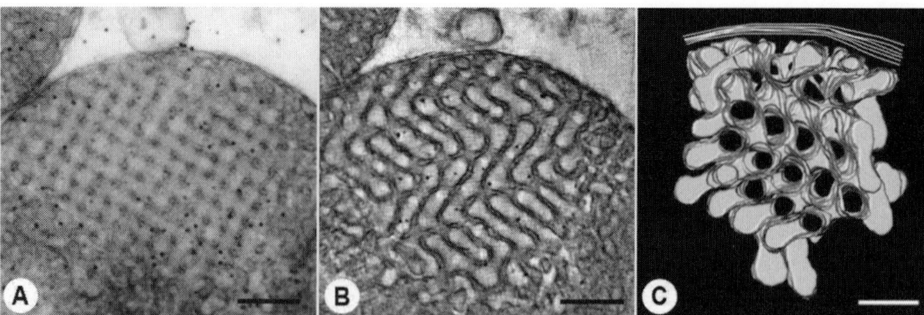

Fig. 5 EM tomographic 3D reconstruction of a mitochondrion with cubic membrane arrangement. (A) Projection image recorded at 400 kV of an untilted 250-nm-thick section of a starved amoeba *Chaos* mitochondrion. (B) A 2.9-nm-thick slice from the EM tomographic reconstruction. (C) Surface-rendering model of the cubic membrane region. All Bars = 0.25 μm each (After Deng *et al.* (1999) with permission).

c. Tomographic double-tilt series are recorded at 400 kV on an electron microscope (JEM-4000FX, JEOL Ltd.). A total of 122 images are recorded covering a tilt range ±60° around each axis, at a tilt interval of 2°.
d. Images are aligned and the reconstruction is computed using SPIDER software (Frank *et al.*, 1996).
e. Surface rendering model of cubic membranes within an amoeba mitochondrion is obtained by tracing membrane contours in successive 2.9-nm tomogram z-slices from the reconstruction (Silicon Graphics' Scene Viewer; Deng *et al.*, 1999).

In this reconstruction, only TEM micrographs containing mitochondria with cubic membrane arrangements were selected. The membranes on this thick section (250 nm) show distinct signs of cubic morphology (Fig. 5A). Simple visual inspection of the mitochondrial inner membrane profile based on thinner sections with individual axial slices that are 2.9-nm thick further reveals that membranes are organized as a D-type cubic structure (Fig. 6B). At viewing angles close to the [1 0 0] lattice direction, the section of highly curved membrane surfaces exhibits a convoluted system of

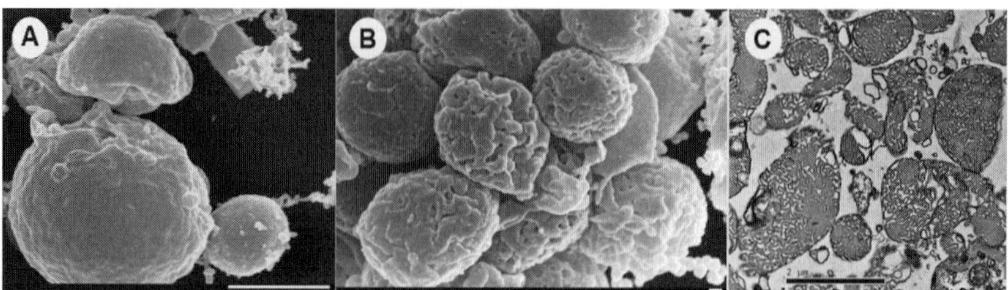

Fig. 6 SEM and TEM images of isolated mitochondria from amoeba *Chaos*. (A) SEM micrograph of isolated mitochondria from fed *Chaos* cell. Bar = 1 μm. (B) SEM micrograph of isolated mitochondria from 7d-starved *Chaos* cells with cubic membrane transformation of their inner mitochondrial membranes. Bar = 100 nm. (C) TEM micrograph of isolated mitochondria with cubic morphology from starved amoeba *Chaos*. Bar = 2 μm.

interwoven compartments with a unique pattern (Fig. 5B). This characteristic 3D periodic arrangement is clearly evident in the surface rendered model (Fig. 5C).

Interestingly, signs of connectivity of internal compartments within the cubic mitochondria are revealed by the membrane surfaces in the reconstruction. The cristae form large, undulating labyrinths (diameter of approximately 80 nm) while in the cubic regions, the matrix is highly condensed and appears to be restricted to a continuous and small space (25 nm) between adjacent cristae membranes (Fig. 5B, C). EM tomographic 3D reconstruction of a mitochondrion in *Chaos* cell provides conclusive evidence that mitochondrial inner membranes are organized into a cubic crystalline form.

C. Discussion

1. Effect of Primary Fixation on the Preservation of Highly Ordered Membrane Organizations

In order to view biological samples under the electron microscope, the sample has to be properly fixed in order to preserve all (membrane) structures as close as possible to their state in living cells. Good fixation is particularly important for viewing fine structures (Palade, 1952). Therefore, fixation is a crucial step to prepare samples with highly ordered membrane arrangements such as cubic membranes to avoid destruction of the native membrane structure or introduction of artifacts.

Glutaraldehyde (GA) is chosen as the primary fixative in our studies. It is important to note the slow penetration rate of GA at approximately 0.7 mm/h (Bozzola and Russell, 1999), which limits its use in thick biological samples. Ensuring that the living specimen is free of excess surrounding medium/buffer (without drying them) before primary fixation contributes significantly to preserving cellular organelle structures in the described cell models. Any excess liquid or medium present might affect the infiltration rate of the fixative and may lead to autolysis of the specimen due to osmotic changes before fixation is completed. The osmotic volume response is further attenuated with the fixative prepared in buffer.

This fixation protocol works very well for certain eukaryotes described here with complex membrane organizations, such as amoeba *Chaos* and green algae *Zygnema*. In mammalian cells, we found that this method of chemical fixation is successful in preserving cER in UT-1 cells and TRS in B-lymphocytes (Fig. 1E, F). Although it might also be possible to fix mammalian cells in suspension culture with a two-fold concentration of fixative (e.g., GA) in an equal volume, this method was not successful for the preservation of highly ordered membranes in amoeba *Chaos* and may need to be individually determined.

It is paramount to note a few basic issues before interpreting a TEM micrograph containing highly ordered membrane structures. Critical parameters to consider are the approximate magnification, overall basic features and subcellular organelles of the specimen, recognition of artifacts and other technical limitations of the micrograph (Bozzola and Russell, 1999). An aptly processed cell or tissue sample is usually indicated by well-preserved mitochondria.

2. Alternative Technique for TEM Sample Preparation – Cryo-techniques

Cryo-techniques for TEM are gaining more and more popularity but require more sophisticated instrumentation, which currently limits their widespread use. The advantage of these techniques is the study of samples at cryogenic temperatures without the need for primary chemical fixation, thus avoiding potential induced artifacts. Related sample preparation techniques are based on fast-freeze fixation followed by freeze-substitution. Quick freezing at extremely low temperatures allows water to form vitreous ice, which, in contrast to secondary ice crystals that may form during freezing, does not destroy the cellular ultrastructure. Several freezing methods and underlying principles of structure preservation are described in detail by Robards and Sleytr (1985), Gilkey and Staehelin (1986), and Steinbrecht and Zierold (1987). Fast-freeze fixation is followed by freeze substitution during which miscible organic solvent is used to replace vitreous ice in the frozen sample at low temperature to prevent ice crystal(s) formation and devitrification of the amorphous ice of the specimen (Nicolas and Bassot, 1993). Although cryo-fixation appears to be a less error prone alternative to chemical fixation for studying convoluted membranes, this technique is also technically highly demanding. Poor freezing (fast-freeze fixation is to be completed in milliseconds) and accidental warming of the sample may lead to ice crystal formation, inducing artifacts of deformed membrane structures. Ice crystals may also form in the sample when they are not adequately infiltrated with organic solvent (Bassot and Nicolas, 1987; Bozzola and Russell, 1999; Nicolas and Bassot, 1993).

Freeze fracture is another technique that utilizes low temperature. This method replicates fractured surfaces of frozen samples – typically occurring between the layers of a biological membrane – which are rendered visible for SEM by using evaporated metal (Bozzola and Russell, 1999). Since this technique provides information on membrane-delineated surfaces, its use to characterize convoluted 3D membrane arrangements is rather limited.

3. Potential Causes of Artifacts in TEM Results

Good fixation of the sample is a key to provide valuable and reliable information on highly ordered membrane arrangements in the cells. In fact, cubic membranes are often seen in TEM micrographs of cells in diseased states or upon viral infections but this remarkable phenomenon is apparently unexplored. Perhaps, cubic membranes were often considered as artifacts that may be caused by primary fixation or other preparation procedures of TEM.

Although all electron micrographs are "artifacts" by definition, it is often a dilemma for a researcher to judge the extent to which the sample has been modified by the preparation technique. Signs of leaching of components, osmotic damage, structural disorder, and discontinuous electron densities are all used to judge the preservation technique. By fixing cells of the same type under various conditions it has to be estimated as to which depiction bears the closest resemblance to their natural state. Therefore, a good judgment of the well-preserved specimen very much depends on the experience of the researcher. As discussed above, the quality of the appearance of mitochondria in TEM images is frequently a good indication and measure of the quality of the sample fixation and processing (refer to Fig. 3B, D).

In addition to the fixation step, there are other potential causes of artifacts that may be induced during dehydrating the sample or slicing the polymerized Araldite plastic block with the embedded samples. Such cases of artifacts are well documented, and the micrographs are typically less ambiguous (Fig. 3). The type of sample itself might also incur artifacts during various preparation steps. For instance, isolated cubic mitochondria might be more susceptible to dehydration or osmotic damage than those contained in whole intact cells. In the dehydration stage, ethanol is the preferred dehydrating agent over acetone. This is because acetone is a more powerful extractor of lipids (Bozzola and Russell, 1999). As such, acetone might destroy the membrane organization by removing excess lipids from the membranes.

Is it possible that highly ordered membranes are indeed simply induced by the preparation technique? In this respect, sample preparation for TEM is most likely destructive rather than constructive to give rise to higher ordered structures. Given that cubic membranes exhibit well-characterized cubic symmetry, it is rather difficult to imagine that they are simply induced by the harsh TEM sample preparation techniques. A strong line of reasoning for the detrimental effects of sample preparation is that cubic membranes are best preserved in the inner zone of the specimen and are commonly surrounded by noticeably disturbed outer membrane regions (Landh, 1996). In this regard, we reiterate that the primary fixative requires a decent amount of time to penetrate the specimen, and inner regions of the cells that are yet to be fixed are susceptible to autolysis. A TEM micrograph containing cubic membranes usually displays varying degrees of destruction and the extent of the region of cubic membranes is occasionally preserved fortuitously. Thus, the key to differentiate these possible artifacts is to make use of the periodic and continuous nature of cubic membranes. An abrupt discontinuity, for instance, might suggest the presence of perturbation caused by inappropriate sample preparation.

IV. A Closer Look at Cubic Membrane Surface Contours through Scanning Electron Microscopy (SEM)

With SEM it is possible to perceive the surface impression of isolated cubic membrane structures, rather than information on their inner 3D arrangement, which limits its use in cubic membranes research. A major obstacle is the requirement for purified cubic membrane-containing organelles (e.g., mitochondria from starved *Chaos* cells). Since the 3D morphology strongly depends on environmental factors such as osmolarity, pH, certain types of ions, such isolation protocols – in addition to the problems arising from sample fixation and preparation for SEM – are crucial determinants of the resulting information quality. Nevertheless, SEM provides valuable information about the surface arrangement and continuity of cubic membranes

(Almsherqi *et al.*, 2008). Preparation of isolated membrane organelles for SEM is similar to that for TEM with respect to the initial fixation and dehydration stages.

Summarized below is the scheme for SEM sample preparation for isolated amoeba mitochondria:

a. **Primary fixation:** Samples are fixed with 2–2.5% GA at 4 °C for 4 h or overnight followed by washing 3–4 times using PBS buffer.
b. **Secondary fixation:** Post fixation by incubation with 1% OsO_4 for 1 h at RT.
c. **Immobilization of sample on cover slips:** Post-fixed sample are incubated on 0.01% poly-L-lysine (Sigma–Aldrich) pre-coated cover slips for 20 min followed by washing with deionized water.
d. **Dehydration:** The sample is subjected to sequential dehydration where it is immersed in a graded series of ethanol with increasing strength of dehydration (25, 50, 75, and 100%) at 10 min each.
e. **Critical point drying:** Dehydrated and fixed samples (isolated mitochondria pellet) that are immobilized on a cover slip are immersed into a chamber filled with pure dehydrant (ethanol) and transferred to a cooled vessel (7–10 °C) of a critical point dryer (CPD) (Balzers: model CPD30) with 99.8% purified compressed carbon dioxide (CO_2) (SOXAL). The initial pressure is 0 bar. Pressurized liquid CO_2 gradually displaces the ethanol in which the sample is soaked. After the exchange of ethanol to CO_2, pressure is changed to around 45–55 bar. This procedure of medium exchange is repeated every 15 min for 2 h for complete displacement of ethanol. Temperature is slowly raised and eventually, the transitional fluid (carbon dioxide) reaches a critical point where transition occurs at 38 °C and 70–90 bar of pressure. The density of the CO_2 in liquid and vapor phase at this point will be the same and this CO_2 is released into the atmosphere.
f. **Specimen mounting and coating:** The cover slip carrying the dried "isolated amoeba mitochondria" is mounted onto a carbon specimen stub of 10-mm diameter (Agar Scientific). The exposed surfaces of the isolated amoeba mitochondria are sputtered with a thin layer of gold particles (Baltec). After coating is completed, the sample is viewed under SEM (JSM6701F, JEOL Ltd.).

A. Potential Factors that Perturb Isolated Cubic Membranes

Similar to TEM, preservation of cubic membrane is accentuated by the initial fixation procedure in SEM sample preparation. Fixed samples are subsequently "immobilized" on a cover slip coated with poly-L-lysine. Poly-L-lysine is a small natural homo-polymer of the hydrophilic and positively charged amino acid L-lysine. This charge might interact electrostatically with the biological membranes and might disturb membrane structures (Shima and Sakai, 1977). However, cubic membranes that are post-fixed by OsO_4 are already structurally preserved; thus, the membrane morphology should not be affected by such electrostatic interaction. In addition, OsO_4 aids in stabilizing unsaturated lipids during the dehydration process (Bozzola and Russell, 1999).

SEM allows to visualize the exterior profile of mitochondria containing inner cubic membrane arrangement. In order to prevent artifacts, one of the critical choices in SEM sample preparation procedures includes the mode of drying of the fixed and dehydrated specimens. In certain cases, air-drying the sample after each step of dehydration might cause a collapsing, flattening or shrinking of the fine structures, especially for biological membranes. The underlying reason for this problem is the surface tension forces associated with retreating air-liquid interfaces (Murphy and Roomans, 1984). Such phenomena can be prevented by using a CPD that makes use of liquid displacement method to dry the isolated cubic membranes (Bozzola and Russell, 1999).

After the cubic mitochondria are dried via CPD, the sample is coated with a thin layer of metal coatings (e.g., gold). When viewing the sample under SEM, the gold coat prevents high voltage charges from building up on the specimen and conduct away the excess damaging heat. As such, the quality and structure of the samples can be preserved for the subsequent viewing and examination. In addition, the absorption of water from the environment might cause swelling of the organelle membranes and further damage the delicate features of cubic structures. To ensure no further artificial changes to the samples, they should be stored in a container in a dry box (Jouan) (Bozzola and Russell, 1999).

B. Perceiving 3D Topology of Mitochondria with Cubic Morphology

For the first time, the surface profile of cubic membranes in biological system was revealed with suitable preparation of isolated mitochondria from the starved amoeba *Chaos* (Fig. 6B). In the control sample (fed amoeba mitochondria without cubic membrane arrangement), the surface profile appears rather smooth with slight folding (Fig. 6A). Interestingly, starvation-induced cubic membrane transition in the amoeba mitochondria, on the other hand, results in distinct membrane invaginations on the surface (Fig. 6B). In TEM micrographs, the outer circumferences of cubic membranes often appear to be perturbed or partially peeled off (Fig. 6C). SEM, in contrast, revealed the appearance of isolated mitochondria with intact outer membrane coverage (Fig. 6B). In fact, SEM data suggest the mitochondrial morphological change in the outer membrane appearance when the inner membranes transform from random tubular to cubic arrangement.

This surface profile of cubic membrane in SEM is supported by computer simulation data (Fig. 7B, provided by Felix Margadant, MechanoBiology Institute, National University of Singapore). Previous TEM work (Deng and Mieczkowski, 1998) has simulated the starved amoeba *Chaos* mitochondria with cubic membrane organizations in great detail through mathematical modeling and computer simulation. Figure 7A shows the mathematical surface model of cubic membrane arrangement in 3D. These surfaces are used to describe cubic membranes and to generate unique patterns in 2D that allow the recognition of cubic membranes in TEM micrographs. Cubic membranes in the mitochondria of starved amoeba commonly

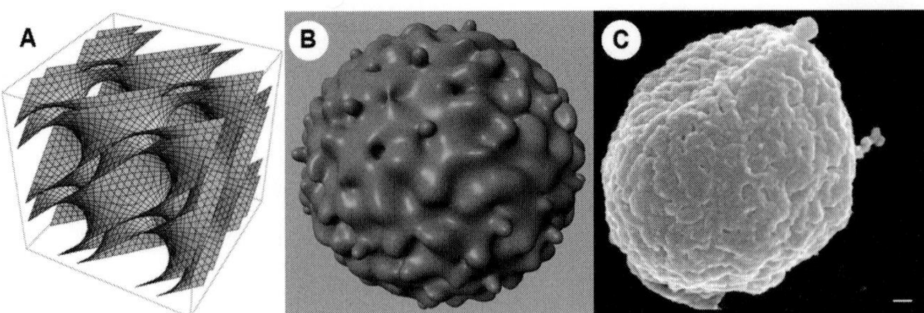

Fig. 7 Computer simulation of SEM view of cubic membrane (isolated mitochondria from starved amoeba *Chaos*). (A) 3D mathematical model of cubic surfaces used to simulate and generate TEM and SEM image of cubic membrane arrangement. (B) Computer simulation of 3D surface contour of a particle (mitochondrion) with inner cubic arrangement. (C) SEM image of a starved amoeba mitochondrion containing inner cubic membrane arrangement. Bar = 100 nm. (For color version of this figure, the reader is referred to the web version of this book.)

appear as double membrane-based D-type cubic structures (Fig. 7A) (Deng and Mieczkowski 1998).

Using the same mathematical 3D model, we show how these highly curved 3D "infinite periodic minimal surfaces" (Hyde *et al.*, 1997) with 2-parallel membranes may appear with a boundary of spherical-shaped mitochondrial outer membrane (Fig. 7B). The computer simulated cubic mitochondrion (Fig. 7B) shows remarkable resemblance to the isolated cubic mitochondria viewed under SEM (Fig. 7C). Indeed, the combination of SEM and computer simulation has provided greater insights to further understand the 3D global topology of cubic membranes organization.

V. Summary

The application of the described methods and protocols has led to the identification and characterization of the 3D arrangements of cubic membranes, through TEM, EM tomography and SEM techniques in conjunction with direct template matching method and computer simulation. The combination of these techniques has provided valuable information on cubic membrane organization. In fact, computer simulation data offered greater scientific acumen to study phospholipid bilayer, membrane-based cell architecture through cubic membranes. Although EM is currently the only available tool to study cubic membranes, emerging techniques such as atomic force microscopy (AFM) and second or third harmonic generation microscopy may also hold the potential to study cubic membrane architecture in biological system(s) from different perspectives. Clearly, obtaining a better understanding of structure-function relationships of cell membrane organizations is a major challenge for future research.

Acknowledgments

We thank Sepp D. Kohlwein for critical reading of the manuscript and Zakaria Almsherqi for helpful discussions. We also thank Felix Margadant for kindly providing computer simulation of SEM image, Mark Mieczkowski for the "Cubic Membrane Simulation Projection" program (QMSP). The technical support by Mei Yin Shoon and Chwee Wah Low is greatly appreciated. We acknowledge the use of the core facilities of Electron Microscopy Unit (Yong Loo Lin School of Medicine, National University of Singapore). Research in the authors' laboratory is supported by research grants NMRC (R-185-000-058-213) and BMRC (R-185-000-065-305) from Singapore to Deng Y.

References

Almsherqi, Z., Hyde, S., Ramachandran, M., and Deng, Y. (2008). Cubic membranes: a structure-based design for DNA uptake. *J. R. Soc. Interface* **5**, 1023–1029.

Almsherqi, Z. A., Kohlwein, S. D., and Deng, Y. (2006). Cubic membranes: a legend beyond the *Flatland** of cell membrane organization. *J. Cell Biol.* **173**, 839–844.

Almsherqi, Z. A., Landh, T., Kohlwein, S. D., and Deng, Y. (2009). Chapter 6: cubic membranes the missing dimension of cell membrane organization. *Int. Rev. Cell Mol. Biol.* **274**, 275–342.

Bassot, J. M., and Nicolas, M. T. (1987). An optional dyadic junctional complex revealed by fast-freeze fixation in the bioluminescent system of the scale worm. *J. Cell Biol.* **105**, 2245–2256.

Bouligand, Y. (1991). Geometry and topology in cell membranes. *In* "Geometry in Condensed Matter Physics," (J. F. Sadoc, ed.), pp. 193–231. World Scientific Press, London.

Bozzola, J. J., and Russell, L. D. (1999). Specimen Preparation for Transmission Electron Microscopy. *In* "Electron Microscopy: Principles and Techniques for Biologists," (D. W. Jones, and A. Bartlett, eds.), pp. 21–31. Sudbury, Mass.

Brenner, N., and Rader, C. (1976). A new principle for Fast Fourier Transformation. *IEEE Acoust., Speech Signal Process* **24**, 264–266.

Brown, W. J., Chambers, K., and Doody, A. (2003). Phospholipase A2 (PLA2) enzymes in membrane trafficking: mediators of membrane shape and function. *Traffic* **4**, 214–221.

Chin, D. J., Luskey, K. L., Anderson, R. G. W., Faust, J. R., Goldstein, J. L., and Brown, M. S. (1982). Appearance of crystalloid endoplasmic reticulum in compactin-resistant Chinese hamster cells with a 500-fold increase in 3-hydroxy-3-methylglutaryl-coenzyme A reductase. *Proc. Natl. Acad. Sci. U. S. A.* **79**, 1185–1189.

Deng, Y. (1998). Transmission electron microscopy studies of cubic membrane morphologies in chloroplasts of the green algae *Zygnema* and mitochondria of the amoeba *Chaos carolinensis*. Ph.D Thesis. State University of New York at Buffalo, USA.

Deng, Y., and Mieczkowski, M. (1998). Three-dimensional periodic cubic membrane structure in mitochondria of amoebae *Chaos carolinensis*. *Protoplasma* **203**, 16–25.

Deng, Y., Marko, M., Buttle, K. F., Leith, A., Mieczkowski, M., and Mannella, C. A. (1999). Cubic membrane structure in amoeba (*Chaos carolinensis*) mitochondria determined by electron microscopic tomography. *J. Struct. Biol.* **127**, 231–239.

de Kruijff, B. (1997). Biomembranes. Lipids beyond the bilayer. *Nature* **386**, 129–130.

Fernandez-Gomez, F. J., Galindo, M. F., Gómez-Lázaro, M., Yuste, V. J., Comella, J. X., Aguirre, N., and Jordán, J. (2005). Malonate induces cell death via mitochondrial potential collapse and delayed swelling through an ROS-dependent pathway. *Br. J. Pharmacol.* **144**, 528–537.

Frank, J., Radermacher, M., Penczek, P., Zhu, J., Li, Y., Ladjadj, M., and Leith, A. (1996). SPIDER and WEB: processing and visualization of images in 3D electron microscopy and related fields. *J. Struct. Biol.* **116**, 190–199.

Gilkey, J. C., and Staehelin, L. A. (1986). Advances in ultrarapid freezing for preservation of cellular ultrastructure. *J. Electron Microsc. Tech.* **3**, 177–210.

Goldsmith, C. S., Tatti, K. M., Ksiazek, T. G., Rollin, P. E., Comer, J. A., Lee, W. W., Rota, P. A., Bankamp, B., Bellini, W. J., and Zaki, S. R. (2004). Ultrastructural characterization of SARS coronavirus. *Emerg. Infect. Dis.* **10**, 320–326.

Grimley, P. M., and Schaff, Z. (1976). Significance of tubuloreticular inclusions in the pathobiology of human diseases. *In* "Pathobiology Annual," (H. L. Ioachim, ed.), Vol. 6, Appleton-Century-Crofts, New York.

Hyde, S., Andersson, S., Larsson, K., Blum, Z., Landh, T., Lidin, S., and Ninham, B. W. (1997). *The Language of Shape: The Role of Curvature in Condensed Matter: Physics, Chemistry and Biology.* Elsevier, Amsterdam.

Kooijman, E. E., Chupin, V., Fuller, N. L., Kozlov, M. M., de Kruijff, B., Burger, K. N., and Rand, P. R. (2005). Spontaneous curvature of phosphatidic acid and lysophosphatidic acid. *Biochemistry* **44**, 2097–2102.

Landh, T. (1995). From entangled membranes to eclectic morphologies: cubic membranes as subcellular space organizers. *FEBS Lett.* **369**, 13–17.

Landh, T. (1996). Cubic cell membrane architectures. Taking another look at membrane.bound cell spaces. Ph.D Thesis. Lund University, Sweden.

Larsson, K. (1989). Cubic lipid–water phases: structures and biomembrane aspects. *J. Phys. Chem.* **93**, 7304–7314.

Lee, M. C., Orci, L., Hamamoto, S., Futai, E., Ravazzola, M., and Schekman, R. (2005). Sar1p N-terminal helix initiates membrane curvature and completes the fission of a COPII vesicle. *Cell.* **122**, 605–617.

Lewis, R. N. A. H., Mannock, D. A., and McElhaney, R. N. (1997). Membrane lipid molecular structure and polymorphism. *In* "Lipid Polymorphism and Membrane Properties," (R. M. Epand, ed.), pp. 25–102. Academic Press, San Diego.

McLean, R. J., and Pessoney, G. F. (1970). A large scale quasi-crystalline lamellar lattice in chloroplasts of the green alga *Zygnema*. *J. Cell Biol.* **45**, 522–531.

Murphy, J. A., and Roomans, G. M. (1984). *In* "The Preparation of Biological Specimens for Scanning Electron Microscopy," (JA. Murphy, and GM. Roomans, eds.), pp. 304. Scanning Electron Microscopy, Inc. (AMF O'Hare, II), Chicago.

Nicolas, M. T., and Bassot, J. M. (1993). Freeze substitution after fast-freeze fixation in preparation for immunocytochemistry. *Microsc. Res. Tech.* **24**, 474–487.

Palade, G. E. (1952). A study of fixation for electron microscopy. *J. Exp Med.* **95**, 285–298.

Pfeiffer, D. R., Gudz, T. I., Novgorodov, S. A., and Erdahl, W. L. (1995). The peptide mastoparan is a potent facilitator of the mitochondrial permeability transition. *J. Biol. Chem.* **270**, 4923–4932.

Robards, A. W., and Sleytr, U. B. (1985). Freeze-substitution and low temperature embedding. *In* "Practical Methods in Electron Microscopy," (A. M. Glauert, ed.), pp. 461–499. Elsevier, Amsterdam, New York.

Robertson, J. D. (1959). The ultrastructure of cell membranes and their derivatives. *Biochem. Soc. Symp.* **16**, 3–43.

Robertson, J. D. (1960). The molecular structure and contact relationships of cell membranes. *Prog. Biophys. Mol. Biol.* **10**, 343–418.

Shima, S., and Sakai, H. (1977). Polylysine produced by *Streptomyces*. *Agric. Biol. Chem.* **41**, 1807–1809.

Starr, R. C., and Zeikus, J. A. (1993). UTEX—the culture collection of algae at the University of Texas at Austin, list of cultures. *J. Phycol.* **92**, 1–106.

Steinbrecht, R. A., and Zierold, K. (1987). *Cryotechniques in Biological Electron Microscopy.* Springer-Verlag, Berlin, Hiedelberg, New York.

Tan, O. L. -L., Almsherqi, Z. A., and Deng, Y. (2005). A simple mass culture of the amoeba *Chaos carolinense*: revisit. *Protistology* **4**, 185–190.

Tinari, A., Ammendolia, M. G., Superti, F., and Donelli, G. (1996). Tubuloreticular structure induced by Rotavirus infection in HT-29 cells. *Ultrastruct. Pathol.* **6**, 571–576.

Unwin, P. N., and Henderson, R. (1975). Molecular structure determination by electron microscopy of unstained crystalline specimens. *J. Mol. Biol.* **94**, 425–440.

von Schnering, H. G., and Nesper, R. (1991). Nodal surfaces of Fourier series: fundamental invariants of structured matter. *Zeitschrift für Physik B: Condensed Matter* **83**, 407–412.

CHAPTER 16

Quantitative Imaging of Lipid Metabolism in Yeast: From 4D Analysis to High Content Screens of Mutant Libraries

Heimo Wolinski*, Kristian Bredies† and Sepp D. Kohlwein*,†

*Institute of Molecular Biosciences, University of Graz, Graz, Austria

†Institute of Mathematics and Scientific Computing, University of Graz, Graz, Austria

METHODS IN CELL BIOLOGY, VOL 108
Copyright 2012, Elsevier Inc. All rights reserved.

0091-679X/10 $35.00
DOI 10.1016/B978-0-12-386487-1.00016-X

Abstract

Due to their central role in cellular fat storage and lipid homeostasis, lipid droplets (LD) have attracted great interest in biomedical research. The integration of both biochemical and genetic tools and the use of model organisms have greatly contributed to the understanding of LD metabolism and its regulation. However, many important aspects such as LD biogenesis, intracellular dynamics, or their potential degradation by autophagy are still poorly understood. Microscopic techniques, in particular fluorescence microscopy using LD specific dyes or fluorescent protein tagging, represent excellent experimental tools to study the dynamic nature of both the protein and lipid content of LD. Single cell systems in culture are particularly suited to identify and characterize proteins required for LD formation and turnover, using genetic knock-down or gene deletion strategies. Here we describe experimental setups to investigate LD dynamics and turnover in yeast, using various labeling techniques suitable for three-dimensional imaging over time (4D imaging), quantitative microscopy and imaging-based screens of mutant libraries. Also, implementation of coherent anti-Stokes Raman scattering (CARS) microscopy as an emerging tool for label-free lipid imaging in living cells will be discussed.

I. Introduction

The worldwide pandemic dimensions of severe lipid-associated metabolic diseases, such as obesity and type 2 diabetes mellitus, have put cellular and organismal lipid homeostasis into the spotlight of biomedical research. The central organelle involved in cellular fat storage, the lipid droplet (LD), is therefore subject to detailed biochemical and molecular studies regarding its protein and lipid content, metabolic and cellular dynamics, and interaction with other cell compartments (Guo *et al.*, 2009; Martin and Parton, 2006; Welte, 2009). Due to the highly dynamic behavior of LD and their associated proteins, noninvasive *in vivo* techniques are of great advantage. Specifically, fluorescence microscopic techniques have proven to yield valuable insight into protein localization as well as the dynamics of lipid storage and lipolysis in a large array of cell types and tissues. Due to the conserved nature of the central cellular processes governing lipid homeostasis, numerous model organisms including the fruit fly *Drosophila melanogaster* or the nematode, *Caenorhabditis elegans* have become attractive experimental systems in lipid research. An unprecedented repertoire of biochemical and genetic tools has also established the unicellular eukaryote yeast, *Saccharomyces cerevisiae,* as an excellent reference organism in biomedical research, in particular, lipid-associated disorders (Athenstaedt and Daum, 2006; Kohlwein, 2010). Many of the fundamental mechanisms controlling cell growth (e.g., cell division cycle control), cell signaling (e.g., TOR), protein secretion through the secretory pathway, and autophagy were first discovered in yeast and were subsequently found to be highly conserved in mammalian cells. Specific labeling techniques make yeast also an excellent system for fluorescence-based screens of mutant libraries, in particular in the context of altered LD homeostasis and inheritance. LD are believed to form between the leaflets of the endoplasmic reticulum (ER) membrane from where they eventually bud off into the cytoplasm (Walther and Farese, 2009). Although this model has been discussed for many years, as yet no direct experimental proof exists. Theoretical considerations suggest that LD in yeast may form presumably in the ER as 20 nm structures (Zanghellini *et al.,* 2010) that eventually fuse and give rise to the typical 300–500 nm structures found in a growing or resting cell. Only few of the enzymes involved in storage lipid synthesis are indeed localized to the LD, and the majority of triacylglycerol (TAG) and all the steryl esters that make up the core of the LD are synthesized in the ER. Once formed, LD appear to remain in close contact to the ER membrane throughout their life cycle (Fei *et al.,* 2008; Wolinski *et al.*, 2011). It is believed that lipid trafficking occurs via these interaction sites, but the specific mechanisms remain obscure. On the other hand, LD harbor several enzymes required for TAG degradation, such as the TAG lipases Tgl3, Tgl4, and Tgl5 (Athenstaedt and Daum, 2003; Athenstaedt and Daum, 2005; Kurat *et al.,* 2006) and the steryl ester hydrolases Tgl1 and Yeh2 (Jandrositz *et al.,* 2005; Koffel *et al.,* 2005).

LD are readily detectable in living cells by labeling with lipophilic (fluorescent) dyes. Commonly used dyes include Oilred O (which is frequently used in histological analyses), BODIPY® 493/503 and Nile Red (see Table I). LD in yeast are also recognized by differential interference contrast (Nomarski) in transmission microscopy due to their high refractive index. The use of various fluorescent dyes in imaging-based screens of *Drosophila* (RNAi) knock-down cells (Guo *et al.,* 2008) as well as of yeast mutant collections (Fei *et al.,* 2008; Szymanski *et al.,* 2007) has

Table I
Frequently used fluorescence dyes for detection of LD in live and fixed specimens and their fluorescence properties.

Label	Typical $\lambda_{ex}/\lambda_{em}$ settings	Notes	Reference
Nile Red	488 or 543/550-560nm: LD 488 or 543/600-650nm: LD and phospholipids	Shows solvatochromism and broad excitation and emission bands; strong fluorescence bleaching	Greenspan *et al.,* 1995 Wolinski *et al.,* 2009b Petschnigg *et al.,*2009
BODIPY 493/503	488/500-535 nm	High quantum yield	Invitrogen, Inc.
LD540	543/550–600 nm 561/565–600 nm	Compatible with GFP detection; high quantum yield	Spandl *et al.,* 2009
BODIPY 558/568 C_{12}	543/550–600 nm 561/570–600 nm	Compatible with GFP detection; incorporation may depend on cell physiology	Invitrogen, Inc.
LipidTox®	LT Green: 488/500–550 nm LT Red: 561/580–630 LT Far Red: 633/640-690nm	LT Red and Far Red compatible with GFP detection; requires cell fixation. Not yet tested for yeast	Invitrogen, Inc.

yielded valuable information on cellular components involved in LD biogenesis and morphology. However, due to the highly dynamic nature of the LD that responds to alterations in growth conditions (Kurat *et al.*, 2006) there is currently very little consensus as to the major regulators for LD homeostasis and morphology, and it is clear that the currently performed screens are far from being saturated.

A. Rationale

The most common strategy for analyzing LD morphology and cellular content involves incubation of mutant cells with the fluorescent dye over defined periods of time and subsequent microscopic imaging and inspection of the results by independent researchers (Vizeacoumar *et al.*, 2010; Wolinski *et al.*, 2009b). Since LD turnover responds very quickly to the nutritional and growth state of the cell, it is instrumental to control the conditions during microscopy of (live) mutant collections to obtain comparable results. Furthermore, commonly fluorescent dyes are substrates for pleiotropic drug resistance pumps (Ivnitski-Steele *et al.*, 2009; Wolinski and Kohlwein, 2008) thus increased staining efficiency observed between mutants may simply be the result of impaired dye efflux. Thus, the elaboration of defined conditions for labeling and imaging is instrumental to obtain reliable results and to eliminate false positive and false negative hits from the screening.

In this chapter we describe efficient methods for both live and fixed yeast cell labeling and imaging for the analysis of LD in small-scale assays and in entire mutant collections, using confocal microscopy. A setup is described to perform 3D imaging over extended periods of time to follow LD morphology, cellular dynamics and inheritance over several generations under the microscope. In addition, we present an approach for quantification of neutral lipids in yeast cell populations. The procedure includes automated extraction of imaged, fluorescently labeled LD and a software tool for automated recognition of yeast cells in high-density cell arrays, which allows for a per cell quantification of neutral lipid and thus reliable quantitative data on cellular neutral lipid metabolism.

Finally, we describe the implementation of label-free lipid imaging of yeast mutants using coherent anti-Stokes Raman scattering (CARS) microscopy. The potential and limitations of this emerging technology for LD imaging in yeast will be discussed.

II. Choice of Fluorescence Dyes for LD Labeling

Several fluorescence dyes are available for labeling of neutral lipids. LD in living cells can either be directly labeled with fluorescence dyes or can be stained by incorporation of fluorescently labeled fatty acid analogues. Table I shows frequently used fluorescence dyes for detection of LD in live specimens, including their fluorescence properties. Although the dyes usually show high specificity for neutral lipids they also label phospholipids and thus subcellular membranes to some extent.

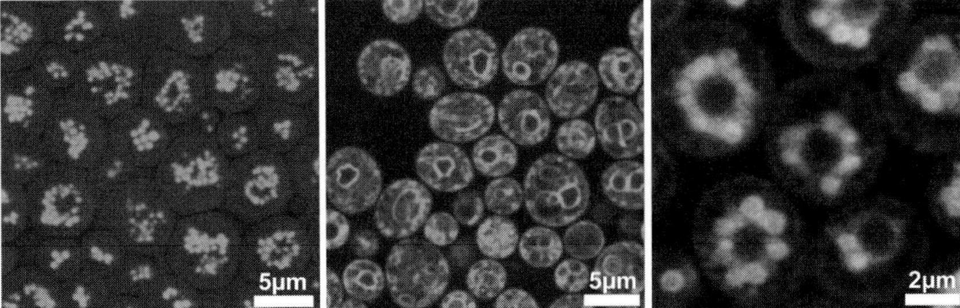

Fig. 1 Live cell imaging of neutral lipid-specific fluorescence dyes in yeast cells. LD of yeast cells that were cultivated to stationary growth phase are labeled with BODIPY 493/503. Maximum-intensity projection of a z-stack. Overlay of fluorescence image and differential interference contrast (DIC) image (image at left). BODIPY 493/503 labeled phospholipid membranes of a yeast quadruple deletion mutant strain lacking LD *(are1are2dga1lro1;*Petschnigg *et al.,* 2009), (middle image). Simultaneous detection of chromosomally integrated Faa4-GFP (long chain fatty acyl-CoA synthetase) localized to the periphery of LD and the endoplasmic reticulum (green) as well as of neutral lipid cores of LD, labeled with LD540 (red), (image at right). (See color plate.)

In this respect, it has to be considered that the affinity of the dyes for phospholipids increases with increasing hydrophobicity of the acyl-chains and also with the decrease of the amount of LD and neutral lipids in a cell. For instance, Nile red displays significant solvatochromism (Greenspan and Fowler, 1985; Greenspan *et al.,* 1985). Particularly, membrane accumulations may be mistakenly interpreted as LD in standard epifluorescence microscopes using TRITC or Rhodamine filter cubes (Petschnigg *et al.,* 2009; Wolinski and Kohlwein, 2008). In addition, numerous fluorescent dyes may be efficiently pumped out of actively growing cells due to the activity of pleotropic drug resistance pumps, indicated by an inhomogeneous staining of cell populations and preferred labeling of sick or dead cells (Ivnitski-Steele *et al.,* 2009; Wolinski and Kohlwein, 2008). In addition, fluorescence dyes such as Nile Red are susceptible to photobleaching, which, however, can usually be compensated by optimizing the imaging settings. Fluorescence dyes such as BODIPY 493/503 or LD540, a recently introduced neutral lipid specific dye emitting red fluorescence (Spandl *et al.,* 2009), show high quantum yield and less photobleaching, making them particularly suitable for long-time microscopic observation of LD (Fig. 1).

III. Four-Dimensional Live Cell Imaging of Yeast LD During Cellular Growth

Four-dimensional (4D) live cell imaging (time-lapsed three-dimensional imaging) is a powerful tool to study the dynamic behavior and interactions of subcellular structures over time (Gerlich *et al.,* 2001; Gerlich and Ellenberg, 2003). Methodical challenges are to keep the cells vital during long-term observation and during repeated illumination. Using suitable equipment, yeast cells can be cultivated directly on the microscope stage. Thus, 4D live cell imaging experiments can be performed during cellular growth and over several cell generations. The following experimental setup makes use of BODIPY 493/503 as an LD label. However, the protocol can also be applied with the red emitting LD540 fluorescence dye as well.

A. Cell Preparation for Microscopy

1. Preparation of the Growth Medium

a. Yeast extract/peptone/dextrose (YPD): Bacto™-yeast extract (1% w/v) and Bacto™-peptone (2% w/v) are weighed in an appropriate flask and filled with distilled water to 9/10 of the final volume. Autoclave at 121 °C for 20 min.
b. Add 1/10 volume of 10X glucose solution (2% w/v final concentration) by sterile filtration to the media, mix well and dispense into sterile flasks or glass tubes.
c. For solid media plates, 2% Bacto™-agar is added to the solution prior to autoclaving.

2. Cell Cultivation

a. Prepare a preculture by cultivating cells that are typically maintained on storage agar plates in culture flasks containing 5 mL YPD for 12 h at 30 °C on a rotary shaker (180 rpm).
b. To obtain yeast cells in stationary phase, inoculate100 mL fresh YPD in a 500-mL flask with 50 μl of the preculture and cultivate cells for 72 h, to stationary phase in a rotary shaker (180 rpm). To obtain logarithmically growing yeast cells inoculate fresh YPD medium with 1/100 volume of a stationary-phase culture and cultivate for 4–8 h at 30 °C on a rotary shaker.
c. Alternatively, for small-scale cultivation, cells are cultivated in 6- or 12-well microtiter plates in 5–2 mL of the medium under shaking and temperature control (30 °C) on a thermomixer. The use of volumes below 2 mL is not recommended since aeration is limiting, and cells tend to pellet in small microtiter wells.

3. Fluorescence Labeling of LD for 4D Live Cell Imaging

a. Grow cells to the desired growth stage in at least 2 mL of growth medium.
b. Transfer 1 mL of the cell culture into the well of a 12-well microtiter plate.
c. Add 1.2 μL of the BODIPY 493/503 stock solution (final concentration 1.2 μg/mL) to the cell culture.
d. Label the cells for 15 min in the microtiter plate at 30 °C and under continuous shaking.

4. Mounting Fluorescently Labeled Cells on Solid Media Slides for Microscopy

a. Melt previously prepared media agar platesin a microwave oven.
b. Pour 5 mL of the liquid agar into a 10 mL Falcon tube.
c. Equilibrate the temperature of the agar to 65 °C in a water bath.
d. Add 6 μL of a BODIPY 493/503 stock solution to the agar (final concentration 1.2 μg/mL) and vortex.
e. Pour the agar containing BODIPY 493/503 on a standard microscope slide (76 × 26 mm). Take care that the edges of the slide are fully covered with agar.
f. Allow polymerization of the agar for 20 min at RT.
g. Place the slide in an oven at 30 °C for 5 min.
h. Immediately, apply 1 μL of the fluorescently labeled cell suspension to the middle of the agar sheet and mount the cells with a large coverslip (50 × 24 mm) suitable for confocal imaging; place the slide on the microscope stage. Alternatively, cut vertical and horizontal slits in distances of ca. 5 mm into the agar. Apply cell suspensions to each resulting field. In this way, multiple samples can be mounted on one single agar slide and imaged under identical incubation conditions. The slits avoid cross contamination of individual strains. This setup with slight modifications enables also high-content screenings of yeast deletion mutant collections (Wolinski *et al.*, 2009a). Transfer the cells as fast as possible onto the fields. If the temperature significantly deviates from the optimal growth temperature of 30 °C cellular growth will be slowed down, which affects physiology and LD metabolism.

If required, dilute the cell suspension before mounting onto the agar slide. Budding of yeast cells leads to movement of the individual cells. If the initial density is too high the cells may overlap during cellular growth, which interferes with the microscopic analysis of subcellular structures in individual cells. Especially when using robot devices (Singer RoToR; Wolinski *et al.*, 2009a) cell colonies are "diluted" by repetitive replicating steps; typically, the second plate contains sufficiently dispersed cells of appropriate density for imaging.

B. Cultivation of Yeast Cells on the Microscope Stage

Controlled heating of the cultivation setup on the microscope is required to obtain morphological information that is comparable to any biochemical or growth assays that are typically performed at the optimal temperature of 30 °C. At our lab two different setups have been tested to maintain the optimal growth temperature of 30 °C for cultivating yeast cells on the microscope stage.

1. Objective Heater System

Immersion media (e.g., oil or water) required for the use of high numerical aperture immersion objectives, act as thermal coupling media between the specimen and the objective. As a consequence heat is drawn away from the specimen. By heating the objective by using a ring heating device yeast cells can easily be cultivated on the microscope stage without the need to control the environment of the cells and microscope slide using a closed chamber. By using the agar sheet immobilization method the yeast cells are immobilized in a liquid medium film directly under the coverslip. This liquid film is continuously heated by the temperature-controlled objective. Some minor shrinkage of the agar may occur over time that is caused by evaporation, but when using a large 50 × 24 mm coverslip this shrinkage is homogeneous and only requires adjustment of the focal plane. This setup is useful both for upright and inverted microscope systems. In our protocols a Bioptechs Inc. (USA) objective heater system is used.

2. Incubation Chamber

Modern incubation chambers enable not only the control of the temperature of the cell's environment but also allow the control of humidity and CO2 concentration during microscopic observation of mammalian cells in culture. Since yeast cells do not need CO2 for cell proliferation the latter is not required to cultivate the cells on the microscope stage. In our lab a small chamber including a heating block suitable for inverted microscopes is used (Pecon Inc., Germany). The device is compatible with the agar sheet immobilization technique and enables long-time observations using an autofocus system and the application of automated imaging approaches (see below). However, heating devices typically require a rather heavy heating block, which is not compatible with standard galvo-stages on confocal microscopes. Thus, for precise axial imaging we prefer the use of an objective heater system.

C. Image Acquisition

The setup described below is based on Leica SP2 (upright) or SP5 (inverted) confocal microscopes (Leica Microsystems, Germany). However, the yeast cell preparation techniques described above are compatible with any other imaging device (confocal or spinning disk). Due to the small size of yeast cells of 5–8 μm, special attention needs to be paid to the objectives and also to the image acquisition speed, to avoid photo damage upon extended 4D imaging and image distortions due to rapid intracellular movements. For high-resolution imaging of yeast cells in the aqueous medium we use a high numerical aperture oil immersion objective (e.g., 63x HCX PL APO CS 1.4 OIL). Typically, for imaging of cells in aqueous medium high numerical aperture water immersion objectives with a correction collar to compensate for errors in the thickness of the coverslip are used (e.g., 62x HCX PL APO CS63 1.2 WATER). However, particularly for long-time observation of different positions within a sample, water as an immersion medium may rapidly evaporate. Thus, when using a water lens for long-term 4D live cell imaging additional equipment for continuous water supply to the lens/coverslip interface may be required. To avoid phototoxicity and cell damage the cells are imaged with high scanning speed (line frequency of 700–1000 Hz), 512 × 512 pixels and typically in bidirectional scanning mode. Laser (illumination) power needs always to be reduced to a minimum. Using this setup acquired images and z-stacks will contain image noise, which, however, can be efficiently reduced in subsequent image processing steps (see II D, 1). The specimens are typically undersampled (f:~1.7) using a sampling rate of 78 × 78 × 240 nm, which reduces exposure time to avoid cell damage during long time observation. During image acquisition, a look-up table indicating over- and underexposed pixels is used to record fluorescently labeled LD within the dynamic range of the detector; the corresponding optimization of the detector gain at each time point is performed manually. Transmission images are recorded simultaneously with the fluorescence images.

Modern microscope stages enable the definition of user-defined positions within the sample. In this way, several sample positions should be sequentially imaged and observed over time, which significantly improves the statistical significance of the analysis. In addition to the common problems of inhomogeneous staining with lipophilic dyes and rapid bleaching, which limits their use for extended 4D imaging, there are also inherent problems to LD imaging that need to be considered during image acquisition. LD undergo a significant directed and oscillating movement, particularly in actively growing cells (Wolinski *et al.*, 2011). Thus, a slow scanning speed may result in improper imaging of moving LD. Also, LD are often closely associated, whereby the distance between the organelles may be below the resolution limit of the optical system. In such cases, detection and counting of LD as separate objects is not possible.

During planning of the experiment, the number of sample positions and the scanning and manipulation time need to be considered. Selection of proper time intervals between two records in 4D imaging experiments depends on the properties of the used strains and cultivation conditions. Thus, the number of time intervals should first be determined empirically. For long-term experiments, initially 30 min steps are defined and the behavior and growth rate of the cells are observed over a total time frame of 6 h. Subsequently, the time intervals may be reduced to a minimum.

D. Data Processing and Visualization

1. Prefiltering of 3D Data

Three-dimensional image data acquired by optical systems are generally degraded. Sources of image degradation are "noise" (signal dependent noise, and noise arising from the imaging system manifested as "salt-and-pepper" grainy noise) and "blur" (caused mainly by diffraction). Image blur particularly degrades image objects in axial direction; the shape and contour of the resulting diffraction pattern is a function of the optical system that is described by the point-spread-function (PSF) of the microscope (Wallace *et al.,* 2001). As a consequence, particularly visualization and analysis of 3D image data are compromised.

A frequently used strategy to decrease image noise in z-stacks is the application of nonlinear and linear 3D image filters such as 3D Gaussian and 3D Median filters. Such filters are included in many commercial and noncommercial software packages that are capable of processing 3D image data, and efficiently suppress image noise.

A more computing intensive strategy to remove both image blur and image noise in 3D image data is the application of image restoration (deconvolution) algorithms. In this respect, the iterative Maximum-Likelihood Estimation (MLE) method provided by Huygens2 Professional software (Scientific Volume Imaging B.V., The Netherlands) used in our lab improves the quality of the acquired 3D image data significantly. Although image deconvolution methods aim primarily on correcting image blur the used algorithm also efficiently reduces image noise.

Application of iterative image deconvolution may result in changes of the relative grey level intensities of acquired image objects (Swedlow, 2007). However, we found that a low number of 5–7 iterations and by using a theoretical PSF the contrast of imaged yeast LD is drastically improved without significant changes of relative object properties. The used algorithm is robust against undersampling with our standard setup (undersampling up to f: $\sim$ 1.7; Scientific Volume Imaging, Inc. 2010). Three-dimensional images improved by deconvolution are ideally suited for subsequent 3D visualization of LD. If due to practical reasons higher sampling distances are required the application of image deconvolution may be critical. In such cases, 3D Gaussian filtering is applied as an alternative method to image deconvolution. However, images processed with this linear filter do not reach the quality level obtained by full-scale deconvolution (Centonze and Pawley, 2006; Landmann, 2002).

2. Visualization of 4D Image Data

At different stages of cellular growth yeast LD are often closely associated and their size is frequently at the limit of optical resolution. This fact has to be considered when choosing a proper visualization method for the analysis of complex 4D image data. The use of polygon-based reconstruction methods, such as isosurface reconstructions based on the marching cubes algorithm (William *et al.,* 1987) and similar techniques are often inefficient to "resolve" all imaged LD in a 3D data set and usually lead to a loss of information. In contrast, the application of voxel rendering techniques avoids artifacts that are created by using polygon-based reconstructions. Modern texture-based volume rendering methods as implemented in amira® (Visage Imaging, Inc.) enable a direct 3D visualization of subcellular structures and fast interactive modification of view angles and view distances to facilitate the analysis of complex 3D and 4D image data. Particularly, features for variation of color and transparency within a defined pixel value range using editable color maps in combination with virtual lighting effects enables more precise rendering of complex spatial structures. As a consequence, the visualization of fine image details such as of closely associated and small LD is enhanced. Thus, volume rendering techniques

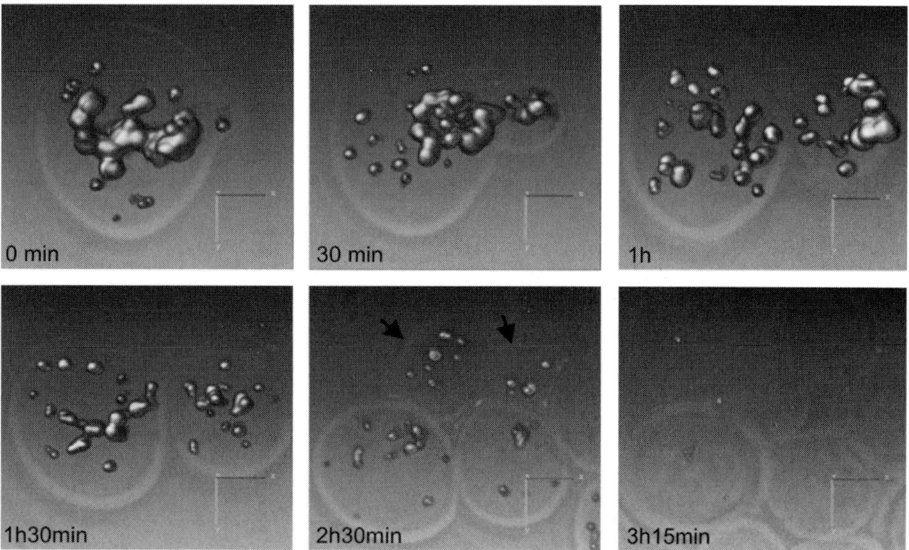

Fig. 2 Tracking BODIPY 493/503 labeled yeast LD during cellular growth and lipolysis (degradation of LD). To induce lipolysis, labeled yeast cells were inoculated into fresh medium in the presence of cerulenin (Kurat *et al.*, 2006), a drug efficiently blocking *de novo* synthesis of fatty acids and the formation of new LD. 4D live cell imaging series. Voxel rendering representations of selected time-points of the 4D data set. Particularly, specular components of a user defined virtual light source enables the visualization also of small LD and fine image detail. The scenes can be interactively viewed from any perspective. Notably, the 4D image series shows that a fraction of LD is inherited into growing daughter cells. The second generation of daughter cells is indicated by black arrows. The neutral lipid stocks are almost depleted 3 h 15 min after induction of lipolysis. Simultaneously acquired transmission images are projected with a transparency transfer function. (For interpretation of the references to color in this figure legend, the reader is referred to the web version of this book.)

allowing the interactive modification of a number of transfer parameters are highly useful for the visualization and analysis of yeast LD. To analyze the different records of a 4D data set we display volume rendering representation to separate viewers. Using a customized script user defined scenes and movies can be started simultaneously, which greatly facilitates comparative analysis of acquired z-stacks (Fig. 2).

IV. Imaging-Based Quantitative Analysis of Yeast LD in Large Cell Populations

The neutral lipid content of cells is typically analyzed qualitatively and quantitatively by chromatographic methods, such as thin layer chromatography (TLC) or high-performance liquid chromatography (HPLC) coupled to mass spectrometry. However, these approaches only provide average values of the lipid content in the cell population. As a consequence of the methodology, important information for the identification of malfunctions of cellular processes involved in neutral lipid homeostasis, such as the morphology and inheritance of LD, or the distribution of LD within a specific cell population is lost. Imaging-based quantification approaches for individual cells can provide this information and, therefore, complement population based analytical methods. Here we present a method for statistically relevant quantitative analysis of neutral lipid mass and distribution in heterogeneous yeast cell populations. The quantification procedure includes a highly efficient method for fluorescence labeling of yeast LD, automated extraction of LD from 3D image data sets using an algorithm, which is rather robust to changes of the detector gain settings, and a method for automated registration of yeast cells in conventional transmission images.

A. Cell Preparation for Microscopy

1. Preparation of Growth Medium and Cell Cultivation

Preparation of cells for quantitative LD imaging follows the same protocol as used for imaging based screens; see **III A, 1–2**.

2. Fixation and Fluorescence Labeling of LD

a. Grow cells to the desired growth stage in at least 2 mL of growth medium.
b. Remove 1 mL of the cell culture and add the cell culture into a 1.5 mL tube.
c. Spin at 3500 rpm for 3 min using a tabletop centrifuge and wash the pellet 2x with 1 mL of 0.5 M sorbitol/Tris.HCL pH 7.2). Sorbitol prevents the formation of large vacuoles during fixation, which may change the subcellular localization of LD.
d. Fix the cells for 15 min using 1 mL of 2% formaldehyde/0.5 M sorbitol/Tris.HCL pH 7.2 (final concentration). For fixation of log-phase cells reduce the concentration of formaldehyde to 1%–1.5%. Typical fixation artifacts are elongated and "fused" LD. In such cases it is recommended to further reduce the concentration of the fixative.
e. Add 1.2 μl of the BODIPY 493/503 stock solution (final concentration 1.2 μg/mL) to the cells in the fixation solution and label for 15 min in a rotating wheel.

3. Preparation of Agar Slides for Microscopy

a. Add Bacto™-agar (2% w/v) to 500 mL of distilled water in an appropriate flask. Stir for 15 min. To dissolve the agar completely autoclave at 121 °C for 20 min.
b. Melt previously prepared agar at 600 W in a microwave oven.
c. Pour 5 mL of the liquid agar in a 10 mL Falcon tube.
d. Equilibrate the temperature of the agar to 65 °C in a water bath.
e. Add 6 μl of a BODIPY 493/503 stock solution to the agar (final concentration 1.2 μg/mL) and vortex.

f. Pour the agar (~3.5 mL) containing BODIPY 493/503 on a standard microscope slide (76 × 26 mm). Take care that the edges of the slide are fully covered with agar.
g. Allow polymerization of the agar for 20 min at RT.
h. Immediately apply 1 μl of the fixed cells to the middle of the agar sheet and mount the cells with a large coverslip (50 × 24 mm) suitable for confocal imaging; place the slide on the microscope stage.
i. Alternatively, prepare a slide containing multiple sample fields (see III A, 4.).

B. Image Acquisition Setup

To acquire a larger number (~50–250) of cells at once yeast samples are imaged with a 40x NA1.25 oil immersion objective (e.g., HCX PL APO CS 40x 1.25 OIL) and a line scanning frequency of 500 Hz (1024 × 1024 pixels, 12 bit) in unidirectional scanning mode. Sampling is performed at 90 nm × 90 nm × 240 nm. A lookup table indicating over- and underexposed pixels is used to ensure imaging of fluorescently labeled LD within the dynamic range of the detector. Depending on the number of imaged cells, one to two image stacks are recorded per sample. Transmission images are recorded simultaneously. For detection of yeast cells in transmission images (see IV E) an optical section is selected showing maximum cell contrast, which is typically indicated by a homogeneous black and sharp cell boundary (usually the middle section of a z-stack). Correct Köhler illumination needs to be checked prior to an experiment to ensure an even illumination of the cells (Leica Microsystems Manual, 2010).

C. Data processing

Since the acquired data is undersampled (f: ~ 1.7) 3D Gaussian filtering is applied instead of using full-scale 3D deconvolution (see II D, 1.). Due to the high contrast of imaged LD additional image processing routines for contrast enhancement typically are not required.

D. Image Segmentation and Quantification of Image Objects

Quantitative analysis of fluorescence images is particularly challenging for the analysis of LD for several reasons. First, the distribution of neutral lipids is typically highly heterogeneous both in individual cells and within a cell population Secondly, fluorescent dyes may be efficiently pumped out of actively growing cells as a result of highly active pleotropic drug resistance pumps (Wolinski and Kohlwein, 2009a). In addition, differences in the labeling efficiency of LD between strains and also dependent on the growth media and growth phase may exist. Moreover, the output power particularly of older gas lasers used in the microscope may fluctuate over time and between different experiments. These phenomena result in significant differences of the grey level intensities of labeled LD in individual cells and in a cell population, resulting in variable fluorescence intensities and inaccurate quantification. Furthermore, since lipophilic dyes that are typically used to label LD also interact with phospholipids, this signal may add to the "background" but is unrelated to the neutral lipid content. As a consequence, imaging of LD for the comparative analysis of neutral lipid content in different yeast strains often requires optimization of the detector sensitivity to acquire most LD in a cell without loss of significant information.

In our approach, cells are first fixed and subsequently labeled with the neutral lipid-specific dye. Fixation enables a more homogeneous labeling of LD with high contrast, and overcomes staining limitations caused by the cell physiology. Secondly, z-stacks of a larger number of cells are imaged with high resolution, whereby LD are strictly imaged within the dynamic range of the detector. Subsequently, we apply an automated binarization algorithm, which is based on the detection of the variance of the intensity distribution of image objects. Such an algorithm is, for instance, implemented as a "factorization" algorithm in the quantification package of amira^{TM}. The algorithm is typically robust against changes in the absolute intensity values of acquired image objects and provides flexibility in optimizing the detector gain when different yeast samples are imaged.

Based on the binary image derived from the microscope, a variety of object properties can be computed. Amira^{TM} calculates 45 parameters to characterize image objects, including shape parameters. For quantitative analysis of neutral lipids or of the LD "mass" in a cell population we compute the sum of voxels of detected image objects in individual cells.

E. Cell Detection

For statistically relevant analysis of LD mass and distribution within a cell population the detection of individual cells is required. Several methods have been established for labeling yeast cell boundaries or the cytoplasm of the cells using fluorescence dyes and fluorescent protein fusions. For segmentation and registration of cells several commercial and noncommercial software tools are available (Carpenter *et al.*, 2006; Chen *et al.*, 2007; Ohtani *et al.*, 2004, 2005; Saito *et al.*, 2004; Suzuki *et al.*, 2004). In our lab, however, an alternative method for detection of yeast cells is applied. We use an image processing method for automated registration of yeast cells in simultaneously acquired transmission images (Bredies and Wolinski, submitted). Thus, possible limitations of fluorescence-based methods such as fluorescence quenching of the labeled molecules of interest, changes of the cell physiology or limitation of the number of spectrally separable fluorescence labels in multi-labeling experiments are avoided. In contrast to other related methods (Gordon *et al.*, 2007; Kvarnstrom *et al.*, 2008) the procedure described here enables detection of yeast cells also in very dense cell populations and without the need for differential interference contrast optics. The software tool runs on Matlab$^{®}$ (MatWorks, Inc.) platforms (tested for Matlab release R2010A) and includes a user-interface for handling of files and for optimization of registration parameters. The software assigns extracted 3D image objects to individual cells and provides statistical values (e.g., mass of LD in individual cells, perimeter of cells). In addition, the software tool includes a feature to create plots for the comparative analysis of different yeast strains. Notably, a feature for convenient processing of images in batch mode makes the software suitable for processing image data that are acquired in high throughput screening approaches, such as for the phenotypic characterization of the entire yeast deletion mutant collection (Wolinski *et al.*, 2009a; 2009b). The principle of the quantification procedure and a sample application are shown in Fig. 3.

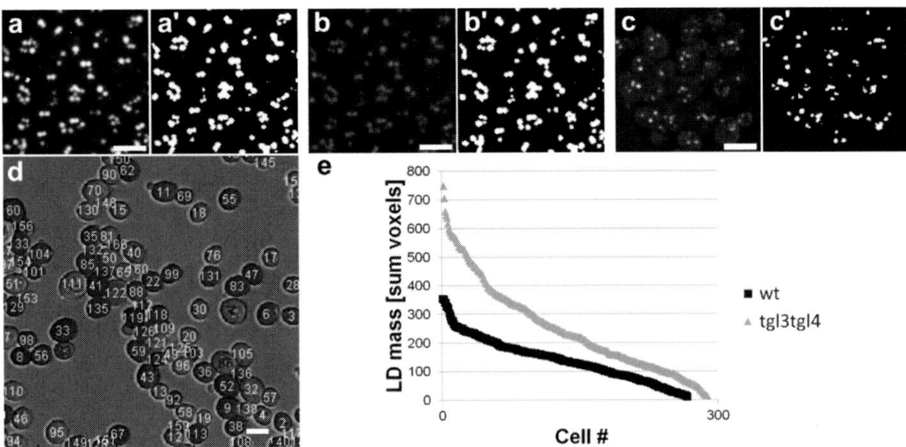

Fig. 3 Quantitative analysis of the amount of neutral lipids in yeast cell populations. BODIPY 493/503 labeled LD of yeast cells cultivated to stationary growth phase (a). Corresponding binary image of LD segmented using an algorithm that is based on detecting the variance of intensity distribution of an image object (a'). Same field of view but imaged with decreased detector gain settings (b). The algorithm is robust against changes of the detector settings, indicated by a similar extraction efficiency of LD (b'). The differences between the computed values for the LD surface and volume of LD between (a) and (b) are 3.7% and 13.2%. Exponentially growing wild-type yeast cells show small LD and a "background" of additionally labeled phospholipid membranes (c). Since the background is homogenous, a background correction in combination with automated segmentation enables precise extraction also of small LD (c'). Images represent maximum-intensity projections of z-stacks and were cropped from larges data sets. After extraction of image objects an algorithm is applied for the registration of yeast cells in conventional transmission images and for assigning extracted image objects (LD) to individual cells. The algorithm detects yeasts cells with high efficiency also in very dense populations of cells (d). As a proof of principle the quantification approach was applied to a comparative analysis of the LD mass in wild-type cells and in a double deletion mutant strain lacking the main yeast lipases *(tgl3 tgl4,* Athenstaed and Daum, 2003, 2005; Kurat *et al.*, 2006). In accordance with published lipid analysis data (Kurat *et al.*, 2006), the plot shows that the double deletion mutant strain accumulates significant more neutral lipids within a cell population compared to the wild-type. The LD mass (sum of voxels) of each detected cell was sorted and then plotted. Due to the high number of imaged cells (~280 per strain), only one image stack per strain was analyzed (e). Bar = 5 μm. (For color version of this figure, the reader is referred to the web version of this book.)

V. Label Free Imaging of yeast LD using CARS Microscopy

CARS microscopy enables the detection of molecules without the need for labels such as fluorescence dyes. A CARS signal is generated by excitation of the sample typically with two laser beams. If the frequency difference between the laser beams matches the frequency of a vibrational transition of a molecule of interest, scattered photons may be emitted with an energy increase that corresponds to the vibrational energy of the molecule (shorter wavelength; anti-Stokes shift). Due to the high density of molecular C-H vibrations, CARS microscopy is especially powerful for the detection of neutral lipids that are highly concentrated in LD (Debarre and Beaurepaire, 2007; Debarre *et al.,* 2006; Evans and Xie, 2008; Brackmann *et al.,* 2009; Le *et al.,* 2010b).

In CARS microscopy, two detection schemes are applied: forward generated CARS (F-CARS), and backward generated CARS (E(epi)-CARS). There are significant differences between these signals. The F-CARS signal is relatively insensitive to sample size or shape and significantly stronger than the E-CARS signal. Biological imaging based on F-CARS may be limited by a nonresonant background signal, which may overshadow weak signals of interest. In contrast, E-CARS is very sensitive to the size and shape of the sample and can be highly suppressed when imaging large objects, due to destructive interference. However, E-CARS shows less nonresonant background compared to F-CARS (Volkmer *et al.,* 2001; Zumbusch *et al.,* 1999).

A. CARS System Configuration

Our CARS system is based on a Leica SP5 confocal microscope (Leica Microsystems, Inc.). To generate a CARS signal a picosecond laser source with integrated OPO is used (picoEmerald; APE, Germany; HighQ Laser, Austria). The microscope is equipped with two nondescanned detectors (NDDs) for acquisition of signals in epi(E-)- and forward(F-) CARS modes (Volkmer *et al.,* 2001; Zumbusch *et al.,* 1999). Detection of the CARS signal is performed using suitable emission filters (650/210, 770SP). For imaging of LD the F-CARS mode is used in our experiments. To detect neutral lipids and LD the laser is tuned to 2845 cm^{-1}, thus enabling imaging of CH2 symmetric stretching vibration. The wavelength is set via a user interfaced that is included in the SP5 microscope control software.

B. Cell Preparation for CARS Microscopy

CARS microscopy enables the detection of a resonant CARS signal that is generated by a specifically excited chemical bond. For the detection of neutral lipids, the symmetric CH2 molecular vibration of fatty acids is excited at $\sim$2845 cm^{-1}. However, the signal from the molecules of interest may be impaired by the nonresonant signal of a solvent or in case of live cell imaging of yeast by the aqueous growth media (Chen *et al.*, 2011; Xie *et al.*, 2006). Fortunately, the nonresonant background produced by yeast media (e.g., rich medium or synthetic medium) does not significantly interfere with the detection of LD in F-CARS mode. Thus, yeast cells can be imaged directly in the particular growth media. For critical experiments and for imaging of very small LD cells are typically washed 3x with distilled water, which further reduces the nonresonant background. For detection of CARS signals both in E-CARS or F-CARS modes, the cells are mounted on standard microscope slides or on agar slides (see II A, 3). We found that the agar does not interfere with the detection of CH2 molecular vibrations at 2845 cm^{-1} in F-CARS mode.

C. Image Acquisition Setup

For CARS imaging of yeast LD we apply a 63x NA1.4 oil immersion objective (e.g., HCX PL APO CS 1.4 OIL) or a 63x NA1.2 water immersion lens with a correction collar to compensate for errors in the thickness of the coverslip (e.g., HXC PL APO CS63 1.2 W CORR).

For 2D imaging of yeast cells with optimized sampling rate a line frequency of 400 Hz (512×512 pixels) and uni-directional scanning mode is applied. Optionally, 8x or 16x line averaging is performed. Slower scan speed may result in light-driven artificial movement of LD or may lead to cell destruction. For 3D imaging of yeast cells an increased line frequency of 700 Hz (512×512 pixels) and uni-directional scanning mode is applied. With this setup no visible cell damage was observed. Notably, using this setup also small LD (~200 nm) can be imaged multi-dimensionally (Fig. 4). However, in 4D experiments over extended periods of time (see II.) significantly impaired cell growth, most likely caused by cell stress, has been observed, using CARS. Yeast LD can also be imaged at video-rate using a resonant scanner at 8 kHz line scanning frequency (Jüngst *et al.*, 2011). However, by using this fast scanning mode imaging of smaller LD (<200 nm) is limited.

One application of CARS microscopy is to use the resonant signal at 2845 cm^{-1} as a control for neutral lipids in fluorescently labeled specimens. Fluorescence dyes for labeling LD also stain phospholipids and membrane accumulations, which may occur in specific mutants. In such cases, the differentiation, based on fluorescence, between the "true" neutral lipid signal and the "artificial" membrane signal is difficult. However, the fluorescence dyes may also be excited by a two-photon process. To test whether the fluorescence or the CARS signal is detected the two lasers are alternately switched on and off, which ablates the CARS signal. Since the CARS signal is exponentially dependent on the concentration of CH_2 molecular vibrations at 2845 cm^{-1} the CARS signal is represented as its square root value for comparative quantitative analysis of signal heterogeneity within a cell population (Le *et al.*, 2010a). A limitation of quantitative CARS imaging of LD is the occurrence of "shadow" effects. The incident light beam may be scattered strongly at objects located above another one in the specimen, resulting in a dark shadows on

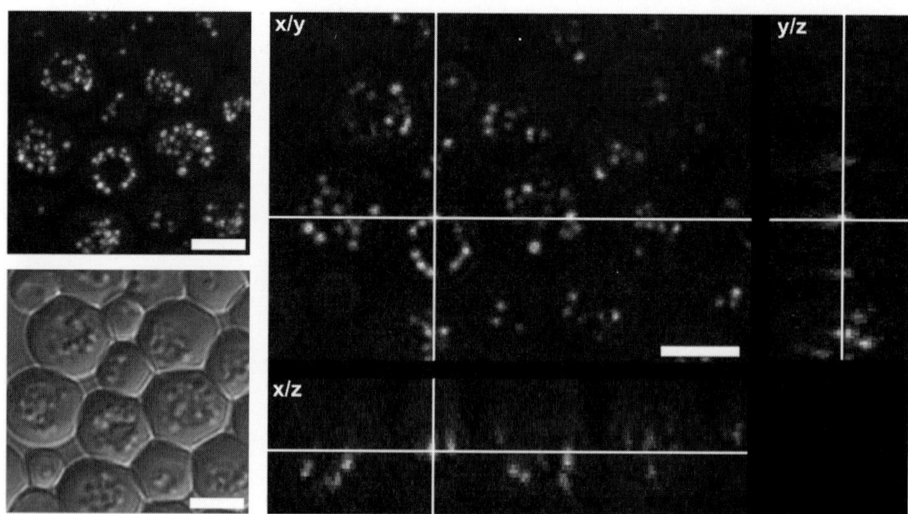

Fig. 4 Label-free detection of neutral lipids in yeast cells. CARS signal of CH2 molecule vibrations detected at 2845 cm^{-1} in a wild-type cell population. Maximum-intensity projection of an acquired z-stack (image top left) and sequentially acquired differential interference contrast (DIC) image (image bottom left). Axial view of a z-stack (image at right). Note the heterogeneity of the CARS signal within individual cells, indicating chemical heterogeneity, which is not observed in conventional fluorescence labeling experiments. Images were cropped from larger data sets. Bar = 5 μm. (For color version of this figure, the reader is referred to the web version of this book.)

the deeper localized object along the optical axis. Thus, consideration of this phenomenon and the analysis of single optical sections is critical for the evaluation of CARS signal heterogeneity (Chen *et al.,* 2011). CARS is an emerging imaging technology and currently largely restricted to optics laboratories; the increasing popularity in biological applications will soon provide more detailed experiences about applications and limitations of this technology in lipid imaging.

VI. Summary and Conclusions

The powerful genetics and ease of biochemical manipulation and cultivation make yeast an attractive reference system for biomedical studies. Its well-established role as a model organism is based on highly conserved biochemical and signaling pathways, up to mammalian cells. The small size of yeast cells, which for a long time has hampered microscopic approaches for detailed subcellular analyses, is not limiting anymore for today's highly sophisticated imaging instrumentation. On the contrary, the large number of yeast cells that can be imaged at once at and beyond the limits of optical resolution provides an excellent data pool for statistically significant quantitative imaging approaches. Especially, the quantitative and qualitative analysis of LD by imaging-based techniques, as a measure of physiological or pathophysiological processes of cellular lipid metabolism, has tremendous potential. Imaging-based screens of mutant collections under various physiological conditions will significantly contribute to better understanding the molecular basis of LD formation and degradation. These analyses are expected to yield fundamental insight into the medically highly relevant problem of cellular lipid homeostasis and associated diseases.

Acknowledgments

The authors thank the members of our laboratory for critically reading the manuscript and helpful suggestions. Work in the authors' laboratories is supported by grants from the Austrian Science Funds FWF (Project LIPOTOX) and the Austrian Federal Ministry for Science and Research (Project GOLD, in the framework of the Austrian Genome Program, GEN-AU).

References

Athenstaedt, K., and Daum, G. (2003). YMR313c/TGL3 encodes a novel triacylglycerol lipase located in lipid particles of Saccharomyces cerevisiae. *J. Biol. Chem.* **278**, 23317–23323.

Athenstaedt, K., and Daum, G. (2005). Tgl4p and Tgl5p, two triacylglycerol lipases of the yeast Saccharomyces cerevisiae are localized to lipid particles. *J. Biol. Chem.* **280**, 37301–37309.

Athenstaedt, K., and Daum, G. (2006). The life cycle of neutral lipids: synthesis, storage and degradation. *Cell Mol. Life Sci.* **63**, 1355–1369.

Brackmann, C., Norbeck, J., Akeson, M., Bosch, D., Larrson, C., Gustafsson, L., and Enejder, A. (2009). CARS microscopy of lipid stores in yeast: The impact of nutritional state and genetic background. *J. Raman Spectrosc.* **40**, 748–756.

Carpenter, A. E., Jones, T. R., Lamprecht, M. R., Clarke, C., Kang, I. H., Friman, O., Guertin, D. A., Chang, J. H., Lindquist, R. A., Moffat, J., Golland, P., and Sabatini, D. M. (2006). CellProfiler: image analysis software for identifying and quantifying cell phenotypes. *Genom. Biol.* **7**, R100.

Chen, B. C., Sung, J., Wu, X., and Lim, S. H. (2011). Chemical imaging and microspectroscopy with spectral focusing coherent anti-Stokes Raman scattering. *J. Biomed. Opt.* **16**, 021112.

Chen, S. C., Zhao, T., Gordon, G. J., and Murphy, R. F. (2007). Automated image analysis of protein localization in budding yeast. *Bioinformatics* **23**, i66–i71.

Centonze, V., and Pawley, J. B. (2006). *In* "Handbook of Biological Confocal Microscopy," (J. P. Pawley, ed.), pp. 627–649. Springer, New York.

Debarre, D., and Beaurepaire, E. (2007). Quantitative characterization of biological liquids for third-harmonic generation microscopy. *Biophys. J.* **92**, 603–612.

Debarre, D., Supatto, W., Pena, A. M., Fabre, A., Tordjmann, T., Combettes, L., Schanne-Klein, M. C., and Beaurepaire, E. (2006). Imaging lipid bodies in cells and tissues using third-harmonic generation microscopy. *Nat. Methods* **3**, 47–53.

Evans, C. L., and Xie, X. S. (2008). Coherent anti-stokes Raman scattering microscopy: Chemical imaging for biology and medicine. *Annu. Rev. Anal. Chem. (Palo Alto Calif)* **1**, 883–909.

Fei, W., Shui, G., Gaeta, B., Du, X., Kuerschner, L., Li, P., Brown, A. J., Wenk, M. R., Parton, R. G., and Yang, H. (2008). Fld1p, a functional homologue of human seipin, regulates the size of lipid droplets in yeast. *J. Cell Biol.* **180**, 473–482.

Gerlich, D., Beaudouin, J., Gebhard, M., Ellenberg, J., and Eils, R. (2001). Four-dimensional imaging and quantitative reconstruction to analyse complex spatiotemporal processes in live cells. *Nat. Cell Biol.* **3**, 852–855.

Gerlich, D., and Ellenberg, J. (2003). 4D imaging to assay complex dynamics in live specimens. *Nat. Cell Biol.* S14–S19.

Gordon, A., Colman-Lerner, A., Chin, T. E., Benjamin, K. R., Yu, R. C., and Brent, R. (2007). Single-cell quantification of molecules and rates using open-source microscope-based cytometry. *Nat. Methods* **4**, 175–181.

Greenspan, P., and Fowler, S. D. (1985). Spectrofluorometric studies of the lipid probe, nile red. *J. Lipid Res.* **26**, 781–789.

Greenspan, P., Mayer, E. P., and Fowler, S. D. (1985). Nile red: a selective fluorescent stain for intracellular lipid droplets. *J. Cell Biol.* **100**, 965–973.

Guo, Y., Cordes, K. R., Farese Jr., R. V., and Walther, T. C. (2009). Lipid droplets at a glance. *J. Cell Sci.* **122**, 749–752.

Guo, Y., Walther, T. C., Rao, M., Stuurman, N., Goshima, G., Terayama, K., Wong, J. S., Vale, R. D., Walter, P., and Farese, R. V. (2008). Functional genomic screen reveals genes involved in lipid-droplet formation and utilization. *Nature* **453**, 657–661.

Ivnitski-Steele, I., Holmes, A. R., Lamping, E., Monk, B. C., Cannon, R. D., and Sklar, L. A. (2009). Identification of Nile red as a fluorescent substrate of the *Candida albicans* ATP-binding cassette transporters Cdr1p and Cdr2p and the major facilitator superfamily transporter Mdr1p. *Anal. Biochem.* **394**, 87–91.

Jandrositz, A., Petschnigg, J., Zimmermann, R., Natter, K., Scholze, H., Hermetter, A., Kohlwein, S. D., and Leber, R. (2005). The lipid droplet enzyme Tgl1p hydrolyzes both steryl esters and triglycerides in the yeast, Saccharomyces cerevisiae. *Biochim. Biophys. Acta* **1735**, 50–58.

Jüngst, C., Winterhalder, M. J., and Zumbusch, A. (2011). Fast and long term lipid droplet tracking with CARS microscopy. *J. Biophotonics* **4**, 435–441.

Koffel, R., Tiwari, R., Falquet, L., and Schneiter, R. (2005). The Saccharomyces cerevisiae YLL012/YEH1, YLR020/YEH2, and TGL1 genes encode a novel family of membrane-anchored lipases that are required for steryl ester hydrolysis. *Mol. Cell Biol.* **25**, 1655–1668.

Kohlwein, S. D. (2010). Triacylglycerol homeostasis: Insights from yeast. *J. Biol. Chem.* **285**, 15663–15667.

Kurat, C. F., Natter, K., Petschnigg, J., Wolinski, H., Scheuringer, K., Scholz, H., Zimmermann, R., Leber, R., Zechner, R., and Kohlwein, S. D. (2006). Obese yeast: Triglyceride lipolysis is functionally conserved from mammals to yeast. *J. Biol. Chem.* **281**, 491–500.

Kvarnstrom, M., Logg, K., Diez, A., Bodvard, K., and Kall, M. (2008). Image analysis algorithms for cell contour recognition in budding yeast. *Opt. Express* **16**, 12943–12957.

Landmann, L. (2002). Deconvolution improves colocalization analysis of multiple fluorochromes in 3D confocal data sets more than filtering techniques. *J. Microsc.* **208**, 134–147.

Le, T. T., Duren, H. M., Slipchenko, M. N., Hu, C. D., and Cheng, J. X. (2010a). Label-free quantitative analysis of lipid metabolism in living Caenorhabditis elegans. *J. Lipid Res* **51**, 672–677.

Le, T. T., Yue, S., and Cheng, J. X. (2010b). Shedding new light on lipid biology with coherent anti-Stokes Raman scattering microscopy. *J. Lipid Res.* **51**, 3091–3102.

Martin, S., and Parton, R. G. (2006). Lipid droplets: a unified view of a dynamic organelle. *Nat. Rev. Mol. Cell Biol.* **7**, 373–378.

Ohtani, M., Saka, A., Sano, F., Ohya, Y., and Morishita, S. (2004). Development of image processing program for yeast cell morphology. *J. Bioinform. Comput. Biol.* **1**, 695–709.

Petschnigg, J., Wolinski, H., Kolb, D., Zellnig, G., Kurat, C. F., Natter, K., and Kohlwein, S. D. (2009). Good fat, essential cellular requirements for triacylglycerol synthesis to maintain membrane homeostasis in yeast. *J. Biol. Chem.* **284**, 30981–30993.

Saito, T. L., Ohtani, M., Sawai, H., Sano, F., Saka, A., Watanabe, D., Yukawa, M., Ohya, Y., and Morishita, S. (2004). SCMD: Saccharomyces cerevisiae Morphological Database. *Nucleic Acids Res.* **32**, D319–D322.

Scientific Volume Imaging, Inc. (2010). Huygens Professional User Manual.

Spandl, J., White, D. J., Peychl, J., and Thiele, C. (2009). Live cell multicolor imaging of lipid droplets with a new dye, LD540. *Traffic* **10**, 1579–1584.

Suzuki, M., Asada, Y., Watanabe, D., and Ohya, Y. (2004). Cell shape and growth of budding yeast cells in restrictive microenvironments. *Yeast* **21**, 983–989.

Swedlow, J. R. (2007). Quantitative fluorescence microscopy and image deconvolution. *Methods Cell Biol.* **81**, 447–465.

Szymanski, K. M., Binns, D., Bartz, R., Grishin, N. V., Li, W. P., Agarwal, A. K., Garg, A., Anderson, R. G., and Goodman, J. M. (2007). The lipodystrophy protein seipin is found at endoplasmic reticulum lipid droplet junctions and is important for droplet morphology. *Proc. Natl. Acad. Sci. U.S.A.* **104**, 20890–20895.

Vizeacoumar, F. J., van Dyk, N., F, S. V., Cheung, V., Li, J., Sydorskyy, Y., Case, N., Li, Z., Datti, A., Nislow, C., Raught, B., Zhang, Z., Frey, B., Bloom, K., Boone, C., and Andrews, B.J. (2010). Integrating high-throughput genetic interaction mapping and high-content screening to explore yeast spindle morphogenesis. *J. Cell Biol.* **188**, 69-81.

Volkmer, A., Cheng, J. X., and Xie, X. S. (2001). Vibrational imaging with high sensitivity via epidetected coherent anti-Stokes Raman scattering microscopy. *Phys. Rev. Lett.* **87**, 23901–23904.

Wallace, W., Schaefer, L. H., and Swedlow, J. R. (2001). A workingperson's guide to deconvolution in light microscopy. *Biotechnique.* **31**, 1076–1078 1080, 1082 passim.

Walther, T. C., and Farese Jr., R. V. (2009). The life of lipid droplets. Biochim. *Biophys. Acta* **1791**, 459–466.

Welte, M. A. (2009). Fat on the move: intracellular motion of lipid droplets. *Biochem. Soc. Trans.* **37**, 991–996.

Wolinski, H., and Kohlwein, S. D. (2008). Microscopic analysis of lipid droplet metabolism and dynamics in yeast. *Methods Mol. Biol.* **457**, 151–163.

Wolinski, H., Natter, K., and Kohlwein, S. D. (2009a). The fidgety yeast: Focus on high-resolution live yeast cell microscopy. *Methods Mol. Biol* **548**, 75–99.

Wolinski, H., Petrovic, U., Mattiazzi, M., Petschnigg, J., Heise, B., Natter, K., and Kohlwein, S. D. (2009b). Imaging-based live cell yeast screen identifies novel factors involved in peroxisome assembly. *J. Proteome Res* **8**, 20–27.

Wolinski, H., Kolb, D., Hermann, S., Koning, R. I., and Kohlwein, S. D. (2011). A role for seipin in lipid droplet dynamics and inheritance in yeast. *J Cell Sci.* **124**, 3894–3904.

Xie, X. S., Cheng, J. X., and Potma, E. (2006). Coherent Anti-Stokes Raman Scattering Microscopy. *In* "Handbook of Biological Confocal Microscopy," (J. P. Pawley, ed.), pp. 595–606. Springer, New York.

Zanghellini, J., Wodlei, F., and von Grunberg, H. H. (2010). Phospholipid demixing and the birth of a lipid droplet. *J. Theor. Biol.* **264**, 952–961.

Zumbusch, A., Holtom, G. R., and Xie, X. S. (1999). Three-dimensional vibrational imaging by coherent anti-Stokes Raman scattering. *Phys. Rev. Lett.* **82**, 4142–4145.

CHAPTER 17

Analysis of Cholesterol Trafficking with Fluorescent Probes

Frederick R. Maxfield* **and Daniel Wüstner**†

*Department of Biochemistry, Weill Cornell Medical College, New York, USA

†Department of Biochemistry and Molecular Biology, University of Southern Denmark, Odense M, Denmark

Abstract

Cholesterol plays an important role in determining the biophysical properties of biological membranes, and its concentration is tightly controlled by homeostatic processes. The intracellular transport of cholesterol among organelles is a key part of the homeostatic mechanism, but sterol transport processes are not well understood. Fluorescence microscopy is a valuable tool for studying intracellular transport processes, but this method can be challenging for lipid molecules because addition of a fluorophore may alter the properties of the molecule greatly. We discuss the use of fluorescent molecules that can bind to cholesterol to reveal its distribution in cells. We also discuss the use of intrinsically fluorescent sterols that closely mimic cholesterol, as well as some minimally modified fluorophore-labeled sterols.

Copyright 2012, Elsevier Inc. All rights reserved.

0091-679X/10 $35.00
DOI 10.1016/B978-0-12-386487-1.00017-1

Methods for imaging these sterols by conventional fluorescence microscopy and by multiphoton microscopy are described. Some label-free methods for imaging cholesterol itself are also discussed briefly.

ABBREVIATIONS

BHK, baby hamster kidney; BODIPY, Bora-diaza-indacene; Bchol, BODIPY-cholesterol; CHO, Chinese hamster ovary; CARS, coherent anti-Stokes Raman scattering; CTL, cholestatrienol; DHE, dehydroergosterol; EMCCD, electron multiplying charge couple device; ERC, endocytic recycling compartment; GUV, giant unilamellar vesicle; LD, lipid droplet; LSO, lysosomal storage organelle; MβCD, methyl-β-cyclodextrin; MP, multiphoton; NPC, Niemann-Pick disease type C; ORP, oxysterol-binding protein-related protein; PBS, phosphate buffered saline; SIMS, Secondary ion mass spectrometry

I. Introduction

Cholesterol is an essential component of mammalian cell membranes, and it plays an important role in determining the biophysical characteristics of these membranes (Mesmin and Maxfield, 2009; Wüstner, 2009). The concentration of cholesterol varies greatly among organelles with levels around 30% of the lipid molecules in the plasma membrane and about 5% in the endoplasmic reticulum. Cholesterol is transported among organelles by a mixture of vesicular and non-vesicular transport processes, and the mechanisms regulating this transport are only partially understood. Cholesterol transport in cells can be studied by following radiolabeled cholesterol, but this requires stringent purification of organelles under conditions in which the sterol does not redistribute. Contamination with a small fraction of membranes with high cholesterol content can lead to significant errors in measurements of cholesterol content in organelles such as the endoplasmic reticulum.

Fluorescence microscopy has been a powerful tool for studying the intracellular transport of proteins. A difficulty in studying transport of cholesterol (and other lipids) using microscopy is that coupling to a fluorophore can dramatically change the properties of the molecule and its interactions with other components of the membrane. Recently, two types of approaches for studying fluorescent sterols have been implemented. One approach has been to use an added fluorophore selected to minimize the perturbation as compared to cholesterol itself. Bora-diaza-indacene (BODIPY)-cholesterol (Fig. 1) has been used as a cholesterol probe in model membranes and in trafficking studies in living cells (Hölttä-Vuori *et al.*, 2008), (Shaw *et al.*, 2006), (Chiantia *et al.*, 2008), (Ariola *et al.*, 2009), (Wüstner *et al.*, 2011a). The second approach takes advantage of intrinsically fluorescent sterols (Fig. 1) (Hao *et al.*, 2002), (Yeagle *et al.*, 1982), (Fischer *et al.*, 1974), (Wüstner, 2010), (McIntosh *et al.*, 2008), (Mukherjee *et al.*, 1998), including dehydroergosterol (DHE; a natural sterol found in yeast) and cholestatrienol (CTL; a synthetic

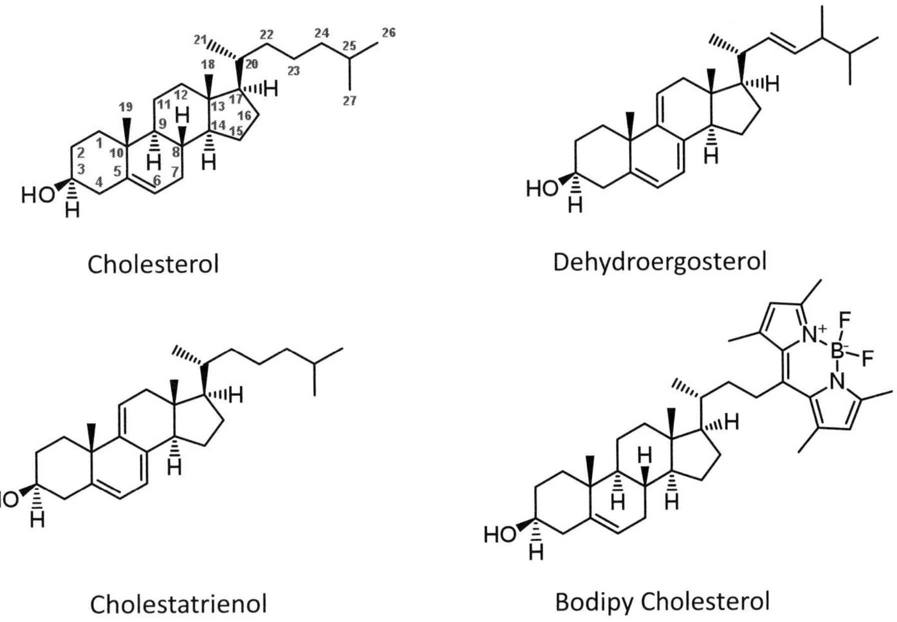

Cholesterol

Dehydroergosterol

Cholestatrienol

Bodipy Cholesterol

Fig. 1 Chemical structures of cholesterol, dehydroergosterol, cholestatrienol, and Bodipy-cholesterol. (For color version of this figure, the reader is referred to the web version of this book.)

sterol that is close to cholesterol in its structure and in many of its biophysical properties). Both of these fluorescent sterols have two additional double bonds in the steroid rings, which creates the fluorophore. Fluorescent sterols can be added to cells either by delivery to the plasma membrane or incorporated into reconstituted lipoproteins. The movement of the sterols through the cells can then be observed directly, or they can be used for techniques such as fluorescence recovery after photobleaching to determine transport kinetics (Hao *et al.*, 2002; Wüstner *et al.*, 2005; Wüstner *et al.*, 2002).

In this chapter, we discuss the relative merits of these fluorescent sterols, and we discuss the methodology for their use. We also discuss some recent developments in the label-free detection of sterols in cells. We begin by discussing the use of fluorescent sterol-binding molecules that are useful for analyzing the distribution of cholesterol in fixed cells or tissues.

A. Visualization of Cholesterol Using Sterol-Binding Probes

Sterol-binding natural products have been used to localize cholesterol in cells, and their use has been described in detail (Gimpl, 2010). Filipin is a naturally fluorescent polyene antibiotic that binds to cholesterol but not to esterified sterols. Thus, it is useful for detecting free (i.e., unesterified) cholesterol in biological membranes. Filipin fluorescence is observed with UV excitation around 360 nm and emission around 480 nm. Filipin binding perturbs the bilayer structure, so filipin cannot be used on living cells. Beyond the free 3'-OH group, the exact basis for filipin's specificity is not fully understood. There are some important practical considerations in the use of filipin. Stock solutions in DMSO must be rigorously dried with Molecular Sieves to remove residual water. Filipin is rapidly photobleached with the UV light intensity available in most fluorescence microscopes. However, with a good camera, we have been able to attenuate the incident light using neutral density filters to 1–10% of the full brightness and obtain good images that photobleach slowly. Excessive photobleaching without the use of a neutral density filter is one of the greatest problems in reproducibility of results using filipin. Since the basis for filipin binding is not fully understood, it must be considered that there can be interfering effects. Nevertheless, we have been able to correlate brightness of filipin labeling with chemically measured cholesterol levels in cells in which cholesterol content was altered by incubation with methyl-β-cyclodextrin (MβCD) or with cholesterol:MβCD complexes (Qin *et al.*, 2006). Filipin treatment causes dimpling of the membrane that can be observed by electron microscopy (Orci *et al.*, 1981). It should be noted, however, that filipin deformation of the membrane can be affected by membrane:protein interactions (Steer *et al.*, 1984). A recent paper discussed the use of filipin to label brain sections from mice with lysosomal storage disorders (Arthur *et al.*, in press). It was found that in addition to cholesterol, filipin was also labeling the GM1 ganglioside. This points to the importance of verifying the validity of sterol-labeling reagents in each experimental system. In addition, particular care should be taken in cells that express large amounts of GM1.

Pore-forming cytolysins that bind to sterols have also been used to visualize cholesterol in cells. These polypeptide bacterial toxins bind to cholesterol in membranes and self-associate to form pores in the bilayer. For fluorescence microscopy the toxins can be labeled with a fluorescent dye and imaged with filter sets appropriate for the dye. Some investigators have preferred the use of these toxins, including perfringolysin-O or a biotinylated derivative of this called BC-toxin, because the fluorescent dyes (or the labeled avidin for BC-toxin) are more photostable than filipin. One issue is that the perfringolysin-O binding to membranes increases very nonlinearly as cholesterol content increases (Sokolov and Radhakrishnan, 2010). It appears that the binding is preferentially to cholesterol molecules with a high chemical activity coefficient. The cholesterol content at which the rapid increase in binding is observed depends on the lipid content of the membrane. For ER lipids, this occurs at about 5% cholesterol, but in other membranes binding is seen above 20% cholesterol (Gimpl, 2010).

B. Automated Microscopy for Cholesterol High Throughput Screens

Filipin labeling has been used for many years for diagnosis of Niemann-Pick Disease type C (NPC), an inherited disorder leading to cholesterol accumulation in lysosomal storage organelles (LSOs), which are modified late endosomes and lysosomes (Peake and Vance, 2010). Recently, filipin labeling of the cholesterol accumulation in LSOs has been used for high throughput automated microscopy screening of compounds that might reduce the cholesterol accumulation (Pipalia *et al.*, 2006; Pipalia *et al.*, 2011). In this application, imaging is essential because the total cholesterol level in the NPC mutant cells does not change greatly. Nevertheless, differences caused by the mutation are easily quantified by microscopy because the filipin-labeled LSOs accumulate in the peri-nuclear region, while the rest of the cell actually has lower cholesterol levels than normal cells (Pipalia *et al.*, 2006).

Cells are plated in 384-well plates, treated with various concentrations of test compounds, fixed with 1.5% paraformaldehyde, and labeled with 50 μg/mL filipin (from a 25 mg/mL stock in dry DMSO) in phosphate buffered saline (PBS) for 45 min. The cells are then rinsed and imaged on an automated microscopy system (ImageXpress^{Micro} from Molecular Devices or equivalent) using UV excitation and a dry 10× objective. Automated image analysis is used to quantify the fluorescence power per cell in areas that are above a threshold brightness corresponding to the brightness in LSOs of untreated cells. Further details are provided elsewhere (Pipalia *et al.*, 2006; Pipalia *et al.*, 2011). Screens based on this procedure have identified compounds that reduce the cholesterol levels in NPC mutant cells by various mechanisms (Pipalia *et al.*, 2011), (Rosenbaum *et al.*, 2009), (Rujoi *et al.*, 2010), (Rosenbaum and Maxfield, 2011).

C. Biophysical Properties of Cholesterol and its Fluorescent Analogs

Valid fluorescent analogs of cholesterol should reproduce its biophysical interactions with lipids in the bilayer as closely as possible. Cholesterol has a small headgroup (the 3'-β-hydroxyl group) that can create hydrogen bonds to other lipid headgroups in the interfacial region of the bilayer, while its hydrophobic steroid ring system and isooctyl side chain are buried in the hydrophobic region of the membrane (Ohvo-Rekila *et al.*, 2002), (Nagle and Tristram-Nagle, 2000), (Hofsass *et al.*, 2003). Cholesterol molecules are shielded under the phospho- and sphingolipid headgroups, thereby minimizing contact with water molecules. This shielding, together with attractive van der Waals interactions with neighboring acyl chains, causes lipid membranes to have a reduced area in the presence of cholesterol. This property is described as cholesterol's condensing effect (Hofsass *et al.*, 2003), (Lindahl and Edholm, 2000), (Henriksen *et al.*, 2006).

When cholesterol is added to a membrane consisting of phospholipids with saturated acyl chains, (e.g., dipalmitoylphosphatidylcholine) below the phase transition temperature of the phospholipid, the sterol will perturb the high structural order in the liquid-crystalline state (Vist and Davis, 1990). In contrast, when added to membranes in the liquid-disordered (l_d) phase, cholesterol has an ordering effect, and it can induce a liquid-ordered (l_o) phase at high sterol mole fractions (approx. above 30 mol% – that is, 30% of the lipid molecules) (Ipsen *et al.*, 1987). The l_o phase is characterized by high lateral lipid mobility, straightened acyl chains and tight lipid packing. The l_o phase has attracted interest, due to similar properties observed in cellular membranes (Ahmed *et al.*, 1997), (Mukherjee and Maxfield, 2004), (Munro, 2003). Despite this resemblance, it has to be emphasized that the l_o phase is strictly defined only for simple two and three-component lipid mixtures at thermodynamic equilibrium. Partitioning of fluorescent sterols between l_o and l_d phase as well as their potential to induce the l_o phase can be used as criteria to assess their potential as cholesterol mimic (Wüstner *et al.*, 2011a; Garvik *et al.*, 2008; L. Solanko, D. Wüstner *et al.*, manuscript in preparation).

D. Properties of Fluorescent Sterols

1. Dehydroergosterol and Cholestatrienol

Both intrinsically fluorescent sterols, DHE and CTL, contain three conjugated double bonds in the steroid ring system giving these probes their slight fluorescence in the near UV-region of the spectrum. Since they contain an identical chromophore, the photo-physical properties of DHE and CTL are nearly identical (Smutzer *et al.*, 1986), (Rogers *et al.*, 1979), (Chong and Thompson, 1986), (Yeagle *et al.*, 1990). Both sterols have excitation and emission maxima in membranes of $\lambda_{ex} = 320$ nm and $\lambda_{em} = 370$–400 nm (Fischer *et al.*, 1974; Rogers *et al.*, 1979; Schroeder *et al.*, 1988; Hyslop *et al.*, 1990). The fluorescence lifetime of both sterols is short, around $\tau_f = 0.3$–0.8 ns in various organic solvents and lipid membranes at room temperature (Smutzer *et al.*, 1986; Chong and Thompson, 1986; Hyslop *et al.*, 1990; Fischer *et al.*, 1985a). Low extinction coefficient ($\varepsilon \approx 11{,}000$ $M^{-1}\cdot cm^{-1}$) and quantum yield ($\Phi_f = 0.04$ in ethanol) results in low fluorescence brightness of these sterols. The environmental sensitivity of DHE and CTL is low since the molecular dipole does not change greatly during electronic transition from the ground S_0, state to the excited S_1 state (Chong and Thompson, 1986; Yeagle *et al.*, 1990; Hyslop *et al.*, 1990; Fischer *et al.*, 1985a). In dimyristoylphosphatidylcholine liposomes, the fluorescence lifetime, τ_f, of DHE decreases in a sigmoidal fashion from $\tau_f = 2$ ns at 10 °C to $\tau_f = 0.5$ ns at 42 °C (Chong and Thompson, 1986). It has been reported that DHE and CTL self-quench at increasing mole fractions in phosphatidylcholine model membranes, as indicated by a drop in fluorescence intensity (Schroeder *et al.*, 1988; Schroeder *et al.*, 1987). Since no change in fluorescence lifetime was observed in these studies, it is likely that the proposed self-quenching is caused by static quenching (Lakowitz, 2006). By measuring fluorescence intensity of DHE in giant unilamellar vesicles (GUVs), we found no evidence for self-quenching but instead a linear relationship between DHE's emission and the mole fraction in the membranes (Garvik *et al.*, 2008). Thus, quenching effects, which could complicate analysis of sterol distribution by fluorescence microscopy of DHE or CTL are unlikely, but systematic studies with varying lipid composition might be necessary to rule out any sterol self-quenching at concentrations used for studies in cells. Fluorescence anisotropy and quantum yield of DHE are very temperature-sensitive and much lower in membranes above than below the phase transition temperature (Smutzer *et al.*, 1986; Chong and Thompson, 1986; Fischer *et al.*, 1985a). Accordingly, under live cell imaging conditions the fluorescence brightness of both sterols should be further reduced. Another unfortunate property of DHE and CTL, at least for cellular imaging, is the high photobleaching propensity of these probes. How this can be avoided or used to advantage for microscopic investigation will be discussed later in this chapter.

Due to their minimal chemical alterations, DHE and CTL resemble the natural sterols ergosterol and cholesterol very closely. DHE differs from ergosterol only in one double bond, while CTL contains two more double bonds in the ring system than cholesterol. Accordingly, DHE is a naturally occurring sterol in yeast and red sponges, and it is the ideal mimic of ergosterol. CTL is the closest analog of cholesterol (Rogers *et al.*, 1979; Wüstner, 2007a). The close resemblance of DHE and CTL to ergosterol and cholesterol is reflected by their biophysical properties in model membranes. DHE can, like ergosterol and cholesterol, induce the l_o phase in ternary lipid mixtures and partitions with high preference into that phase (Garvik *et al.*, 2008). CTL also prefers the l_o over the l_d phase in GUV membranes (Baumgart *et al.*, 2007). CTL has a very similar potential to order acyl chains in membranes as cholesterol even at high sterol mole fraction, while DHE is slightly less efficient (Scheidt *et al.*, 2003). Similarly, DHE (like ergosterol, but in contrast to cholesterol) shows a concentration saturation effect in various biophysical effects on lipid membranes. That means that bilayer properties such as bending elasticity and ordering of phospholipid acyl chains depend on sterol mole fraction only up to about 10 mol% DHE in the membranes (Henriksen *et al.*, 2006; Garvik *et al.*, 2008; Scheidt *et al.*, 2003; Henriksen *et al.*, 2004). In contrast, acyl chain ordering was found to depend linearly on cholesterol and CTL concentrations in the membrane, even at high sterol mole fractions (Henriksen *et al.*, 2006; Scheidt *et al.*, 2003). The less flexible side chain of DHE and ergosterol as compared to CTL and cholesterol is likely responsible for this difference. We found that DHE and CTL have very similar intracellular trafficking itineraries to cholesterol, suggesting that the minor differences in biophysical properties of the analogs are not determining their overall transport in mammalian cells (Wüstner and Færgeman, 2008a), (Hartwig Petersen *et al.*, 2008), (Mondal *et al.*, 2009). Interestingly, in a study in yeast cells, we observed different metabolism of DHE and ergosterol compared to cholesterol, indicating that DHE is the most suitable sterol analog in organisms containing ergosterol (Kohut *et al.*, 2011).

2. BODIPY-Cholesterol

Recently, cholesterol analogs containing a BODIPY fluorophore have been synthesized by Bittman and co-workers (Li *et al.*, 2006). The BODIPY moiety is particularly suitable for fluorescent lipid probes since it is relatively non-polar, allowing for insertion of the analogs into the hydrophobic interior of lipid membranes (Marks *et al.*, 2008). The BODIPY dye is electrically neutral, has a low environmental sensitivity, low Stokes shift (typically, λ_{ex} = 505 nm and λ_{em} = 515 nm), high extinction coefficient, and high quantum yield ($\Phi_f \approx 0.9$ in organic solvents) (Bergström *et al.*, 2002). Several BODIPY dyes exhibit a red-shifted excited-state dimer (excimer) at high concentration in membranes (Chen *et al.*,

1997), (Puri *et al.*, 2001), (Pagano *et al.*, 1991). This has been used to determine lateral clustering and sorting of these lipid probes in cells by quantitative fluorescence microscopy (Puri *et al.*, 2001), (Puri *et al.*, 1999), (Sharma *et al.*, 2003), (Sharma *et al.*, 2004). BODIPY-cholesterol (BChol) with the dye at carbon 24 of the sterol side chain is a promising new cholesterol analog, which can supplement the well-established intrinsically fluorescent probes DHE and CTL (Hölttä-Vuori *et al.*, 2008), (Shaw *et al.*, 2006), (Chiantia *et al.*, 2008), (Ariola *et al.*, 2009), (Wüstner *et al.*, 2011a). BChol partitions preferentially into the l_o phase compared to the l_d phase in various ternary lipid mixtures (Shaw *et al.*, 2006; Ariola *et al.*, 2009), though a direct comparison with DHE revealed that the latter sterol has an even higher preference for the l_o phase (Wüstner *et al.*, 2011a). Since Bchol is more than 500-fold brighter than DHE, it can be used at very low concentrations (about 0.1–0.5 mol% of lipids) (Hölttä-Vuori *et al.*, 2008; Wüstner *et al.*, 2011a). In contrast, visualization of DHE in cells may require replacing up to 5% of the sterols (Wüstner *et al.*, 2011a; Wüstner *et al.*, 2002). BChol has been used to investigate cholesterol mobility in the l_o and l_d phase in GUV's made of dioleoylphosphatidylcholine, egg yolk sphingomyelin, and cholesterol (Ariola *et al.*, 2009). The BODIPY-moiety of this BChol is oriented perpendicular to the bilayer normal (and the acyl chains), as measured by two-photon fluorescence polarization (Ariola *et al.*, 2009; L. Solanko, D. Wüstner *et al.*, manuscript in preparation), and the strength of this orientation is enhanced in the presence of cholesterol in the membranes (L. Solanko, D. Wüstner *et al.*, manuscript in preparation). BChol self-quenches at concentrations above 3 mol% in lipid membranes, but it does not form red-shifted excimers (Wüstner *et al.*, 2011a). Instead, dark ground state dimers seem to be responsible for the self-quenching effect, as reported for some other BODIPY-tagged probes (Wüstner *et al.*, 2011a; Bergström *et al.*, 2002; Qin *et al.*, 2005). Molecular dynamics simulations of BChol indicate that it has a higher molecular tilt compared to cholesterol (Hölttä-Vuori *et al.*, 2008). This can affect the local lipid structure surrounding the probe and might contribute to its lower partition into the l_o phase compared to DHE (Hölttä-Vuori *et al.*, 2008; Wüstner *et al.*, 2011a). The molecular tilt has been proposed to be a main determinant of the ability of a sterol to order phospholipid acyl chains and condense lipid bilayers (Aittoniemi *et al.*, 2006). For the intrinsically fluorescent sterols, DHE and CTL, the order parameter, S, which is inversely related to the molecular tilt, could be measured by fluorescence polarization spectroscopy, since the transition dipole of these sterols lies along the long molecular axis (Chong and Thompson, 1986; Fischer *et al.*, 1985a). The order parameter was found to be only a little lower for DHE compared to cholesterol ($S = 0.67$ for DHE and $S = 0.85$ for cholesterol in dimyristoylphosphatidylcholine at 37 °C)(Fischer *et al.*, 1985a; Fischer *et al.*, 1985b; Marsan *et al.*, 1999).

3. NBD- and Dansyl-cholesterol

Cholesterol tagged with a 7-nitrobenz-2-oxa-1,3-diazole (NBD)-group at carbon 22 or carbon 25 has been used in model membrane and cellular trafficking studies in yeast and in mammalian cells (Shaw *et al.*, 2006; Mukherjee *et al.*, 1998; Scheidt *et al.*, 2003; Craig *et al.*, 1981; Reiner *et al.*, 2005). A problem with these probes is their up-side down orientation in model membranes compared to cholesterol and intrinsically fluorescent sterols, as well as their low ordering capacity and partitioning into the l_d phase in ternary model membranes (Scheidt *et al.*, 2003; Loura *et al.*, 2001). NBD-cholesterol with the fluorophore at carbon 25 has been shown to be mistargeted in cells to mitochondria (Mukherjee *et al.*, 1998). Dansyl-cholesterol is another fluorescent cholesterol analog used in cellular studies (Wiegand *et al.*, 2003; Huang *et al.*, 2010). The Dansyl moiety was linked to carbon 6 of the steroid ring system, and recent fluorescence studies found that the Dansyl-group of this sterol is localized on average 1.56 nm from the bilayer center (Shrivastava *et al.*, 2009). This should significantly affect the lipid acyl chain packing in proximity of this probe when inserted into membranes. Partitioning of Dansyl-cholesterol between l_o and l_d phases in model membranes has not been reported. Quantitative studies of intracellular sterol distribution based on fluorescence of NBD- and Dansyl-cholesterol are also hampered by the high environmental sensitivity of the attached fluorophores. Accurate measurement of sterol distribution requires that emitted fluorescence is proportional to probe concentration, and this is not likely for these cholesterol probes (Benson *et al.*, 1985). Further details about NBD- and Dansyl-cholesterol can be found elsewhere (Gimpl, 2010; Wüstner, 2007a).

E. Transport of Fluorescent Cholesterol Probes in Cells

1. Live-Cell Imaging of Intrinsically Fluorescent Sterols

Both DHE and CTL are suitable for studies in living cells. DHE is widely available from commercial sources, although concerns have been expressed about the purity of this material, which may vary depending on the supplier and the lot number (McIntosh *et al.*, 2008). CTL is not available commercially, but the method for synthesis has been published (Fischer *et al.*, 1984; Fellmann *et al.*, 1994). Like many lipids, these fluorescent sterols are subject to oxidation, so they should be protected from exposure to air (e.g., by purging solvents with argon). They are also sensitive to light and should be stored in the dark. The purity can be checked by HPLC (McIntosh *et al.*, 2008). Of particular concern, oxidized DHE or CTL may affect the structure of lipid bilayers.

Several methods have been used to incorporate DHE or CTL into cells. For simplicity, we will describe methods for DHE, but the same methods would apply for CTL. The simplest method is to inject DHE in an ethanolic stock solution into the culture medium. The DHE is very poorly soluble in water; some of it will adsorb to serum proteins, but most will form microcrystals. These microcrystals may be taken up by the cells and slowly dissolved to allow the DHE to distribute into cell membranes. In our experience, this procedure results in very heterogeneous labeling of cells and incomplete breakup of the microcrystals. A much better procedure involves preparation of DHE complexes with MβCD, which solubilizes the sterol

Box 1
Labeling cells with filipin

1. Prepare a stock solution of filipin (25 mg/mL) in dimethylsulfoxide that has been treated with Molecular Sieves to remove water. This solution can be stored as aliquots at −20 °C in tightly capped containers in a dessicated box. Thaw each aliquot in a dessicated container. Discard after use; do not refreeze.
2. Rinse cells with buffered saline, and fix with paraformaldehyde (1.5%).
3. Rinse again with buffered saline.
4. Incubate cells with filipin at a final concentration of 50 μg/mL in buffered saline for 45 minutes at room temperature.
5. Rinse cells three times with buffered saline.
6. Acquire images using 360/40 nm excitation and 480/40 nm emission filters with a 365 nm dichroic long pass filter. (Other similar UV filters should work.) It is very important to use a low level of excitation. Depending on lamp, filters, etc., you can attenuate excitation by 90–99% with a neutral density filter. This reduced excitation will slow photobleaching.

and allows it to be rapidly exchanged into the plasma membrane (Box 2). Sterol: MβCD complexes form at a 1:2 ratio (Breslow and Zhang, 1996), but these complexes can dissociate rapidly. Thus, it is necessary to maintain an excess of the MβCD in order to store sterol: MβCD complexes without precipitation of the DHE.

In normal cell physiology, cholesterol is delivered to cells as cholesteryl esters in the core of lipoproteins. DHE-oleate esters can be synthesized and incorporated into

Box 2
Formation of DHE: MβCD complexes and labeling of cells

Dissolve 5 mg of DHE in 2.5 ml of ethanol to give a 5 mM stock solution. Transfer to a 30 ml clean glass vial and evaporate the ethanol under argon to produce a thin film. Add 2.5 ml of 25 mM MβCD in buffered saline to get a DHE/MβCD ratio of 1:5. Vortex repeatedly to resuspend the DHE film. (Warming to 37 °C may help to release the dried film.) Sonicate for 10 min, and then shake at 37 °C overnight. Centrifuge for 10 min at 21,000 x g to remove undissolved DHE, and then aliquot and store at 4 °C under argon. The solution can be used for labeling the cells without further dilution. The labeling solution can be stored at 4 °C, but it should be centrifuged to remove DHE crystals just before use.

Cells are rinsed and incubated with the labeling solution at 37 °C for 0.5-1 minute to allow exchange of the DHE into the plasma membrane. The cells are then rinsed and returned to culture medium.

the core of reconstituted LDL (Wüstner *et al.*, 2005). Established procedures for reconstituting LDL can be modified to incorporate DHE-oleate into the core of LDL or acetylated-LDL (Krieger, 1986). We have used this method to incorporate a mixture of DHE-oleate and cholesteryl-oleate into acetylated-LDL at a molar ratio of 1:10, using 6 mg of total neutral lipid per 1.9 mg of AcLDL protein (Wüstner *et al.*, 2005).

The LDL can be taken into cells by receptor-mediated endocytosis via the LDL receptor. For uptake by macrophages, the reconstituted LDL can be acetylated, and the acetylated-LDL is taken up via scavenger receptor type A. The esters are hydrolyzed in late endosomes and lysosomes, releasing the unesterified DHE. We have verified that DHE esters are hydrolyzed, transported out of the digestive organelles, and distributed among cellular membranes (Wüstner *et al.*, 2005). DHE and CTL are also effective substrates for the esterifying enzyme, ACAT, which forms steryl esters that are incorporated into cytoplasmic lipid droplets, and these can be visualized in cells (Wüstner *et al.*, 2005).

Although in this chapter we emphasize studies of mammalian cells in culture, *Caenorhabditis elegans* can be labeled with DHE by dietary feeding (Matyash *et al.*, 2001; Wüstner *et al.*, 2010). The yeast, *Saccharomyces cerevisiae,* can also take up DHE when grown under hypoxic conditions (Georgiev *et al.*, 2011), and the distribution and transport of the DHE can be analyzed by microscopy. For yeast, DHE is a naturally formed sterol, and it is also a natural component of the diet of *Caenorhabditis elegans*.

2. Measurement of Transport Kinetics by Fluorescence Recovery after Photobleaching

Photobleaching of DHE can be a problem when several frames of the same field of view are acquired (e.g., in time-lapse imaging or when determining the effect of a drug or fluorescence quencher). Photobleaching of DHE can be used to advantage, however, for dynamic measurements of sterol transport. After selectively destroying DHE fluorescence by illumination of only a small region with closed field aperture on a wide field microscope, fluorescence recovery of DHE into that region was measured over time with the field aperture opened to image the whole cell (Hao *et al.*, 2002; Wüstner *et al.*, 2005; Wüstner *et al.*, 2002; Wüstner, 2007b). To quantify fluorescence recovery, DHE intensity in the bleached region was normalized to total cell intensity, thereby correcting for fluorescence loss during repeated acquisition.

3. Transbilayer Distribution of Fluorescent Sterols

Transbilayer asymmetry is a general feature of most lipids in the plasma membrane and many other organelles. This asymmetry has important consequences for membrane physical properties and cell signaling. Although cholesterol is a major lipid in these membranes, its transbilayer distribution is not well understood. Fluorescent sterols such as DHE and CTL can be used with fluorescence quenchers that do not cross the bilayer to determine the sterol asymmetry in membranes (McIntosh *et al.*, 2008; Mondal *et al.*, 2009; Hale and Schroeder, 1982). In most such studies, it has been found that the abundance of DHE is greater on the cytoplasmic leaflet of the plasma membrane than on the exofacial leaflet. Using imaging methods described here, it is also possible to examine the DHE transbilayer distribution in organelles. When the membrane impermeant quencher, trinitrobenzene sulfonic acid, was microinjected into DHE-labeled cells, about 60% of the fluorescence in both the endocytic recycling compartment and the plasma membrane was quenched (Hartwig Petersen *et al.*, 2008). When the trinitrobenzene sulfonic acid was added outside the cell, only 20–30% of the fluorescence was quenched.

4. Metabolism of DHE in Cells

DHE-oleate that is delivered to lysosomes in the core of LDL particles is hydrolyzed, presumably by lysosomal acid lipase, the lysosomal enzyme that hydrolyzes cholesteryl esters (Wüstner *et al.*, 2005). DHE is also esterified by the endoplasmic reticulum enzyme, ACAT, and stored in lipid droplets. Esterification of DHE can be verified by extracting cellular neutral lipids and using HPLC to quantify free and esterified DHE (Hölttä-Vuori *et al.*, 2008; Wüstner *et al.*, 2005). Incorporation into lipid droplets can be measured by fluorescence microscopy by determining the colocalization of DHE with a lipid droplet vital stain such as Nile Red or Lipidtox (Wüstner *et al.*, 2011a; Wüstner *et al.*, 2005; Wüstner and Færgeman, 2008b). While DHE mimics many aspects of cholesterol in cells, it is not recognized by the regulatory proteins, SCAP and Insig, in the endoplasmic reticulum (A. Radhakrishnan, personal communication).

F. Imaging Modalities for Visualization of Intrinsically Fluorescent Sterols

1. UV-Sensitive Wide Field Imaging

Excitation of DHE requires UV wavelengths around 330 nm, and the emission is around 400 nm. Both of these wavelengths impose special requirements as compared to typical epi-fluorescence microscopy. Most fluorescence microscopes contain glass elements that block transmission of light below 340 nm, and this greatly diminishes the excitation of DHE unless modifications are made. The most important modification is in the lamp housing. Conventional collecting lenses are a few cm thick, and they completely block transmission at 330 nm. Most research grade microscopes can be modified to incorporate UV-transmitting collecting lenses. An issue with single element collecting lenses is that the focal points for UV and visible wavelengths are very different, so optimal Köhler illumination cannot be obtained for the visible and UV simultaneously. We have obtained a multi-lens collector from Leica that corrects illumination adequately for wavelengths from 335 to 630 nm. If fiber optical systems are used for excitation, it is essential to verify that they have high transmittance at 330 nm.

Dichroic filter cubes for DHE imaging are available from various suppliers. We have used 335 nm (20 nm bandpass) excitation filter/365 nm longpass dichromatic filter/405 nm (40 nm bandpass) emission filter from Chroma (Brattleboro, VT). A microscope objective must also be chosen that has significant transmission below 340 nm. Finally, various glass elements in the optical path must be checked for transmission at 330 nm and replaced if necessary. For example, some infrared-blocking (heat) filters have poor transmission at 330 nm and must be replaced.

Because DHE fluorescence is weak and in the near UV, it places special requirements on the detector. The sensitivity of many CCD cameras falls off sharply around 400 nm, but detectors are available with coatings that extend sensitivity into the near UV. An important development is the availability of near-UV sensitive electron multiplying CCD (EMCCD) cameras. These cameras can provide a large increase in signal-to-noise ratio for detecting DHE as compared to previous versions of commercially available cameras. Using these near-UV sensitive EMCCD cameras, excitation light intensity can be reduced by 80–90% while still obtaining high quality images with less than 1 s exposure time. This allows repeated observations of the same cell for time course studies or for obtaining images at multiple focal positions.

2. Image Post-Processing for UV-Sensitive Wide Field Microscopy of DHE and CTL

Despite the advances in imaging technology described above, reliable detection of sub-cellular DHE and CTL distribution remains a challenge due to the weak fluorescence of these sterol probes. One potential problem of wide field imaging is a significant contribution of out-of-focus light (Swedlow, 2007). Image restoration methods based on deconvolution can improve both axial and lateral resolution in wide field fluorescence images (Swedlow, 2007), (Young, 1989), (Sibarita, 2005). The performance of any deconvolution algorithm is limited by spherical aberration, image noise and by photobleaching of the fluorophore as multiple images are obtained at different focal planes (Sibarita, 2005; Markham and Conchello, 2001). We have shown that iterative deconvolution in combination with photobleaching correction can significantly improve the performance of DHE imaging (see Fig. 2) (Wüstner and Færgeman, 2008b).

Chromatic aberration caused by the wavelength dependence of the microscope optical glass elements is a potential problem in multicolor imaging with DHE and organelle markers. While good research objectives are normally corrected for chromatic aberration in the visible range of the spectrum, they fail to collect UV light at the same focal position as red and green light. This can be corrected for by acquiring z-stacks of multicolor fluorescent beads, measuring the axial and lateral off-set between the UV channel and red or green channels and correcting for these off-sets in a post-processing step (Wüstner and Færgeman, 2008b). Software has been developed that performs a nonlinear regression of various decay models to photobleaching-induced intensity loss on a pixel-by-pixel basis (Wüstner *et al.*, 2010). This analysis revealed that DHE bleaches homogenously throughout cells, in stark contrast to other fluorescently labeled sterols, for example NBD-cholesterol

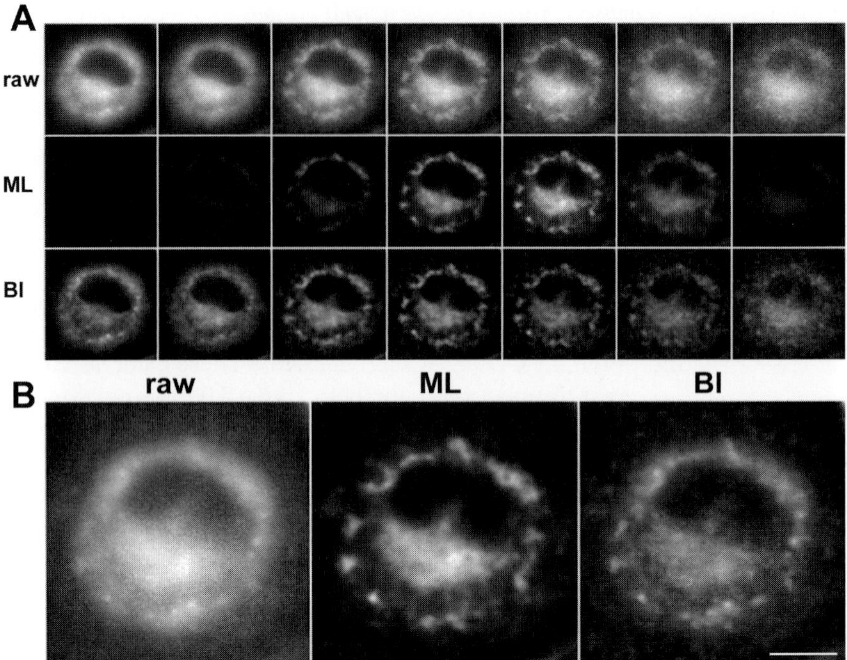

Fig. 2 Image deconvolution improves analysis of DHE distribution in living cells. J774 macrophages were incubated with 10 μg/mL acetylated LDL for 6 h to induce foam cell formation, washed and labeled for 5 min at 37 °C with DHE/ MβCD, as described in Box 1 (Wüstner et al., 2005). Cells were washed, chased for 30 min and imaged on a UV-sensitive wide field microscope. Ten images were acquired along the optical axis with 0.5 μm distance between the frames. The images were corrected for photobleaching along the stack, as described (Wüstner and Færgeman, 2008b). A, central 7 frames of the raw image stack, upper panel ("raw"), after applying maximum likelihood deconvolution implemented in Huygens software (Scientific Volume Imaging, Hilversum, The Netherlands) with a theoretical point spread function (PSF), middle panel ("ML"), the bleach-corrected image stack after applying a blind deconvolution as implemented in Autoquant (Media Cybernetics, Inc., Silver Spring, MD, USA), lower panel ("BI"). The latter method estimates the PSF directly from the data. B, maximum intensity projection of the image stack shown in A for all three methods. Deconvolved images appear less blurred than the raw data set. Note that the ML deconvolution is most noise-suppressive and gives the best results for deconvolution of DHE image sets (Wüstner and Færgeman, 2008b). Bar, 5 μm.

(Benson *et al.*, 1985; Wüstner *et al.*, 2010). Moreover, DHE bleaching follows a simple mono-exponential decay, and the measured decay rate is comparable in model and cellular membranes and quite independent of DHE intensity (Wüstner *et al.*, 2010). Accordingly, bleach-rate imaging can be used as a fingerprint to detect DHE in the presence of other spectrally indistinguishable fluorophores. For example, we showed that bleach-rate imaging of DHE allows for selective detection of sterol enriched tissue in *Caenorhabditis elegans* despite the presence of significant autofluorescence (Wüstner *et al.*, 2010; Wüstner and Sage, 2010).

3. Multiphoton Microscopy

As an alternative to conventional epifluorescence microscopy, intrinsically fluorescent sterols can be monitored in cells using multiphoton (MP) excitation. In this imaging modality, almost simultaneous absorption of two or more photons causes an electronic transition in the fluorophore from the ground to the first excited state. The absorbed photons have approximately one-half (for two-photon excitation) or one-third of the energy (for three-photon excitation) compared to the one-photon excitation process. Thus, as a rough rule of thumb, excitation of a fluorescent probe by two- or three-photon microscopy can be stimulated using twice or three times the wavelength used for one-photon excitation. For example, DHE is excited around 320 nm by a one-photon process, and it can be excited with 920 nm light for MP microscopy (McIntosh *et al.*, 2008), (Wüstner *et al.*, 2010), (McIntosh *et al.*, 2003), (Frolov *et al.*, 2000), (Zhang *et al.*, 2005), (Wüstner *et al.*, 2011b). The longer excitation wavelength has several advantages for UV microscopy including less cytotoxicity and photobleaching, deeper specimen penetration and less light scattering. Apart from these advantages, the key benefit of using MP microscopy is its intrinsic sectioning capability since multiphoton events only occur in the focal plane (Wüstner, 2010), (Zipfel *et al.*, 2003), (Masters and So, 2008), (Xu and Zipfel, 2008), (Schrader *et al.*, 1997).

Schroeder, Webb, Gratton and colleagues, pioneered MP excitation microscopy of DHE (Frolov *et al.*, 2000). Using an excitation wavelength of $\lambda ex = 920$–930 nm, these authors visualized DHE in L-cell fibroblasts and found the sterol mainly in the plasma membrane and in lipid droplets (LDs) (Frolov *et al.*, 2000). The published images were, however, of low quality (i.e., low signal-to-noise ratio, SNR), and no direct comparison to non-stained cells was made. The extremely low signal of DHE in MP microscopy is mostly a consequence of the low propensity of DHE for multiphoton excitation (Wüstner, 2010; Wüstner *et al.*, 2011b). Accordingly, high laser powers are required to obtain the necessary photon density at the focal point. We compared wide field and MP microscopy of DHE directly in the same samples and found that both methods provide similar information about sterol distribution (Wüstner *et al.*, 2011b). Average laser powers in the range of 40–70 mW were required for MP excitation of DHE at $\lambda ex = 920$ nm in CHO and HepG2 cells, and even then the images of DHE-stained cells were of low SNR. For comparison, two-photon excitation of enhanced green fluorescent protein at the same excitation wavelength required a laser power of only 0.3 mW (Wüstner *et al.*, 2011b). One could also try to excite DHE more efficiently by a two-photon process using excitation wavelengths $\leq$700 nm. Our attempts to do that in cells, however, were unsuccessful due to overwhelming autofluorescence. Photobleaching of DHE is restricted to the focal plane in MP microscopy. Accordingly, we were able to acquire many frames (typically between 50 and 100) without significant intensity loss in MP microscopy of DHE. Although individual frames were of low SNR, averaging many frames increased the SNR dramatically. The frame-averaged MP image of

DHE has a higher lateral resolution than a corresponding single-frame image of the same specimen acquired on an epifluorescence microscope.

As has been reported for other fluorescent lipids, we did not find evidence for lateral domains of DHE in the plasma membrane of living cells by deconvolution of wide field microscopy images or by MP microscopy (Wüstner and Færgeman, 2008a; Wüstner, 2007b; Wüstner *et al.*, 2011b). By both methods, fine cell protrusions in various cell types and sterol-containing nanotubes could be observed by microscopy of DHE. The need for frame averaging dramatically lowers the time-resolution of MP microscopy of DHE. While one image of DHE stained cells with good SNR can be acquired in less than 0.5 s by widefield microscopy, a comparable image consists of about 30 averaged frames in MP microscopy, which requires between 1 and 2 min total acquisition time (depending on the pixel-dwell time) (Wüstner *et al.*, 2011b). Accordingly, dynamic events occurring on a shorter time scale cannot be monitored and will result in motion-blurring. Despite the slow acquisition, this strategy has been used to record z-sections of DHE labeled CHO cells, and the sterol distribution in three dimensions could be observed (Wüstner *et al.*, 2011b).

In MP microscopy of DHE, a photomultiplier tube operating in photon-counting mode is used as the detector. In photon-counting the main source of noise in detection is the statistical uncertainty in the number of detected photons in a fixed time interval (Yazdanfar and So, 2008). Computational progress in denoising routines allowed us to post-process individual frames of MP sequences of DHE stained cells achieving dramatic improvement in image quality (F. Lund, D.Wüstner *et al.*, manuscript in preparation; Luisier, 2010). After denoising MP microscopy images, we could follow the dynamics of individual vesicles containing DHE in living CHO cells with a time-resolution of 4 s (F. Lund, D. Wüstner *et al.*, manuscript in preparation). We found similar types of vesicle dynamics by wide field microscopy of DHE stained HepG2 and J774 cells (Wüstner *et al.*, 2002; Wüstner and Færgeman, 2008a). Since intracellular vesicles must contain significant amounts of DHE (about 3–5 mol%) to be detectable by either wide field or MP microscopy, both imaging modalities will only catch dynamics of sterol-enriched vesicles.

4. Analysis of Intracellular Sterol Transport Using Fluorescence Microscopy of Bodipy-Cholesterol

As discussed earlier, BChol is the only fluorescent cholesterol analogue with an extrinsic fluorescence moiety that partitions preferentially into the l_o phase in model membranes (Ariola *et al.*, 2009; Wüstner *et al.*, 2011a). Thus, it fulfills an essential requirement of any suitable fluorescent analog of cholesterol. Co-labeling of M19 cells, a partial sterol-auxotroph Chinese hamster ovary (CHO) cell-line, with ^{3}H-cholesterol and BChol followed by sucrose density fractionation indicated a comparable distribution of cholesterol and BChol among cell membranes of varying equilibrium density (Hölttä-Vuori *et al.*, 2008). Efflux of BChol from CHO cells to extracellular acceptors like BSA or apoA1 is significantly higher than that of cholesterol (Hölttä-Vuori *et al.*, 2008). This is likely a consequence of the less efficient packing of phospholipid acyl chains around the BODIPY-moiety, as suggested by molecular dynamics simulations (Hölttä-Vuori *et al.*, 2008) and by studies in model membranes (Shaw *et al.*, 2006; Ariola *et al.*, 2009; Wüstner *et al.*, 2011a). BChol becomes esterified by Raw264.7 macrophages after incubation with AcLDL, though with a lower efficiency than cholesterol (Hölttä-Vuori *et al.*, 2008). By directly comparing transport of DHE and BChol, we found that both sterols are targeted to the endocytic recycling compartment (ERC), a major cellular sterol pool, with identical kinetics in baby hamster kidney (BHK) and in HeLa cells (Wüstner *et al.*, 2011a). Uptake of DHE and BChol from the cell surface was strongly reduced in BHK cells overexpressing a dominant-negative clathrin heavy chain, and co-internalization of BChol with fluorescent transferrin, a marker for this uptake pathway, could be demonstrated (Wüstner *et al.*, 2011a). These observations suggest that some plasma membrane sterol is internalized by clathrin-dependent endocytosis in these cells. By fluorescence lifetime imaging, we demonstrated that BChol's emission is constant throughout the cell, such that the fluorescence of this probe is a reliable measure of sterol concentration (Wüstner *et al.*, 2011a).

BChol has also been used to study sterol trafficking in CHO cells with defective Niemann-Pick C1 protein and to determine the role of oxysterol-binding protein-related proteins (ORPs) in cholesterol transport (Hölttä-Vuori *et al.*, 2008; Jansen *et al.*, 2011). In the latter study, an impact of ORPs on transport of BChol from the plasma membrane to the ER and to LDs has been suggested using HeLa cells treated with oleic acid to stimulate droplet formation (Jansen *et al.*, 2011). Importantly, we found that BChol is preferentially targeted to LDs in oleic-acid treated HeLa and BHK cells compared to DHE (Wüstner *et al.*, 2011a). Low free cholesterol content of oleic-acid induced LDs in HeLa cells was recently confirmed by filipin staining by another group (Du *et al.*, 2011). Thus, the preferred targeting of BChol to LDs compared to cholesterol or DHE indicates that the BODIPY-fluorophore affects the intracellular transport of BChol in cells with elevated fat content (Wüstner *et al.*, 2011a).

Based on our experience, we suggest using BChol exclusively for analysis of endocytic sterol trafficking, for example between plasma membrane and ERC. Moreover, we recommend performing co-localization studies of BChol with DHE by multi-color wide field microscopy, and only under conditions in which the probe distributions coincide, can one use the much better fluorescence properties of BChol for studies of sterol transport (Wüstner *et al.*, 2011a). This is exemplified in Fig. 3, where cells were labeled with BChol using similar protocols as described for DHE in Box 2, and imaged on a multiphoton microscope. The intrinsic sectioning capability of this technique combined with negligible bleaching propensity outside the focal region enabled us to follow vesicles containing BChol over several hundred frames (F. Lund, D. Wüstner *et al.*, manuscript in preparation). In this particular example, two dimly fluorescent vesicles form close to the brightly stained plasma membrane and fuse over a time course of a few minutes (Fig. 3B). If DHE would have been used

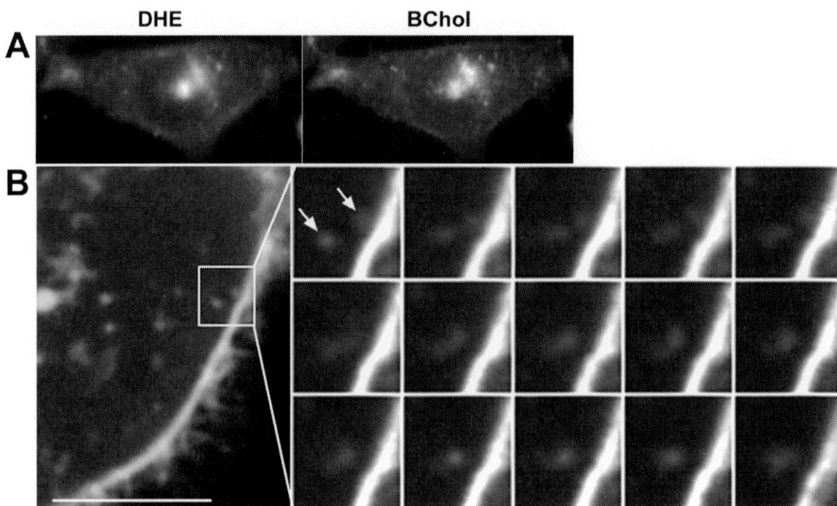

Fig. 3 Multi-color sterol imaging and multiphoton time-lapse microscopy of BChol. A, CHO cells were pulse-labeled with DHE and BChol using a sterol/ MβCD complex, washed and chased for 30 min at 37 °C before imaging at a UV-sensitive wide field microscope, as described (Wüstner *et al.*, 2011a). Left panel, DHE; middle panel, BChol; right panel, color overlay with DHE in red and BChol in green, respectively. B, HeLa cells were labeled with BChol/ MβCD for 1 min at 37 °C, washed and chased for 20 min at 37 °C and placed on the nitrogen-floated stage of a home-built two-photon microscope. Images were acquired every 1 sec with an excitation wavelength of 930 nm for a total of 425 frames. Zoomed box in the large panel (left) shows two small BChol-containing vesicles (arrows), the upper just formed at the cell membrane (bright line on right half of the zoomed region). Over a time course of 15 sec, these two vesicles fuse (time montage on right panel). Bar, 10 μm in A, and 5 μm in B. (See color plate.)

instead of BChol, such vesicles would be hard to detect. Due to the much higher molecular brightness of BChol, we can detect it in intracellular membranes even after several rounds of vesicle fusion and fission (F. Lund, D. Wüstner *et al.*, manuscript in preparation). In summary, BChol is useful as supplement to imaging of intrinsically fluorescent sterols in cells. In addition to its much better photophysical properties, fluorescence imaging of BChol can be combined with electron microscopy studies making use of the diaminobenzidine reaction (Rohrl *et al.*, 2011). Upon illumination of BChol in glutaraldehyde or formaldehyde fixed cells, reactive oxygen species cause photo-oxidation of diaminobenzidine, thereby creating an electron-dense precipitate in sub-cellular regions where BChol is located (Rohrl *et al.*, 2011; Ellinger *et al.*, 2010). By this method, BChol was found in tubular endosomes, multivesicular bodies and the trans-Golgi network in human HepG2 hepatoma cells. The same approach has been used to follow the intracellular fate of BChol-oleate incorporated into high density lipoprotein (Rohrl *et al.*, 2011). A potential problem is the necessary fixation step, which, depending on the fixative and experimental conditions, can alter the intracellular sterol distribution.

G. Label-Free Imaging of Sterols in Model Membranes and Living Cells

Ideally, one would like to visualize intracellular sterol distribution without the need for introducing labeled sterol probes. Two approaches for this that are being developed are reviewed here.

1. Sterol Imaging by Mass Spectrometry

Imaging by mass spectrometry has been used to determine the steady state distribution of cholesterol and other membrane lipids based on chemically specific ionization patterns. Secondary ion mass spectrometry (SIMS) allows for detecting cholesterol and other biomolecules based on characteristic fragmentation patterns. Matrix-enhanced SIMS and time-of-flight SIMS were used to detect phospholipids and cholesterol in neuroblastoma cells and in tissue slices and to visualize cholesterol in cell attachment sites (Altelaar *et al.*, 2006; Nygren and Malmberg, 2004). Chemical mapping is typically limited by the low signal intensity from various biomolecules. Fragmentation of cholesterol in SIMS imaging typically provides several peaks in the mass range of the fragment ion m/z = 366–370 which can be used to visualize and quantify cellular cholesterol. Recently, a characteristic fragment of cholesterol of mass-charge ratio m/z = 147 has been used as a fingerprint to visualize cholesterol accumulation in J774 murine macrophages (Piehowski *et al.*, 2008). J774 cells pre-treated with MβCD-cholesterol (Zidovetzki and Levitan, 2007) had about two times more cholesterol in the plasma membrane than control cells (Piehowski *et al.*, 2008; Ostrowski *et al.*, 2007). For cellular cholesterol imaging by mass spectrometry, two major limitations need to be overcome; SIMS can be used only in freeze-dried samples, and the method has a spatial resolution approximating the cell diameter (i.e., at present no sub-cellular sterol distribution can be revealed). The technique might currently be useful for determining sterol distribution in model membranes and lipid monolayers, since domain sizes in these systems can be quite large (Sostarecz *et al.*, 2004), (Zheng *et al.*, 2007), (Boxer *et al.*, 2009), (Anderton *et al.*, 2011).

2. Sterol Imaging by Vibrational Microscopy

Raman scattering is due to inelastic scattering of light, which results in frequency shifts compared to the incident light. Scattered light of lower frequency ($\omega_S = \omega_p - \omega_R$) or increased frequency ($\omega_{AS} = \omega_p + \omega_R$) compared to the incident light are called Stokes and anti-Stokes components, respectively (Potma and Xie, 2008). The frequencies ω_{R} are caused by molecular vibrations, which can be used to identify and characterize (bio)molecules. Since the Raman signal is very weak, application of this technique to bioimaging is severely limited. Nevertheless, the method combined with confocal detection has been used to characterize cholesterol esters in HeLa cells and macrophage foam cells (van Manen and Otto, 2009). Moreover, Raman spectroscopy has been used as diagnostic tool to detect breast cancer, and combined Raman and fluorescence microscopy were applied to characterize atherosclerotic plaques (Haka *et al.*, 2009; Haka *et al.*, 2011).

A major technical improvement is coherent anti-Stokes Raman scattering (CARS). In CARS, two coherent light beams of frequency ω_1 and ω_2 are used to drive a Raman scattering of frequency ω_R, which results in scattering signals about 10^5-fold stronger than spontaneous Raman scattering (Potma and Xie, 2008; Cheng *et al.*, 2002). The coherent addition of anti-Stokes Raman scattering radiation from many molecular oscillators in the focal volume is responsible for the sensitivity of the technique. This paved the way for biomedical applications of CARS, and this technique has been used for analysis of neutral lipids (i.e., cholesteryl and triacylglycerol esters) in cytoplasmic lipid droplets of mammalian cells (Nan *et al.*, 2003), *Drosophila melanogaster* (Chien *et al.*, 2011) and *Caenorhabditis elegans* (Yen *et al.*, 2010) with chemical specificity. Neutral lipids are particularly suitable for CARS analysis due to the abundance of C-H stretching vibrations (Nan *et al.*, 2003). Characteristic vibrations of saturated versus unsaturated fatty acyl chains allow monitoring of lipid remodeling during nutritional or genetic perturbations (Yen *et al.*, 2010; Mörck *et al.*, 2009). Another advantage of CARS is its ease combination with nonlinear fluorescence microscopy. For example in *C. elegans*, fat storage organelles have been visualized by CARS, while multiphoton fluorescence microscopy provided complementary information on autofluorescent gut granules in the same samples (Yen *et al.*, 2010). Quantitative CARS microscopy revealed heterogeneity in acyl chain saturation of LDs in individual adipocytes (Rinia *et al.*, 2008). Despite the improvements in CARS, vibrational microscopy is limited to cellular regions with large stores of similar lipid species (e.g., lipid droplets), and CARS remains largely confined to detection of neutral lipids due the contribution of the C-H stretching and the C=C vibrations to the CARS signal. CARS microscopy has great potential to unravel the physico-chemical basis of membrane phase separation, with CARS combined with fluorescence microscopy to visualize cholesterol-mediated lipid demixing in supported membranes (Li *et al.*, 2005). Finally, CARS microscopy has been combined with other nonlinear optical imaging methods, like two-photon excitation and sum-frequency generation to monitor atherosclerotic lesions and plaque formation in aortas of ApoE-deficient mice (Wang *et al.*, 2009; Lim *et al.*, 2010).

II. Concluding remarks

There has been considerable progress over the past decade in the ability to observe the distribution of sterols in cells and to measure their transport between organelles using optical microscopy methods. CTL and DHE are reasonably good analogs of cholesterol and it seems unlikely that fluorescent analogs that are better cholesterol analogs will be developed. However, these molecules are not identical to cholesterol and this caveat must always accompany studies using them in mammalian cells. (Of course, DHE is a natural yeast sterol, so it can be used with confidence in these cells and in sterol auxotrophs for which DHE and other ergosterols are suitable dietary sterols.) With improvements in instrumentation, particularly blue-sensitive EMCCD cameras, DHE and CTL can be imaged reasonably well even for time lapse studies. Multiphoton microscopy opens the possibility of determining the three-dimensional distribution of these fluorescent sterols. The weak fluorescence brightness and rapid photobleaching of these fluorescent sterols has led to a long search for sterol derivatives with brighter, more photostable fluorescence. At present, Bchol seems to be the best choice among such sterol derivatives in terms of its distribution and transport in cells. Bchol is less like cholesterol than either DHE or CTL, so the investigator must make a choice among these sterols depending on available instrumentation and the goals of the experiment.

There is still a need for better methods to label endogenous cholesterol in cells or tissues. Both filipin and the cholesterol-binding toxins have been used for such purposes. As discussed in this chapter, within a limited range of conditions filipin can be validated as a linear indicator of cholesterol levels, but there are many instances in which filipin binds to other molecules or has a nonlinear response to cholesterol levels. The sterol binding toxins are generally nonlinear in their sterol response, but they may be useful for detecting high levels of cholesterol in membranes.

The label-free detection of cholesterol has promise for detecting endogenous cholesterol, but at present the methods available do not have sufficient resolution and/or sensitivity for studies of the subcellular distribution of cholesterol.

Acknowledgments

DW acknowledges funding by grants of the Lundbeck Foundation, the Danish Research Agency Forskningsstyrelsen, Forskningsrådet for Natur og Univers (FNU), and the Danish Research Agency Forskningsstyrelsen, Forskningsrådet for Sundhed og sygdom (FSS). FRM received support from NIH Grant R37-DK27083.

References

Ahmed, S. N., Brown, D. A., and London, E. (1997). On the origin of sphingolipid/cholesterol-rich detergent-insoluble cell membranes: physiological concentrations of cholesterol and sphingolipid induce formation of a detergent-insoluble, liquid-ordered lipid phase in model membranes. *Biochemistry* **36**, 10944–10953.

Aittoniemi, J., Róg, T., Niemelä, P., Pasenkiewicz-Gierula, M., Karttunen, M., and Vattulainen, I. (2006). Tilt: major factor in sterols' ordering capability in membranes. *J. Phys. Chem. B* **110**, 25562–25564.

Anderton, C. R., Lou, K., Weber, P. K., Hutcheon, I. D., and Kraft, M. L. (2011). Correlated AFM and NanoSIMS imaging to probe cholesterol-induced changes in phase behavior and non-ideal mixing in ternary lipid membranes. *Biochim. Biophys. Acta* **1808**, 307–315.

Altelaar, A. F., Klinkert, I., Jalink, K., de Lange, R. P., Adan, R. A., Heeren, R. M., and Piersma, S. R. (2006). Gold-enhanced biomolecular surface imaging of cells and tissue by SIMS and MALDI mass spectrometry. *Anal. Chem.* **78**, 734–742.

Ariola, F. S., Li, Z., Cornejo, C., Bittman, R., and Heikal, A. A. (2009). Membrane fluidity and lipid order in ternary giant unilamellar vesicles using a new bodipy-cholesterol derivative. *Biophys. J.* **96**, 2696–2708.

Arthur, J.R., Heinecke, K.A., and Seyfried, T.N. Filipin recognizes both GM1 and cholesterol in GM1 gangliosidosis mouse brain. *J. Lipid Res.* **52**, 1345-1351.

Baumgart, T., Hunt, G., Farkas, E. R., Webb, W. W., and Feigenson, G. W. (2007). Fluorescence probe partitioning between Lo/Ld phases in lipid membranes. *Biochim. Biophys. Acta* **1768**, 2182–2194.

Benson, D. M., Bryan, J., Plant, A. L., Gotto, A. M. J., and Smith, L. C. (1985). Digital imaging fluorescence microscopy: spatial heterogeneity of photobleaching rate constants in individual cells. *J. Cell Biol.* **100**, 1309–1323.

Bergström, F., Mikhalyov, I., Hägglöf, P., Wortmann, R., Ny, T., and Johansson, L. B. -Å. (2002). Dimers of Dipyrrometheneboron difluoride (BODIPY) with light spectroscopic applications in chemistry and biology. *J. Am. Chem. Soc.* **124**, 196–204.

Boxer, S. G., Kraft, M. L., and Weber, P. K. (2009). Advances in imaging secondary ion mass spectrometry for biological samples. *Annu. Rev. Biophys.* **38**, 53–74.

Breslow, R., and Zhang, B. L. (1996). Cholesterol recognition and binding by cyclodextrin dimers. *J. Am. Chem. Soc.* **118**, 8495–8496.

Chen, C. S., Martin, O. C., and Pagano, R. E. (1997). Changes in the spectral properties of a plasma membrane lipid analog during the first seconds of endocytosis in living cells. *Biophys. J.* **72**, 37–50.

Cheng, J. X., Jia, Y. K., Zheng, G., and Xie, X. S. (2002). Laser-scanning coherent anti-Stokes Raman scattering microscopy and applications to cell biology. *Biophys. J.* **83**, 502–509.

Chiantia, S., Ries, J., Chwastek, G., Carrer, D., Li, Z., Bittman, R., and Schwille, P. (2008). Role of ceramide in membrane protein organization investigated by combined AFM and FCS. *Biochim. Biophys. Acta* **1778**, 1356–1364.

Chien, C. H., Chen, W. W., Wu, J. T., and Chang, T. C. (2011). Label-free imaging of Drosophila in vivo by coherent anti-Stokes Raman scattering and two-photon excitation autofluorescence microscopy. *J. Biomed. Optics* **16**, 016012.

Chong, P. L., and Thompson, T. E. (1986). Depolarization of dehydroergosterol in phospholipid bilayers. *Biochim. Biophys. Acta* **863**, 53–62.

Craig, I.F., Via, D.P., Mantulin, W.W., Pownall, H.J., Gotto, A.M.J., and L.C.S. (1981). Low density lipoproteins reconstituted with steroids containing the nitrobenzoxadiazole fluorophore. *J. Lipid Res.***22**, 687-696.

Du, X., Kumar, J., Ferguson, C., Schulz, T. A., Ong, Y. S., Hong, W., Prinz, W. A., Parton, R. G., Brown, A. J., and Yang, H. (2011). A role for oxysterol-binding protein-related protein 5 in endosomal cholesterol trafficking. *J. Cell Biol.* **192**, 121–135.

Ellinger, A., Vetterlein, M., Weiss, C., Meisslitzer-Ruppitsch, C., Neumüller, J., and Pavelka, M. (2010). High-pressure freezing combined with in vivo-DAB-cytochemistry: a novel approach for studies of endocytic compartments. *J. Struct. Biol.* **169**, 286–293.

Fischer RT, Stephenson FA, Shafiee A, and F., S. (1974). Delta 5,7,9(11)-Cholestatrien-3 beta-ol: a fluorescent cholesterol analogue. *Chem. Phys. Lipids* **36**, 1-14.

Fischer, R. T., Stephenson, F. A., Shafiee, A., and Schroeder, F. (1984). Delta 5,7,9(11)-Cholestatrien-3 beta-ol: a fluorescent cholesterol analogue. *Chem. Phys. Lipids* **36**, 1–14.

Fischer, R. T., Stephenson, F. A., Shafiee, A., and Schroeder, F. (1985a). Structure and Dynamic Properties of Dehydroergosterol, delta 5,7,9(11),22-ergostatetraen-3 beta-ol. *J. Biol. Chem.* **13**, 13–24.

Fischer, R. T., Cowlen, M. S., Dempsey, M. E., and Schroeder, F. (1985b). Fluorescence of delta 5,7,9 (11),22-ergostatetraen-3 beta-ol in micelles, sterol carrier protein complexes, and plasma membranes. *Biochemistry* **24**, 3322–3331.

Fellmann, P., Zachowski, A., and Devaux, P. F. (1994). Synthesis and use of spin-labeled lipids for studies of the transmembrane movement of phospholipids. *Methods Mol. Biol.* **27**, 161–175.

Frolov, A., Petrescu, A., Atshaves, B. P., So, P. T., Gratton, E., Serrero, G., and Schroeder, F. (2000). High density lipoprotein-mediated cholesterol uptake and targeting to lipid droplets in intact L-cell fibroblasts. *J. Biol. Chem.* **275**, 12769–12780.

Garvik, O., Benediktsen, P., Ipsen, J. H., Simonsen, A. C., and Wüstner, D. (2008). The fluorescent cholesterol analog dehydroergosterol induces liquid-ordered domains in model membranes. *Chem. Phys. Lipids* **159**, 114–118.

Georgiev, A.G., Sullivan, D.P., Kersting, M.C., Dittman, J.S., Beh, C.T., and Menon, A.K. (2011). Osh Proteins Regulate Membrane Sterol Organization but Are Not Required for Sterol Movement Between the ER and PM. *Traffic* **12**, 1341–1355.

Gimpl, G. (2010). Cholesterol-protein interaction: methods and cholesterol reporter molecules. *Subcell. Biochem.* **51**, 1–45.

Haka, A. S., Volynskaya, Z., Gardecki, J. A., Nazemi, J., Shenk, R., Wang, N., Dasari, R. R., Fitzmaurice, M., and Feld, M. S. (2009). Diagnosing breast cancer using Raman spectroscopy: prospective analysis. *J. Biomed. Optics* **14**, 054023.

Haka, A. S., Kramer, J. R., Dasari, R. R., and Fitzmaurice, M. (2011). Mechanism of ceroid formation in atherosclerotic plaque: in situ studies using a combination of Raman and fluorescence spectroscopy. *J. Biomed. Optics* **16**, 011011.

Hale, J. E., and Schroeder, F. (1982). Asymmetric transbilayer distribution of sterol across plasma membranes determined by fluorescence quenching of dehydroergosterol. *Eur. J. Biochem.* **122**, 649–661.

Hao, M., Lin, S. X., Karylowski, O. J., Wüstner, D., McGraw, T. E., and Maxfield, F. R. (2002). Vesicular and non-vesicular sterol transport in living cells. The endocytic recycling compartment is a major sterol storage organelle. *J. Biol. Chem.* **277**, 609–617.

Hartwig Petersen, N., Færgeman, N. J., Yu, L., and Wüstner, D. (2008). Kinetic imaging of NPC1L1 and sterol trafficking between plasma membrane and recycling endosomes in hepatoma cells. *J. Lipid Res.* **49**, 2023–2037.

Henriksen, J., Rowat, A. C., and Ipsen, J. H. (2004). Vesicle fluctuation analysis of the effects of sterols on membrane bending rigidity. *Eur. Biophys. J.* **33**, 732–741.

Henriksen, J., Rowat, A. C., Brief, E., Hsueh, Y. W., Thewalt, J. L., Zuckermann, M. J., and Ipsen, J. H. (2006). Universal behavior of membranes with sterols. *Biophys. J.* **90**, 1639–1649.

Hofsass, C., Lindahl, E., and Edholm, O. (2003). Molecular dynamics simulations of phospholipid bilayers with cholesterol. *Biophys. J.* **84**, 2192–2206.

Hölttä-Vuori, M., Uronen, R. L., Repakova, J., Salonen, E., Vattulainen, I., Panula, P., Li, Z., Bittman, R., and Ikonen, E. (2008). BODIPY-cholesterol: a new tool to visualize sterol trafficking in living cells and organisms. *Traffic* **9**, 1839–1849.

Huang, H., McIntosh, A. L., Atshaves, B. P., Ohno-Iwashita, Y., Kier, A. B., and Schroeder, F. (2010). Use of dansyl-cholestanol as a probe of cholesterol behavior in membranes of living cells. *J. Lipid Res.* **51**, 1157–1172.

Hyslop, P. A., Morel, B., and Sauerheber, R. D. (1990). Organization and interaction of cholesterol and phosphatidylcholine in model bilayer membranes. *Biochemistry* **29**, 1025–1038.

Jansen, M., Ohsaki, Y., Rita Rega, L., Bittman, R., Olkkonen, V. M., and Ikonen, E. (2011). Role of ORPs in sterol transport from plasma membrane to ER and lipid droplets in mammalian cells. *Traffic* **12**, 218–231.

Ipsen, J. H., Karlstrom, G., Mouritsen, O. G., Wennerstrom, H., and Zuckermann, M. J. (1987). Phase equilibria in the phosphatidylcholine-cholesterol system. *Biochim. Biophys. Acta* **905**, 162–172.

Kohut, P., Wüstner, D., Hronska, L., Kuchler, K., Hapala, I., and Valachovic, M. (2011). The role of ABC proteins Aus1p and Pdr11p in the uptake of external sterols in yeast: Dehydroergosterol fluorescence study. *Biochem. Biophys. Res. Commun.* **404**, 233–238.

Krieger, M. (1986). Reconstitution of the hydrophobic core of low-denstiy lipoprotein. *Methods Enzymol.* **128**, 608–613.

Lakowitz, J. R. (2006). *Principles of Fluorescence Spectroscopy.* Springer, New York, USA.

Lindahl, E., and Edholm, O. (2000). Spatial and energetic-entropic decomposition of surface tension in lipid bilayers from molecular dynamics simulations. *J. Chem. Phys.* **113**, 3882–3893.

Li, L., Wang, H., and Cheng, J. X. (2005). Quantitative coherent anti-stokes Raman scattering imaging of lipid distribution in coexisting domains. *Biophys. J.* **89**, 3480–3490.

Li, Z., Mintzer, E., and Bittman, R. (2006). First synthesis of free cholesterol-BODIPY conjugates. *J. Org. Chem.* **71**, 1718–1721.

Lim, R. S., Kratzer, A., Barry, N. P., Miyazaki-Anzai, S., Miyazaki, M., Mantulin, W. W., Levi, M., Potma, E. O., and Tromberg, B. J. (2010). Multimodal CARS microscopy determination of the impact of diet on macrophage infiltration and lipid accumulation on plaque formation in ApoE-deficient mice. *J. Lipid Res.* **51**, 1729–1737.

Loura, L. M. S., Fedorov, A., and Prieto, M. (2001). Exclusion of a cholesterol analog from the cholesterol-rich phase in model membranes. *Biochim. Biophys. Acta* **1511**, 236–243.

Luisier, F. (2010). The SURE-LET Approach to Image Denoising. PhD thesis. EPFL, Thesis No. 4566.

Marks, D. L., Bittman, R., and Pagano, R. E. (2008). Use of Bodipy-labeled sphingolipid and cholesterol analogs to examine membrane microdomains in cells. *Histochem. Cell Biol.* **130**, 819–832.

Markham, J., and Conchello, J. A. (2001). Artefacts in restored images due to intensity loss in three-dimensional fluorescence microscopy. *J. Microsc.* **204**, 93–98.

Marsan, M. P., Muller, I., Ramos, C., Rodriguez, F., Dufourc, E. J., Czaplicki, J., and Milon, A. (1999). Cholesterol orientation and dynamics in dimyristoylphosphatidylcholine bilayers: a solid state deuterium NMR analysis. *Biophys. J.* **76**, 351–359.

Masters, B. R., and So, P. T. C. (2008). Classical and quantum theory of one-photon and multiphoton fluorescence spectroscopy. *In* "Handbook of Biomedical Nonlinear Optical Microscopy," (B. R. Masters, and P. T. C. So, eds.), pp. 91–152. Oxford University Press, Oxford, United Kingdom.

Matyash, V., Geier, C., Henske, A., Mukherjee, S., Hirsh, D., Thiele, C., Grant, B., Maxfield, F. R., and Kurzchalia, T. V. (2001). Distribution and transport of cholesterol in Caenorhabditis elegans. *Mol. Biol. Cell* **12**, 1725–1736.

McIntosh, A. L., Gallegos, A. M., Atshaves, B. P., Storey, S. M., Kannoju, D., and Schroeder, F. (2003). Fluorescence and multiphoton imaging resolve unique structural forms of sterol in membranes of living cells. *J. Biol. Chem.* **278**, 6384–6403.

McIntosh, A. L., Atshaves, B. P., Huang, H., Gallegos, A. M., Kier, A. B., and Schroeder, F. (2008). Fluorescence techniques using dehydroergosterol to study cholesterol trafficking. *Lipids* **43**, 1185–1208.

Mesmin, B., and Maxfield, F. R. (2009). Intracellular sterol dynamics. *Biochim. Biophys. Acta* **1791**, 636–645.

Mondal, M., Mesmin, B., Mukherjee, S., and Maxfield, F. R. (2009). Sterols are mainly in the cytoplasmic leaflet of the plasma membrane and the endocytic recycling compartment in CHO cells. *Mol. Biol. Cell* **20**, 581–588.

Mörck, C., Olsen, L., Kurth, C., Persson, A., Storm, N. J., Svensson, E., Jansson, J. O., Hellqvist, M., Enejder, A., Faergeman, N. J., and Pilon, M. (2009). Statins inhibit protein lipidation and induce the unfolded protein response in the non-sterol producing nematode Caenorhabditis elegans. *Proc. Natl. Acad. Sci. U.S.A.* **106**, 18285–18290.

Mukherjee, S., Zha, X., Tabas, I., and Maxfield, F. R. (1998). Cholesterol distribution in living cells: fluorescence imaging using dehydroergosterol as a fluorescent cholesterol analog. *Biophys. J.* **75**, 1915–1925.

Mukherjee, S., and Maxfield, F. R. (2004). Membrane domains. *Annu. Rev. Cell Dev. Biol.* **20**, 839–866.

Munro, S. (2003). Lipid rafts: elusive or illusive? *Cell* **115**, 377–388.

Nagle, J. F., and Tristram-Nagle, S. (2000). Structure of lipid bilayers. *Biochim. Biophys. Acta* **1469**, 159–195.

Nan, X., Cheng, J. X., and Xie, X. S. (2003). Vibrational imaging of lipid droplets in live fibroblast cells with coherent anti-Stokes Raman scattering microscopy. *J. Lipid Res.* **44**, 2202–2208.

Nygren, H., and Malmberg, P. (2004). Silver deposition on freeze-dried cells allows subcellular localization of cholesterol with imaging TOF-SIMS. *J. Microsc.* **215**, 156–161.

Ohvo-Rekila, H., Ramstedt, B., Leppimaki, P., and Slotte, J. P. (2002). Cholesterol interactions with phospholipids in membranes. *Prog. Lipid Res.* **41**, 66–97.

Orci, L., Montesano, R., Meda, P., Malaisse-Lagae, F., Brown, D., Perrelet, A., and Vassalli, P. (1981). Heterogeneous distribution of filipin–cholesterol complexes across the cisternae of the Golgi apparatus. *Proc. Natl. Acad. Sci. U.S.A.* **78**, 293–297.

Ostrowski, S. G., Kurczy, M. E., Roddy, T. P., Winograd, N., and Ewing, A. G. (2007). Secondary ion MS imaging to relatively quantify cholesterol in the membranes of individual cells from differentially treated populations. *Anal. Chem.* **79**, 3554–3560.

Pagano, R. E., Martin, O. C., Kang, H. C., and Haugland, R. P. (1991). A novel fluorescent ceramide analogue for studying membrane traffic in animal cells: accumulation at the Golgi apparatus results in altered spectral properties of the sphingolipid precursor. *J. Cell Biol.* **113**, 1267–1279.

Peake, K. B., and Vance, J. E. (2010). Defective cholesterol trafficking in Niemann-Pick C-deficient cells. *FEBS Lett.* **584**, 2731–2739.

Piehowski, P. D., Carado, A. J., Kurczy, M. E., Ostrowski, S. G., Heien, M. L., Winograd, N., and Ewing, A. G. (2008). MS/MS methodology to improve subcellular mapping of cholesterol using TOF-SIMS. *Anal. Chem.* **80**, 8662–8667.

Pipalia, N. H., Huang, A., Ralph, H., Rujoi, M., and Maxfield, F. R. (2006). Automated microscopy screening for compounds that partially revert cholesterol accumulation in Niemann-Pick C cells. *J. Lipid Res.* **47**, 284–301.

Pipalia, N. H., Cosner, C. C., Huang, A., Chatterjee, A., Bourbon, P., Farley, N., Helquist, P., Wiest, O., and Maxfield, F. R. (2011). Histone deacetylase inhibitor treatment dramatically reduces cholesterol accumulation in Niemann-Pick type C1 mutant human fibroblasts. *Proc. Natl. Acad. Sci. U.S.A.* **108**, 5620–5625.

Potma, E. O., and Xie, X. S. (2008). Theory of spontaneous and coherent Raman scattering. *In* "Handbook of Biomedical Nonlinear Optical Microscopy," (B. R. Masters, and P. T. C. So, eds.), pp. 164–190. Oxford University Press, Oxford, United Kingdom.

Puri, V., Watanabe, R., Dominguez, M., Sun, X., Wheatley, C. L., Marks, D. L., and Pagano, R. E. (1999). Cholesterol modulates membrane traffic along the endocytic pathway in sphingolipid-storage diseases. *Nat. Cell Biol.* **1**, 386–388.

Puri, V., Watanabe, R., Singh, R. D., Dominguez, M., Brown, J. C., Wheatley, C. L., Marks, D. L., and Pagano, R. E. (2001). Clathrin-dependent and -independent internalization of plasma membrane sphingolipids initiates two Golgi targeting pathways. *J. Cell Biol.* **154**, 535–547.

Qin, W., Baruah, M., Van der Auweraer, M., De Schryver, F. C., and Boens, N. (2005). Photophysical properties of borondipyrromethene analogues in solution. *J. Phys. Chem. A* **109**, 7371–7384.

Qin, C., Nagao, T., Grosheva, I., Maxfield, F. R., and Pierini, L. M. (2006). Elevated plasma membrane cholesterol content alters macrophage signaling and function. *Arterioscler. Thromb. Vasc. Biol.* **26**, 372–378.

Reiner, S., Micolod, D., and Schneiter, R. (2005). Saccharomyces cerevisiae, a model to study sterol uptake and transport in eukaryotes. *Biochem. Soc. Trans.* **33**, 1186–1188.

Rinia, H. A., Burger, K. N., Bonn, M., and Müller, M. (2008). Quantitative label-free imaging of lipid composition and packing of individual cellular lipid droplets using multiplex CARS microscopy. *Biophys. J.* **95**, 4908–4914.

Rogers, J., Lee, A. G., and Wilton, D. C. (1979). The organisation of cholesterol and ergosterol in lipid bilayers based on studies using non-perturbing fluorescent sterol probes. *Biochim. Biophys. Acta* **552**, 23–37.

Röhrl, C., Meisslitzer-Ruppitsch, C., Bittman, R., Li, Z., Pabst, G., Prassl, R., Strobl, W., Neumüller, J., Ellinger, A., Pavelka, M., and Stangl, H. (2011). Combined light and electron microscopy using diaminobenzidine photooxidation to monitor trafficking of lipids derived from lipoprotein particles. *Curr. Pharm. Biotechnol.* In press. PMID: 21470121.

Rosenbaum, A. I., and Maxfield, F. R. (2011). Niemann-Pick type C disease: molecular mechanisms and potential therapeutic approaches. *J. Neurochem.* **116**, 789–795.

Rosenbaum, A. I., Rujoi, M., Huang, A. Y., Du, H., Grabowski, G. A., and Maxfield, F. R. (2009). Chemical screen to reduce sterol accumulation in Niemann-Pick C disease cells identifies novel lysosomal acid lipase inhibitors. *Biochim. Biophys. Acta* **1791**, 1155–1165.

Rujoi, M., Pipalia, N. H., and Maxfield, F. R. (2010). Cholesterol pathways affected by small molecules that decrease sterol levels in Niemann-Pick type C mutant cells. *PLoS One* **5**, e12788.

Scheidt, H. A., Müller, P., Herrmann, A., and Huster, D. (2003). The potential of fluorescent and spin-labeled steroid analogs to mimic natural cholesterol. *J. Biol. Chem.* **278**, 45563–45569.

Schrader, M., Bahlmann, K., and Hell, S. W. (1997). Three-photon-excitation microscopy: theory, experiment and applications. *Optik* **104**, 116–124.

Schroeder, F., Barenholz, Y., Gratton, E., and Thompson, T. E. (1987). A fluorescence study of dehydroergosterol in phosphatidylcholine bilayer vesicles. *Biochemistry* **26**, 2441–2448.

Schroeder, F., Nemecz, G., Gratton, E., Barenholz, Y., and Thompson, T. E. (1988). Fluorescence properties of cholestatrienol in phosphatidylcholine bilayer vesicles. *Biophys. Chem.* **32**, 57–72.

Sharma, D. K., Choudhury, A., Singh, R. D., Wheatley, C. L., Marks, D. L., and Pagano, R. E. (2003). Glycosphingolipids internalized via caveolar-related endocytosis rapidly merge with the clathrin pathway in early endosomes and form microdomains for recycling. *J. Biol. Chem.* **278**, 7564–7572.

Sharma, D. K., Brown, J. C., Choudhury, A., Peterson, T. E., Holicky, E., Marks, D. L., Simari, R., Parton, R. G., and Pagano, R. E. (2004). Selective stimulation of caveolar endocytosis by glycosphingolipids and cholesterol. *Mol. Biol. Cell* **15**, 3114–3122.

Shaw, J. E., Epand, R. F., Epand, R. M., Li, Z., Bittman, R., and Yip, C. M. (2006). Correlated fluorescence-atomic force microscopy of membrane domains: structure of fluorescence probes determines lipid localization. *Biophys. J.* **90**, 2170–2178.

Shrivastava, S., Haldar, S., Gimpl, G., and Chattopadhyay, A. (2009). Orientation and dynamics of a novel fluorescent cholesterol analogue in membranes of varying phase. *J. Phys. Chem. B* **113**, 4475–4481.

Sibarita, J. B. (2005). Deconvolution microscopy. *Adv. Biochem. Eng. Biotechnol.* **95**, 201–243.

Smutzer, G., Crawford, B. F., and Yeagle, P. L. (1986). Physical properties of the fluorescent sterol probe dehydroergosterol. *Biochim. Biophys. Acta* **862**, 361–371.

Sokolov, A., and Radhakrishnan, A. (2010). Accessibility of cholesterol in endoplasmic reticulum membranes and activation of SREBP-2 switch abruptly at a common cholesterol threshold. *J. Biol. Chem.* **285**, 29480–29490.

Sostarecz, A. G., McQuaw, C. M., Ewing, A. G., and Winograd, N. (2004). Phosphatidylethanolamine-induced cholesterol domains chemically identified with mass spectrometric imaging. *J. Am. Chem. Soc.* **126**, 13882–13883.

Steer, C. J., Bisher, M., Blumenthal, R., and Steven, A. C. (1984). Detection of membrane cholesterol by filipin in isolated rat liver coated vesicles is dependent upon removal of the clathrin coat. *J. Cell Biol.* **99**, 315–319.

Swedlow, J. R. (2007). Quantitative fluorescence microscopy and image deconvolution. *Methods Cell Biol.* **81**, 447–465.

van Manen, H. -J., and Otto, C. (2009). Cholesterol esters are detected by Raman microspectroscopy in HeLa cells. *J. Raman Spectrosc.* **40**, 117–118.

Vist, M. R., and Davis, J. H. (1990). Phase equilibria of cholesterol/dipalmitoylphosphatidylcholine mixtures: 2H nuclear magnetic resonance and differential scanning calorimetry. *Biochemistry* **29**, 451–464.

Wang, H. W., Langohr, I. M., Sturek, M., and Cheng, J. X. (2009). Imaging and quantitative analysis of atherosclerotic lesions by CARS-based multimodal nonlinear optical microscopy. *Arterioscler. Thromb. Vasc. Biol.* **29**, 1342–1348.

Wiegand, V., Chang, T.Y., Strauss, J.F. r., Fahrenholz, F., and Gimpl, G. (2003). Transport of plasma membrane-derived cholesterol and the function of Niemann-Pick C1 Protein. *FASEB J.* **17**, 782–784.

Wüstner, D. (2009). Intracellular cholesterol transport. *In* "Cellular Lipid Metabolism," (C. Ehnholm, ed.), pp. 157–190. Springer press, Heidelberg, Germany.

Wüstner, D. (2010). Following intracellular cholesterol transport by linear and non-linear optical microscopy of intrinsically fluorescent sterols. *Curr. Pharm. Biotechnol.* In press. PMID: 21470123.

Wüstner, D., Brewer, J. R., Bagatolli, L. A., and Sage, D. (2011b). Potential of ultraviolet widefield imaging and multiphoton microscopy for analysis of dehydroergosterol in cellular membranes. *Microsc. Res. Tech.* **74**, 92–108.

Wüstner, D., and Færgeman, N. J. (2008a). Spatiotemporal analysis of endocytosis and membrane distribution of fluorescent sterols in living cells. *Histochem. Cell Biol.* **130**, 891–908.

Wüstner, D., and Færgeman, N. J. (2008b). Chromatic aberration correction and deconvolution for UV sensitive imaging of fluorescent sterols in cytoplasmic lipid droplets. *Cytometry A* **73**, 727–744.

Wüstner, D., Herrmann, A., Hao, M., and Maxfield, F. R. (2002). Rapid nonvesicular transport of sterol between the plasma membrane domains of polarized hepatic cells. *J. Biol. Chem.* **277**, 30325–30336.

Wüstner, D., Landt Larsen, A., Færgeman, N. J., Brewer, J. R., and Sage, D. (2010). Selective visualization of fluorescent sterols in Caenorhabditis elegans by bleach-rate based image segmentation. *Traffic* **11**, 440–454.

Wüstner, D., Mondal, M., Tabas, I., and Maxfield, F. R. (2005). Direct observation of rapid internalization and intracellular transport of sterol by macrophage foam cells. *Traffic* **6**, 396–412.

Wüstner, D. (2007a). Fluorescent sterols as tools in membrane biophysics and cell biology. *Chem. Phys. Lipids* **146**, 1–25.

Wüstner, D. (2007b). Plasma membrane sterol distribution resembles the surface topography of living cells. *Mol. Biol. Cell* **18**, 211–228.

Wüstner, D., Solanko, L. M., Sokol, E., Lund, F. W., Garvik, O., Li, Z., Bittman, R., Korte, T., and Herrmann, A. (2011a). Quantitative assessment of sterol traffic in living cells by dual labeling with dehydroergosterol and BODIPY-cholesterol. *Chem. Phys. Lipids.* **164**, 221–235.

Wüstner, D., and Sage, D. (2010). Multicolor bleach-rate imaging enlightens in vivo sterol transport. *Commun. Integr. Biol.* **3**, 1–4.

Xu, C., and Zipfel, W. R. (2008). Multiphoton excitation of fluorescent probes. *In* "Handbook of Biomedical Nonlinear Optical Microscopy," (B. R. Masters, and P. T. C. So, eds.), pp. 311–333. Oxford University Press, Oxford, United Kingdom.

Yazdanfar, S., and So, P. T. (2008). Signal detection and processing in nonlinear optical microscopes. *In* "Handbook of Biomedical Nonlinear Optical Microscopy," (B. R. Masters, and P. T. C. So, eds.), pp. 283–309. Oxford University Press, Oxford, United Kingdom.

Yeagle, P. L., Albert, A. D., Boesze-Battaglia, K., Young, J., and Frye, J. (1990). Cholesterol dynamics in membranes. *Biophys. J.* **57**, 413–424.

Yeagle, P. L., Bensen, J., Boni, L., and Hui, S. W. (1982). Molecular packing of cholesterol in phospholipid vesicles as probed by dehydroergosterol. *Biochim. Biophys. Acta* **692**, 139–146.

Yen, K., Le, T. T., Bansal, A., Narasimhan, S. D., Cheng, J. X., and Tissenbaum, H. A. (2010). A comparative study of fat storage quantitation in nematode Caenorhabditis elegans using label and label-free methods. *PLoS One* **5**, e12810.

Young, I. T. (1989). Image fidelity: characterizing the imaging transfer function. *Methods Cell Biol.* **30**, 2–45.

Zhang, W., McIntosh, A. L., Xu, H., Wu, D., Gruninger, T., Atshaves, B., Liu, J. C., and Schroeder, F. (2005). Structural analysis of sterol distributions in the plasma membrane of living cells. *Biochemistry* **44**, 2864–2884.

Zheng, L., McQuaw, C. M., Ewing, A. G., and Winograd, N. (2007). Sphingomyelin/phosphatidylcholine and cholesterol interactions studied by imaging mass spectrometry. *J. Am. Chem. Soc.* **129**, 15730–15731.

Zidovetzki, R., and Levitan, I. (2007). Use of cyclodextrins to manipulate plasma membrane cholesterol content: evidence, misconceptions and control strategies. *Biochim. Biophys. Acta* 1768.

Zipfel, W. R., Williams, R. M., and Webb, W. W. (2003). Nonlinear magic: multiphoton microscopy in the biosciences. *Nat. Biotechnol.* **21**, 1369–1377.

CHAPTER 18

Fluorescence Correlation Methods for Imaging Cellular Behavior of Sphingolipid-Interacting Probes

Rachel Kraut[*], **Nirmalya Bag**[†] **and Thorsten Wohland**[†]

[*]School of Biological Sciences, Nanyang Technological University, Singapore

[†]Department of Chemistry, National University of Singapore, Singapore

METHODS IN CELL BIOLOGY, VOL 108
Copyright 2012, Elsevier Inc. All rights reserved.

0091-679X/10 $35.00
DOI 10.1016/B978-0-12-386487-1.00018-3

Abstract

For cell biologists interested in the properties of cell membranes, their composition, and dynamics, the realization that sphingolipids and cholesterol have the capacity to self-organize into ordered domains has given rise to a need to visualize these lipids in actual living cell membranes. In order to find out how various classes of lipids distribute in the membrane and what their behaviors are, it is extremely useful to apply fluorescent probes that either interact with these lipids, or that themselves behave like naturally occurring lipids. At the same time, imaging modalities to observe their behaviors require the appropriate spatial and temporal resolution, on the milli- or microsecond timescale. Knowledge of membrane organization and how it changes during processes like cell signaling and invasion by pathogens will undoubtedly be relevant to our understanding of the action of infectious diseases, bacterial toxins, and even disease pathologies like prion and Alzheimer's disease, where glycosphingolipids (GSL) and sphingolipid-rich domains act as membrane receptors or docking sites. In such cases, membrane composition and dynamics will clearly play a role in infectivity. The present challenge is in coming up with high-resolution and non-invasive approaches to observing these dynamic and structural features of sphingolipid-containing membrane domains. This chapter will discuss several variants and applications of fluorescence correlation spectroscopy as well as probes that can be used to study sphingolipid dynamics.

I. Introduction

Through the application of novel lipid-interacting probes and fluorescence spectroscopic methods, some of which we discuss in this chapter, membrane biologists have discovered that nano-scaled sphingolipid-rich domains exhibit biophysical properties distinct from the surrounding disordered regions. For various cell types that have been observed, the plasma membrane appears to be composed of a mosaic of fluid, disordered regions (the liquid disordered or l_d phase), interspersed with more ordered, less fluid domains (the liquid ordered or l_o phase). Statistically, the frequency of occurrence of the ordered regions seems to be on the order of 25% at any given time, although it is generally agreed that the domains are extremely short-lived and dynamic, probably existing only for milliseconds at a time, when a cell is in a resting state (Lingwood *et al.*, 2009).

The behavior of sphingolipid-rich regions in natural cell membranes has been likened to mixtures of phase-separating lipids that are at a critical point, where domain formation can be triggered by various perturbations such as ligand binding to a receptor, pathogen invasion, or cytoskeletal organization (Feigenson, 2009; Veatch *et al*., 2007; Zhang *et al*., 2009). The exact composition of domains and the extent of heterogeneity that might exist in the membrane remain unknown, mainly due to the dearth of easily traceable probes that interact with sphingolipid rich membrane domains, or sphingolipid analogs that behave naturally. However, it is known that sphingolipids and cholesterol are required for detectable ordered domains to form. In order to observe the behavior of sphingolipids in the membrane and the domains they form, one needs to use visible probes that mimic the behavior of sphingolipids themselves, or that bind to them and follow their movements. For this purpose, fluorescently tagged versions of gangliosides (sialylated glycosphingolipids) and sphingomyelin (a sphingolipid with a phosphocholine headgroup) are available, and a number of groups have described variants of such lipid analogs coupled to fluorescent tags at different positions either at the headgroup or on the acyl tail (see Table I)(Eggeling *et al*., 2008).

As a complement to studying the behavior of sphingolipids themselves via fluorescently tagged analogs, sphingolipid-interacting peptides and lipid-linked proteins can be observed at the membrane, and their dynamics interpreted as a proxy for the behavior of the actual lipids in the membrane. This approach is schematically depicted in Fig. 1, and the structures of the represented lipids are shown in Fig. 2. In this chapter, we discuss several types of fluorescent sphingolipid and non-sphingolipid analogs and peptide probes that we and others have used to analyze membrane dynamics and heterogeneity.

Since the dimensions of membrane domains is thought to be in the nanometer range, which roughly corresponds to anywhere from a few tens of lipid molecules to hundreds of them, they cannot be visualized by conventional microscopy techniques due to the diffraction limit. The very dynamic nature of these domains with characteristic times possibly as fast as milliseconds renders also many super-resolution techniques incapable of observing the domains and their dynamics. Therefore, new methods are needed which can observe cell membranes with high spatio-temporal resolution. Fluorescence Correlation Spectroscopy, or FCS, is one of the favored methods of extracting information about the average diffusion behavior of mobile fluorescent molecules, and can also be used to measure the concentration of those molecules. The conventional method of FCS, performed in a confocal microscope, does not scan the confocal light beam, but keeps it at a stationary spot on the region to be measured (in this case the cell membrane), and records an intensity trace of the fluorescence emitted as individual probe molecules pass through (depicted in Fig. 3). A main advantage of FCS in general is that it is capable of generating detailed, quantitative information about molecular processes by averaging over the behavior of thousands of molecules in a short space of time (Haustein and Schwille, 2007; Kim *et al*., 2007).

In this chapter, we describe a method that uses FCS adaptations of conventional confocal and total internal reflection fluorescence (TIRF) microscopes to analyze the behavior of sphingolipid-tracing probes in terms of their simple diffusion characteristics. FCS when applied in combination with TIRF, which we call Imaging TIR-FCS (ITIR-FCS) (Fig. 3), is particularly useful for probing the anisotropy in

Table I
Sphingolipid and non-sphingolipid fluorescent analogs for labeling ordered and disordered domains in the membranes of live cells; fluorescently labeled peptide probes that can be used for detecting sphingolipid domains of different composition, for example, mixed SM/ganglioside (SBD), SM alone (lysenin-GFP), or GM1 (CtxB-Alexa). Locations of the fluorophore on the headgroup, acyl chain, or backbone of the lipid are shown in the schematic figures, color coded as in Fig. 2. Representative citations documenting the diffusion and/or phase behavior of the probes are given.

Probe name	molecule/source	D/phase	intended target	location of label	reference	source
LIPIDS						
Atto647N-PE	n-(Atto647N)-1,2-dipalmitoyl-sn-glycero-3 phosphoethanolamine	free	DPPE		Manzo, van Zanten *et al*, 2011; Eggeling *et al*, 2008	Atto-Tec
BODIPY-FL-DHPE (aka DPPE)	n-(BODIPY-FL-3-propionyl)-1, 2-dihexadecanoyl-sn-glycero-3-phosphoethanolamine	disordered	DPPE		Burns *et al*, 2005	Invitrogen
BODIPY 530/550-HPC	2-(BODIPY)-C5/12-1-hexadecanoyl-*sn*-glycero-3-phosphocholine	disordered	DPPC		Burns *et al*, 2005	Invitrogen
BODIPY-FL-SM	n-(BODIPY-FL)-C5/C12-sphingosyl-phosphocholine	free	SM		Hebbar *et al*, 2008; Eggeling *et al*, 2008	Invitrogen
Atto647N-SM	n-(Atto647N)-sphingosyl-phosphocholine	confined, but disordered (see Honigmann *et al*, 2010)	SM		Eggeling *et al*, 2008; Honigmann *et al*, 2010	Atto-Tec
DiI-C18	1,1′-dioctadecyl-3,3,3′,3′-tetramethylindocarbocyanine perchlorate	free (on neuroblastoma plasma membrane)	depends on lipid composition	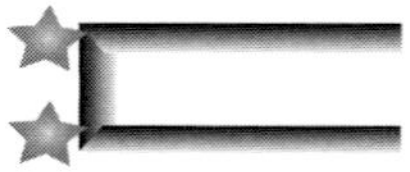	Hebbar *et al*, 2008; Sankaran *et al*, 2009	Invitrogen
BODIPY-C5-cer	n-(BODIPY-FL or TR)-C5-ceramide	?	ceramide		Pagano & Chen, 1998	Invitrogen
BODIPY-Gal-cer	n-(BODIPY-FL)-C12-sphingosyl 1-β-D-galactopyranoside	?	GSL		Simons, Kraemer *et al*, 2002	Invitrogen

BODIPY-C5-lac-cer	n-(BODIPY-FL)-C5-sphingosyl 1-β-D-lactoside	confined?	GSL		Pagano & Chen, 1998; Homchaudhuri, Wohland & Kraut, unpublished	Invitrogen
BODIPY-GM1	n-(BODIPY)-C5 ganglioside GM1 or head-GM1	disordered	GM1		Burns *et al*, 2005; Maruschak *et al*, 2007; Mikhalyov *et al*, 2009	Invitrogen; L.B.-A. Johansson lab
Atto647N-GM1	n-(Atto647N)-C4 ganglioside GM1, head-GM1, or additional acyl chain	confined	GM1		Eggeling *et al*, 2008; Polyakova *et al*, 2009	Atto-Tec; S. W. Hell lab
PEPTIDES						
SBD	Amyloid peptide Aβ 1-25 aa	confined	GD1a, GT1b, GM1, SM	-(AEEAc)2-DAEFRHDSGYEVHHQELVFFAEDVG	Hebbar *et al*, 2008; Steinert *et al*, 2009; Zhang *et al*, 2009; Sankaran *et al*, 2009	Kraut lab
lysenin	Earthworm toxin	ordered?	SM	297aa full length GFP fusion	Ishitsuka & Kobayashi, 2004	Peptide Institute Japan
CtxB-Alexa	Cholera toxin B subunit	confined	GM1	full length B subunit, Alexa tagged	Bacia *et al*, 2004; Hebbar *et al*, 2008	Invitrogen

*BODIPY = 4,4-difluoro-5,7-dimethyl-4-bora-3a,4a-diaza-s-indacene
**AEEAc = Amino-ethoxy-ethoxy-acetyl

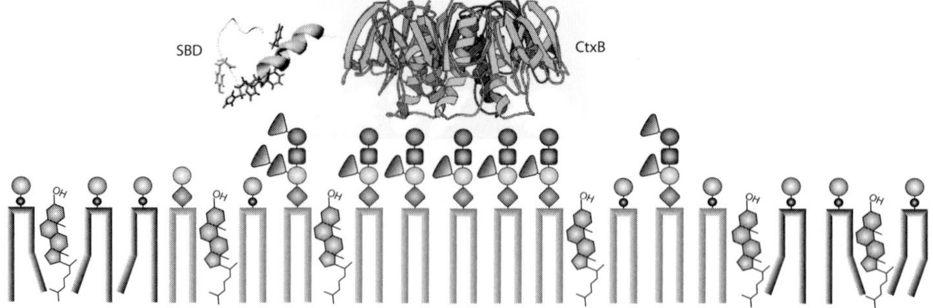

Fig. 1 Tracing sphingolipid behavior in the membrane with labeled sphingolipid-interacting peptides. An idealized outer plasma membrane leaflet containing a mixture of saturated straight-chain sphingolipids (orange backbones), including sphingomyelin (pink headgroup), gangliosides with carbohydrate headgroups (colored shapes), cholesterol (orange molecules), and unsaturated glycerophospholipids (gray backbones). Particular glycosphingolipids were chosen to represent those that bind to the shown sphingolipid-tracing peptides SBD and CtxB. SBD requires SM, cholesterol, and ganglioside to bind effectively to the membrane, while CtxB binds as a pentamer to five molecules of GM1 ganglioside. Key to colored shapes is shown in Fig. 2. SBD and CtxB were modified from Fantini *et al*., 2002, Expert Reviews in Molecular Medicine, and Merritt *et al*, Protein Science, 1994, Cambridge University Press. (See color plate.)

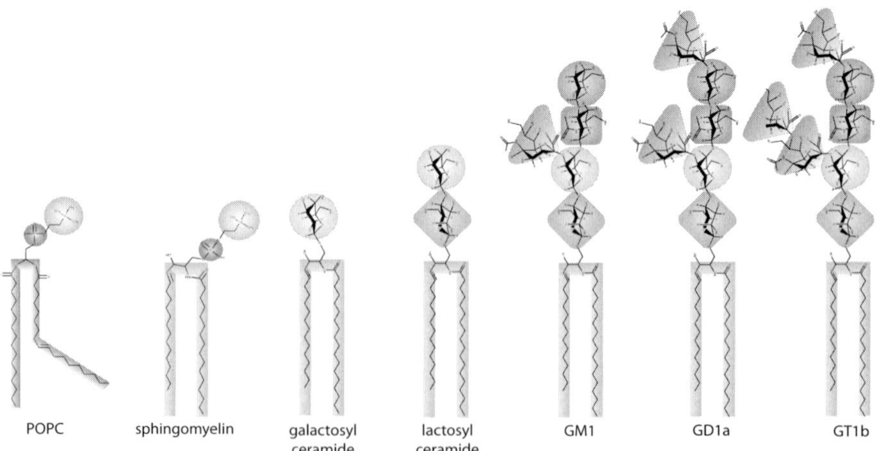

Fig. 2 Structures of a typical outer-leaflet phospholipid (phosphatidylcholine; POPC) and representative sphingolipids that are depicted graphically in the outer leaflet shown in Fig. 1. Headgroups are color-coded as follows: phosphocholine: gray and pink balls; galactose: aqua balls; glucose: orange diamonds; N-acetyl-galactosamine-green squares; sialic acid-purple triangles. Saturated straight acyl chains of the sphingolipids are colored orange; saturated and unsaturated acyl chains of the glycerophospholipid POPC are gray. (See color plate.)

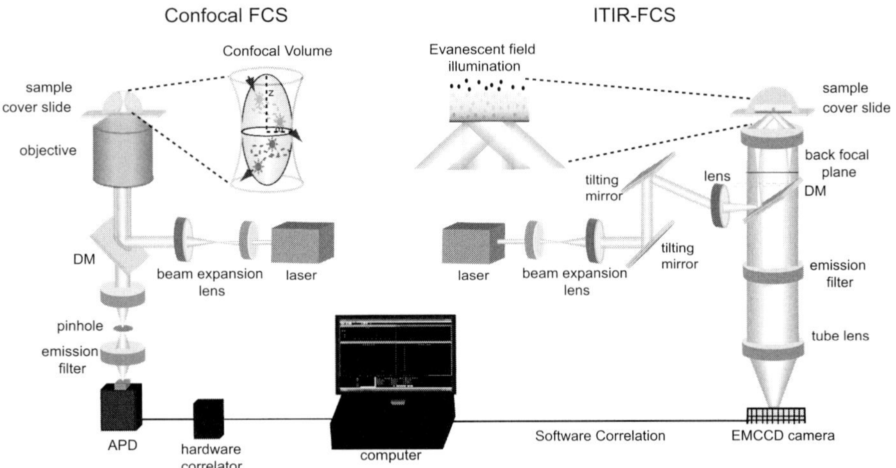

Fig. 3 Set ups of different FCS modalities. Confocal FCS (left) and Imaging Total Internal Reflection FCS (ITIR-FCS) (right). In confocal mode, a pinhole mediated focal volume is created and the emitted light is detected using an avalanche photodiode, a point detector. However, in TIR mode an evanescent field creates a 2D surface-bound observation volume that is imaged by EMCCD or sCMOS camera. The pixels of the CCD or sCOMS chips can be regarded as confocal volumes and thus the image is being multiplexed. (For color version of this figure, the reader is referred to the web version of this book.)

the movement of sphingolipid-tracing probes in the membrane, using a novel parameter called the ΔCCF (Sankaran *et al.*, 2009). The ΔCCF measurement (depicted in Fig. 5), referring to the "difference in cross correlation functions," is especially suited to distinguishing between random diffusion processes versus transport phenomena, because these tend to show differences in directionality over large areas. It gives a "gestalt" measure of the overall state of the membrane under a given condition, and thus can reveal features of membrane organization that are invisible to other methods that rely on single-point measurements, like confocal FCS or single particle tracking (SPT). ΔCCF is related to another method for detecting spatial heterogeneities, called the spatial pair-correlation function. Devised by Digman and Gratton (2009), this takes advantage of the observation that the maximum correlation between a probe observed at any two points in the membrane is proportional to the average transit time of that probe between the two points. Both of these methods are ideal for detecting diffusion barriers which cannot be seen by conventional microscopy.

II. Rationale

A goal in studying the biology and biophysics of sphingolipids is to be able to trace their behaviors in the cell membrane, and characterize what is thought to be a dynamic heterogeneity in the distribution and the composition of sphingolipid-containing domains. To achieve that goal, sphingolipid or microdomain-binding sequences from proteins that are known to interact with certain sphingolipids have been used as probes for labeling sphingolipid rich domains, or particular species of sphingolipid in the membrane (Bakrac *et al.*, 2010; Ishitsuka and Kobayashi, 2004; Lencer and Saslowsky, 2005; Steinert *et al.*, 2008). This approach works because sphingolipids occupy the outer leaflet of the plasma membrane, and therefore are accessible to exogenously added probes (a schematic drawing of the outer leaflet with representative sphingolipids and peptide probes is presented in Fig. 1).

Using synthesized peptides coupled with a small molecule fluorophore is a somewhat more expensive but much more direct way of labeling these domains than using transgenically expressed fluorescent protein fusions, or heterologously expressed GFP fusions. Transgenically produced fluorescent protein fusions suffer from the inherent problem that they must first be made in the biosynthetic pathway and then extruded across the membrane; GFP is a relatively large polypeptide (238 aa) that may also interfere with the interaction of the probe with the membrane.

Sphingolipid analogs have been used extensively, mainly as tracers of sphingolipid trafficking within cells (Pagano, 2003). Until recently, this has been limited to N-linked acyl chain substituted BODIPY- or NBD-linked sphingomyelin, ceramide and mono- or di-glycosylated ceramides like lactosylceramide (Fig. 2, Table I). The more complex gangliosides are very expensive and difficult to synthesize, limiting the available sphingolipid analog probes to the simpler species GM1, which is comparatively cheap and abundant. Acyl-chain substituted sphingolipid probes are commercially available (Invitrogen), but are often of limited usefulness for membrane dynamics studies because of their non-physiological behavior—for example, the BODIPY moiety in some cases flips up inside the membrane toward the polar headgroups of the lipids, giving it obviously very different biophysical properties from the endogenous lipids (Kaiser and London, 1998; Wang and Silvius, 2003). Notably, this appears not to be the case with acyl chain substitutions of the large and relatively lipophilic dye Atto647N (see Eggeling *et al.*, 2008).

That the probe itself may interfere with the physiological process or phenomenon to be studied is a potential hazard with all probes. This is particularly problematic with sphingolipid analogs that have substitutions of the fluorophore at the acyl chain, because these generally (with few exceptions (Eggeling *et al.*, 2008; Polyakova *et al.*, 2009)) don't behave in membranes like natural lipids. Head labeled or backbone-labeled probes have been used with more success (Burns *et al.*, 2005; Eggeling *et al.*, 2008; Schwarzmann *et al.*, 2005) (Table I), but are not commercially available and require expertise in organic chemistry to synthesize.

Lipid anchored probes like glycosyl-phosphatidylinositol (GPI)-linked proteins display partially ordered domain behavior, but their segregation is variable, and depending on the context seems to not always correspond to the behavior of sphingolipid containing liquid ordered domains (Glebov and Nichols, 2004; Kenworthy *et al.*, 2004; Sharma *et al.*, 2004). Rather, GPI-linked probes' behavior may depend more on the size of the headgroup, rather than the GPI-linkage (Bhagatji *et al.*,

2009). So what are the alternatives, if we wish to trace the behaviors of endogenous sphingolipids?

A. Using Peptides to Look at Sphingolipid Behavior

In order to characterize the diffusion behaviors of sphingolipids – or any membrane lipid for that matter – and how they interact with each other, ideally one would like to be able to observe individual molecules of a probe, or multiple probes, each binding to or mimicking a particular membrane component. At present, the main obstacle to doing this is that there are an extremely limited number of probes targeting different membrane lipids, including many different varieties of sphingolipid and glycolipid. Most probes currently being measured, as mentioned above, are either GPI-linked or transmembrane proteins that have been characterized as being associated with detergent-resistant membrane fractions of cells, along with the one commonly available glycolipid interacting probe, the Cholera Toxin B subunit (CtxB), coupled with an Alexa dye (Fig. 1 and Table I; see (Lagerholm *et al.*, 2005; Simons and Gerl, 2010)).

A different approach from using lipid analogs, lipid-linked proteins, or intact raft-associated proteins to analyze lipid behavior in the membrane, is to use short lipid binding peptides. Here, probes consisting of short glycolipid or sphingolipid-interacting peptides, like CtxB, which binds very tightly and specifically to ganglioside clusters as a pentamer, are labeled with a fluorophore and applied exogenously to cells. The probes bind to sphingolipid-containing domains in the outer leaflet of the plasma membrane, and their behavior can be followed. Other such probes, for example lysenin and perfringolysin, Shiga Toxin (StxB) and others, have also been developed as fluorescent protein fusions to label specific subtypes or components of membrane domains (sphingomyelin, cholesterol, and glycolipid Gb3 respectively) (Waheed *et al.*, 2001; Ishitsuka and Kobayashi, 2004; Romer *et al.*, 2007). It is important to keep in mind, and often neglected in the literature documenting microdomain behavior, that peptide probes may affect membrane organization upon binding, as has been shown to occur when CtxB, Shiga Toxin, and SV40 virus bind to the membrane. It has been proposed that these toxins and viruses, because they bind GM1 as pentamers, lead to coalescence and capping of GM1-containing domains, and strong curvature of the membrane, creating membrane-deformation and tubulating effects (Ewers *et al.*, 2009; Johannes and Mayor, 2010; Lingwood *et al.*, 2008; van Zanten *et al.*, 2010).

Analysis of the diffusion behavior of a probe is a means to deduce the location of that probe in either a liquid ordered or disordered state, or some other state (e.g., a gel state), and therefore to give information about the distribution in the membrane of the lipids to which the probe is binding. The underlying assumption is that because sphingolipid containing domains both in artificial membranes and in real cell membranes are known to exist in a *more* ordered state, whereas phospholipids exist in a *less* ordered state in the membrane, this will be detectable as a difference in both the order of the membrane itself, and in the fluidity of molecules that adhere to the membrane in some manner. The state of disorder in the membrane is often measured by changes in fluorescence anisotropy when labeled molecules that are close together in domains transfer their excited state energies and depolarize the signal (a method which will not be discussed here, but see (Goswami *et al.*, 2008; Reyes Mateo *et al.*, 1993)).

Lipid order is also influenced by the presence of proteins either embedded in the membrane, or associated with the membrane, such as cytoskeletal elements (Engelman, 2005; Lenne *et al.*, 2006; Chichili and Rodgers, 2007). Protein barriers in the membrane effectively reduce the degrees of freedom of the lipids and reduce diffusion, as well as possibly engaging in long-range protein–protein interactions that will affect the dynamics of the lipids around them (Daumas *et al.*, 2003; Destainville and Dumas, 2008). Actin polymerization has even been shown to induce phase separation in vesicles (Liu and Fletcher, 2006). Another nice example of membrane order being influenced by proteins is the mechanism by which Synaptotagmin may help to trigger vesicle fusion by de-mixing phosphatidyl-serine and induce buckling of the membrane (Lai *et al.*, 2011).

The degree of fluidity can also be effectively measured by the diffusion coefficient, since the lipids contained in less fluid, more ordered membranes diffuse at a slower rate. In fact, this scenario appears also to be true in cell membranes, as the diffusion rate of ordered-domain associated molecules has been shown by us and a number of other groups to be strongly influenced by membrane content of lipids that are thought to belong to ordered domains (Eggeling *et al.*, 2008; Hebbar *et al.*, 2008; Lasserre *et al.*, 2008; Lenne *et al.*, 2006; Sankaran *et al.*, 2009).

In this chapter, we describe one example of relatively short (<50 aa) peptide probes that are designed to interact with certain sphingolipid or glycolipid species, and that are derived from various toxins and pathogens (Fantini *et al.*, 2006; Mahfoud *et al.*, 2002). We have used the SBD probe, for Sphingolipid Binding Domain (SBD), a peptide of 25 aa, which interacts with a subset of gangliosides (see Figs. 1 and 2), as a tracer of sphingolipid containing domains.

B. How is Diffusion Behavior Determined? Comparison of Imaging Methods

A number of different spectroscopic, or fluorescence imaging and tracking methods can be used to measure diffusion rates. Two of the most popular are single fluorescent molecule tracking (or SPT) and confocal FCS (Kusumi *et al.*, 2010), and variants thereof, which we will discuss below.

1. FCS-based Methods

While probes are still sparse, methodologies of tracking the few probes that are available are relatively numerous and are quickly developing. Many of these involve spectroscopic approaches to measuring diffusion rates, such as FCS, which most often uses conventional confocal microscopy, to generate an illuminated area that generally stays fixed at the membrane, while individual molecules of probe move through it. Since the detection of single molecules demands either very low concentrations or a small observation volume, but at the same time many biological processes can be observed only above a certain concentration threshold, FCS records typically in a small femtoliter sized observation volume(s) at intermediate nanomolar to micromolar concentrations (Fig. 3).

For describing the behavior of fluorescent molecules diffusing on the membrane at the surface of a cell, FCS is an ideal method (see review by (Kim *et al.*, 2007; Machan and Hof, 2010)). It is capable of identifying diffusion characteristics for many thousands of individual molecules in a very short time, and therefore has the advantage of good statistics. However, measurements of immobile or very slowly diffusing particles are difficult in this FCS mode. Therefore, Weissman and Petersen introduced scanning FCS ((Petersen, 1986; Petersen *et al.*, 1986; Weissman *et al.*, 1976) see also (Petrasek and Schwille, 2008) and (Shi and Wohland, 2010) and references therein). Scanning FCS has multiple advantages. Characteristics of immobile or slow moving particles can now be determined along the scanning path since the transition time of the particle is determined by the scanning speed. Photobleaching is reduced since the scanning process distributes excitation energies along the scanning trajectory. Flow velocities can be measured and one can correct for sample movement by extracting the appropriate points in which the laser resided on the sample from the recorded scans.

Recently, stimulated emission-depletion (STED) microscopy, or TIRF microscopy have also been adapted to great effect for obtaining FCS measurements on very small, sub-resolution-size confocal areas (in STED), or over many pixels on a high-sensitivity CCD chip in TIRF. While confocal and STED both use a confocal volume, which ranges from ~30 to 500 nm, ITIR-FCS treats each pixel of the CCD camera as an individual "observational volume" (depicted schematically in Figs. 3 and 4). The size of this observation volume is determined by the convolution of the point spread function of the microscope and the pixel size and is similar in lateral size to a confocal volume but is limited in the axial direction by the penetration of the evanescent wave into the sample (~100 nm) (Fig. 3). The range of time resolution for these methods is nanoseconds for FCS to fractions of a millisecond for TIRF-FCS. Since transit times through the confocal volume are usually in the range of milliseconds for membrane bound molecules (with a diffusion rate of ~0.1–5 $\mu m^2/s$), this is generally sufficient to accurately describe membrane diffusion behavior.

2. FCS versus SPT

The detailed understanding of molecular processes and organization requires the investigation of single molecules in space and time. The statistics needed to draw conclusions from the single molecule data mandates the measurements of hundreds of single molecule events with the concomitant need for large scale data processing. This is, for instance, performed in SPT and single molecule spectroscopy (SMS) where a multiplicity of single molecule tracks or traces are evaluated individually to extract parameters which then can be statistically analyzed. Practitioners of single particle techniques argue that SPT, in contrast to spectroscopic techniques, gives possibly more accurate, and certainly more specific knowledge about the behavior of single molecules because they are observed individually over long periods of time, even when they are not moving (Kusumi *et al.*, 2010). However, due to the laborious measurement process, only a limited number of molecules can be analyzed, and the picture might therefore be biased or incomplete. Clearly, SPT would require unrealistic time and effort to yield the same quantity of data "by hand" as FCS. FCS uses fast, single molecule sensitive detection to record fluorescence fluctuations in time over many molecules and uses Fourier transforms or correlation functions to automatically analyze the data in a statistical manner. In this way, FCS circumvents the time consuming analysis of individual traces and provides better statistics, however at the cost of providing only ensemble data.

3. ITIR-FCS and ΔCCF: Advantages of ITIR-FCS

Conventional FCS has the drawback that it measures instantaneous diffusion behavior at individual points on the membrane. Clearly, this method by itself will have difficulty describing a membrane that is a dynamic and fluid structure consisting of many interacting parts. To study membrane dynamics, what is needed is a method that is capable of observing an entire region of the membrane simultaneously, and detecting features of its interconnectedness. TIRF microscopy is ideally suited to do this, since it sees only the surface of the cell (the membrane), images of which are acquired essentially simultaneously over the entire CCD chip, with region sizes up to hundreds of square microns, depending on the region of interest and the desired time resolution. By the same token, a CCD camera can capture hundreds of "snapshots" of *diffusion rates* over entire regions of the cell membrane simultaneously (see Figs. 3 and 4) (Kannan *et al.*, 2006, 2007). Information can be obtained in this way about fluorescently labeled sphingolipid-interacting peptides, or fluorescent sphingolipid analogs as markers of sphingolipid containing domains, or analogs of other lipid markers which may not be coincident with the sphingolipid containing domains (Sankaran *et al.*, 2009).

There are principally two different correlations on TIRF data that can be calculated individually or in combination. On the one side are temporal correlations collected from a single spot as done in the confocal modality of FCS. Information about molecular processes that cause temporal fluorescence fluctuations can be extracted from the fluorescence correlation functions. For instance, FCS will provide information on rotational and translational diffusion coefficients, directed translational motion (flow), concentrations, and photophysical properties of molecules, all of which can be used to derive secondary properties of the sample including chemical reactions, aggregation, binding, changes in size and/or viscosity etc. (see Fig. 4). On the other hand, spatial autocorrelation functions can also be calculated from images in so called image correlation spectroscopy (ICS) to provide information about directed movement of particles and their clustering behavior. ICS is a powerful

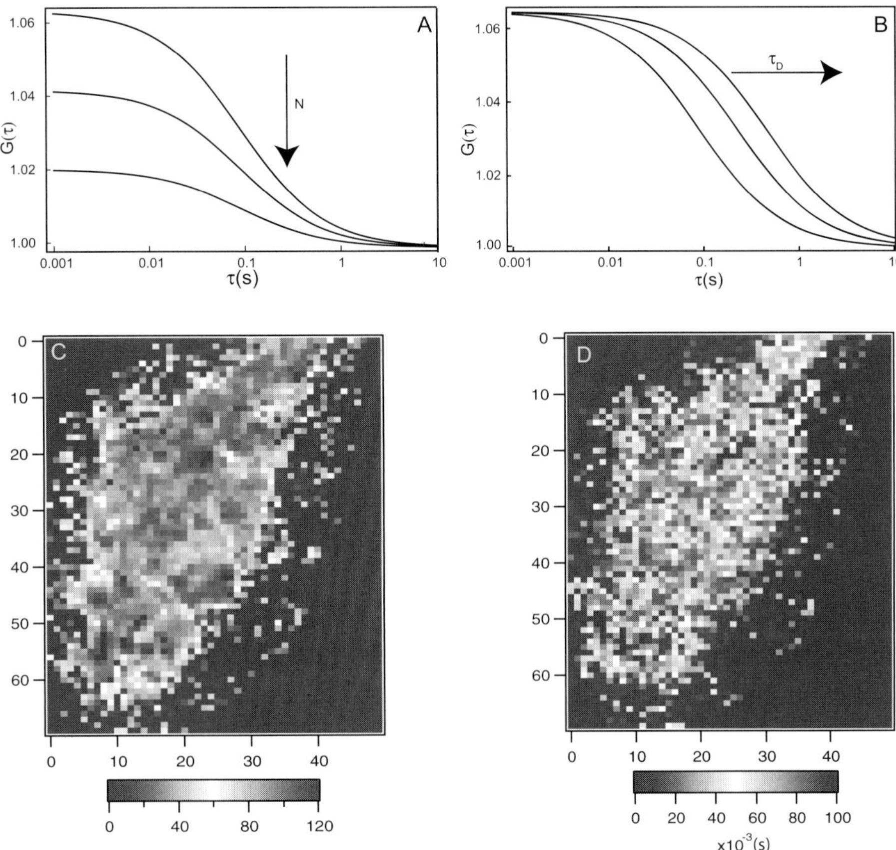

Fig. 4 Information that can be extracted from the characteristics of correlations functions and multiplexed images. (A) Autocorrelation curve as a function of particle number. The arrow shows the direction of increasing particle number (*N*). The amplitude of the autocorrelation function is inversely proportional to particle number. (B) Autocorrelation curve as a function of diffusion time. The arrow shows the direction of increasing diffusion time (τ_D). τ_D is defined as the time point of the full-width at half-max of the autocorrelation curve. (C) Heat-map image of particle numbers (*N*) measured at each pixel in the image of a whole cell. The color bar corresponds to particle numbers (*N*). (D) Heat-map image of diffusion times (τ_D) measured at each pixel of the same cell. The color bar corresponds to diffusion times (τ_D). (See color plate.)

tool to understand cell membrane dynamics, because the organization of the membrane represents the coordinated activity of multiple regions and domains, not just individual points. ICS has been extended to the temporal domain by collecting images with better time resolution and then observing the temporal development of ICS in temporal ICS (TICS). The final generalization to spatio-temporal correlations then came in the form of spatio-temporal ICS (STICS, (Hebert *et al.*, 2005)).

Similarly FCS was multiplexed by collecting multiple spots at the same time and thus extending the temporal correlations of FCS into space (Burkhardt and Schwille, 2006; Kannan *et al.*, 2006). This resulted then in the first contiguous FCS images being taken by Sisan *et al.* (2006) using a spinning disk confocal microscope. However, due to the spinning disk the measurement time per spot is only a fraction of the total measurement time and light collection efficiency is limited. In our case, the combination of spatial and temporal resolution is accomplished by a technique we developed called Imaging Total Internal Reflection Fluorescence Correlation Spectroscopy (ITIR-FCS) and a method of analyzing the ITIR data, called ΔCCF, which we describe below and represent graphically in Fig. 5 (see related methods in (Kolin and Wiseman, 2007; Petersen *et al.*, 1993)). This technique overcomes the problem of spinning disk FCS since the measurement time is larger than for the spinning disk. However, this comes at the price of being able to observe only samples close to the cover slip, as for example membranes. A solution to this is offered by single plane illumination microscopy based FCS (SPIM-FCS) which creates an illuminated plane anywhere within a sample. The thickness of this illuminated plane is ~10 times larger than the penetration depth of the evanescent wave but it allows the acquisition of FCS images anywhere in a 3D sample (Sankaran *et al.*, 2010; Wohland *et al.*, 2010).

The combination of the spatial and temporal correlations allows the extraction of information that is not evident from either one of them alone. In particular, the ITIR-FCS approach that we will describe in this chapter provides information about the

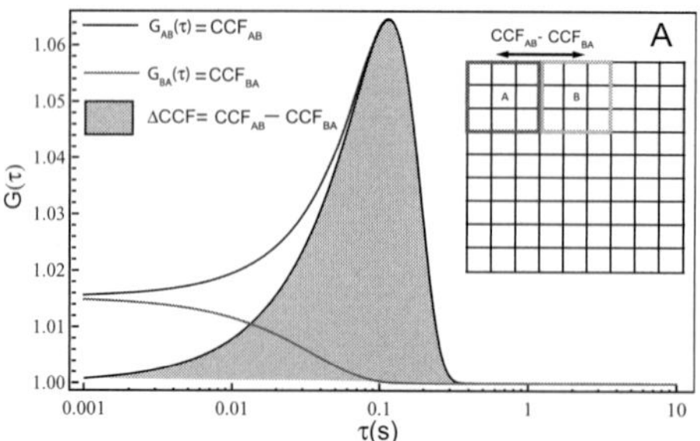

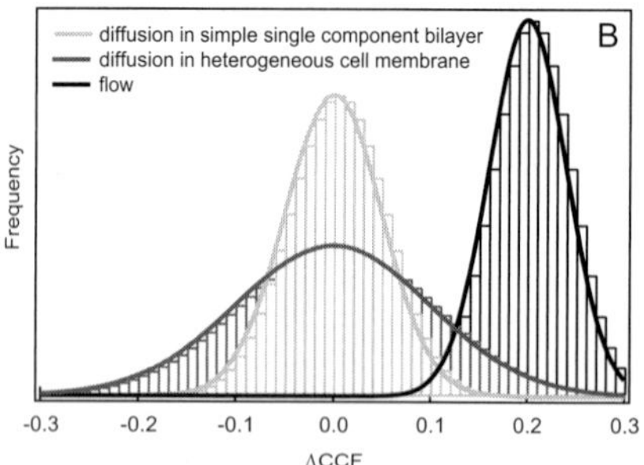

Fig. 5 Spatial cross correlation and ΔCCF. (A) Chip grid representation to show both the forward and backward spatial cross correlations between 3 x 3 pixel regions A (red) and B (green). For a typical flow process from region A to B, a forward CCF (AB) will have a peak at the transit time required to flow from A to B (blue curve) while a backward CCF (BA) will not have any peak (red curve). ΔCCF is represented by the green area which is the difference between the areas under the forward and backward correlations. (B) ΔCCF distributions for different dynamic processes in different systems. An anisotropic process like flow will have a ΔCCF distribution centered at some non-zero value (black). However, for an isotropic process like diffusion, the ΔCCF distribution will be centered at zero. Furthermore, more heterogeneity in the system will cause broadening in the ΔCCF distribution as can be seen from the distribution of homogeneous single lipid bilayers (green) and that of a cell membrane (red). (See color plate.)

transport of molecules from one area to another in a direction dependent manner. In the following, we will discuss the two methods which we have employed recently to investigate membrane dynamics and organization, namely confocal FCS and ITIR-FCS.

4. ΔCCF analysis on ITIR data: ΔCCF detects anisotropic movements

The last method we will discuss for observing membrane dynamics is the so-called ΔCCF function. A particular advantage of the ITIR methodology is that the spatiotemporal data of ITIR-FCS provides information about the transport of molecules from one area to another in a direction dependent manner. Therefore, one can use the information about the movement of particles between two areas A and B to characterize any anisotropy in a membrane. For instance, if particles flow from pixel A to pixel B then the temporal correlation between the two pixels will be different when calculated from A to B (C_{AB}) in comparison to the calculation from B to A (C_{BA}) (Fig. 5). By subtracting the two correlation functions from each other ($C_{AB} - C_{BA}$) and calculating the area under this curve one obtains a so-called Difference in Cross-Correlation Functions value, giving rise to the term ΔCCF (Sankaran *et al.*, 2009). By calculating this value for all neighboring pixel pairs in an FCS image one can create a ΔCCF histogram that shows characteristic changes depending on the state of the membrane (schematically shown in Fig. 5, and examples of actual histograms in Fig. 6).

On average there should be no directional bias of flow in a completely homogeneous membrane, so the ΔCCF between forward and backward correlations over many measurements tends to cluster near zero for membranes in which ordered domains have been disrupted, and which are therefore more homogeneous. In other words, a completely disordered membrane should result in a ΔCCF distribution of zero mean. In fact, for freely diffusing lipids in an artificial membrane, this is what is seen (Fig. 5). However, when there is flow or translational movement in the membrane, the mean of the ΔCCF distribution takes on a non-zero value. In addition, the width of the distribution will also be influenced by the degree of membrane order. Small local variations in flows or diffusion coefficients or obstacles to movements, while still random in nature, will leave the average of the ΔCCF distribution at zero but will lead to an increase in the range of non-zero values seen and thus a broadening of the ΔCCF distribution. While a completely disordered membrane should have a small width (e.g., lipid probes in the l_d phase), a membrane with several phases should exhibit a more complex behavior and a wider ΔCCF distribution. Thus, this method enables one to describe the degree of anisotropy (or heterogeneity) of dynamical properties in the membrane. ΔCCF should in principle be good at detecting borders between different regions, since borders will demarcate areas with different properties (see schematics in Figs. 5 and 6)(see (Feigenson, 2009) for a discussion of phase behavior in membranes).

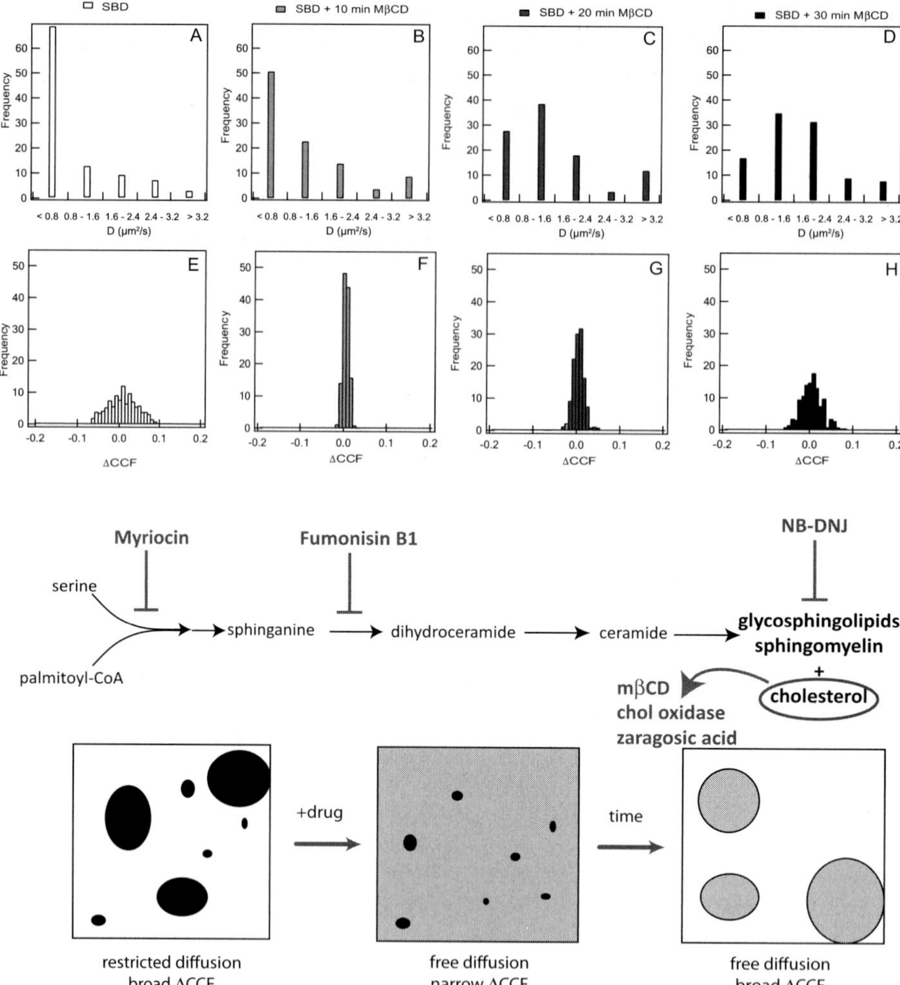

Fig. 6 Effects of sphingolipid and cholesterol disrupting drugs on diffusion and ΔCCF of the same cell over time. Distributions of diffusion D and ΔCCF for SBD: (A and E) before addition of mβCD, (B and F) 10 min. after addition of mβCD, (C and G) 20 min. after addition of mβCD, and (D and H) 30 min. after addition of mβCD, respectively. With the addition of mβCD, which depletes cholesterol from the cell membrane, the diffusion rate D of SBD increases with time (up to 30 min.), reflected in an increasing fraction of measurements on the histogram showing a high diffusion coefficient. In contrast to D, which remains severely altered in the presence of the drug over time, the distribution of ΔCCF reverts to a similar pattern to that seen before addition of the drug (compare H after 30' with E initially). This indicates that the cell membrane may regain some form of heterogeneity not reflected in D, but measurable by ΔCCF (depicted graphically in the lower part of the figure). The sphingolipid biosynthetic pathway (highly simplified) is shown in the flowchart, with drugs blocking individual steps in red. Each of these drugs disrupts both the D and the ΔCCF of SBD (Manna, Bag, Wohland, Kraut, in preparation), in a manner similar to that shown in the top histograms. The dark shapes in the leftmost block at the bottom of the figure represent restricted diffusion domains affecting both D and ΔCCF. The center block represents post-drug disrupted domains, while the block on the right shows unknown domains or some other source of heterogeneity which is detectable only by ΔCCF, but not by D. (For interpretation of the references to color in this figure legend, the reader is referred to the web version of this book.)

A very heterogeneous membrane should give a wide range of ΔCCF values. In other words, the width of the distribution and possibly its shape can give empirical evidence of membrane changes. Experimentally, we observe the distribution of ΔCCF values to be markedly different for a variety of membrane markers in the absence of any membrane perturbations; after drug treatments, depending on which phase the marker partitions into, which type of lipid it associates with, and how a given treatment affects the membrane composition, the ΔCCF changes in characteristic ways (Fig. 6) (Sankaran *et al.*, 2009); Manna, Bag, Wohland, Kraut, in preparation).

Interestingly, ΔCCF detects different properties of the membrane from diffusion alone. This is clearly demonstrated by the observation that the increase in the spot diffusion rate of a disordered marker like short-chain DiI is only marginal, or undetectable, after treatments that should produce ceramide, for example, by cleaving sphingomyelin or by inhibiting glycosylation of ceramides (see section heading "perturbation of lipid order" below). In contrast, ΔCCF detects very strong changes in anisotropic distribution of DiI diffusion when ceramides are produced by these methods (Manna, Wohland, Kraut, in preparation), indicating that borders of domains perhaps consisting of gel-state ceramide are being detected (Chiantia *et al.*, 2008). The difference between the properties of the membrane measured by simple diffusion versus ΔCCF are also made clear by the fact that the membrane can simultaneously recover a broad, heterogeneous ΔCCF distribution, while retaining the faster diffusion reflective of a raft-disrupted membrane, fo example, during mβCD treatment (see Fig. 6, and unpublished results).

III. Methods

A. Short sphingolipid binding domain (SBD) Probes

As discussed above, a sphingolipid- or glycolipid-binding probe that can follow the dynamic behaviors of sphingolipids in the outer leaflet of the cell, without interfering with the normal behaviors of those lipids, would be a valuable tool for cell biologists interested in sphingolipids, biophysicists who study membrane dynamics, and clinical researchers interested in diagnostics for disorders that perturb sphingolipid metabolism. Of course, anyone familiar with the Heisenberg uncertainty principle would wonder how one gets around the problem of the probe itself interfering with the process to be observed. Although any probe will in some way affect its environment, it is certainly possible to generate a frame of reference by comparing different probes' degrees of interference, by measuring the distribution or other properties (diffusion, anisotropy, quenching, etc.) of a second fluorescent molecule in the presence of the first probe (e.g., Bakht and London, 2007; Pinaud *et al.*, 2009). Rather than incorporating a fluorescent lipid analog into cell membranes, or expressing a fluorescent protein fusion, a strategy where the synthetic fluorescent probe can be applied to cells exogenously can be advantageous. This avoids the need to express a fusion protein in cells, as well as the obstacle of getting the probe extruded from the cell such that it can bind to the outer leaflet, where sphingolipids are found.

B. SBD Probe From Aβ

Based on the idea of a sphingolipid binding domain, or SBD, we developed a small peptide probe, derived from the putative sphingolipid interacting domain proposed by Fantini and coworkers in 2002 (Mahfoud *et al.*, 2002), that is present in the N-terminal ~25 amino acids of the Aβ peptide, the endogenously produced, secreted fragment of the Amyloid Precursor protein (App), the causative agent in Alzheimer's disease. This sequence was selected because of the extensive characterization of the full length, naturally occurring peptides Aβ1-40 or 1-42, which are known to interact with gangliosides in artificial membranes, and can be isolated with ganglioside (particularly GM1) in brain extracts (Yanagisawa, 2007; Yanagisawa *et al.*, 1995).

The molecule used to link the peptide with a fluorophore seems to be critical to its use as a probe, perhaps because ganglioside-interacting peptides would be expected to aggregate readily, due to the hydrophobic nature of their aromatic-rich regions (Fantini, 2007). For example, a poly-Glycine linker attached to the SBD from Aβ led to strong aggregation, and rendered the probe unusable. Polyethyleneglycol (PEG)-based linkers were used with greater success: two tandem copies of the PEG-like sequence amino-ethoxy-ethoxy-acetyl (AEEAc; Table I; described in (Steinert *et al.*, 2008)) and a multiple PEG linker were both tried. PEG is thought to be inert and non-interacting with the membrane, since it is uncharged and not particularly hydrophobic. One should note that in an otherwise identical peptide tagged with the same fluorophore, the AEEAc linker appears to give fewer problems with aggregation than the PEG linker (Lauterbach, Kraut, Wohland, unpublished), so this slight difference in composition of the linker may have relatively strong effects on probe behavior.

C. Charge Variation in the Probe

Probes interacting with glycosphingolipids (GSL), especially gangliosides with large negatively charged headgroups (see Figs. 1 and 2), would presumably encounter a sea of negative charges upon approaching the membrane surface, assuming that like-charged GSL cluster together (which is an unsafe assumption; see (Prinetti *et al.*, 2009; Westerlund and Slotte, 2009)). The "native distribution" of GSL in the unperturbed membrane is as yet completely unclear, precisely because there are extremely few probes available to examine their distribution directly. Lateral segregation into domains will be affected by the size and charge of the particular glycolipid headgroup (Longo and Blanchette, 2007), which in the case of complex gangliosides and sulfatides are bulky and negatively charged. While factors like hydration or cations in the surrounding medium may play a role in masking these charges (Westerlund and Slotte, 2009), the charge of the probe is probably quite important in influencing the ability of the probe to approach and bind, as has been suggested for Prion protein's interaction with the membrane (Critchley *et al.*, 2004). This may be the reason for the observed pH sensitivity of Aβ's binding to GM1 as well as other gangliosides (it binds more strongly at low pH, when the sialic acids would be partially protonated (Ikeda and Matsuzaki, 2008; Valdes-Gonzalez *et al.*, 2003) (see also (Kamata *et al.*, 1997)). As is the case with Aβ, the binding of SBD probe to gangliosides seems to be increased at low pH close to its pKa, with lower pH increasing the relative binding to GM1 over other more sialylated gangliosides (Steinert *et al.*, 2008).

D. Sphingolipid Analogs

The primary choice for directly examining the behavior of sphingolipids in membranes would of course be to use fluorescently substituted sphingolipid analogs. The problem with this approach is the non-physiological behavior of the N-acyl chain substituted sphingolipid analogs with fluorophores, usually BODIPY or NBD, that are commercially available (see above, and Table I). Although BODIPY is more lipophilic and therefore a better choice than NBD for acyl-chain substitutions, there have been many reports that acyl chain substitutions in general, also with BODIPY, interfere with the normal phase behavior and membrane positioning of sphingolipids (Kaiser and London, 1998; Wang and Silvius, 2003). Head-labeling of for example GM1 with Alexa fluorophores (Burns *et al.*, 2005), or sphingomyelin and GM1 with the more lipophilic dye Atto647N (Eggeling *et al.*, 2008) have yielded far more promising results.

Using combined AFM and fluorescence, Burns *et al.* favorably compared Alexa488 head-labeled GM1 with commercial acyl-chain BODIPY-substituted GM1, showing that the head-labeled GM1 preferred ordered dipalmitoyl-phosphatidylcholine (DPPC) domains, whereas the acyl-chain BODIPY-GM1 clearly did not. Notably, however, trapping and anomalous diffusion behavior expected for ordered domain-associated molecules is seen even with N-acyl-substituted Atto647N labeled sphingolipids, including GM1 and sphingomyelin, in the STED-FCS measurements of Eggeling *et al.* Their analogs with Atto dyes suggest – not surprisingly – that lipophilic Atto647N is a better choice than the more hydrophilic Atto532 for acyl chain labeling of SM. These conclusions on the trapping behavior of Atto-substituted sphingolipids were based on the FCS diffusion laws discussed below, but it should be noted that the l_o/l_d partitioning and rotational diffusion of Atto647N-lipids in phase separating mixtures clearly indicated incorporation of the same probes into *disordered* domains (Honigmann *et al.*, 2010). It is important that the effects on biophysical behavior resulting from any manipulation or change in choice of fluorophore be tested, in the best case using several different approaches. Given the multitude of choices for fluorescent labels and constructions of the analogs, there do not seem to be as yet any hard and fast rules by which one could predict how a lipid analogue will behave after substitutions.

The addition of sphingolipid analogs to cell membranes can be achieved easily by coupling of the lipid to bovine serum albumin (BSA), as described in Martin and Pagano (1994). Briefly, the BSA-coupled lipid is added in a Hanks Buffered Salt Solution (HBSS)/HEPES buffer to between 1 and 5 μM for approximately 30 minutes, on cells that have either been cooled to between 4 and 18 °C, or warmer, depending on whether trafficking is to be synchronized as in a pulse-chase experiment (see (Steinert *et al.*, 2008)). Cells to be imaged should be seeded on coverslip-bottomed dishes or 8-well plates. 8-well plates have the advantage that they have a small capacity (~200 μL), and therefore use less probe, if scarcity or expense is an issue. Imaging by TIRF-FCS or FCS has to be carried out with a lens of high numerical aperture (NA). TIRF lenses are generally NA 1.42 or higher.

E. Perturbations of Lipid Order and Cytoskeleton as a Means to Test Domain Localization

When examining the diffusion behavior of sphingolipid analogs or probes that putatively associate with sphingolipid domains in cells, an important test for association with these domains is to eradicate them pharmacologically and then observe the changes in diffusion behavior, for example, (Eggeling *et al.*, 2008; Lasserre *et al.*, 2008; Sankaran *et al.*, 2009). Since the diffusion behavior of a probe will also be influenced by interactions of its target lipids with other proteins and lipids in/on the membrane (Epand, 2004) it is informative to test the behavior of the probe in an artificial system such as supported lipid bilayers, which lack proteins and have a known lipid composition (Bacia, Scherfeld, BPJ 04; Kahya & Schwille 06). In this way, a comparison can be made to parallel conditions in drug-treated cells, for example by simply leaving out or substituting different components. Common methods of altering the sphingolipid composition of cells to which the probes are applied include treating the cells with inhibitors of sphingolipid biosynthesis such as the fungal antibiotic Fumonisin B1 (FB1), which inhibits dihydroceramide synthase, and myriocin, which inhibits serine palmitoyl transferase (Merrill, 2002; Merrill *et al.*, 1993). We have also applied bacterial sphingomyelinase and N-butyl-deoxynojirimycin (NB-DNJ) (Platt *et al.*, 1994; Yu *et al.*, 2005) to convert membrane sphingomyelin into ceramide, and to inhibit the production of GSL by glucosylceramide synthase (Fig. 6 shows the schematized pathway with different inhibitors). It should be kept in mind when using these treatments that inhibiting any biosynthetic step will probably lead to a buildup of the precursor to that step, for example, NB-DNJ may give abnormally high levels of ceramide, and FB1 may raise sphinganine levels (Merrill, 2002).

Cholesterol extraction by methyl-β-cyclodextrin (mβCD) has been commonly used as a means to disrupt sphingolipid containing ordered domains. While low concentrations of this drug (~1 mM) can be used, higher concentrations have been found to strongly disrupt the actin cytoskeleton, such that it is hard to draw conclusions about the origin of changes in diffusion behavior, and whether they are traceable to cholesterol at all, as opposed to for example cholesterol-directed cytoskeletal processes (Chadda *et al.*, 2007; Goswami *et al.*, 2008). We observe strong increases in the diffusion of SBD and other l_o markers already at 1 mM mβCD (Fig. 6; Sankaran *et al.*, 2009). Interference with cholesterol biosynthesis or structure have been used as alternatives to mβCD extraction, for example, zaragosic acid (Lasserre *et al.*, 2008) or cholesterol oxidase (Eggeling *et al.*, 2008).

As a reference to observe the diffusion behavior of a non-ordered (l_d) phase lipid, we have used DiI with a chain length of 18 carbons (DiI-C18). DiI-C18 diffuses about 3–4 times faster in live cells than SBD or other raft markers and the diffusion rate does not change appreciably after sphingolipid perturbations, although cytoskeletal disruption also increases its diffusion slightly (Sankaran *et al.*, 2009).

F. Methods for Examining Probe Behavior

1. Principle of Confocal FCS

The principle of FCS rests on the analysis of fluorescence fluctuations from a small observation volume. The fluctuations can be caused by any processes which influence the fluorescence intensity in the observation volume, in particular the transition of fluorescent molecules through the observation volume (diffusion, flow), but as well changes in the fluorescence properties of the molecules (photophysical changes, molecular rotations if polarized lasers or detection schemes are used). The correlation functions are calculated from the recorded photon count traces, typically recorded over a few seconds to minutes, and give access to the number of molecules observed on average within and their transition time through the observation volume (Figs. 3 and 4). Since practically it is often difficult to establish the exact size of the observation volume, FCS is calibrated with a fluorescent dye standard of known concentration and diffusion coefficient (Ries *et al.*, 2010). Since molecules free in solution and bound to membranes diffuse at rates which differ by about two orders of magnitude they can be easily distinguished by their transition times through the observation volume (~20–100 μs in solution vs. ~5–100 ms in the membrane). Confocal FCS typically measures different spots on the membrane sequentially, in which case one has to position the focal volume repeatedly at different places by moving the sample or the laser beam, although multiplexed detection for confocal FCS has also been used (Brinkmeier *et al.*, 1999; Gösch *et al.*, 2004; Kannan *et al.*, 2006; Sisan *et al.*, 2006).

2. Instrumentation for Doing FCS

Confocal FCS can be performed either with commercial systems that are offered by many of the microscope manufacturers or companies dedicated to SMS, or it can be performed on customized instruments (Pan *et al.*, 2007). The standard detectors used in FCS are avalanche photodiodes (APDs) with single photon counting capabilities and a quantum efficiency of 40% or more in the range between 400 and 700 nm. The advantage of confocal FCS is its time resolution (nanoseconds) and the free placement of the observation volume in a 3D environment, as for example in a tissue or organism (Bacia *et al.*, 2006; Shi *et al.*, 2009). Correlation of the raw data is performed by hardware or software correlators. Software correlators are the cheaper alternative and offer more flexibility but hardware correlators have typically the better time resolution, down to nanoseconds. The time resolution of the software correlators, typically on the order of microseconds, depends on the computer system used but is in most cases, and in particular for membrane measurements, sufficient for almost all cases. The data collected can be analyzed by software delivered with commercial systems or by a range of self-written software tools which have been previously published (Sengupta *et al.*, 2003; Skakun *et al.*, 2005) (Fig. 3).

G. ITIR-FCS Analysis of Probe Behavior Over Large Areas of the Cell Surface

1. Why ITIRF-FCS is a Good Way of Looking at the Membrane & How it Works

Imaging Total Internal Reflection-FCS (ITIR-FCS) uses a standard TIRF microscope to obtain diffusion measurements as with confocal FCS, but collects those measurements simultaneously on large membrane areas. This is made possible by two important facts. First, the observation volume in ITIR-FCS is created by a thin, exponentially decaying evanescent layer which can be created within the solution side of a glass water interface (Axelrod, 2003). The thickness of this evanescent layer can be controlled to some extent but is typically on the order of 100 nm. This provides a very thin z-section from which fluorescence can be detected. In combination with the limited pixel size of the camera, and the small point spread function of the microscope, this leads to very small observation volumes per pixel and in addition reduces significantly the background signal since no light enters the sample expect within the layer observed. Therefore, every camera pixel has a small observation volume in which single molecules can be easily detected. Since the size of the camera pixels are exactly known in ITIR-FCS for membrane measurements no calibration is needed. As a matter of fact SPT, ICS and single molecule imaging use the same illumination approach. Second, the difference in ITIR-FCS is the use of high-speed sensitive cameras which allow recording with a time resolution on the order of milliseconds or below, which is sufficient to observe membrane dynamics. The cameras used in this approach have been to date either electron multiplying CCD (EMCCD) cameras or scientific CMOS (sCMOS) cameras due to their high quantum efficiencies of 60–95% and their fast read-out speeds.

2. Data Analysis and Curve Fitting

The data analysis and interpretation is arguably the most difficult part in FCS. From the observation volume characteristics and the expected molecular behavior, a pre-determined model has to be fit to the FCS raw data. However, the choice and correctness of the model crucially depends on the knowledge of the experimenter. For example, in some cases a fit with a model assuming one or two different particles can be difficult to differentiate and extra confirmation of a second existing particle might be needed from a complementary method before correct fitting can be performed. In praxis, the experimenter fits models of increasing complexity as long as a statistical test confirms that the more complex model is significantly better. This can be performed by different statistical tests (e.g., *F*-test, or runs test). When the more complex model fails the search has finished. Here in the applications to membranes in general we assume two differently moving particles, for example, one bound to the membrane and one free in solution. In addition, for confocal FCS, which provides the necessary time resolution below the μs range, the models will include photophysical behavior (Shi and Wohland, 2010; Widengren *et al.*, 1995, 1999). For ITIR-FCS, with a time resolution of only milliseconds (limited by the camera speed), the photophysical characteristics of the probes cannot be detected and simpler models, which only take into account diffusion and transport, can be employed (Sankaran *et al.*, 2009).

H. FCS Diffusion Laws

In 2005, Wawrezinieck *et al.* (2005) proposed the so-called FCS diffusion laws. Since the diffusion coefficient is a constant it should not vary in a homogeneous membrane, and the time a particle needs to cross a certain area (the diffusion time) should be linearly proportional to the area. Therefore a plot of the diffusion time versus the various area sizes observed in FCS should give a linear plot through the origin. However, if the membrane is heterogeneous the plot, while still linear, should exhibit positive y-intercepts for transient membrane domains which trap the probe molecules, or negative intercepts due to an underlying meshwork which inhibits free diffusion. These measurements can be performed by enlarging the confocal spot size in steps to test the linearity of the relationship between diffusion time (τD) and spot diameter. In the past, various different techniques have been used to vary the spot size (Masuda *et al.*, 2005; Ruprecht *et al.*, 2011; Wawrezinieck *et al.*, 2004, 2005). Either the confocal pinhole and/or the beam expansion of the exciting laser beam have been varied to change the size of the observation volume, or in STED-FCS the spot size can be controlled by the light intensity of the laser causing stimulated emission (Eggeling *et al.*, 2008). However, they required consecutive measurements for each individual spot size. In camera-based FCS, this test can be readily performed in a single measurement. The data in an image can be analyzed either pixel by pixel or any number of pixels can be binned to create various effective spot sizes from which correlations und thus the diffusion coefficients can be determined. A comparison of the results over increasing spot sizes can then be used to test the FCS diffusion laws, and to deduce the presence or absence of trapping domains or other barriers to diffusion.

IV. Materials

A. Fluorophores-General Features

FCS experiments rely upon the use of fluorescence labels as reporters of the molecular dynamics of a tagged particle, protein, or molecule. Since the signal-to-noise ratio in FCS crucially depends on the molecular brightness, that is, the number of photons detected per particle and unit time, it is important to select the optimal probes which maximize molecular brightness but are at the same time compatible with the sample and its integrity. For example, a good membrane label should have a high partition coefficient for the membrane, or specific regions or phases of the membrane, and at the same time should not alter membrane dynamics or organization. An exhaustive list of fluorescent membrane labels, including DiI of different chain lengths and other fluorescent lipid mimetics, for example, polycyclic aromatic hydrocarbons such as perylene, and their segregation behavior into ordered membrane domains is given in Baumgart *et al.* (2007).

There are some main characteristics of a fluorophore necessary for a satisfactory FCS signal. First, the excitation and emission wavelengths must be matched as closely as possible to the lasers and optical filters of the system at hand to allow optimal excitation and detection of the fluorophores. Second, the molar absorptivity or extinction coefficient, which determines how well a flurophore absorbs light and thus how easily it is excited, should be as large as possible. This is very important in light sensitive samples so that a good signal-to-noise ratio can be reached with the lowest possible excitation intensity. Third, the fluorescence quantum yield determines how many of the absorbed photons are re-emitted. Fourth, the probe should be photostable and have a low triplet state yield (a microsecond or longer lived state in which the fluorophore will not emit light), so that molecules can be observed over long times with minimal time in dark states (Widengren *et al.*, 1995). These factors are the main determinants of the molecular brightness of a fluorophore and are important for the signal-to-noise ratio in FCS. For a more comprehensive discussion of fluorophores, see (Bacia *et al.*, 2006; Maier *et al.*, 2002).

B. Lipid Analogs

BODIPY-labeled sphingolipid analogs of ceramide, sphingomyelin, GM1, and lactosylceramide with the fluorophore attached to the N-linked acyl chain at different positions (C5, C6, or C12) are available from Invitrogen. Avanti Polar Lipids and Atto-tec also synthesize GM1 analogs with acyl chain substitutions of NBD or the Atto dyes, which have different lipophilicities depending on the dye (see Eggeling *et al.*, 2008). For lipid analogs, we find that BODIPY-sphingomyelin and a ceramide derivative (Manna, Jennings, Kraut, Wohland, unpublished) behave like a l_d phase probe in that they are fast-diffusing (Hebbar *et al.*, 2008). Lactosyl ceramide diffuses with a D intermediate between the sphingolipid-associated peptides CtxB and SBD, but more slowly than DiI-C18 or BODIPY-SM, for example (Homchaudhuri, Kraut, Wohland, unpublished).

C. Peptides

CtxB coupled with Alexa dyes can be obtained from Invitrogen; fluorescently labeled SBD peptide, and other peptides have been produced by individual laboratories, cited within the text (see Table I). For tracing sphingolipid domains, we have obtained consistent measurements with SBD, CtxB, and a raftophilic sterol-derivative, which all have slow, bimodal diffusion rates typical of raft markers, and are sensitive to sphingolipid depleting agents.

D. TIRF Hardware and ITIR-FCS Analysis Software

Although to our knowledge no commercial software exists for the evaluation of ITIR-FCS and ΔCCF data to date, a free version which uses image stacks as input and correlates and provides tools to fit the data is available at http://staff.science.nus.edu.sg/~chmwt/ImFCS.html. The advantage of ITIR-FCS is that it requires no customized instrumentation and can be implemented at any TIRF microscope by using a fast and highly sensitive camera.

E. Experimental Considerations for Diffusion Analysis

Since the diffusion coefficient of particles, but also the composition and phase behavior of cell membranes is temperature-dependent, one should make sure to measure membrane dynamics at a fixed temperature. Ideally this temperature is physiological; however, in some cases probe internalization on live cells can be very fast and it can be advantageous to measure the behavior of the probe at lower temperatures.

The position of a membrane measurement can be important. It has been shown by an FCS modality called z-scan FCS, which allows more precise diffusion coefficient measurements than confocal FCS by precisely controlling the position of the membrane within the confocal volume, that membrane in contact with a glass surface exhibits slower diffusion by a factor of about 1.5–2 due to sticking behavior of the lipids in artificial and cell membranes to the surface (Benda *et al.*, 2003). Therefore one should select membranes not in contact with glass surfaces for FCS measurements, if possible. While this is not possible with ITIR-FCS, a variant of imaging FCS based on single plane imaging microscopy (SPIM-FCS) allows imaging FCS to be performed in 3D. A comparison of different methods for the determination of diffusion coefficients (z-scan FCS, FCS, ITIR-FCS, FRAP, and SPT) showed the same effect of sticking but demonstrated that overall, the different methods obtain broadly similar diffusion coefficients, although one should be clearly aware of the different capabilities and uncertainties of the different approaches (Guo *et al.*, 2008).

Lastly, it should be mentioned that the diffusion behavior in the inner and outer leaflets of a cell membrane can also differ (Golebiewska *et al.*, 2008). The specific interactions of proteins and lipids in different layers of cell membrane cause the difference in their diffusion behavior.

V. Discussion

A. Probing Sphingolipid Organization in Membranes

That sphingolipid (or any other lipid)-binding fluorescent probes can be used to assay membrane organization relies on the phenomenon that these probes interact with a particular lipid species, or combination of lipids, or a lipid membrane surface with certain characteristics—that is, order, size, and charge of head group, height, etc. Given this interaction, our methods to look at the movement of those probes in the membrane and interpret diffusion behavior rests on the assumption that the diffusion of the probe molecules mirrors the diffusion and organization of the lipid molecules beneath them. While it seems reasonable, in reality, we do not know to what extent this assumption is valid, since only an indirect description of membrane composition and dynamics can be achieved by following non-lipid probes. There are two main considerations.

Exogenous peptide-based probes may not give an accurate measure of actual lipid behavior: that is, the diffusion of the peptide presumably reports on the presence of its target lipid, but certainly does not give a precise description of the entire composition of the interacting domain, its size, mixing with other domains, or its lifetime.

The problem of perturbing the system by observing it: interaction of the probe with the membrane may itself influence lipid distribution in the membrane. As mentioned earlier, this is known to occur with CtxB, which clusters small domains into larger aggregates, but it is not clear if all membrane-interacting peptides (or all sphingolipid interacting peptides) would do this. Does it depend on the binding strength of the probe? Or the fact that the probe multimerizes? As yet, no controlled and systematic test measuring the effects of these parameters in different binding sequences has been carried out, although this should in principle be possible. Toxins like Tetanus and the Botulinum toxins have been thoroughly studied, and their likely ganglioside targets are known (Chen *et al.*, 2009; Fotinou *et al.*, 2001; Fu *et al.*, 2009; Kitamura *et al.*, 1980; Stenmark *et al.*, 2008). The different serotypes of Botulinum and tetanus toxin are related by sequence homology to each other (Ginalski *et al.*, 2000), but have different affinities for gangliosides (Rummel *et al.*, 2004). It has also been established that a relatively short fragment of Tetanus still can interact with its favored target, GT1b (Halpern and Loftus, 1993; Louch *et al.*, 2002; Shapiro *et al.*, 1997).

The SBD peptide studied in our laboratories appears to have an affinity in the range of one order of magnitude lower for its target ganglioside GT1b, than these toxin sequences, and a relatively loose specificity for GT1b versus other gangliosides such as GD1a. SBD can bind equally well to GM1, but only at low pH. Binding preference was determined by Surface Plasmon Resonance assay on dextran-modified L1 chips specifically designed for immobilizing lipid surfaces (GE Healthcare), and by liposome-capture experiments comparing lipid mixtures of different composition (Steinert *et al.*, 08; and unpublished). SBD may also differ from the toxins discussed above in that it requires the presence not only of cholesterol but also sphingomyelin for optimum binding, in both artificial membranes and in cell membranes (Steinert *et al.*; unpublished). This range of properties in the different sphingolipid-interacting candidate proteins and peptides presents an interesting opportunity to compare the effects of such features as binding strength, multimerization, and glycolipid specificity on behavior, since some of the toxin sequences have closely related structures, but different affinities for target lipids. Conversely, peptides like SBD may have similar lipid specificities (in this case for GT1b or GD1a), but completely unrelated sequences. Although the clustering behavior of SBD has not been studied, clustering would be expected to affect probe dynamics as well as perhaps exerting organizing effects on the membrane it binds to. Multimerization or more haphazard types of clustering would be an interesting parameter to test systematically for its effects on membrane organization and diffusion of the peptides. This could be achieved by engineering artificially multimerized probes from existing monomeric sequences such as the clostridial toxins. To make probe design easier, even though these toxin sequences are relatively long and may interact with multiple glycolipid molecules, individual interacting domains could potentially be isolated.

B. How SBD Domains Might Bind

The diversity and range of proteins that interact with sphingolipids, in particular glycolipids, at the cell surface is enormous. Because of this diversity, it may seem counterintuitive that a single short sequence could mediate any sort of specific interaction. An overall motif for sphingolipid, in particular ganglioside-interacting sequences, although by necessity very poorly defined, appears to include a high density of aromatic amino acids near or surrounding a turn (in some cases induced by proline or glycine), and in the vicinity of one or more basic amino acids, like Arg or Lys (Fantini, 2007). According to Fantini's model, sphingolipid (or more specifically, glycolipid)-interacting peptides do so by virtue of the presence of these strategically placed aromatic amino acids that can engage in pi-CH bonding with the sugar rings that are presented at the termini of the glycolipids, analogous to lectin binding to carbohydrates (Taieb *et al.*, 2004;

Weis and Drickamer, 1996). While this mechanism has not as yet been strictly proven, it does seem to be the case that certain aromatic amino acids clustered near the binding region are important for the interaction of some glycolipid-associated proteins, like bacterial toxins Tetanus and Botulinum, and Aβ (Fotinou *et al.*, 2001; Louch *et al.*, 2002; Rummel *et al.*, 2004). Sphingolipid interacting domains of proteins often have a low affinity for the lipids themselves, but are nonetheless necessary for function, for example, in the case of virus infectivity. In these cases, the binding may act more like a surveyor of the membrane landscape, before high affinity binding to a protein receptor takes place. For the purposes of designing probes, this may not be a bad thing, since one would like to follow the behaviors of lipids rather than force them into a certain configuration, for example, by virtue of tight binding, which might lead to aggregation behavior and distortion of the membrane.

VI. Summary and Outlook

A. How Far the Study of Sphingolipid Organization in Membranes has Come

Ever since the discovery in the 1980 s that sphingolipids phase separate and form coexisting liquid phases with glycerol-lipids in artificial membranes (Brown, 1998; Thompson and Tillack, 1985), and the suggestion shortly thereafter that sphingolipids and GSL cluster into microdomains in actual cell membranes (van Meer *et al.*, 1987), there has been great excitement and interest in being able to visualize these sphingolipid domains in cells. Although tantalizing, this goal has proven extremely challenging for a number of reasons: sphingolipid-containing domains themselves are well below the optical diffraction limit, rendering conventional fluorescence microscopy useless for visualization of any meaningful dynamic or structural features at the molecular level. Secondly, these domains are not only mobile, but ephemeral, probably forming and dissipating on a millisecond time-scale (Simons and Gerl, 2010). Finally, until very recently there were only very few specific probes available that could recognize specific sphingolipids or clusters of sphingolipids. The advent of spectroscopic techniques with sufficiently fast temporal resolution such as FCS and TIRF-FCS, and novel adaptations of these methods, for example by combining spatial correlation methods as we have outlined here, begins to solve the issues of rapid and complex temporal and spatial dynamics at the membrane. As to the choice of probes for examining sphingolipid behavior in membranes, we have given examples of fluorescent probes based on sphingolipid-interacting peptides and lipid analogs that can be used to monitor the dynamics of sphingolipids in live cell membranes. With more such probes currently being developed and intensive activity in refinement of fast spectroscopic techniques combined with super-resolution microscopy (as for example (Eggeling *et al.*, 2008; Sahl *et al.*, 2010; van Zanten *et al.*, 2010)), an accurate picture of sphingolipid dynamics and organization in cell membranes is in sight.

Acknowledgements

We are very grateful to the members of our laboratories and to our colleagues in the field for discussions, and to Dr. Kamila Oglecka for the lipid structures shown in Fig. 2. The work discussed in this chapter was supported in part by the Association for Science, Technology, and Research, Singapore (A*STAR), the Singapore Ministry of Education, and the Biomedical Research Council, Singapore.

References

Axelrod, D. (2003). Total internal reflection fluorescence microscopy in cell biology. *Methods Enzymol.* **361**, 1–33.

Bacia, K., and Kim, S. A., *et al.* (2006). Fluorescence cross-correlation spectroscopy in living cells. *Nat. Methods* **3**(2), 83–89.

Bakht, O., and London, E. (2007). Detecting ordered domain formation (lipid rafts) in model membranes using Tempo. *Methods Mol. Biol.* **398**, 29–40.

Bakrac, B., and Kladnik, A., *et al.* (2010). A toxin-based probe reveals cytoplasmic exposure of golgi sphingomyelin. *J. Biol. Chem.* **285**, 22186–22195.

Baumgart, T., and Hammond, A. T., *et al.* (2007). Large-scale fluid/fluid phase separation of proteins and lipids in giant plasma membrane vesicles. *Proc. Natl. Acad. Sci.* **104**(9), 3165–3170.

Benda, A., and Benes, M., *et al.* (2003). How to determine diffusion coefficients in planar phospholipid systems by confocal fluorescence correlation spectroscopy. *Langmuir* **19**(10), 4120–4126.

Bhagatji, P., and Leventis, R., *et al.* (2009). Steric and not structure-specific factors dictate the endocytic mechanism of glycosylphosphatidylinositol-anchored proteins. *J. Cell Biol.* **186**(4), 615–628.

Brinkmeier, M., and Dorre, K., *et al.* (1999). Two-beam cross-correlation: a method to characterize transport phenomena in micrometer-sized structures. *Analyt. Chem.* **71**(3), 609–616.

Brown, R. (1998). Sphingolipid organization in biomembranes: what physical studies of model membranes reveal. *J. Cell Sci.* **111**(1), 1–9.

Burkhardt, M., and Schwille, P. (2006). Electron multiplying CCD based detection for spatially resolved fluorescence correlation spectroscopy. *Opt. Express* **14**(12), 5013–5020.

Burns, A. R., and Frankel, D. J., *et al.* (2005). Local mobility in lipid domains of supported bilayers characterized by atomic force microscopy and fluorescence correlation spectroscopy. *Biophys. J.* **89**(2), 1081–1093.

Chadda, R., and Howes, M. T., *et al.* (2007). Cholesterol-sensitive Cdc42 activation regulates actin polymerization for endocytosis via the GEEC pathway. *Traffic* **8**(6), 702–717.

Chen, C., and Fu, Z., *et al.* (2009). Gangliosides as high affinity receptors for tetanus neurotoxin. *J. Biol. Chem.* **284**(39), 26569–26577.

Chiantia, S., and Ries, J., *et al.* (2008). Role of ceramide in membrane protein organization investigated by combined AFM and FCS. *Biochim. Biophys. Acta* **1778**(5), 1356–1364.

Chichili, G. R., and Rodgers, W. (2007). Clustering of membrane raft proteins by the actin cytoskeleton. *J. Biol. Chem* **282**(50), 36682–36691.

Critchley, P., and Kazlauskaite, J., *et al.* (2004). Binding of prion proteins to lipid membranes. *Biochem. Biophys. Res. Commun.* **313**(3), 559–567.

Daumas, F., and Destainville, N., *et al.* (2003). Confined diffusion without fences of a G-protein-coupled receptor as revealed by single particle tracking. *Biophys J.* **84**(1), 356–366.

Destainville, N., and Dumas, F., *et al.* (2008). What do diffusion measurements tell us about membrane compartmentalisation? Emergence of the role of interprotein interactions. *J. Chem. Biol.* **1**(1–4), 37–48.

Digman, M. A., and Gratton, E. (2009). Imaging barriers to diffusion by pair correlation functions. *Biophys. J.* **97**(2), 665–673.

Eggeling, C., and Ringemann, C., *et al.* (2008). Direct observation of the nanoscale dynamics of membrane lipids in a living cell. *Nature* **457**(7233), 1159–1162.

Engelman, D. M. (2005). Membranes are more mosaic than fluid. *Nature* **438**(7068), 578–580.

Epand, R. M. (2004). Do proteins facilitate the formation of cholesterol-rich domains? *Biochim. Biophys. Acta* **1666**(1–2), 227–238.

Ewers, H., and Romer, W., *et al.* (2009). GM1 structure determines SV40-induced membrane invagination and infection. *Nat. Cell Biol.* **12**(1), 11–18.

Fantini, J. (2007). Interaction of proteins with lipid rafts through glycolipid-binding domains: biochemical background and potential therapeutic applications. *Curr. Med. Chem.* **14**(27), 2911–2917.

Fantini, J., and Garmy, N., *et al.* (2006). Prediction of glycolipid-binding domains from the amino acid sequence of lipid raft-associated proteins: application to HpaA, a protein involved in the adhesion of helicobacter pylori to gastrointestinal cells. *Biochemistry* **45**(36), 10957–10962.

Feigenson, G. W. (2009). Phase diagrams and lipid domains in multicomponent lipid bilayer mixtures. *Biochim. Biophys. Acta* **1788**(1), 47–52.

Fotinou, C., and Emsley, P., *et al.* (2001). The crystal structure of tetanus toxin Hc fragment complexed with a synthetic GT1b analogue suggests cross-linking between ganglioside receptors and the toxin. *J. Biol. Chem.* **276**(34), 32274–32281.

Fu, Z., and Chen, C., *et al.* (2009). Glycosylated SV2 and gangliosides as dual receptors for botulinum neurotoxin serotype F. *Biochemistry* **48**(24), 5631–5641.

Ginalski, K., and Venclovas, C., *et al.* (2000). Structure-based sequence alignment for the beta-trefoil subdomain of the clostridial neurotoxin family provides residue level information about the putative ganglioside binding site. *FEBS Lett.* **482**(1–2), 119–124.

Glebov, O. O., and Nichols, B. J. (2004). Lipid raft proteins have a random distribution during localized activation of the T-cell receptor. *Nat. Cell Biol.* **6**(3), 238–243.

Golebiewska, U., and Nyako, M., *et al.* (2008). Diffusion coefficient of fluorescent phosphatidylinositol 4,5-bisphosphate in the plasma membrane of cells. *Mol. Biol. Cell* **19**(4), 1663–1669.

Gösch, M., Serov, A., Anhut, T., Lasser, T., Rochas, A., Besse, P. A., Popovic, R. S., Blom, H., and Rigler, R. (2004). Parallel single molecule detection with a fully integrated single-photon 2×2 CMOS detector array. *J. Biomed. Opt.* **9**(5), 913–921.

Goswami, D., and Gowrishankar, K., *et al.* (2008). Nanoclusters of GPI-anchored proteins are formed by cortical actin-driven activity. *Cell* **135**(6), 1085–1097.

Guo, L., and Har, J. Y., *et al.* (2008). Molecular Diffusion Measurement in Lipid Bilayers over Wide Concentration Ranges: A Comparative Study. *ChemPhysChem* **9**(5), 721–728.

Halpern, J. L., and Loftus, A. (1993). Characterization of the receptor-binding domain of tetanus toxin. *J. Biol. Chem.* **268**(15), 11188–11192.

Haustein, E., and Schwille, P. (2007). Fluorescence correlation spectroscopy: novel variations of an established technique. *Annu. Rev. Biophys. Biomol. Struct.* **36**, 151–169.

Hebbar, S., and Lee, E., *et al.* (2008). A fluorescent sphingolipid binding domain peptide probe interacts with sphingolipids and cholesterol-dependent raft domains. *J. Lipid Res.* **49**(5), 1077–1089.

Hebert, B., and Costantino, S., *et al.* (2005). Spatiotemporal image correlation spectroscopy (STICS) theory, verification, and application to protein velocity mapping in living CHO cells. *Biophys. J.* **88**(5), 3601–3614.

Honigmann, A., and Walter, C., *et al.* (2010). Characterization of horizontal lipid bilayers as a model system to study lipid phase separation. *Biophys J* **98**(12), 2886–2894.

Ikeda, K., and Matsuzaki, K. (2008). Driving force of binding of amyloid [beta]-protein to lipid bilayers. *Biochem. Biophys. Res. Commun.* **370**(3), 525–529.

Ishitsuka, R., and Kobayashi, T. (2004). Lysenin: a new tool for investigating membrane lipid organization. *Anat. Sci. Int.* **79**(4), 184–190.

Johannes, L., and Mayor, S. (2010). Induced domain formation in endocytic invagination. *Lipid Sorting, and Scission.* **142**(4), 507–510.

Kaiser, R. D., and London, E. (1998). Determination of the depth of BODIPY probes in model membranes by parallax analysis of fluorescence quenching. *Biochim. Biophys. Acta* **1375**(1–2), 13–22.

Kamata, Y., and Yoshimoto, M., *et al.* (1997). Interaction between botulinum neurotoxin type a and ganglioside: ganglioside inactivates the neurotoxin and quenches its tryptophan fluorescence. *Toxicon* **35**(8), 1337–1340.

Kannan, B., and Guo, L., *et al.* (2007). Spatially resolved total internal reflection fluorescence correlation microscopy using an electron multiplying charge-coupled device camera. *Anal. Chem.* **79**(12), 4463–4470.

Kannan, B., and Har, J. Y., *et al.* (2006). Electron multiplying charge-coupled device camera based fluorescence correlation spectroscopy. *Anal. Chem.* **78**(10), 3444–3451.

Kenworthy, A. K., and Nichols, B. J., *et al.* (2004). Dynamics of putative raft-associated proteins at the cell surface. *J. Cell Biol.* **165**(5), 735–746.

Kim, S. A., and Heinze, K. G., *et al.* (2007). Fluorescence correlation spectroscopy in living cells. *Nat. Methods* **4**(11), 963–973.

Kitamura, M., and Iwamori, M., *et al.* (1980). Interaction between Clostridium botulinum neurotoxin and gangliosides. *Biochim. Biophys. Acta* **628**(3), 328–335.

Kolin, D. L., and Wiseman, P. W. (2007). Advances in image correlation spectroscopy: measuring number densities, aggregation states, and dynamics of fluorescently labeled macromolecules in cells. *Cell Biochem. Biophys.* **49**(3), 141–164.

Kusumi, A., and Shirai, Y. M., *et al.* (2010). Hierarchical organization of the plasma membrane: investigations by single-molecule tracking vs. fluorescence correlation spectroscopy. *FEBS Lett.* **584**(9), 1814–1823.

Lagerholm, B. C., and Weinreb, G. E., *et al.* (2005). Detecting microdomains in intact cell membranes. *Annu Rev. Phys. Chem.* **56**, 309–336.

Lai, A. L., and Tamm, L. K., *et al.* (2011). Synaptotagmin 1 Modulates Lipid Acyl Chain Order in Lipid Bilayers by Demixing Phosphatidylserine. *J. Biol. Chem.* **286**(28), 25291–25300.

Lasserre, R., and Guo, X. J., *et al.* (2008). Raft nanodomains contribute to Akt/PKB plasma membrane recruitment and activation. *Nat. Chem. Biol.* **4**(9), 538–547.

Lencer, W. I., and Saslowsky, D. (2005). Raft trafficking of AB5 subunit bacterial toxins. *Biochim. Biophys. Acta* **1746**(3), 314–321.

Lenne, P. F., and Wawrezinieck, L., *et al.* (2006). Dynamic molecular confinement in the plasma membrane by microdomains and the cytoskeleton meshwork. *EMBO J.* **25**(14), 3245–3256.

Lingwood, D., and Kaiser, H. -J., *et al.* (2009). Lipid rafts as functional heterogeneity in cell membranes. *Biochem. Soc. Trans.* **037**(5), 955–960.

Lingwood, D., and Ries, J., *et al.* (2008). Plasma membranes are poised for activation of raft phase coalescence at physiological temperature. *Proc. Natl. Acad. Sci. U. S. A.* **105**(29), 10005–10010.

Liu, A. P., and Fletcher, D. A. (2006). Actin Polymerization Serves as a Membrane Domain Switch in Model Lipid Bilayers. *Biophys. J.* **91**(11), 4064–4070.

Longo, M. L., and Blanchette, C. D. (2007). Imaging cerebroside-rich domains for phase and shape characterization in binary and ternary mixtures. *Biochimica et Biophysica Acta (BBA) - Biomembranes* **1798**(7), 1357–1367.

Louch, H. A., and Buczko, E. S., *et al.* (2002). Identification of a binding site for ganglioside on the receptor binding domain of tetanus toxin. *Biochemistry* **41**(46), 13644–13652.

Machan, R., and Hof, M. (2010). Recent Developments in Fluorescence Correlation Spectroscopy for Diffusion Measurements in Planar Lipid Membranes. *Int. J. Mol. Sci.* **11**(2), 427–457.

Mahfoud, R., and Garmy, N., *et al.* (2002). Identification of a common sphingolipid-binding domain in Alzheimer, prion, and HIV-1 proteins. *J. Biol. Chem.* **277**(13), 11292–11296.

Maier, O., and Oberle, V., *et al.* (2002). Fluorescent lipid probes: some properties and applications (a review). *Chem. Phys. Lipids* **116**(1–2), 3–18.

Martin, O. C., and Pagano, R. E. (1994). Internalization and sorting of a fluorescent analogue of glucosylceramide to the Golgi apparatus of human skin fibroblasts: utilization of endocytic and nonendocytic transport mechanisms. *J. Cell Biol.* **125**(4), 769–781.

Masuda, A., and Ushida, K., *et al.* (2005). New fluorescence correlation spectroscopy enabling direct observation of spatiotemporal dependence of diffusion constants as an evidence of anomalous transport in extracellular matrices. *Biophys. J.* **88**(5), 3584–3591.

Merrill Jr., A. H. (2002). De Novo sphingolipid biosynthesis: a necessary, but dangerous, pathway. *J. Biol. Chem.* **277**(29), 25843–25846.

Merrill Jr., A. H., and van Echten, G., *et al.* (1993). Fumonisin B1 inhibits sphingosine (sphinganine) N-acyltransferase and de novo sphingolipid biosynthesis in cultured neurons in situ. *J. Biol. Chem.* **268** (36), 27299–27306.

Pagano, R. E. (2003). Endocytic trafficking of glycosphingolipids in sphingolipid storage diseases. *Philos. Trans. R. Soc. Lond. B Biol. Sci.* **358**(1433), 885–891.

Pan, X., and Foo, W., *et al.* (2007). Multifunctional fluorescence correlation microscope for intracellular and microfluidic measurements. *Rev. Sci. Instrum.* **78**(5), 053711.

Petersen, N. O. (1986). Scanning fluorescence correlation spectroscopy. I. Theory and simulation of aggregation measurements. *Biophys. J.* **49**(4), 809–815.

Petersen, N. O., and Hoddelius, P. L., *et al.* (1993). Quantitation of membrane receptor distributions by image correlation spectroscopy: concept and application. *Biophys. J.* **65**(3), 1135–1146.

Petersen, N. O., and Johnson, D. C., *et al.* (1986). Scanning fluorescence correlation spectroscopy. II. Application to virus glycoprotein aggregation. *Biophys. J.* **49**(4), 817–820.

Petrasek, Z., and Schwille, P. (2008). Scanning fluorescence correlation spectroscopy. *In* "Single Molecules and Nanotechnology," (R. R. Rigler and H. Vogel, eds.), pp. 83–105. Springer, Berlin.

Pinaud, F., and Michalet, X., *et al.* (2009). Dynamic partitioning of a glycosyl-phosphatidylinositol-anchored protein in glycosphingolipid-rich microdomains imaged by single-quantum dot tracking. *Traffic* **10**(6), 691–712.

Platt, F. M., and Neises, G. R., *et al.* (1994). N-butyldeoxynojirimycin is a novel inhibitor of glycolipid biosynthesis. *J. Biol. Chem.* **269**(11), 8362–8365.

Polyakova, S. M., and Belov, V. N., *et al.* (2009). New GM1 ganglioside derivatives for selective single and double labelling of the natural glycosphingolipid skeleton. *Eur. J. Org. Chem* **2009**(30), 5162–5177.

Prinetti, A., and Loberto, N., *et al.* (2009). Glycosphingolipid behaviour in complex membranes. *Biochim. Biophys. Acta* **1788**(1), 184–193.

Reyes Mateo, C., and Brochon, J. C., *et al.* (1993). Lipid clustering in bilayers detected by the fluorescence kinetics and anisotropy of trans-parinaric acid. *Biophys. J.* **65**(5), 2237–2247.

Ries, J., Petrasek, Z., Garcia-Saez, A. J., and Scwille, P. (2010). A comprehensive framework for fluorescence cross-correlation spectroscopy. *New J. Phys.* **12**, (November 2010), 113009.

Romer, W., and Berland, L., *et al.* (2007). Shiga toxin induces tubular membrane invaginations for its uptake into cells. *Nature* **450**(7170), 670–675.

Rummel, A., and Mahrhold, S., *et al.* (2004). The HCC-domain of botulinum neurotoxins A and B exhibits a singular ganglioside binding site displaying serotype specific carbohydrate interaction. *Mol. Microbiol.* **51**(3), 631–643.

Ruprecht, V., and Wieser, S., *et al.* (2011). Spot variation fluorescence correlation spectroscopy allows for superresolution chronoscopy of confinement times in membranes. *Biophys. J.* **100**(11), 2839–2845.

Sahl, S. J., and Leutenegger, M., *et al.* (2010). Fast molecular tracking maps nanoscale dynamics of plasma membrane lipids. *Proc. Natl. Acad. Sci.* **107**(15), 6829–6834.

Sankaran, J., and Manna, M., *et al.* (2009). Diffusion, transport, and cell membrane organization investigated by imaging fluorescence cross-correlation spectroscopy. *Biophys. J.* **97**(9), 2630–2639.

Sankaran, J., and Shi, X., *et al.* (2010). ImFCS: a software for imaging FCS data analysis and visualization. *Opt. Express* **18**(25), 25468–25481.

Schwarzmann, G., and Wendeler, M., *et al.* (2005). Synthesis of novel NBD-GM1 and NBD-GM2 for the transfer activity of GM2-activator protein by a FRET-based assay system. *Glycobiology* **15**(12), 1302–1311.

Sengupta, P., and Garai, K., *et al.* (2003). Measuring size distribution in highly heterogeneous systems with fluorescence correlation spectroscopy. *Biophys. J.* **84**(3), 1977–1984.

Shapiro, R. E., and Specht, C. D., *et al.* (1997). Identification of a ganglioside recognition domain of tetanus toxin using a novel ganglioside photoaffinity ligand. *J. Biol. Chem.* **272**(48), 30380–30386.

Sharma, P., and Varma, R., *et al.* (2004). Nanoscale organization of multiple GPI-anchored proteins in living cell membranes. *Cell* **116**(4), 577–589.

Shi, X., and Teo, L. S., *et al.* (2009). Probing events with single molecule sensitivity in zebrafish and Drosophila embryos by fluorescence correlation spectroscopy. *Dev. Dyn.* **238**(12), 3156–3167.

Shi, X., and Wohland, T. (2010). Fluorescence Correlation Spectroscopy. *In* "Nanoscopy and Multidimensional Optical Fluorescence Microscopy," (A. Diaspro, ed.), CRC Press, Boca Raton.

Simons, K., and Gerl, M. (2010). Revitalizing membrane rafts: new tools and insights. *Nat. Rev. Mol. Cell Bi*

al. (2006). Spatially resolved fluorescence correlation spectroscopy using
oscope. *Biophys. J.* **91**(11), 4241–4252.

et al. (2005). Global analysis of fluorescence fluctuation data. *Eur.*

08). A fluorescent glycolipid-binding peptide probe traces cholesterol
trafficking pathways. *PLoS ONE* **3**(8), e2933.

08). Crystal structure of botulinum neurotoxin type A in complex with
nsight into the toxin-neuron interaction. *PLoS Pathog.* **4**(8), e1000129.

. Rafts and related glycosphingolipid-enriched microdomains in the
s linked to nutrient absorption. *Adv. Drug Deliv. Rev.* **56**(6), 779–794.

1985). Organization of glycosphingolipids in bilayers and plasma
nu. Rev. Biophys. Chem. **14**, 361–386.

et al. (2003). Characterization of the interactions of ß-amyloid
surface plasmon resonance. *Spectroscopy* **17**(2), 241–254.

37). Sorting of sphingolipids in epithelial (Madin-Darby canine
3–1635.

0). Direct mapping of nanoscale compositional connectivity on
Sci. **107**(35), 15437–15442.

Critical fluctuations in domain-forming lipid mixtures. *Proc.*
17655.

2001). Selective binding of perfringolysin O derivative to
(rafts). *PNAS* **98**(9), 4926–4931.

olipid partitioning into ordered domains in cholesterol-free
ophys. J. **84**(1), 367–378.

04). Fluorescence correlation spectroscopy to determine
cell membrane. *Proc. SPIE* **5462**, 92–102.

Rigneault, H., *et al.* (2005). Fluorescence correlation spectroscopy diffusion laws
the submicron cell membrane organization. *Biophys. J.* **89**(6), 4029–4042.

Weis, W. I., and Drickamer, K. (1996). Structural basis of lectin-carbohydrate recognition. *Annu. Rev. Biochem.* **65**, 441–473.

Weissman, M., and Schindler, H., *et al.* (1976). Determination of molecular weights by fluctuation spectroscopy: application to DNA. *Proc. Natl. Acad. Sci. U. S. A.* **73**(8), 2776–2780.

Westerlund, B., and Slotte, J. P. (2009). How the molecular features of glycosphingolipids affect domain formation in fluid membranes. *Biochim. Biophys. Acta* **1788**(1), 194–201.

Widengren, J., and Mets, U., *et al.* (1995). Fluorescence correlation spectroscopy of triplet-states in solution – a theoretical and experimental-study. *J. Phys. Chem.* **99**(36), 13368–13379.

Widengren, J., and Mets, U., *et al.* (1999). Photodynamic properties of green fluorescent proteins investigated by fluorescence correlation spectroscopy. *Chem. Phys.* **250**(2), 171–186.

Wohland, T., and Shi, X., *et al.* (2010). Single plane illumination fluorescence correlation spectroscopy (SPIM-FCS) probes in homogeneous three-dimensional environments. *Opt. Express* **18**(10), 10627–10641.

Yanagisawa, K. (2007). Role of gangliosides in Alzheimer's disease. *Biochim. Biophys. Acta* **1768**(8), 1943–1951.

Yanagisawa, K., and Odaka, A., *et al.* (1995). GM1 ganglioside-bound amyloid beta-protein (A beta): a possible form of preamyloid in Alzheimer's disease. *Nat. Med.* **1**(10), 1062–1066.

Yu, C., and Alterman, M., *et al.* (2005). Ceramide displaces cholesterol from lipid rafts and decreases the association of the cholesterol binding protein caveolin-1. *J. Lipid Res.* **46**(8), 1678–1691.

Zhang, Y., and Li, X., *et al.* (2009). Ceramide-enriched membrane domains – structure and function. *Biochim. Biophys. Acta* **1788**, 178–183.

CHAPTER 19

Monitoring Phospholipid Dynamics during Phagocytosis: Application of Genetically-Encoded Fluorescent Probes

Helen Sarantis and Sergio Grinstein

Program in Cell Biology, The Hospital for Sick Children, Toronto, Canada

METHODS IN CELL BIOLOGY, VOL 108
Copyright 2012, Elsevier Inc. All rights reserved.

0091-679X/10 $35.00
DOI 10.1016/B978-0-12-386487-1.00019-5

Abstract

The internalization of foreign or unwanted particles by cells, a process that is important in many aspects of immunity and development, is known as phagocytosis. Rearrangement of cellular actin enables the phagocytic cell to gradually wrap itself around and ultimately engulf the target particle. Phagocytosis is initiated by receptor engagement that in turn triggers multiple signaling pathways, including extensive lipid remodeling. Lipid modification not only generates a variety of second messengers, but is important for the redistribution of key proteins involved in phagocytosis. Lipids can associate with proteins bearing stereospecific association domains. In addition the phosphoinositides and phosphatidylserine (PS), which are anionic, can also recruit proteins electrostatically by generating a considerable negative surface charge. We describe a method whereby lipids can be monitored dynamically in live cells performing phagocytosis. This method involves the expression by phagocytic cells of genetically encoded fluorescent lipid-binding probes, which can be monitored using confocal fluorescence microscopy.

I. Introduction

Phagocytosis is the term used to describe the process by which a cell engulfs a particle larger than 0.5 μm. In mammals this is typically carried out by specialized cell types, mainly neutrophils, dendritic cells, and macrophages (the 'professional' phagocytes); however, other cells also have the inherent capacity to internalize particles, which can be unmasked by heterologous expression of phagocytic receptors (Downey *et al.*, 1999). The induction of phagocytosis is triggered by the binding of surface receptors to ligands on the target particle. Ligation of receptors occurs progressively as the membrane advances along the particle surface, until the leading edges of the membrane meet and fuse, thus completing the engulfment process (Swanson, 2008). The newly formed sealed compartment, the "phagosome," then undergoes a series of fission and fusion reactions with endosomes and later with lysosomes. This remodeling process, termed phagosome maturation, allows the intra-phagosomal milieu to undergo a marked change in composition and properties, from an initial environment resembling the innocuous and neutral extracellular space, to an acidic compartment rich in active proteases and other degradative enzymes that act to effectively eliminate and break down the internalized target (Huynh *et al.*, 2007). By enabling the host cells to sequester and subsequently degrade foreign bodies – which include bacteria, fungi, and apoptotic cells – phagocytosis plays a key role in immunity and development.

Phagocytic receptors can directly recognize and bind ligands on the surface of the target particle. For example, receptors such as SR-A, MARCO, and Dectin-1 interact directly with components of the bacterial or fungal cell wall. However, binding is often indirect, mediated by a host protein that deposits on the surface of the foreign particle. In this instance, the bound serum protein or opsonin is the ligand recognized by the phagocytic receptor (Aderem and Underhill, 1999). Examples of opsonic phagocytosis include those driven by the binding of the Fc-gamma receptors (FcγR) to IgG-coated targets, or of complement receptors to C3b- or iC3b-coated particles.

Work over the past couple of decades has defined roles for numerous proteins in phagocytosis (Swanson and Hoppe, 2004). The reactions induced by FcγR have been studied in greatest detail. An immunoreceptor tyrosine-based activation motif (ITAM) in the cytoplasmic domains of the FcγR is critical for signaling, which is initiated when these receptors undergo lateral clustering upon binding immunoglobulin-coated particles. Upon activation, the receptors are phosphorylated on the tyrosine residues in the ITAM, thus providing a docking site for Src-Homology 2 (SH2) domains such as those found in the protein tyrosine kinase Syk, and the lipid kinases of the phosphatidylinositol (PI) 3-kinase (PI3K) family. Class I PI3Ks are critical for FcγR-mediated phagocytosis, producing the key second messenger phosphatidylinositol 3,4,5-*tris*phosphate ($PI(3,4,5)P_3$) from phosphatidylinositol 4,5-*bis*phosphate ($PI(4,5)P_2$). Indeed, inhibition of PI3K activity using wortmannin or LY294002 results in the arrest of phagocytosis, particularly in the case of large ($\geq 3\ \mu m$) particles (Araki *et al.*, 1996; Cox *et al.*, 1999). In addition, class III PI3Ks contribute to phagosome maturation through the production of phosphatidylinositol 3-phosphate (PI3P) from phosphatidylinositol (PI) (Vieira *et al.*, 2001). Accordingly, when added immediately after particle ingestion, wortmannin and LY294002 arrest phagosome maturation.

$PI(4,5)P_2$ is not only important as a substrate for the synthesis of $PI(3,4,5)P_3$, but also serves as a direct regulator of F-actin dynamics during phagocytosis. $PI(4,5)P_2$ is present in the resting plasma membrane and undergoes a biphasic change during phagocytosis: it initially undergoes a modest accumulation in pseudopods, followed by an abrupt disappearance (Botelho *et al.*, 2000).

$PI(4,5)P_2$ and $PI(3,4,5)P_3$ act in large part by recruiting a variety of proteins to sites of phagocytosis. They interact with proteins bearing lipid-recognition domains such as the Pleckstrin Homology (PH) domains found in proteins like AKT and phospholipase C (Liao *et al.*, 1992). However, in addition to their association with stereospecific recognition domains, it is becoming increasingly appreciated that lipids can also recruit proteins electrostatically; phospholipids with anionic headgroups generate a surface charge that attracts and retains polycationic proteins. This has been documented for Rac1, a small GTPase that is crucial for FcγR-mediated phagocytosis; the cationic C-terminus of Rac1 is essential for its association with the plasmalemma (Yeung *et al.*, 2008).

From the preceding considerations it should be apparent that phagocytosis is an important biological process and that lipids play a central role in both phagosome formation and maturation. Some of these insights were obtained using the method described below, which should also be applicable to future studies using more powerful and versatile probes.

II. Rationale

Many of the proteins known to be involved in phagocytosis were identified by biochemical and immunochemical means. Phagocytic receptor complexes have been immunoprecipitated to identify associated proteins, or immunostained for visualization by light or electron microscopy. Unfortunately, these procedures are not readily applicable to lipids: the detergent solubilization required for immunoprecipitation generally strips lipids from the complexes, and the fixation and permeabilization procedures used for immunostaining fail to preserve the native lipid architecture. Clearly, more conservative approaches are needed to study lipid distribution and dynamics.

Phosphoinositides can be monitored using biochemical assays that typically involve labeling of cells with isotopic phosphate or myo-inositol, followed by analysis of cell extracts by chromatography coupled with detection of radioactivity (Rusten and Stenmark, 2006). However these are usually end-point determinations made in large populations of cells and are therefore insensitive to localized, transient, and asynchronous changes such as those that occur during phagocytosis. Local, continuous monitoring of phosphoinositide changes during phagocytosis is not feasible by biochemical means.

Biochemical analysis of phosphatidylserine (PS) is similarly complicated by limited spatial and temporal resolution. On the other hand, PS distribution can be studied using fluorescent probes. Indeed, PS tagged with NBD [6-(7-nitrobenz-2-oxa-1,3-diazol-4-yl)aminocaproyl]-PS (NBD-PS) has been used extensively. However, NBD-PS is highly susceptible to photobleaching and, more worrisome, the distortion introduced by the bulky, polar NBD moiety alters the behavior of the probe such that its subcellular localization does not accurately reflect that of endogenous PS (Kobayashi and Arakawa, 1991). Annexin V, a protein that has been used extensively for the measurement of PS on the outer cell surface, is not helpful in the case of intracellular PS. The association of annexin V with PS requires calcium at concentrations that far exceed those present in the cytosol of normal cells.

Clearly, better approaches are required to monitor the distribution of anionic lipids in live cells with suitable spatial and temporal resolution. Herein we describe the application of a sensitive method that can be used in single cells to monitor lipids in real time during phagocytosis. The method is based on the generation of chimeric constructs consisting of a specific lipid-binding domain attached to a fluorescent protein that serves as a beacon (Balla and Varnai, 2002). cDNA encoding the chimera is transfected into the cells of choice and the location and dynamics of the lipid of interest are monitored by fluorescence microscopy. While conventional epifluorescence can be used, laser scanning or spinning-disc confocal microscopy are recommended for improved resolution.

The general concept is applicable to a variety of phosphoinositides and also to PS, because specific probes have been developed in recent years for a number of lipid species. This has been made possible by the systematic identification and characterization of lipid-binding domains of a large array of proteins. A partial list of such domains and their cognate lipids are presented in Table I. Salient examples include the PH domain of PLCδ, which is widely used to monitor PI $(4,5)P_2$, and the C2 domain of lactadherin that was introduced recently as a probe for PS.

The method described below uses RAW264.7 cells. These are murine macrophage-like cells that are adherent, differentiated, and can be maintained in culture without requiring special, often expensive additives such as cytokines or growth factors. Importantly, these cells express FcγR and complement receptors, making them suitable for studies of phagocytosis of opsonized particles. RAW264.7 cells can be transfected by lipofection with modest efficiency, which is nevertheless more than sufficient to perform single-cell studies, and the transfectants can then be challenged with opsonized targets, while being monitored in real time using confocal fluorescence microscopy. The localization and redistribution of the fluorescent chimeric probes is a useful index of the fate of lipids during phagocytosis.

Table I
Lipid-binding probes that can be used to monitor changes during phagocytosis

Lipid	Protein-domain	References
PI3P	EEA1-[FYVE] × 2[a]	(Gillooly *et al.*, 2000)
	Hrs-FYVE	(Raiborg *et al.*, 2001)
	$p40^{phox}$-PX	(Ellson *et al.*, 2001)
PI4P	FAPP1-PH	(Godi *et al.*, 2004)
	OSH2-PH	(Roy and Levine, 2004)
PI$(3,4)P_2$	TAPP1-PH	(Kimber *et al.*, 2002)
PI$(3,5)P_2$	Centaurin–b2-PH	(Dowler *et al.*, 2000)
	Svp1p-PH	(Dove *et al.*, 2004)
PI$(4,5)P_2$	PLCδ-PH	(Stauffer *et al.*, 1998)
	Tubby	(Santagata *et al.*, 2001)
PI$(3,4,5)P_3$	GRP1-PH	(Gray *et al.*, 1999)
	ARNO-PH	(Venkateswarlu *et al.*, 1998)
	Btk-PH	(Varnai *et al.*, 1999)
	Gab2-PH	(Gu *et al.*, 2003)
PI$(3,4)P_2$ and	AKT-PH	(Gray *et al.*, 1999)
PI$(3,4,5)P_3$	PDK1-PH	(Komander *et al.*, 2004)
	CRAC-PH	(Dormann *et al.*, 2002)
DAG	PKCδ-C1	(Colon-Gonzalez and Kazanietz, 2006)
	PKD-C1	(Chen *et al.*, 2008)
PS	Annexin V[b]	(Andree *et al.*, 1990)
	Lactadherin-C2	(Yeung *et al.*, 2008)

[a] Must be used in tandem to detect PI3P in living cells.

[b] Cannot be used as a probe in the cytoplasm of live cells due to inappropriate calcium concentrations.

III. Materials

A. Cell Culture

1. Tissue culture flasks (Sarstedt, Montreal, QC, Canada)
2. Dulbecco's modified Eagle's medium (DMEM; Wisent, St. Bruno, QC, Canada)
3. Fetal bovine serum (FBS; Wisent, St. Bruno, QC, Canada)
4. Sterile cell scrapers (Sarstedt, Montreal, QC, Canada)
5. 18 mm glass coverslips (Fisher, Pittsburgh, PA)
6. 12-well tissue culture-treated plates (Becton Dickinson, Mississauga, ON, Canada)

B. Transfection

1. FuGENE HD (Roche Applied Science, Mississauga, ON, Canada)
2. Opti-MEM (Invitrogen, Burlington, ON, Canada)

C. Preparation of Target Particles

1. Sheep red blood cells (RBCs; MP Biomedicals, Solon, OH)
2. Rabbit anti-sheep RBC IgG (ICN Pharmaceuticals, Costa Mesa, CA)
3. Phosphate-buffered saline (PBS; Wisent, St. Bruno, QC, Canada)
4. Anti-sheep RBC IgM (Cedarlane Laboratories, Burlington, ON, Canada)
5. $CaCl_2$, 1 M sterile solution (Bioshop, Burlington, ON, Canada)
6. $MgCl_2$, 1 M sterile solution (Bioshop, Burlington, ON, Canada)
7. Human C5-deficient serum (Sigma Aldrich, Oakville, ON, Canada)

D. Phagocytosis Assay

1. Phorbol-12-myristate-13-acetate (PMA; Bioshop, Burlington, ON, Canada)
2. HEPES-buffered medium RPMI-1640 (HPMI; Wisent, St. Bruno, QC, Canada)
3. Live-cell imaging chamber for 18 mm coverslip (Invitrogen, Burlington, ON, Canada)

E. Imaging

1. A spinning-disc confocal microscope equipped with a 63x magnifying objective, suitable light source(s) and filter sets that that are appropriate for the fluorescent proteins in use, and chamber- and objective-heating devices and temperature controllers.

F. Image Analysis

1. ImageJ (rsbweb.nih.gov/ij/), or other suitable analysis software.

IV. Methods

A. Cell Culture

RAW264.7 cells obtained from the American Tissue Culture Collection are grown in tissue culture flasks in DMEM supplemented with 5% FBS. To passage, remove old medium and replace with fresh DMEM supplemented with 5% FBS, then scrape the cells into the medium using a sterile cell scraper. Pipette up and down to create a homogenous suspension and split into a new flask (a 1:20 passage can typically grow for 3–4 days before it needs to be re-split). To prepare cells for phagocytosis assays, place a sterile 18 mm coverslip in each well of a 12-well tissue culture plate and add 2×10^5 cells per well. Allow cells to adhere before transfecting.

B. Transfection

We routinely use FuGENE HD to introduce plasmids encoding tagged lipid-binding domains into RAW264.7 cells. For one well of a 12-well plate, mix 100 μL Opti-MEM, 3 μL FuGENE HD, and 1 μg plasmid DNA (if co-transfecting two plasmids, use 1 μg of each). Vortex and incubate the mixture at room temperature for 15 min before adding to the well; pipette up and down to mix. Incubate at least 6 h at 37 °C under 5% CO_2, or until expression of the fluorescent proteins is observed. Overnight transfection is convenient, yielding cells that are ready for use early the following morning.

C. Preparation of Target Particles

We typically use sheep red blood cells (RBCs) as our prey for phagocytosis. To opsonize for FcγR-mediated phagocytosis, sediment 200 μL of RBCs (e.g., by spinning at 6000 rpm for 2 min in a microcentrifuge), wash with 500 μL PBS, re-centrifuge, and resuspend in 200 μL PBS. Add 4 μL of anti-sheep RBC IgG (1:50), and incubate at room temperature for 1 h with constant mixing. Before use, wash RBCs 3–4 times with 500 μL PBS.

To opsonize for complement receptor-mediated phagocytosis, wash and pellet the RBCs as described above, then incubate with sub-agglutinating (1:10) concentrations of anti-sheep RBC IgM in PBS supplemented with 0.5 mM $CaCl_2$ and 0.5 mM $MgCl_2$ for 1 h at room temperature under constant agitation. Wash RBCs as described above, then incubate with human C5-deficient serum (1:6) for 20 min at 37 °C with frequent mixing. Wash RBCs 3–4 times with PBS before using and suspend in 500 μL.

D. Phagocytosis Assay

To initiate phagocytosis, 10 μL of the washed RBCs are added to a coverslip placed at the bottom of a well containing 1 mL of DMEM. This results in an approximate 10:1 ratio of RBCs per macrophage (a more precise determination can be obtained by counting the cells electronically using a Coulter device or with a hemocytometer). Before adding the RBCs to the RAW264.7 cells, it is advisable to wash the latter at least twice with DMEM without serum, to ensure that serum components do not interfere with the assay. Macrophages need to be primed to perform complement-mediated phagocytosis. To this end the RAW264.7 cells are treated with 100 nM PMA for 20 min before addition of the complement-opsonized RBCs. Priming is not required for FcγR-mediated phagocytosis.

After addition to the well, the RBCs sediment slowly by gravity and make contact with the phagocytic cells. To promote the association of RBCs with the macrophages – and to synchronize phagocytosis – the plate can be subjected to centrifugation at 270 x *g* for 1 min. When studying the earliest stages of phagocytosis, it is preferable to add the opsonized RBCs to the macrophages while on the microscope stage, after mounting the coverslips in the temperature-controlled chamber (see below).

E. Imaging

Place the coverslip in a temperature-regulated holder, and add a suitable volume of HPMI that has been pre-warmed to the desired temperature. If imaging is to be performed at 37 °C, the stage holder and objective must be heated prior to imaging to avoid heat loss from the sample; this also minimizes the extent of focal drift during imaging. Monitor cells for a suitable period of time, acquiring images as required. For fast processes, such as the engulfment of opsonized RBCs, images should be acquired frequently, for example, 3–4 time points per min, for 5–10 min; for slower processes, such as phagosome maturation, images can be acquired once every 1–3 min, for 30 min.

F. Image Analysis

Images can be exported from the acquisition software and analyzed with suitable analysis software. We use ImageJ for this purpose, which is available for download without charge from: rsbweb.nih.gov/ij/. A number of basic tools are available within this software, including subtraction of background, and measurements of intensities, both based on the definition of a region of interest. In the case of phagocytosis, where thickness of membrane can confound the measurements (i.e., during pseudopod extension, membrane is essentially doubled around the particle), we can co-transfect the cells with a vector encoding a generic plasma membrane marker fused to a separate fluorophore (e.g., the plasma membrane-targeting sequence from Lyn, fused to GFP; PM-GFP) (Teruel *et al.*, 1999) and normalize acquisition of our probe of interest to the amount of membrane marker, to see if a true increase in intensity has occurred.

While data obtained with this method are typically qualitative, quantification of results can also be made in specific instances. In cases of acquisition of a probe, we can count the number of phagosomes that are positive for the marker, typically for a least 15 cells per experiment, in at least three independent experiments. Per cent co-localization with a probe can be expressed as the average of the three (or more) means, also expressing the standard error of the mean; statistical tests can be conducted on paired samples when analyzing different conditions, allowing a calculation of significance. At least five separate fields should be acquired for each sample, representing various areas of the coverslip.

Similarly, timing of lipid generation at sites of phagocytosis/phagosome maturation can also be quantified; in this case we can plot the time when a probe is acquired and lost, again using an average of at least 15 cells per experiment, to apply similar statistical analyses as suggested above. Images from single movies or multiple samples acquired at fixed time points can be quantified, and phagosomes can be scored as positive or negative for the probe in question.

V. Considerations when Designing an Experiment

While the procedure described above has proven to be extremely useful for us and others, a number of considerations must be kept in mind when designing an experiment or analyzing experimental data obtained with this method. Some of these considerations are discussed below.

A. Accessibility of Probes to Cellular Compartments

For a probe to effectively detect its intended substrate, it must be able to access the compartment where the lipid is located.Unless otherwise targeted, the lipid-binding probes are expressed in the cytosol. As such, they can be readily recruited to the leaflet of the membrane that faces the cytoplasm, and this applies to the plasmalemma as well as to organellar membranes. However, lipids on the extracellular leaflet of the plasma membrane or on the leaflet facing the lumen of organelles are not accessible and will not be detected. This concern is minimal in the case of phosphoinositides, which, with the exception of internal vesicles of multivesicular bodies, are thought to be exposed cytosolically. However, it must be borne in mind that a considerable fraction of PS and other lipids like diacylglycerol is likely to be inaccessible to cytosolic probes and therefore invisible to the investigator.

To circumvent such problems, the chimeric probe can be engineered to include a targeting determinant that will deliver it to the compartment of choice. Sequences that target proteins to the lumen of the endoplasmic reticulum (Choy *et al.*, 1999; Andersson *et al.*, 1999), mitochondria (Li *et al.*, 2008), peroxisomes (Hill and Walton, 1995), or other organelles have been identified and can be added to the structure of the probe of interest. It must be borne in mind, however, that the resolution of the optical microscope may be insufficient to discern whether the probe is bound to the membrane or free in the lumen of a small or narrow organelle. More sophisticated techniques, such as fluorescence correlation spectroscopy, may be required to ascertain whether the probe is bound and hence immobilized.

B. Interference with Biological Processes

Whenever a foreign component is introduced into a cell, the investigator must be mindful of the possibility of interference with normal cellular function. This maxim applies also to the phospholipid-sensing probes. By binding to their headgroups the fluorescent probes not only detect the phospholipids, but also scavenge a fraction of them. The fraction of lipids occluded will depend on the concentrations of the probe and lipid, and on the affinity of their interaction. Possible alterations of the normal physiology can be minimized by reducing the level of expression of the probe to the lowest level compatible with reliable detection. Nevertheless, it is advisable to compare the biological responsiveness of the transfected cells with that of cells not expressing the probe, to ensure that no significant changes occurred. An even better control consists of comparison of cells transfected with comparable amounts of the lipid-binding probe of interest and of a closely related mutant that is incapable of binding. For instance, the R40L mutation renders the PLCδ-PH domain probe inactive and serves as an ideal control; other examples can be found in (Balla and Varnai, 2002).

C. Specificity and Sensitivity Considerations

While some proteinaceous lipid-binding domains are very specific for their target (e.g., the PH domain of PLCδ), others are more promiscuous. The PH domain from AKT, which is commonly used as a marker for PI(3,4,5)P_3 (Gray *et al.*, 1999), also has the capacity to bind to PI(3,4)P_2. When this domain associates with a membrane, either one or both of these lipids may be recruiting it. More specific probes are required to distinguish between these two species (see Table I). Similarly, in some cases lipid-binding domains can bind both a lipid and protein; one such case is the PH domain of FAPP, which binds to PI4P but also to the small GTPase Arf (Godi *et al.*, 2004). Whenever possible, it is advisable to confirm the results using multiple probes that recognize the same lipid.

In addition to specificity, the sensitivity of the probe is also an issue of concern. Often protein domains recognize lipids with high selectivity, but their affinity is insufficient to recruit a sufficient fraction of the molecules to the membrane containing the target, which is often scarce. In these instances, constructs can be engineered to express two or more lipid-binding domains in tandem. This strategy was used successfully to detect the rare lipid phosphatidylinositol 3-phosphate using the FYVE domain of EEA1 (Gillooly *et al.*, 2000). Two tandem PH domains of PLCδ were also used to detect low levels of PI(4,5)P_2 which are not revealed by a single PH domain (Bohdanowicz *et al.*, 2010; Gillooly *et al.*, 2000). However, when increasing the avidity of the probe in this manner, care must be taken not to obstruct a significant fraction of the headgroups of the scarce lipid that is being investigated.

In cases where small amounts of the target lipid are generated at the plasma membrane, or if the probe is not very sensitive, total internal reflection microscopy (TIRF) (Axelrod, 1981; Mattheyses *et al.*, 2010) may be used. In this method, the focal depth is restricted to 100–150 nm from the surface of the coverslip, thus monitoring preferentially the adherent, ventral membrane of the cell while minimizing background from the bulk cytoplasm or other organelles.

D. Type of Data Obtained

When using the method described above it must be kept in mind that the data generated are typically qualitative; unlike biochemical methods that can provide quantitative, absolute determinations of lipid concentrations, the fluorescence microscopy determinations offer only relative estimates. The fluorescence intensity measured in any particular location of the cell is influenced by multiple parameters, including the level of expression of the probe (which varies between cells), the degree of photobleaching (which is a function of light intensity, duration of exposure and the sensitivity of the particular probe to photoconversion), the local environment surrounding the probe (some fluorescent proteins are highly sensitive to the pH, or can be quenched by other cellular components), and the features of the hardware used for image acquisition. These factors must be considered when interpreting the results.

E. Extension of the Method to Primary Cells

One of the limitations of this method is that the tagged lipid-binding domain must be introduced into the cell of interest. While the protocol detailed above generates sufficient numbers of transfected RAW264.7 cells and is satisfactory for a variety of cell lines, it may be inadequate for other cell types, such as primary macrophages and neutrophils that cannot be transfected by lipophilic agents. In this case, a number of options are available. Electroporation (using, for example, the Amaxa nucleofection system) has been shown to successfully introduce plasmids into primary macrophages (Weischenfeldt *et al.*, 2008) and neutrophils (Magalhaes *et al.*, 2007). Even more effective, and markedly less toxic, is the introduction of cDNAs using retroviral- or lentiviral-based vectors, the latter of which can infect even non-dividing cells (Emi *et al.*, 1991). Of course, handling of such vectors is more complicated and involves institution-specific biohazard considerations. Transgenic mouse lines expressing lipid-binding probes such as PH-AKT (Nishio *et al.*, 2007) are also beginning to become available, and have proven useful in the study of cells that are refractory to transfection and lipofection.

F. Combination of the Method with Other Techniques

While the technique we describe herein is relatively simple and straight-forward, the combination of imaging of live cells with confocal microscopy with other cell biology techniques such as TIRF, fluorescence resonance energy transfer (FRET), rapamycin-based heterodimerization systems, and others significantly increases the power of the approach while allowing other types of questions to be addressed. FRET-based biosensors have been used to more specifically and quantitatively detect levels of $PI(3,4,5)P_3$ (Sato *et al.*, 2003), diacylglycerol (Sato *et al.*, 2006), and phosphatidic acid (Nishioka *et al.*, 2010). These biosensors are composed of a lipid-binding domain fused between cyan (CFP) and yellow (YFP) fluorescent protein variants, through a linker region that includes a flexible hinge domain. The molecule is targeted to a membrane of choice by a membrane-localization sequence (MLS). Binding to the target lipid causes the biosensor to adopt a conformation that causes FRET to occur or, in some cases, to decrease, depending on how the probe is generated (Sato *et al.*, 2003). Such systems offer the advantage of

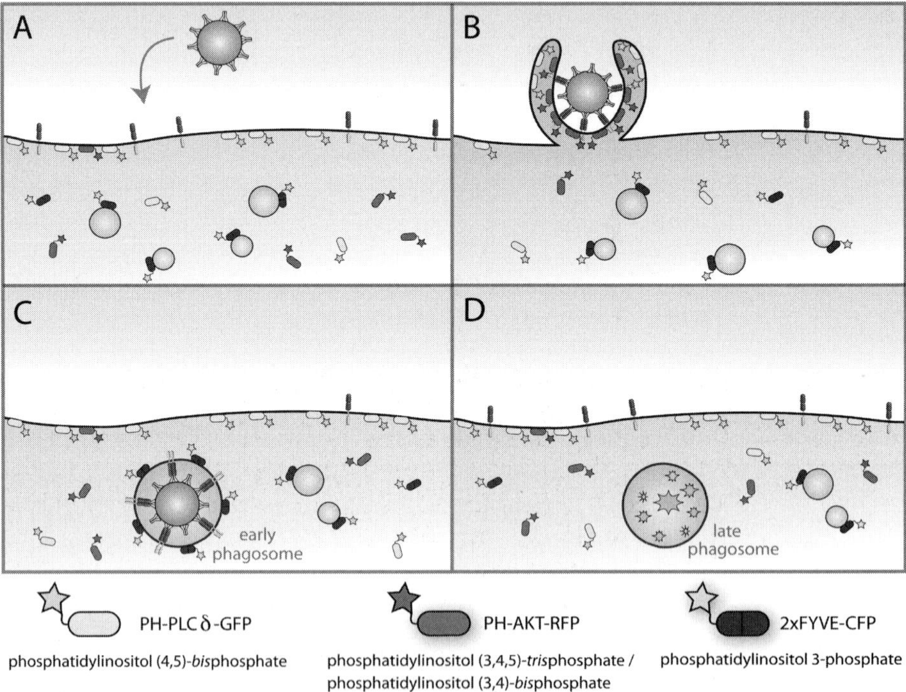

Fig. 1 Schematic Representation of PI probes before, during and after Phagocytosis. (A) In resting cells, the $PI(4,5)P_2$ probe (PH-PLCδ-GFP) is present at the membrane, whereas the PIP_3 probe (PH-AKT-RFP) remains in the cytoplasm as there is no PIP_3 for it to bind to. The PI3P probe (2xFYVE-CFP) is localized to endosomes. (B) Upon initiation of phagocytosis, more PIP_2 is generated, resulting in an accumulation of the $PI(4,5)P_2$ probe at the membrane. Almost as soon as it is made, $PI(4,5)P_2$ begins to be metabolized to PIP_3, through the action of PI3Ks. This results in the recruitment of the PIP_3 probes to the membrane. (C) By the time the phagosome seals, PIP_2 and PIP_3 have been metabolized to other products, resulting in the relocalization of their respective probes. Fusion of endosomes with the phagosome results in appearance of PI3P on the phagosome membrane, thus recruiting the PI3P probe. (D) As the phagosome matures through fusion with lysosomes, PI3P is lost from the phagosome membrane. The phagosome acquires degradative properties, resulting in the digestion of the internalized particle. (See color plate.)

allowing the user to monitor generation of a lipid at a specific membrane, defined by the MLS utilized when the probe is created. This is particularly useful in cases where one lipid is produced at multiple membranes within the cell, as in the case of PI (3,4,5)P_3 (Sato *et al.*, 2003).

The advent of inducible membrane-targeting systems such as those based on rapamycin-binding domains (e.g., Suh *et al.*, 2006) allows timed, directed recruitment of a fluorescently tagged protein of interest during imaging. Thus, one can use such a method to determine the effect of recruiting a lipid kinase or phosphatase (as a general example) to the membrane during phagocytosis or any phenomenon of interest. Selective targeting directly to the phagosome has not yet been described, though an elegant study has recently shown direct targeting of the phosphatase Synaptojanin 1 to endophilin-induced plasma membrane tubules (but not to the rest of the bulk plasma membrane), thus allowing the determination of the effects of phosphoinositide remodeling in a tubulated structure in real time (Chang-Ileto *et al.*, 2011).

VI. Summary

We describe a method that can be used to monitor dynamic changes in lipids during phagosome formation and maturation, with excellent spatial and temporal resolution. This method can be extended to other cell types to study processes in which lipid remodeling is thought to be important for signaling and metabolism.

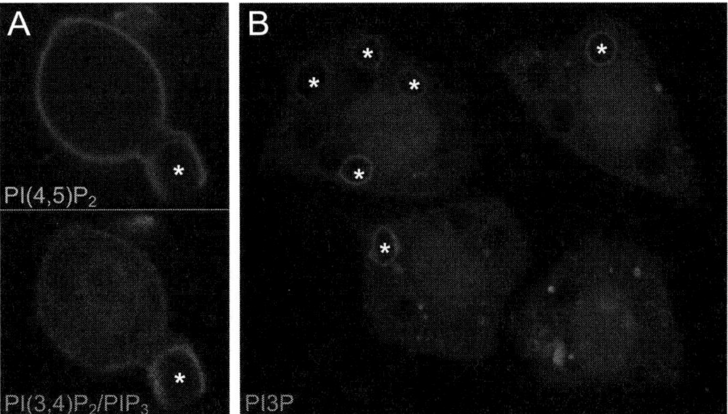

Fig. 2 Imaging of PH-PLCδ-GFP, PH-AKT-RFP, and 2xFYVE-RFP during Fcγ receptor-mediated phagocytosis in RAW264.7 cells. (A) A cell co-expressing the PI(4,5)P_2-binding PH domain from PLCδ fused to GFP (PH-PLCδ-GFP) and the PI(3,4)P_2/PIP_3-binding PH domain from AKT fused to RFP (PH-AKT-RFP) is shown. The asterisk marks a RBC in the process of being engulfed; at this early stage both PIP_2 and PIP_3 are present in the nascent phagosome. (B) A group of cells expressing two tandem PI3P-binding FYVE domains from EEA1 fused to RFP (2xFYVE-RFP) are shown. Asterisks mark phagosomes decorated with PI3P, as judged by recruitment of the fluorescent protein. (For color version of this figure, the reader is referred to the web version of this book.)

Acknowledgements

We thank Paul Paroutis for preparing Fig. 1, and Michal Bohdanowicz for providing us with images for Fig. 2. Original work in the authors' laboratory is supported by the Canadian Cystic Fibrosis Foundation and by the Canadian Institutes of Health Research. S. Grinstein is the current holder of the Pitblado Chair in Cell Biology and is cross-appointed to the Department of Biochemistry of the University of Toronto.

References

Aderem, A., and Underhill, D. M. (1999). Mechanisms of phagocytosis in macrophages. *Annu. Rev. Immunol.* **17**, 593–623.

Andersson, H., *et al.* (1999). Protein targeting to endoplasmic reticulum by dilysine signals involves direct retention in addition to retrieval. *J. Biol. Chem.* **274**, 15080–15084.

Andree, H. A., *et al.* (1990). Binding of vascular anticoagulant alpha (VAC alpha) to planar phospholipid bilayers. *J. Biol. Chem.* **265**, 4923–4928.

Araki, N., *et al.* (1996). A role for phosphoinositide 3-kinase in the completion of macropinocytosis and phagocytosis by macrophages. *J. Cell Biol.* **135**, 1249–1260.

Axelrod, D. (1981). Cell-substrate contacts illuminated by total internal reflection fluoresce'nce. *J. Cell Biol.* **89**, 141–145.

Balla, T., and Varnai, P. (2002). Visualizing cellular phosphoinositide pools with GFP-fused protein-modules. *Sci. STKE* **2002**, pl3.

Bohdanowicz, M., *et al.* (2010). Class I and class III phosphoinositide 3-kinases are required for actin polymerization that propels phagosomes. *J. Cell Biol.* **191**, 999–1012.

Botelho, R. J., *et al.* (2000). Localized biphasic changes in phosphatidylinositol-4,5-bisphosphate at sites of phagocytosis. *J. Cell Biol.* **151**, 1353–1368.

Chang-Ileto, B., *et al.* (2011). Synaptojanin 1-mediated PI(4,5)P2 hydrolysis is modulated by membrane curvature and facilitates membrane fission. *Dev. Cell* **20**, 206–218.

Chen, J., *et al.* (2008). Selective binding of phorbol esters and diacylglycerol by individual C1 domains of the PKD family. *Biochem. J.* **411**, 333–342.

Choy, E., *et al.* (1999). Endomembrane trafficking of ras: the CAAX motif targets proteins to the ER and Golgi. *Cell* **98**, 69–80.

Colon-Gonzalez, F., and Kazanietz, M. G. (2006). C1 domains exposed: from diacylglycerol binding to protein-protein interactions. *Biochim. Biophys. Acta* **1761**, 827–837.

Cox, D., *et al.* (1999). A requirement for phosphatidylinositol 3-kinase in pseudopod extension. *J. Biol. Chem.* **274**, 1240–1247.

Dormann, D., *et al.* (2002). Visualizing PI3 kinase-mediated cell-cell signaling during Dictyostelium development. *Curr. Biol.* **12**, 1178–1188.

Dove, S. K., *et al.* (2004). Svp1p defines a family of phosphatidylinositol 3,5-bisphosphate effectors. *EMBO J.* **23**, 1922–1933.

Dowler, S., *et al.* (2000). Identification of pleckstrin-homology-domain-containing proteins with novel phosphoinositide-binding specificities. *Biochem J.* **351**, 19–31.

Downey, G. P., *et al.* (1999). Phagosomal maturation, acidification, and inhibition of bacterial growth in nonphagocytic cells transfected with FcgammaRIIA receptors. *J. Biol. Chem.* **274**, 28436–28444.

Ellson, C. D., *et al.* (2001). PtdIns(3)P regulates the neutrophil oxidase complex by binding to the PX domain of p40(phox). *Nat. Cell Biol.* **3**, 679–682.

Emi, N., *et al.* (1991). Pseudotype formation of murine leukemia virus with the G protein of vesicular stomatitis virus. *J. Virol.* **65**, 1202–1207.

Gillooly, D. J., *et al.* (2000). Localization of phosphatidylinositol 3-phosphate in yeast and mammalian cells. *EMBO J.* **19**, 4577–4588.

Godi, A., *et al.* (2004). FAPPs control Golgi-to-cell-surface membrane traffic by binding to ARF and PtdIns(4)P. *Nat. Cell Biol.* **6**, 393–404.

Gray, A., *et al.* (1999). The pleckstrin homology domains of protein kinase B and GRP1 (general receptor for phosphoinositides-1) are sensitive and selective probes for the cellular detection of phosphatidylinositol 3,4-bisphosphate and/or phosphatidylinositol 3,4,5-trisphosphate in vivo. *Biochem. J.* **344**(Pt 3), 929–936.

Gu, H., *et al.* (2003). Critical role for scaffolding adapter Gab2 in Fc gamma R-mediated phagocytosis. *J. Cell Biol.* **161**, 1151–1161.

Hill, P. E., and Walton, P. A. (1995). Import of microinjected proteins bearing the SKL peroxisomal targeting sequence into the peroxisomes of a human fibroblast cell line: evidence that virtually all peroxisomes are import-competent. *J. Cell Sci.* **108**(Pt 4), 1469–1476.

Huynh, K. K., *et al.* (2007). Fusion, fission, and secretion during phagocytosis. *Physiology (Bethesda)* **22**, 366–372.

Kimber, W. A., *et al.* (2002). Evidence that the tandem-pleckstrin-homology-domain-containing protein TAPP1 interacts with Ptd(3,4)P2 and the multi-PDZ-domain-containing protein MUPP1 in vivo. *Biochem. J.* **361**, 525–536.

Kobayashi, T., and Arakawa, Y. (1991). Transport of exogenous fluorescent phosphatidylserine analogue to the Golgi apparatus in cultured fibroblasts. *J. Cell Biol.* **113**, 235–244.

Komander, D., *et al.* (2004). Structural insights into the regulation of PDK1 by phosphoinositides and inositol phosphates. *EMBO J.* **23**, 3918–3928.

Li, S. K., *et al.* (2008). Identification of functionally important amino acid residues in the mitochondria targeting sequence of hepatitis B virus X protein. *Virology* **381**, 81–88.

Liao, F., *et al.* (1992). Tyrosine phosphorylation of phospholipase C-gamma 1 induced by cross-linking of the high-affinity or low-affinity Fc receptor for IgG in U937 cells. *Proc. Natl. Acad. Sci. U. S. A.* **89**, 3659–3663.

Magalhaes, M. A., *et al.* (2007). Expression and translocation of fluorescent-tagged p21-activated kinase-binding domain and PH domain of protein kinase B during murine neutrophil chemotaxis. *J. Leukoc. Biol.* **82**, 559–566.

Mattheyses, A. L., *et al.* (2010). Imaging with total internal reflection fluorescence microscopy for the cell biologist. *J. Cell Sci.* **123**, 3621–3628.

Nishio, M., *et al.* (2007). Control of cell polarity and motility by the PtdIns(3,4,5)P3 phosphatase SHIP1. *Nat. Cell Biol.* **9**, 36–44.

Nishioka, T., *et al.* (2010). Heterogeneity of phosphatidic acid levels and distribution at the plasma membrane in living cells as visualized by a Foster resonance energy transfer (FRET) biosensor. *J. Biol. Chem.* **285**, 35979–35987.

Raiborg, C., *et al.* (2001). FYVE and coiled-coil domains determine the specific localisation of Hrs to early endosomes. *J. Cell Sci.* **114**, 2255–2263.

Roy, A., and Levine, T. P. (2004). Multiple pools of phosphatidylinositol 4-phosphate detected using the pleckstrin homology domain of Osh2p. *J. Biol. Chem.* **279**, 44683–44689.

Rusten, T. E., and Stenmark, H. (2006). Analyzing phosphoinositides and their interacting proteins. *Nat. Methods* **3**, 251–258.

Santagata, S., *et al.* (2001). G-protein signaling through tubby proteins. *Science* **292**, 2041–2050.

Sato, M., *et al.* (2003). Production of PtdInsP3 at endomembranes is triggered by receptor endocytosis. *Nat. Cell Biol.* **5**, 1016–1022.

Sato, M., *et al.* (2006). Imaging diacylglycerol dynamics at organelle membranes. *Nat. Methods* **3**, 797–799.

Stauffer, T. P., *et al.* (1998). Receptor-induced transient reduction in plasma membrane PtdIns(4,5)P2 concentration monitored in living cells. *Curr. Biol.* **8**, 343–346.

Suh, B. C., *et al.* (2006). Rapid chemically induced changes of PtdIns(4,5)P2 gate KCNQ ion channels. *Science* **314**, 1454–1457.

Swanson, J. A. (2008). Shaping cups into phagosomes and macropinosomes. *Nat. Rev. Mol. Cell Biol.* **9**, 639–649.

Swanson, J. A., and Hoppe, A. D. (2004). The coordination of signaling during Fc receptor-mediated phagocytosis. *J. Leukoc. Biol.* **76**, 1093–1103.

Teruel, M. N., *et al.* (1999). A versatile microporation technique for the transfection of cultured CNS neurons. *J. Neurosci. Methods* **93**, 37–48.

Varnai, P., *et al.* (1999). Phosphatidylinositol 3-kinase-dependent membrane association of the Bruton's tyrosine kinase pleckstrin homology domain visualized in single living cells. *J. Biol. Chem.* **274**, 10983–10989.

Venkateswarlu, K., *et al.* (1998). Insulin-dependent translocation of ARNO to the plasma membrane of adipocytes requires phosphatidylinositol 3-kinase. *Curr. Biol.* **8**, 463–466.

Vieira, O. V., *et al.* (2001). Distinct roles of class I and class III phosphatidylinositol 3-kinases in phagosome formation and maturation. *J. Cell Biol.* **155**, 19–25.

Weischenfeldt, J., *et al.* (2008). NMD is essential for hematopoietic stem and progenitor cells and for eliminating by-products of programmed DNA rearrangements. *Genes Dev.* **22**, 1381–1396.

Yeung, T., *et al.* (2008). Membrane phosphatidylserine regulates surface charge and protein localization. *Science* **319**, 210–213.

CHAPTER 20

Genetically Encoded Probes for Phosphatidic Acid

Nawal Kassas, Petra Tryoen-Tóth, Matthias Corrotte, Tamou Thahouly, Marie-France Bader, Nancy J. Grant and Nicolas Vitale

Institut des Neurosciences Cellulaires et Intégratives, CNRS UPR3212, Strasbourg, France

Abstract

In addition to forming bilayers to separate cellular compartments, lipids participate in vesicular trafficking and signal transduction. Among others, phosphatidic acid (PA) is emerging as an important signaling molecule. The spatiotemporal distribution of cellular PA appears to be tightly regulated by localized synthesis and a rapid metabolism. Although PA has been long proposed as a pleiotropic bioactive lipid, when and where PA is produced in the living cells have only recently

METHODS IN CELL BIOLOGY, VOL 108
Copyright 2012, Elsevier Inc. All rights reserved.

0091-679X/10 $35.00
DOI 10.1016/B978-0-12-386487-1.00020-1

been explored using biosensors that specifically bind to PA. The probes that we have generated are composed of the PA-binding domains of either Spo20p or Raf1 directly fused to GFP.

In this chapter, we will describe the expression and purification of GST-fusion proteins of these probes, and the use of phospholipid strips to validate the specificity of their interaction with PA. We will then illustrate the use of GFP-tagged probes to visualize the synthesis of PA in the neurosecretory PC12 cells and RAW 267.4 macrophages. Interestingly, the two probes show a differential distribution in these cell types, indicating that they may have different affinities for PA or recognize different pools of PA. In conclusion, the development of a broader choice of probes may be required to adequately follow the complex dynamics of PA in different cell types, in order to determine the cellular distribution of PA and its role in various cellular processes.

I. Phosphatidic Acid: A Rapid Overview

Phosphatidic acid (PA) is the simplest diacyl-glycerophospholipid and consists of an alcohol and a glycerol moiety, which has two fatty acids and a phosphate group esterified at positions 1, 2, and 3, respectively. In general, a saturated fatty acid is bond to carbon 1, a saturated fatty acid to carbon 2, and a phosphate group to carbon 3. PA is found only in limited amounts in biological membranes, yet is crucial for cell survival. It is only the recent development of liquid-chromatography coupled to mass spectrometry techniques that have allowed a more precise idea of the relative levels of the different PA species (Oliveira *et al.*, 2010) and provided some idea of the PA composition in different subcellular compartments (Shulga *et al.*, 2010). The physiological importance of PA is related to its central role in glycerophospholipid synthesis, in addition to diverse functions in membrane dynamics and lipid signaling (Athenstaedt and Daum, 1999). In mammalian two different pathways are known for the de novo synthesis of PA, the glycerol 3-phosphate pathway or the dihydroxyacetone phosphate pathway (Roth, 2008). The signaling PA is produced through three major pathways: the hydrolysis of phosphatidylcholine by phospholipase D (PLD), the acylation of lysophosphatidic acid by lysoPA-acyltransferase (LPAAT), and the phosphorylation of diacylglycerol (DAG) by DAG kinase (Roth, 2008). PA is degraded by conversion into DAG by lipid phosphate phosphohydrolases (LPPs) or into lyso-PA by phospholipase A (PLA) (Roth, 2008). These potent enzymes maintain at extremely low levels PA concentrations in the cell.

PA, like other lipid signals can serve as membrane docking sites, which serve to recruit specific proteins. PA can activate proteins either by directly stimulating their enzymatic activity or by indirectly targeting them to the site where they are required to function. Alternatively, PA can inactivate proteins by depleting them from the place where they are active. In this way, the endoplasmic reticulum (ER) sequesters the yeast transcriptional repressor Opi1p (Loewen *et al.*, 2004). When PA levels decrease, Opi1p is no longer retained in the ER and translocates to the nucleus where it represses target gene expression (Loewen *et al.*, 2004). PA can also directly inhibit enzymes such as the protein phosphatase-1 (Jones and Hannun, 2002). In other cases, PA appears to act in concert with other lipid second messengers. For example, when protein kinase C is stimulated, PA binds to its C2-domain allowing translocation of the kinase to the plasma membrane. Subsequently, DAG binding to the C1-domain then has a synergistic effect on protein kinase C activity (Jose Lopez-Andreo *et al.*, 2003). Similarly, the pleckstrin homology (PH) domain of Son of sevenless was shown to bind both phosphatidylinositol 4,5-bisphosphate (PIP2) and PA (Zhao *et al.*, 2007). Another interesting possibility, not directly involving protein targeting, is that the shape and negative charge of PA itself promote a membrane curvature and thereby, affect membrane topology and fusion properties (Bader and Vitale, 2009; Chernomordik and Kozlov, 2008).

A. Rationale

The PA synthesized by the mammalian PLD isoforms has been proposed to play important roles in many important cellular functions, in particular membrane trafficking. By measuring the production of labeled phosphatidylethanol by PLD-mediated transphosphatidylation in the presence of ethanol, an increase in PLD activity has been reported in numerous processes, including neurosecretion (Caumont *et al.*, 1998; Humeau *et al.*, 2001) and phagocytosis (Iyer *et al.*, 2004). More recently numerous experiments based on the use of selective knockdown of the PLD isoforms, using a small RNA interference approach have confirmed the implication of PLD in many vesicular trafficking events (Bader and Vitale, 2009; Corrotte *et al.*, 2006; Du *et al.*, 2004; Huang *et al.*, 2005). Nevertheless, the precise localization and dynamics of PA in these processes remains a major challenge. Here, we describe the characterization of two genetically encoded probes for PA and their use in studying exocytosis in neurosecretory cells and phagocytosis in macrophages.

II. Choice of the PA-Probes

The specificity of PA-protein interactions appears to largely rely on the unique ionization properties of the phosphomonoester headgroup of PA (Kooijman and Burger, 2009). High affinity PA-protein interactions have been described for over 20 proteins in mammalian, plant and yeast cells (for review, see Stace and Ktistakis, 2006). These include the protein kinase Raf-1 and the yeast SNARE protein Spo20p that we have used as probes to visualize PA in cells. Interestingly, when one compares the PA-binding region of the subset of proteins in which the binding region has been evaluated, no consensus sequence (sequence homology) is apparent. This is in sharp contrast to other lipid binding modules such as the PH, PX, C1, and C2 domains. One general, but not surprising, feature that arises is the presence of basic amino acid residues. In most cases, hydrophobic residues also appear to be important for PA binding. To visualize PA within cells, we have used two probes based on the PA-binding domains of the yeast SNARE protein Spo20p and the protein kinase Raf-1.

In *Saccharomyces cerevisiae*, the developmentally regulated Soluble *N*-ethylmaleimide sensitive factor attachment protein receptor (SNARE) protein Spo20p mediates the fusion of vesicles with the prospore membrane that is required for spore formation. Spo20p is positively regulated through an amphipathic helix, which confers plasma membrane or prospore membrane localization (Nakanishi *et al.*, 2004). Genetic manipulations of phospholipid pools indicate that the likely *in vivo* ligand of this domain is PA (Nakanishi *et al.*, 2004). Accordingly *in vitro* flotation assay have shown that Spo20p binds preferentially to PA when compared to negatively charged phosphoinositol 4,5-bisphosphate (Nakanishi *et al.*, 2004). The minimal PA binding domain consists of a stretch of 40 residues that are predicted to form an amphipathic α-helix. Hydrophobic residues on one side of the helix, as well as basic residues appear to be critical for PA binding (Nakanishi *et al.*, 2004).

The protooncogene Raf-1 kinase plays a crucial role in several normal and pathological cellular processes. A common aspect of Raf-1 activation is its translocation to the plasma membrane. *In vitro* analysis of Raf-1-lipid interaction reveals two distinct phospholipid binding sites within Raf-1 kinase, of which a PA interacting domain is located between residues 390 and 423 of human Raf-1 (Ghosh *et al.*, 2003; Rizzo *et al.*, 2000). The tetrapeptide RKTR (residues 398-401 of human Raf-1) was shown to be critical in Raf-1 for interaction with PA by a combination of alanine-scanning, deletion mutagenesis and liposome assay (Ghosh *et al.*, 2003).

III. Specific Binding of PA to Probes

A. Expression of Recombinant PA Probes

The Spo20p PA-binding domain (PABD) is amplified by PCR from the pEGFP-C1-PABD-Spo20p construct using the forward primer 5'-CGGGATCCCTCG AGCGTCTAGAATGG-3' and reverse primer 5'-GCGAATTCTTAACTAGTCT TAGTGGCGTC3', as previously described (Zeniou-Meyer *et al.*, 2007). The Raf1 PA binding domain is amplified by PCR from the pEGFP-C1-PABD-Raf1 construct using the forward primer 5'-CGGGATCCTCCAGGCCTTCAGGAATGAGG-3' and reverse primer 5'-GCGAATTCGCCCTCGCACCACTGGGTC-3'. The PA-binding fragment is inserted in frame into pGEX4T1 using a procedure described previously (Vitale *et al.*, 2000a). Constructs have been verified by automated DNA sequencing.

Large-scale production of chimeric GST-PABD proteins has been previously described (Vitale *et al.*, 1996, 2001). Briefly, expression is induced at 37 °C for GST-Spo20p-PABD and at 18 °C for GST-Raf1-PABD. Fusion proteins are purified on glutathione-Sepharose (Vitale *et al.*, 1998) and purity is estimated to be >98% by Coomassie blue staining of SDS-PAGE gels. The quantity of purified protein can be determined from this dye-binding assay using bovine serum albumin as a standard (Vitale *et al.*, 2000b). Protein aliquots are stored at −80 °C.

B. *In vitro* Lipid Binding Assay

The PA-binding region of Spo20p was originally characterized in yeast by successive deletions from Spo20p (Nakanishi *et al.*, 2004), whereas the PA binding fragment of Raf1 was mapped *in vitro* (Ghosh *et al.*, 2003). We have previously shown that GFP-PABD-Spo20p in lysates of transfected PC12 cells specifically binds to PA on phospholipid strips (Zeniou-Meyer *et al.*, 2007). The GFP-Raf1-PABD also appears to be specific for PA because mutations of basic residues in the binding domain (Rizzo *et al.*, 2000) eliminate the recruitment of the probe to phagosomes in transfected RAW 264.7 macrophages (Corrotte *et al.*, 2006).

To support the specificity of these probes for PA, lipid overlay assays are performed with recombinant GST-PABD proteins. Briefly, strips containing various phospholipids, as indicated (Molecular Probes), are initially saturated with PBS 3% BSA + 0.1% Tween 20 for 4 h at 4 °C. Then GST-PABD-proteins (1 μg/mL in PBS 3% BSA + 0.1% Tween20) or as a control, GST alone are incubated for 2 h at 4 °C with the phospholipid strips. Subsequently, the strips are processed for Western blot analysis with an anti-GST antibody (Molecular Probes). Images are acquired with a Chemi-smart 5000 (Vilber Lourmat). Our data reveal that GST-PABD-Spo20p and GST-PABD-Raf1 bind to PA whereas they weakly bind to other acidic lipids (Fig. 1). On the other hand GST does not bind any of the lipids tested (Fig. 1).

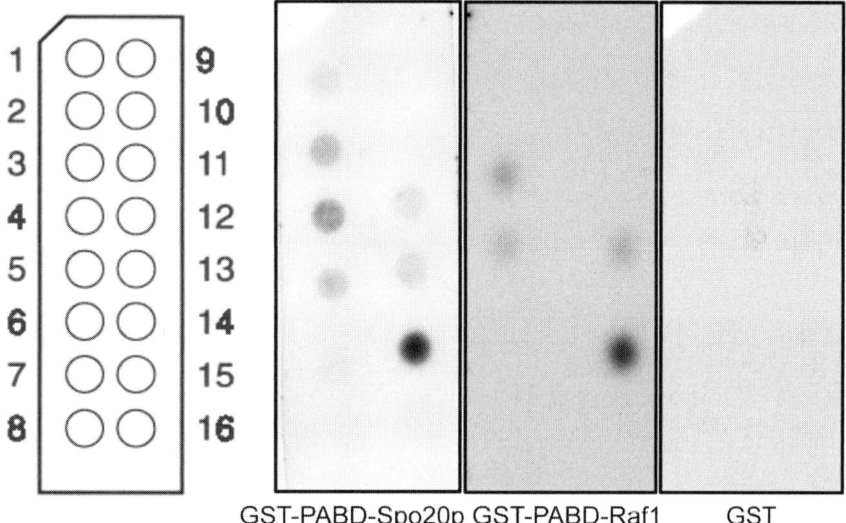

Fig. 1 GST-PABD-Spo20p and GST-PABD-Raf1 specifically binds to PA *in vitro*. (A) Layout of phospholipid strips membranes. 1: Lysophosphatidic acid, 2: Lysophosphatidylcholine, 3: Phosphatidylinositol (PtdIns), 4: PtdIns(3)P, 5: PtdIns(4)P, 6: PtdIns(5)P, 7: Phosphatidylethanolamine, 8: Phosphatidylcholine, 9: Sphingosine 1-phosphate, 10: PtdIns(3,4)P2, 11: PtdIns(3,5)P2, 12: PtdIns(4,5)P2, 13: PtdIns(3,4,5)P3, 14: Phosphatidic acid, 15: Phosphatidylserine, 16: Blank. (B) Recombinant GST-PABD-Spo20p, GST-PABD-raf1 or GST proteins were incubated with phosphlipid strips for 2 h at 4 °C. Strips were subsequently processed for Western blot analysis with an anti-GST antibody. Similar results were obtained with three distinct protein preparations.

IV. Imaging PA in Cells

A. Choice of Cell Types

The catecholamine-secreting PC12 cell line derived from a rat pheochromoytoma originating from chromaffin cells of the adrenal medulla has long been considered a suitable model system for studying neurosecretion and neuronal differentiation. PC12 cells contain a large number of secretory granules for storage of small molecules, processing enzymes, neuropeptides, and peptide hormones. Secretory granule exocytosis in PC12 cells is tightly regulated by calcium and occurs in response to a secretagogue (Bader *et al.*, 2002). Because the PC12 cells are still today easier to transfect than the primary chromaffin cells, they represent a valuable model to determine the dynamics of PA synthesis during neurosecretion using genetically encoded PA probes.

The murine macrophage cell line RAW 264.7 has been used extensively for characterizing the molecular machinery governing phagocytosis. Particle activation of surface receptors triggers signaling cascades that lead to the engulfment of the particle within the phagosome and its subsequent degradation. Accumulating evidence indicates that the PLD isoforms actively participate in this process (Corrotte *et al.*, 2006, 2010; Iyer *et al.*, 2004), and that PA production is necessary for efficient phagocytosis (Corrotte *et al.*, 2006). One major advantage of the RAW 264.7 cell line over primary cultures of professional phagocytes such as macrophages or neutrophiles is that the cell line can be transfected at reasonable rates either by electroporation or lipofection. The use of the PA-biosensors in RAW 264.7 cells for studying PA production during phagocytosis will be described below.

B. PC12 Cell Culture Conditions and Transfection

1. Stock cells are grown in 10 cm cell culture dishes in 10 mL Dulbecco's modified Eagle's medium (DMEM) containing fetal bovine serum (FBS) 5%, horse serum 10%, and antibiotics (complete DMEM medium), in a humidified incubator, 5% CO_2, 37 °C. The dishes should be pre-coated with poly-L-lysine. Dishes are pre-coated with 5 mL poly-L-lysine (100 μg/mL H_2O) solution and incubated 60 min at 37 °C. To eliminate excess of poly-L-lysine dishes are washed with sterile water before the addition of the cells.
2. Split cells at a subcultivation ratio of 1:3 up to 1:7, as needed for experiments. Doubling time is about 30–35 h. Cells should be split when the confluency is between 70 and 80%. Wash cells with 2 mL of pre-warmed trypsin-EDTA solution (dissociation solution) then incubate them with 2 mL of cell dissociation solution at 37 °C for few minutes, before collecting the cells in complete DMEM medium.
3. After centrifugation at 120 g for 5 min, cells are suspended and counted in cell culture medium without antibiotics. For the imaging experiments, cells are plated

in 500 μL of this medium on coverslips precoated with poly-L-lysine in 24-well plates at a density of about 1,00,000 cells per well.

4. The day after cell plating, the GFP-PABD probes are transfected into PC12 cells using Lipofectamine 2000 following the manufacturer's optimized protocol (Invitrogen). Briefly, 0.8 μg of DNA is diluted in 50 μL of OPTI-MEM / well, and 1.6 μL of Lipofectamine is diluted in 50 μL of OPTI-MEM/well. The diluted DNA is then mixed with the Lipofectamine solution and incubated at room temperature for 20 min. Next the DNA-Lipofectamine transfection solution is added to wells containing 500 μL of culture medium as described above. Cells are incubated with the transfection solution for 4–5 h, then medium is replaced by complete DMEM medium and cells are cultured for 24 h.

C. RAW 264.7 Macrophage Culture Conditions and Transfection

1. RAW 264.7 macrophages (American Tissue Culture Collection (ATCC)) are grown in 10 cm Petri dishes containing 12 mL RPMI plus heat-inactivated FBS 10%, HEPES 10 mM, pH7.0, Na pyruvate 1 mM, β-mercaptoethanol 0.05 mM, and supplemented with streptomycin 10^{-6} M, and penicillin $5x10^4$ U/L. The cells are maintained at 37 °C in a 5% CO_2 humidified atmosphere, and split every 2–3 days. To dilute cells, they are first detached mechanically, then diluted 1:6, and seeded at 4 x 10^4 cells/cm^2.
2. The day before transfection, RAW264.7 cells are plated in 10 cm dishes at a dilution of 1:3 to obtain a confluency between 60 and 80% the following day. Cells can be transfected either by electroporation, or by lipofection as described above for the PC12 cells. For electroporation of the PABD plasmids, the Electrobuffer Kit and protocol (Cell Projects) is used. Cells (8 x 10^6) are suspended in wash buffer and collected by centrifugation at 90 g for 5 min. Cells are then suspended in 240 μL electroporation buffer B containing ATP (2.5 mM) and glutathione (4 mM) and 10 μg of plasmid. The cell suspension is electoporated in a 4 mm cuvette at 230 Vand 950 μF in a Biorad Gene Pulser (Xcell). The average pulse time is 40 ms. Cells are immediately diluted with warm culture medium without antibiotics, and then seeded at 4 x 10^4 cells on the glass bottom of 35 mm Microwell dishes (MatTek Corporation). Transfection efficiency by either electroporation or lipofection is about 30%.

D. Imaging PA in PC12 Cells

1. The day after PABD probe transfection cultures are washed twice for 5 min with Locke buffer (HEPES 15 mM; pH 7,5; NaCl 140 mM; KCl 4,7 mM; $CaCl_2$ 2,5 mM; KH_2PO_4 1,2 mM; $MgSO_4$ 1,2 mM; EDTA 0,01 mM). Then cells are incubated in either Locke buffer (resting condition) or in the presence of Locke solution containing K^+ 59 mM (stimulation solution) for 5 min at 37 °C.

2. Immediately after these treatments, cells are fixed with ice-cold 4% paraformaldehyde (PFA) diluted in phosphate buffer for 10 min at room temperature. Cells are then permeabilized by incubating in 0.1% Triton X-100/4% PFA for 10 min and washed thoroughly with PBS to eliminate all fixation residues. To saturate non-specific sites cells are incubated in PBS containing 3% BSA/10% goat serum for 1 h at room temperature.
3. To delineate the plasma membrane in fixed cells, we label the SNARE protein SNAP25 with a monoclonal anti-SNAP25 antibody (Covance), diluted 1:200 in 3% BSA-PBS. A 50-μL drop of diluted antibody is placed on parafilm and coverslips with transfected PC12 cells are placed upside down on the droplets and incubated for 1 h at room temperature. Following incubation, coverslips are returned to wells and washed three times for 5 min in PBS. Incubation with secondary antibody, Alexa 555 conjugated goat anti-mouse IgG (Invitrogen) diluted to 1:1500 is carried out as described for the primary antibody. After three washes in PBS, cells are rinsed with ultrapure water, and coverslips are mounted in Mowiol-Evanol.

E. Imaging PA in RAW 264.7 Macrophage and Phagocytosis Assay

1. For phagocytosis assays, the medium of RAW264.7 cells, cultured 18–24 h on glass bottom Petri dishes is changed to RPMI-BSA (medium without phenol red (Sigma R7509) supplemented with BSA 1% (Fraction V, Sigma A8806), HEPES 10 mM, pH7.0, Na pyruvate 1 mM, and β-mercaptoethanol 0.05 mM).
2. Phagocytosis is stimulated using 3-μm latex microspheres (PS05N, Bangs Laboratories) opsonized with human IgG (Invitrogen). To opsonize particles, 25 μL of microspheres (stock 10% w/v) are washed in PBS, suspended in 250 μL PBS containing 0.2 mg/mL IgG and incubated with rotation 2 h at room temperature. Particles are then washed in PBS and suspended in 250 μL RPMI-BSA to obtain a 1% solution of particles.
3. Before initiating phagocytosis, transfected cells can be identified by epifluorescence using the binoculars. Fresh RPMI-BSA medium (1 mL) pre-heated to 37 °C is then placed in the Petri dishes. To initiate phagocytosis, 5 μL of the IgG-coated particle suspension is added to the dish directly over the cells. When the particles reach the cells (about 5 min), switch to the confocal mode and images can be acquired over the next 20 min. To avoid photobleaching and phototoxicity, illumination of the cells should be minimized.

1. Fluorescent Measurements/Image Acquisition

To examine the cellular distribution of the PA probes, a scanning confocal microscope (Zeiss LSM 510) is used. Samples are examined with an oil immersion 63x objective (Plan Apochromat n.a. = 1.4) using an argon laser at an excitation wavelength of 488 nm for GFP and a helium/neon laser at 543 nm to excite Alexa 555 fluorophore Digital images (of median confocal sections of cells) are acquired with pinhole of 1.0.

F. Image Analysis and Results in PC12 Cells

Multiple confocal image acquisitions (8 bits) are carried out for each experimental condition and digital images are analyzed. To determine the recruitment of the probes at the plasma membrane, the co-localization of GFP-PABD-Spo20p or GFP-PABD-Raf1 (green) with SNAP25 (red) can be visualized by generating co-localization masks and the proportion of co-localization can be quantified using the Zeiss CLSM 3.2 instrument software. Briefly, for each cell, a scatter plot of double-labelled pixels is generated and threshold levels are adjusted for each fluorophore to eliminate background-staining levels. The weighted co-localization percentages, which we use, take into account the number of pixels, the average fluorescence intensity and their frequency. For each condition, 20–30 cells were analyzed and data are expressed as the mean ± SD. Similar observations were obtained from at least three different cell culture.

The two GFP-tagged PA probes do not show the same cellular distribution in PC12 cells; however, their cellular localization is not altered by cell fixation (data not shown). In resting cells, the GFP-PABD-Spo20p probe is concentrated in the cell nucleus (Fig. 2a) and only about 4 ± 1% of the PA probe co-localized with SNAP25. Upon stimulation of exocytosis, a fraction of the Spo20p fluorescent probe is efficiently recruited to the plasma membrane (Fig. 2a). The recruitment kinetics of the PA sensor reaches a maximum level after 5 min of stimulation, showing 30 ± 4% co-localization with the SNARE protein SNAP25 (Béglé *et al.*, 2009; Zeniou-Meyer *et al.*, 2007). The GFP-PABD-Raf1 probe on the other hand, shows a clear cytosolic distribution with faint nuclear localization. In stimulated cells only a modest co-localization of GFP-PABD-Raf1 with SNAP25 is observed (Fig. 2b). All together these results suggest that the PABD of Spo20p appears to be a more sensitive reporter for PA than that of Raf1 in the neurosecreting PC12 cells.

G. Results in RAW 264.7 Macrophages

As we observed in fixed PC12 cells, the PA biosensors PABD-Raf1 and PABD-Spo20 do not show the same distribution in live macrophages. Prior to a phagocytic stimulation, the PABD-Raf1 probe is principally localized in the cytoplasm (Fig. 3a), whereas the PABD-Spo20p is mainly in the nucleus and not in the cytoplasm (Fig. 3d). However, the PABD-Spo20p probe is also observed at the plasma membrane in about 20% of the macrophages under resting conditions (data not shown). As PLD activity has also been implicated in cell adhesion of RAW 264.7 macrophages (Iyer *et al.*, 2006), it is possible that these cells are migrating cells. Following stimulation of phagocytosis, the PABD-Raf1 probe is sometimes observed in the phagocytic cup as the pseudopods extend around the particle (Fig. 3b), but is

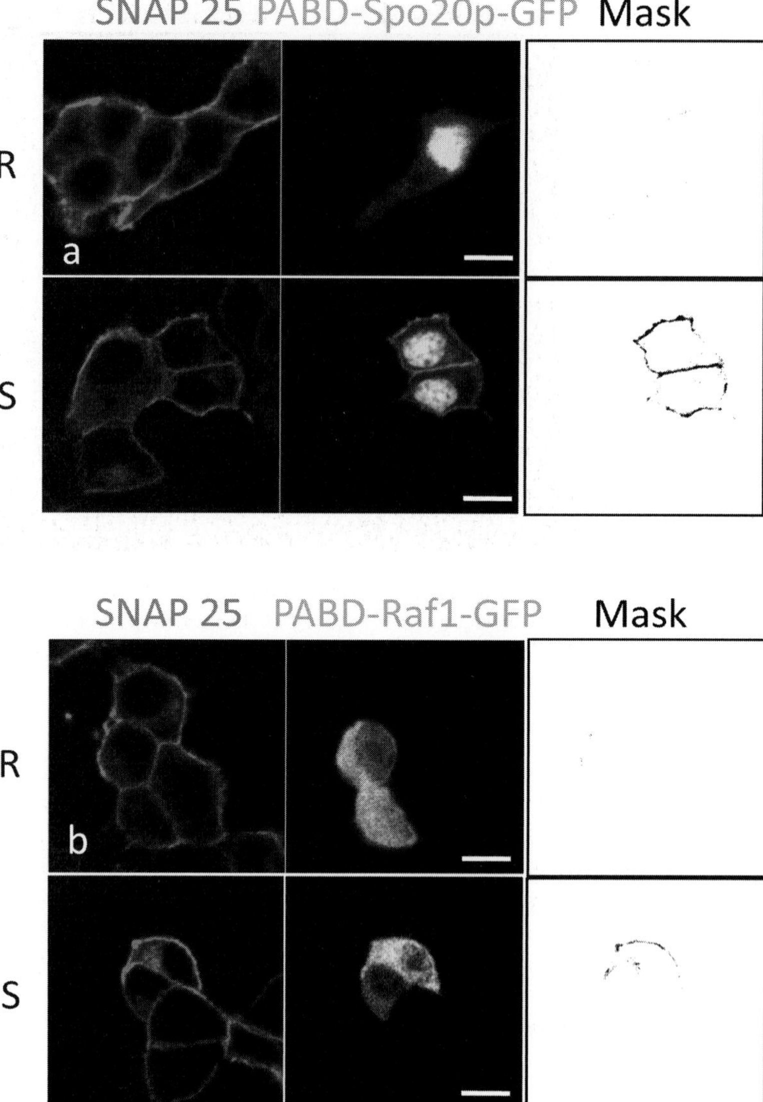

Fig. 2 Distribution of GFP-PABD-Raf1 and GFP-PABD-Spo20p in PC12 cells. Confocal immunofluorescence images showing the subcellular distribution of GFP-PABD-Raf1 and GFP-PABD-Spo20p in PC12 cells. Cells were maintained in Locke's solution under resting conditions, or stimulated for 10 min with a depolarizing concentration of potassium. The plasma membrane-bound protein SNAP25 was visualized using anti-SNAP25 antibody. In the mask images, the black staining indicates the areas of co-localization of the PABD probes with SNAP25 as illustrated by displaying the double-labeled pixels. Bars, 5 μm. (For color version of this figure, the reader is referred to the web version of this book.)

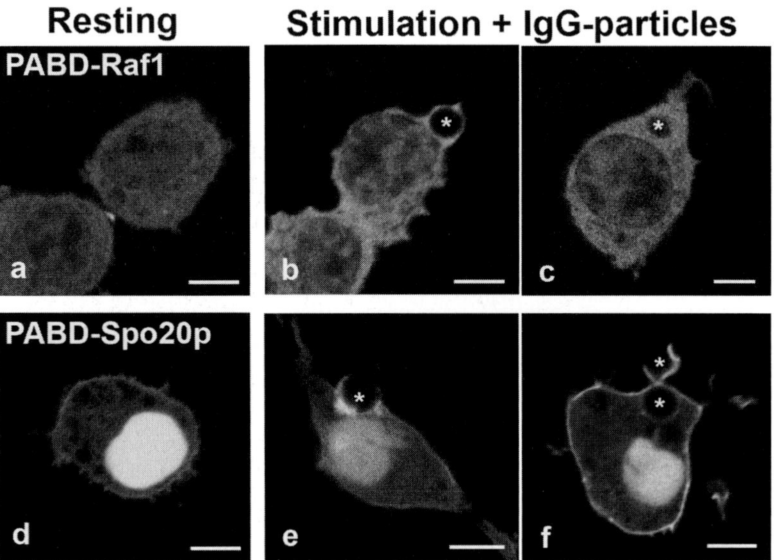

Fig. 3 Distribution of GFP-PABD-Raf1 and GFP-PABD-Spo20p in RAW 267.4 macrophages. Confocal images of live macrophages 12–18 h after transfection with either GFP-PABD-Raf1 (a–c) or GFP-PABD-Spo20p (d–f). Macrophages in resting conditions (a, d) and following stimulation with IgG-coated 3 μm microspheres (b, c, e, f). Stars indicate particles at different stages of phagocytosis. Bars, 5 μm. (For color version of this figure, the reader is referred to the web version of this book.)

rarely found associated with internalized phagosomes (Fig. 3c), in agreement with our earlier observations of PABD-Raf1 localization in fixed RAW264.7 cells (Corrotte *et al.*, 2006). During phagocytosis, the change in distribution of PABD-Spo20p is more evident. In phagocytosing cells, there is a net increase in the probe at the plasma membrane and a corresponding decrease in the nucleus (Fig. 3e, f). As in the case of the PABD-Raf1 probe, the PABD-Spo20p probe is present in the phagosomal cup and on pseudopods during phagosome formation (Fig. 3e, f), but is less evident on internalized phagosomes (Fig. 3f).

V. Summary and Conclusion

Here, we have described the use of two probes for PA that can serve as tools to visualize PA synthesis at exocytotic sites in neurosecreting cells and at the phagocytic cup in macrophages in response to stimulation. In PC12 cells, the secretagogue-induced PA synthesis requires PLD1 activation (Zeniou-Meyer *et al.*, 2007), whereas in macrophages both PLD1 and PLD2 isoforms appear to be involved in PA synthesis at the phagocytic cup (Corrotte *et al.*, 2006). These tools have also proven valuable to validate the implication of several regulators of PLD1 in exocytosis. For instance, depletion of endogenous ARF6 (Béglé *et al.*, 2009), Rac1 (Momboisse *et al.*, 2009), or Rsk2 (Zeniou-Meyer *et al.*, 2009) dramatically reduced the recruitment of GFP-PABD-Spo20p in stimulated PC12 cells. The specificity of these observations has also been confirmed in macrophages and neuroendocrine cells (Corrotte *et al.*, 2006; Zeniou-Meyer *et al.*, 2007) by using probes with point mutants that have been shown *in vitro* to no longer bind PA (Ghosh *et al.*, 2003; Nakanishi *et al.*, 2004).

As mentioned in the introduction several biosynthetic or signaling pathways converge to modulate the level of PA in living cells. In order to define the source of PA visualized by these probes, a number of specific manipulations need to be performed. For instance overexpression or knockdown of PA generating enzymes (PLD, DGK, LPAAT) should increase or prevent the recruitment of the probes if these pathways are implicated in the synthesis of the PA detected by the probes. The implication of these different PA-synthesizing enzymes can also be evaluated by pharmacological inhibitors such as the recently described isoform specific PLD inhibitors (Vitale, 2010), DGK inhibitors such as R59022, and LPAAT inhibitor such as CT-32228. Alternatively, the specificity of these probes can be tested *in situ* by examining their recruitment in cells where PA conversion into DAG is enhanced by overexpressing LPP or PLA or reduced by using pharmacological inhibitors of these enzymes.

One problem that can arise when using these probes to study PA dynamics is the observation that they can interfere with cellular functions by sequestering PA, as in the case of phagocytosis (Corrotte *et al.*, 2006). As well, when the PABD probes are expressed at too high levels or for prolonged periods (>18 h), we have observed that they can affect cell viability both for PC12 cells and macrophages. It is therefore important to minimize the expression of the probe to a level, which permits detection of PA recruitment and minimizes deleterious effects on cellular functions and cell viability.

Biological membranes contain a host of anionic lipids and are therefore generally negatively charged. The negative charge of biological membranes is an important determinant of biomembrane structure and function. It is well established that the negative charge carried by anionic lipids in biomembranes forms an important site of attraction for positively charged (carrying basic amino acids) proteins. Yet, proteins bind to individual anionic lipids with a high degree of specificity, based mostly on specific structural motives. The binding of PA to the probes described here appears to rely mainly on basic residues probably having a specific orientation towards the negative charges of PA, although the exact recognition sites remain to be determined.

The transmembrane topology of phospholipids in cellular compartments remains ill defined except for PS. The presence of PS in the lumenal monolayer of the ER and Golgi complex and its cytosolic exposure at the trans-Golgi network has recently been elegantly demonstrated ultrastructurally on cryosections overlaid with PS probes (Fairn *et al.*, 2011). This transmembrane flipping of PS is thought to contribute to the exit of cargo from the Golgi complex. Whether the PA is also present on the lumenal leaflet of subcellular compartments and the possibility that PA flipping may be involved in regulating biological functions remains to be determined. As the GFP-based PA probes described here are likely to miss a pool of PA present in the luminal/ectoplasmic leaflets of cellular membranes, an ultrastructural study using these PA-probes on cryosections is a line of investigation that may provide some answers to these questions.

The difference in subcellular localization of the two PA probes merits to be mentioned. The PABD of Raf1 is mainly cytosolic, whereas the PABD of Spo20p concentrates in the nucleus. The nuclear localization of the Spo20p probe probably results from the presence of a small nuclear sequence at the amino-terminal part of the PA binding region of Spo20p (Nakanishi *et al.*, 2004). As a probe for PA synthesis, this is however of interest because it facilitates the visualization of the recruitment of the PABD-Spo20p probe to cytoplasmic organelles and the plasma membrane. Conversely, the cytoplasmic distribution of the PABD-Raf1 probe tends to obscure observations of PA generation on organelles and the plasma membrane and this seems to be accentuated in live cells. Indeed, the recruitment of the probes to the phagosome was more easily observed in fixed macrophages (Corrotte *et al.*, 2006) than in live macrophages (Fig. 3). For the moment, it appears that the Spo20p probe is probably be more useful as a PA biosensor than the PABD-Raf1 probe for studying PA dynamics in live cells.

Another possibility to explain the differential recruitment of the two probes is that they do not recognize the same pools of PA. Variations in the length or saturation of the fatty acid chains of PA probably affect the orientation of the negatively charged polar head group of PA in the lipid bilayer. Since this head group interacts with proteins, changes in its orientation could affect the affinity of PA for binding domains of different proteins. This remains an open question that deserves to be investigated. In addition, full characterization of other potential biosensors for PA will undoubtedly prove crucial for dissecting the numerous pathways involving PA.

Acknowledgments

We wish to thank all the members of our laboratory that have contributed to this work. This work was supported by the Agence Nationale de la Recherche (Grant ANR-09-BLAN-0264). N.K. was supported by a grant from the CNRS-Liban.

References

Athenstaedt, K., and Daum, G. (1999). Phosphatidic acid, a key intermediate in lipid metabolism. *Eur. J. Biochem.* **266**, 1–16.

Bader, M. F., and Vitale, N. (2009). Phospholipase D in calcium-regulated exocytosis: lessons from chromaffin cells. *Biochim. Biophys. Acta* **1791**, 936–941.

Bader, M. F., *et al.* (2002). Exocytosis: the chromaffin cell as a model system. *Ann. N. Y. Acad. Sci.* **971**, 178–183.

Béglé, A., *et al.* (2009). ARF6 regulates the synthesis of fusogenic lipids for calcium-regulated exocytosis in neuroendocrine cells. *J. Biol. Chem.* **284**, 4836–4845.

Caumont, A. S., *et al.* (1998). Regulated exocytosis in chromaffin cells – Translocation of ARF6 stimulates a plasma membrane-associated phospholipase D. *J. Biol. Chem.* **273**, 1373–1379.

Chernomordik, L. V., and Kozlov, M. M. (2008). Mechanics of membrane fusion. *Nat. Struct. Mol. Biol.* **7**, 675–683.

Corrotte, M., *et al.* (2006). Dynamics and function of phospholipase D and phosphatidic acid during phagocytosis. *Traffic* **7**, 365–377.

Corrotte, M., *et al.* (2010). Ral isoforms are implicated in Fc gamma R-mediated phagocytosis: activation of phospholipase D by RalA. *J. Immunol.* **185**, 2942–2950.

Du, G., *et al.* (2004). Phospholipase D2 localizes to the plasma membrane and regulates angiotensin II receptor endocytosis. *Mol. Biol. Cell.* **3**, 1024–1030.

Fairn., *et al.* (2011). High-resolution mapping reveals topologically distinct cellular pools of phosphatidylserine. *J. Cell Biol.* **194**, 257–275.

Ghosh, S., *et al.* (2003). Functional analysis of a phosphatidic acid binding domain in human Raf-1 kinase: mutations in the phosphatidate binding domain lead to tail and trunk abnormalities in developing zebrafish embryos. *J. Biol. Chem.* **278**, 45690–45696.

Huang, P., *et al.* (2005). Insulin-stimulated plasma membrane fusion of Glut4 glucose transporter-containing vesicles is regulated by phospholipase D1. *Mol. Biol. Cell* **6**, 2614–2623.

Humeau, Y., *et al.* (2001). A role for phospholipase D in neurotransmitter release. *Proc. Natl. Acad. Sci. U. S. A.* **93**, 1941–1944.

Iyer, S. S., *et al.* (2004). Phospholipases D1 and D2 coordinately regulate macrophage phagocytosis. *J. Immunol.* **173**, 2615–2623.

Iyer, S. S., *et al.* (2006). Phospholipases D1 regulates phagocyte adhesion. *J. Immunol.* **176**, 3686–3696.

Jones, J. A., and Hannun, Y. A. (2002). Tight binding inhibition of protein phosphatase-1 by phosphatidic acid. Specificity of inhibition by the phospholipid. *J. Biol. Chem.* **277**, 15530–15538.

Jose Lopez-Andreo., *et al.* (2003). The simultaneous production of phosphatidic acid and diacylglycerol is essential for the translocation of protein kinase Cepsilon to the plasma membrane in RBL-2H3 cells. *Mol. Biol. Cell* **14**, 4885–4895.

Kooijman, E. E., and Burger, K. N. (2009). Biophysics and function of phosphatidic acid: a molecular perspective. *Biochim. Biophys. Acta* **1791**, 881–888.

Loewen, C. J., *et al.* (2004). Phospholipid metabolism regulated by a transcription factor sensing phosphatidic acid. *Science* **304**, 1644–1647.

Momboisse, F., *et al.* (2009). betaPIX-activated Rac1 stimulates the activation of phospholipase D, which is associated with exocytosis in neuroendocrine cells. *J. Cell Sci.* **122**, 798–806.

Nakanishi, H., *et al.* (2004). Positive and negative regulation of a SNARE protein by control of intracellular localization. *Mol. Biol. Cell* **15**, 1802–1815.

Oliveira, T. G., *et al.* (2010). Phospholipase d2 ablation ameliorates Alzheimer's disease-linked synaptic dysfunction and cognitive deficits. *J. Neurosci.* **30**, 16419–16428.

Rizzo, M. A., *et al.* (2000). The recruitment of Raf-1 to membranes is mediated by direct interaction with phosphatidic acid and is independent of association with Ras. *J. Biol. Chem.* **275**, 23911–23918.

Roth, M. G. (2008). Molecular mechanisms of PLD function in membrane traffic. *Traffic* **9**, 1233–1239.

Shulga, Y. V., *et al.* (2010). Molecular species of phosphatidylinositol-cycle intermediates in the endoplasmic reticulum and plasma membrane. *Biochemistry* **49**, 312–317.

Stace, C. L., and Ktistakis, N. T. (2006). Phosphatidic acid- and phosphatidylserine-binding proteins. *Biochim. Biophys. Acta* **1761**, 913–926.

Vitale, N., *et al.* (1996). ARD1, a 64-kDa bifunctional protein containing an 18-kDa GTP-binding ADP-ribosylation factor domain and a 46-kDa GTPase-activating domain. *Proc. Natl. Acad. Sci. U. S. A.* **93**, 1941–1944.

Vitale, N., *et al.* (1998). Molecular characterization of the GTPase-activating domains of ADP-ribosylation factor domain protein 1 (ARD1). *J. Biol. Chem.* **273**, 2553–2560.

Vitale, N., *et al.* (2000a). Specific functional interaction of human cytohesin-1 and ADP-ribosylation factor domain protein (ARD1). *J. Biol. Chem* **275**, 21331–21339.

Vitale, N., *et al.* (2000b). GIT proteins, A novel family of phosphatidylinositol 3,4,5-trisphosphate-stimulated GTPase-activating proteins for ARF6. *J. Biol. Chem* **275**, 13901–13906.

Vitale, N., *et al.* (2001). Purification of ARD1 an ADP-ribosylation factor (ARF)-related protein with GTPase-activating domain. *Methods Enzymol.* **329**, 324–334.

Vitale, N. (2010). Therapeutic potentials of recently identified PLD inhibitors. *Curr. Chem. Biol.* **4**, 244–249.

Zeniou-Meyer, M., *et al.* (2007). Phospholipase D1 production of phosphatidic acid at the plasma membrane promotes exocytosis of large dense-core granules at a late stage. *J. Biol. Chem.* **282**, 21746–21757.

Zeniou-Meyer, M., *et al.* (2009). The Coffin-Lowry syndrome-associated protein RSK2 controls neuroendocrine secretion through the regulation of phospholipase D1 at the exocytotic sites. *Ann. N. Y. Acad. Sci.* **1152**, 201–208.

Zhaso, C., *et al.* (2007). Phospholipase D2-generated phosphatidic acid couples EGFR stimulation to Ras activation by Sos. *Nat. Cell Biol.* **6**, 706–712.

INDEX

M

R

S

T

U

V

W

Y

Z

VOLUMES IN SERIES

Founding Series Editor
DAVID M. PRESCOTT

Volume 1 (1964)
Methods in Cell Physiology
Edited by David M. Prescott

Volume 2 (1966)
Methods in Cell Physiology
Edited by David M. Prescott

Volume 3 (1968)
Methods in Cell Physiology
Edited by David M. Prescott

Volume 4 (1970)
Methods in Cell Physiology
Edited by David M. Prescott

Volume 5 (1972)
Methods in Cell Physiology
Edited by David M. Prescott

Volume 6 (1973)
Methods in Cell Physiology
Edited by David M. Prescott

Volume 7 (1973)
Methods in Cell Biology
Edited by David M. Prescott

Volume 8 (1974)
Methods in Cell Biology
Edited by David M. Prescott

Volume 9 (1975)
Methods in Cell Biology
Edited by David M. Prescott

Volume 10 (1975)
Methods in Cell Biology
Edited by David M. Prescott

Volume 11 (1975)
Yeast Cells
Edited by David M. Prescott

Volume 12 (1975)
Yeast Cells
Edited by David M. Prescott

Volume 13 (1976)
Methods in Cell Biology
Edited by David M. Prescott

Volume 14 (1976)
Methods in Cell Biology
Edited by David M. Prescott

Volume 15 (1977)
Methods in Cell Biology
Edited by David M. Prescott

Volume 16 (1977)
Chromatin and Chromosomal Protein Research I
Edited by Gary Stein, Janet Stein, and Lewis J. Kleinsmith

Volume 17 (1978)
Chromatin and Chromosomal Protein Research II
Edited by Gary Stein, Janet Stein, and Lewis J. Kleinsmith

Volume 18 (1978)
Chromatin and Chromosomal Protein Research III
Edited by Gary Stein, Janet Stein, and Lewis J. Kleinsmith

Volume 19 (1978)
Chromatin and Chromosomal Protein Research IV
Edited by Gary Stein, Janet Stein, and Lewis J. Kleinsmith

Volume 20 (1978)
Methods in Cell Biology
Edited by David M. Prescott

Advisory Board Chairman
KEITH R. PORTER

Volume 21A (1980)
Normal Human Tissue and Cell Culture, Part A: Respiratory, Cardiovascular, and Integumentary Systems
Edited by Curtis C. Harris, Benjamin F. Trump, and Gary D. Stoner

Volume 21B (1980)
Normal Human Tissue and Cell Culture, Part B: Endocrine, Urogenital, and Gastrointestinal Systems
Edited by Curtis C. Harris, Benjamin F. Trump, and Gray D. Stoner

Volume 22 (1981)
Three-Dimensional Ultrastructure in Biology
Edited by James N. Turner

Volume 23 (1981)
Basic Mechanisms of Cellular Secretion
Edited by Arthur R. Hand and Constance Oliver

Volume 24 (1982)
The Cytoskeleton, Part A: Cytoskeletal Proteins, Isolation and Characterization
Edited by Leslie Wilson

Volume 25 (1982)
The Cytoskeleton, Part B: Biological Systems and In Vitro Models
Edited by Leslie Wilson

Volume 26 (1982)
Prenatal Diagnosis: Cell Biological Approaches
Edited by Samuel A. Latt and Gretchen J. Darlington

Series Editor
LESLIE WILSON

Volume 27 (1986)
Echinoderm Gametes and Embryos
Edited by Thomas E. Schroeder

Volume 28 (1987)
Dictyostelium discoideum: Molecular Approaches to Cell Biology
Edited by James A. Spudich

Volume 29 (1989)
Fluorescence Microscopy of Living Cells in Culture, Part A: Fluorescent Analogs, Labeling Cells, and Basic Microscopy
Edited by Yu-Li Wang and D. Lansing Taylor

Volume 30 (1989)
Fluorescence Microscopy of Living Cells in Culture, Part B: Quantitative Fluorescence Microscopy—Imaging and Spectroscopy
Edited by D. Lansing Taylor and Yu-Li Wang

Volume 31 (1989)
Vesicular Transport, Part A
Edited by Alan M. Tartakoff

Volume 32 (1989)
Vesicular Transport, Part B
Edited by Alan M. Tartakoff

Volume 33 (1990)
Flow Cytometry
Edited by Zbigniew Darzynkiewicz and Harry A. Crissman

Volume 34 (1991)
Vectorial Transport of Proteins into and across Membranes
Edited by Alan M. Tartakoff

Selected from Volumes 31, 32, and 34 (1991)
Laboratory Methods for Vesicular and Vectorial Transport
Edited by Alan M. Tartakoff

Volume 35 (1991)
Functional Organization of the Nucleus: A Laboratory Guide
Edited by Barbara A. Hamkalo and Sarah C. R. Elgin

Volume 36 (1991)
***Xenopus laevis:* Practical Uses in Cell and Molecular Biology**
Edited by Brian K. Kay and H. Benjamin Peng

Series Editors
LESLIE WILSON AND PAUL MATSUDAIRA

Volume 37 (1993)
Antibodies in Cell Biology
Edited by David J. Asai

Volume 38 (1993)
Cell Biological Applications of Confocal Microscopy
Edited by Brian Matsumoto

Volume 39 (1993)
Motility Assays for Motor Proteins
Edited by Jonathan M. Scholey

Volume 40 (1994)
A Practical Guide to the Study of Calcium in Living Cells
Edited by Richard Nuccitelli

Volume 41 (1994)
Flow Cytometry, Second Edition, Part A
Edited by Zbigniew Darzynkiewicz, J. Paul Robinson, and Harry A. Crissman

Volume 42 (1994)
Flow Cytometry, Second Edition, Part B
Edited by Zbigniew Darzynkiewicz, J. Paul Robinson, and Harry A. Crissman

Volume 43 (1994)
Protein Expression in Animal Cells
Edited by Michael G. Roth

Volume 44 (1994)
Drosophila melanogaster: Practical Uses in Cell and Molecular Biology
Edited by Lawrence S. B. Goldstein and Eric A. Fyrberg

Volume 45 (1994)
Microbes as Tools for Cell Biology
Edited by David G. Russell

Volume 46 (1995)
Cell Death
Edited by Lawrence M. Schwartz and Barbara A. Osborne

Volume 47 (1995)
Cilia and Flagella
Edited by William Dentler and George Witman

Volume 48 (1995)
Caenorhabditis elegans: Modern Biological Analysis of an Organism
Edited by Henry F. Epstein and Diane C. Shakes

Volume 49 (1995)
Methods in Plant Cell Biology, Part A
Edited by David W. Galbraith, Hans J. Bohnert, and Don P. Bourque

Volume 50 (1995)
Methods in Plant Cell Biology, Part B
Edited by David W. Galbraith, Don P. Bourque, and Hans J. Bohnert

Volume 51 (1996)
Methods in Avian Embryology
Edited by Marianne Bronner-Fraser

Volume 52 (1997)
Methods in Muscle Biology
Edited by Charles P. Emerson, Jr. and H. Lee Sweeney

Volume 53 (1997)
Nuclear Structure and Function
Edited by Miguel Berrios

Volume 54 (1997)
Cumulative Index

Volume 55 (1997)
Laser Tweezers in Cell Biology
Edited by Michael P. Sheetz

Volume 56 (1998)
Video Microscopy
Edited by Greenfield Sluder and David E. Wolf

Volume 57 (1998)
Animal Cell Culture Methods
Edited by Jennie P. Mather and David Barnes

Volume 58 (1998)
Green Fluorescent Protein
Edited by Kevin F. Sullivan and Steve A. Kay

Volume 59 (1998)
The Zebrafish: Biology
Edited by H. William Detrich III, Monte Westerfield, and Leonard I. Zon

Volume 60 (1998)
The Zebrafish: Genetics and Genomics
Edited by H. William Detrich III, Monte Westerfield, and Leonard I. Zon

Volume 61 (1998)
Mitosis and Meiosis
Edited by Conly L. Rieder

Volume 62 (1999)
Tetrahymena thermophila
Edited by David J. Asai and James D. Forney

Volume 63 (2000)
Cytometry, Third Edition, Part A
Edited by Zbigniew Darzynkiewicz, J. Paul Robinson, and Harry Crissman

Volume 64 (2000)
Cytometry, Third Edition, Part B
Edited by Zbigniew Darzynkiewicz, J. Paul Robinson, and Harry Crissman

Volume 65 (2001)
Mitochondria
Edited by Liza A. Pon and Eric A. Schon

Volume 66 (2001)
Apoptosis
Edited by Lawrence M. Schwartz and Jonathan D. Ashwell

Volume 79 (2007)
Cellular Electron Microscopy
Edited by J. Richard McIntosh

Volume 80 (2007)
Mitochondria, 2nd Edition
Edited by Liza A. Pon and Eric A. Schon

Volume 81 (2007)
Digital Microscopy, 3rd Edition
Edited by Greenfield Sluder and David E. Wolf

Volume 82 (2007)
Laser Manipulation of Cells and Tissues
Edited by Michael W. Berns and Karl Otto Greulich

Volume 83 (2007)
Cell Mechanics
Edited by Yu-Li Wang and Dennis E. Discher

Volume 84 (2007)
Biophysical Tools for Biologists, Volume One: In Vitro Techniques
Edited by John J. Correia and H. William Detrich, III

Volume 85 (2008)
Fluorescent Proteins
Edited by Kevin F. Sullivan

Volume 86 (2008)
Stem Cell Culture
Edited by Dr. Jennie P. Mather

Volume 87 (2008)
Avian Embryology, 2nd Edition
Edited by Dr. Marianne Bronner-Fraser

Volume 88 (2008)
Introduction to Electron Microscopy for Biologists
Edited by Prof. Terence D. Allen

Volume 89 (2008)
Biophysical Tools for Biologists, Volume Two: In Vivo Techniques
Edited by Dr. John J. Correia and Dr. H. William Detrich, III

Volume 90 (2008)
Methods in Nano Cell Biology
Edited by Bhanu P. Jena

Volume 91 (2009)
Cilia: Structure and Motility
Edited by Stephen M. King and Gregory J. Pazour

Volume 92 (2009)
Cilia: Motors and Regulation
Edited by Stephen M. King and Gregory J. Pazour

Volume 93 (2009)
Cilia: Model Organisms and Intraflagellar Transport
Edited by Stephen M. King and Gregory J. Pazour

Volume 94 (2009)
Primary Cilia
Edited by Roger D. Sloboda

Volume 95 (2010)
Microtubules, in vitro
Edited by Leslie Wilson and John J. Correia

Volume 96 (2010)
Electron Microscopy of Model Systems
Edited by Thomas Müeller-Reichert

Volume 97 (2010)
Microtubules: In Vivo
Edited by Lynne Cassimeris and Phong Tran

Volume 98 (2010)
Nuclear Mechanics & Genome Regulation
Edited by G.V. Shivashankar

Volume 99 (2010)
Calcium in Living Cells
Edited by Michael Whitaker

Volume 100 (2010)
The Zebrafish: Cellular and Developmental Biology, Part A
Edited by: H. William Detrich III, Monte Westerfield and Leonard I. Zon

Volume 101 (2011)
The Zebrafish: Cellular and Developmental Biology, Part B
Edited by: H. William Detrich III, Monte Westerfield and Leonard I. Zon

Volume 102 (2011)
Recent Advances in Cytometry, Part A: Instrumentation, Methods
Edited by Zbigniew Darzynkiewicz, Elena Holden, Alberto Orfao, William Telford and Donald Wlodkowic

Volume 103 (2011)
Recent Advances in Cytometry, Part B: Advances in Applications
Edited by Zbigniew Darzynkiewicz, Elena Holden, Alberto Orfao, Alberto Orfao and Donald Wlodkowic

Volume 104 (2011)
The Zebrafish: Genetics, Genomics and Informatics 3rd Edition
Edited by H. William Detrich III, Monte Westerfield, and Leonard I. Zon

Volume 105 (2011)
The Zebrafish: Disease Models and Chemical Screens 3rd Edition
Edited by H. William Detrich III, Monte Westerfield, and Leonard I. Zon

Volume 106 (2011)
Caenorhabditis elegans: Molecular Genetics and Development 2nd Edition
Edited by Joel H. Rothman and Andrew Singson

Volume 107 (2011)
Caenorhabditis elegans: Cell Biology and Physiology 2nd Edition
Edited by Joel H. Rothman and Andrew Singson

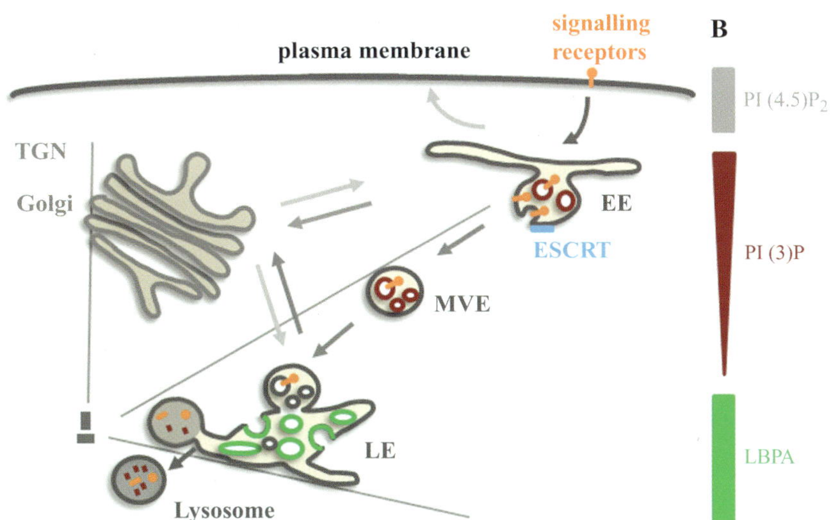

Plate 1 (Figure 2.1 on page 20 of this volume).

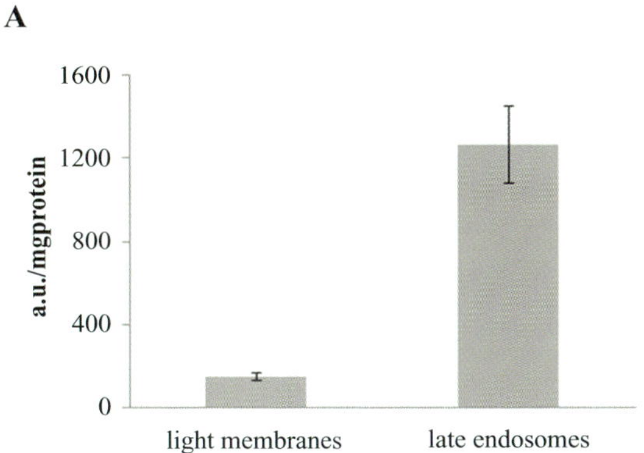

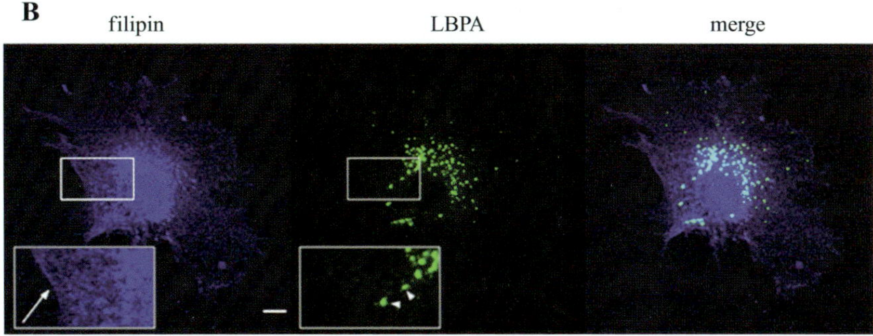

Plate 2 (Figure 2.2 on page 32 of this volume).

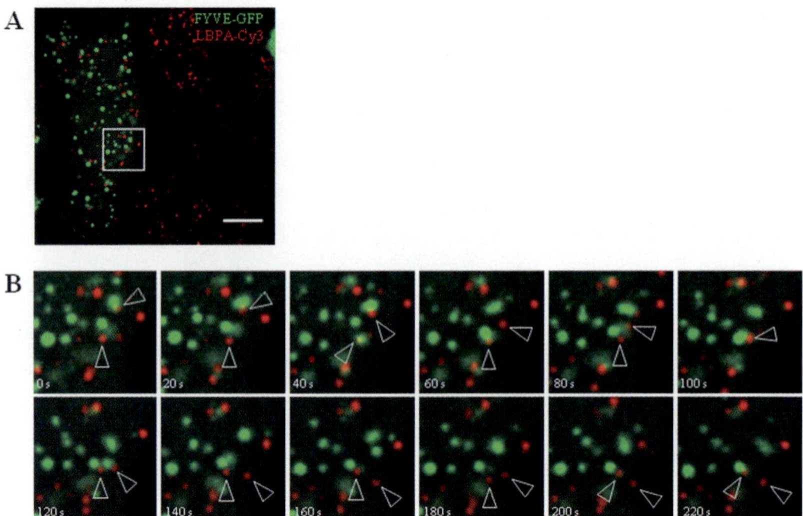

Plate 3 (Figure 2.4 on page 38 of this volume).

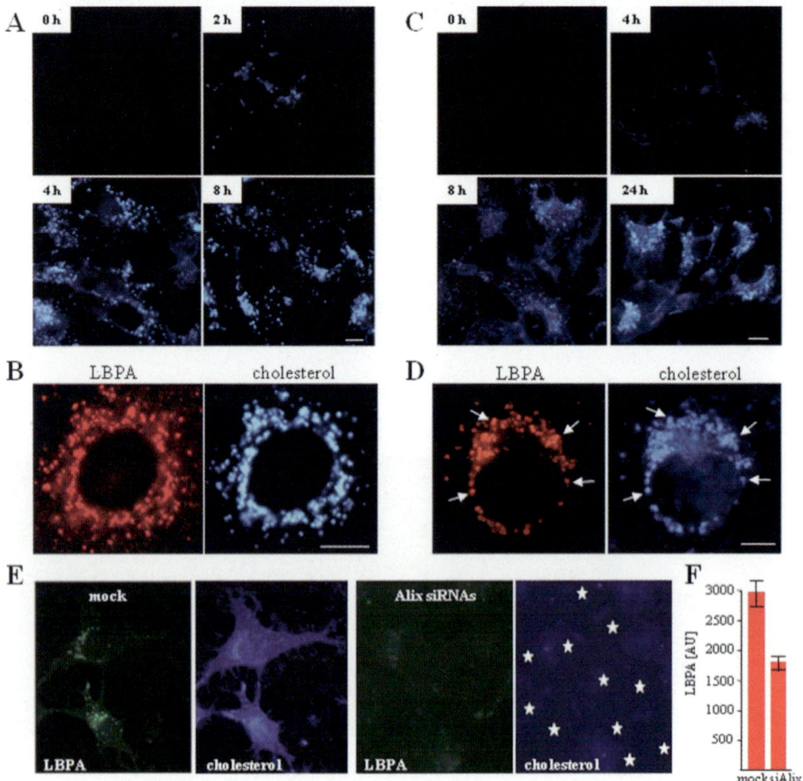

Plate 4 (Figure 2.5 on page 41 of this volume).

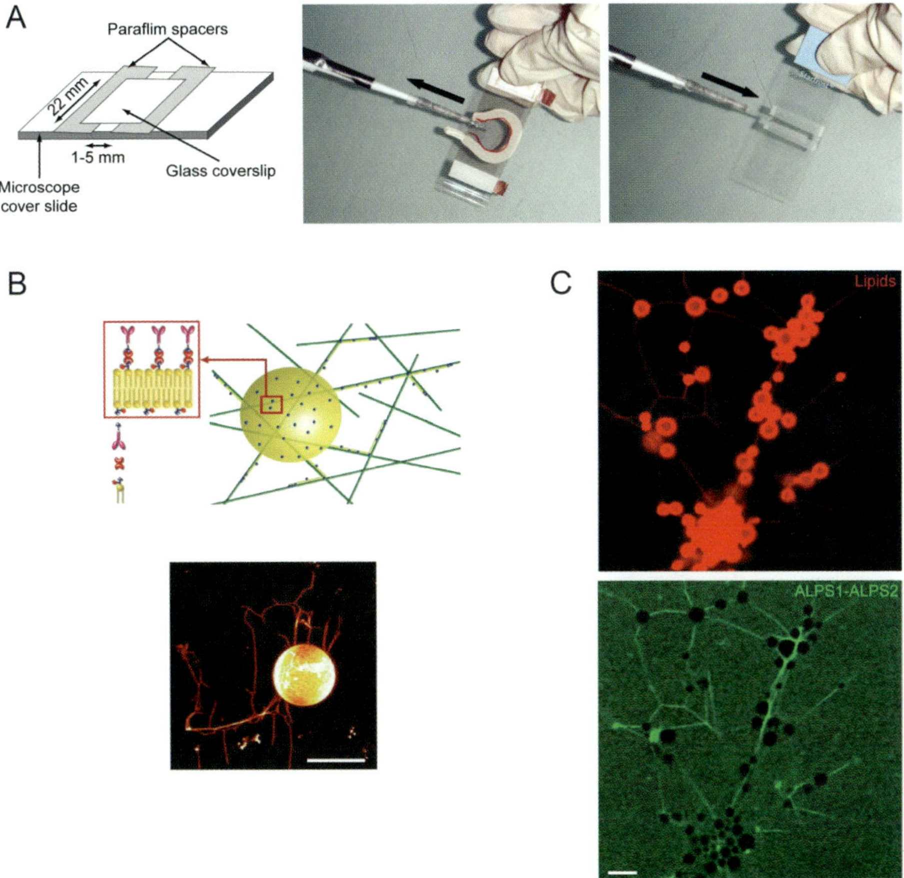

Plate 5 (Figure 3.4 on page 58 of this volume).

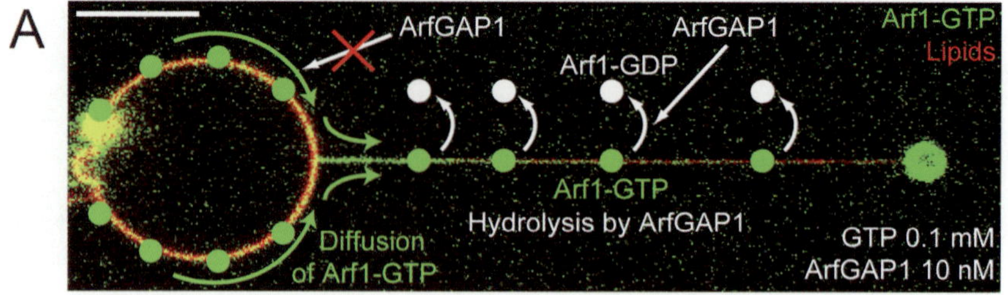

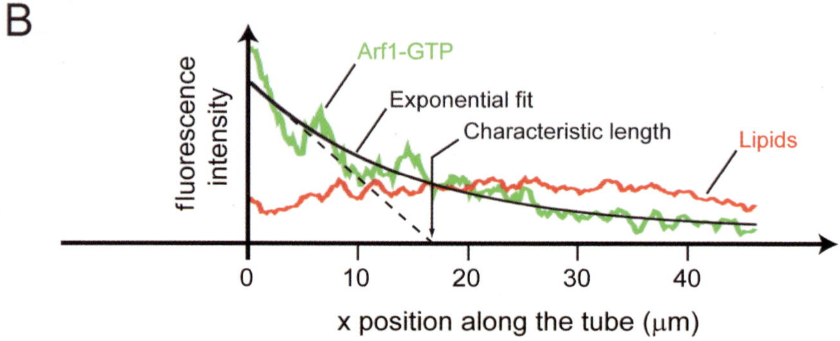

Plate 6 (Figure 3.8 on page 66 of this volume).

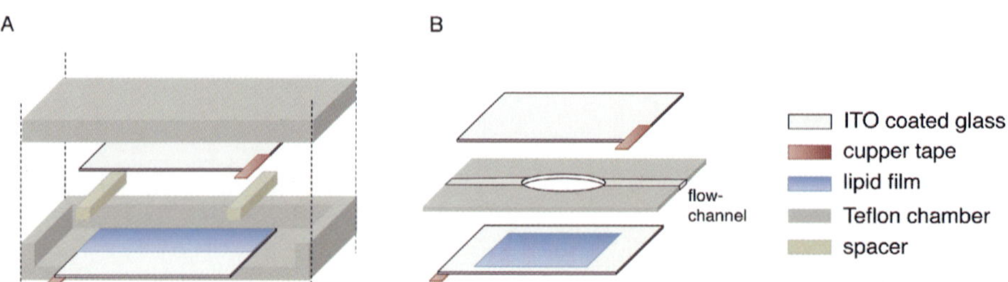

Plate 7 (Figure 4.1 on page 80 of this volume).

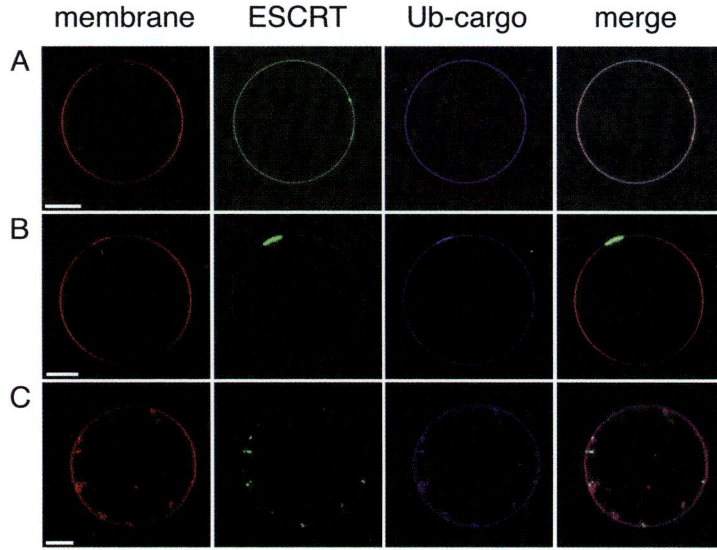

Plate 8 (Figure 4.3 on page 88 of this volume).

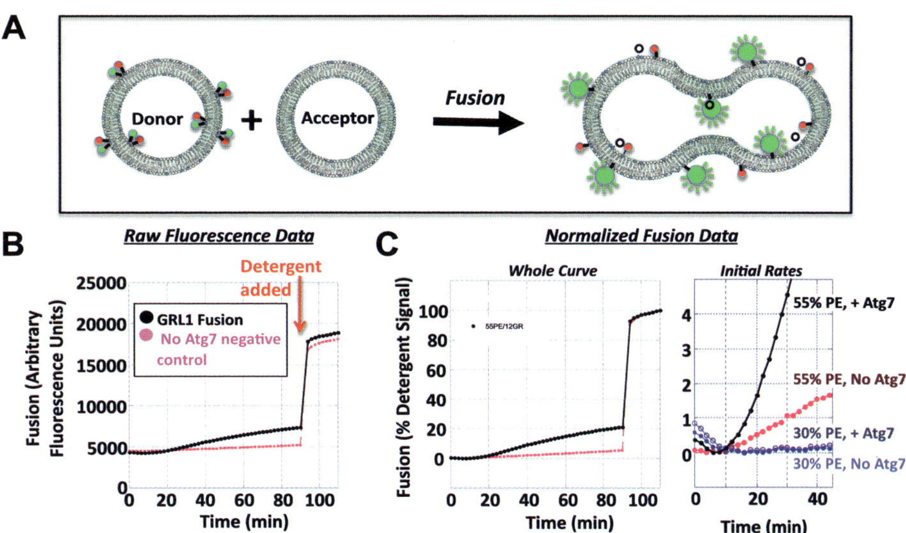

Plate 9 (Figure 5.4 on page 109 of this volume).

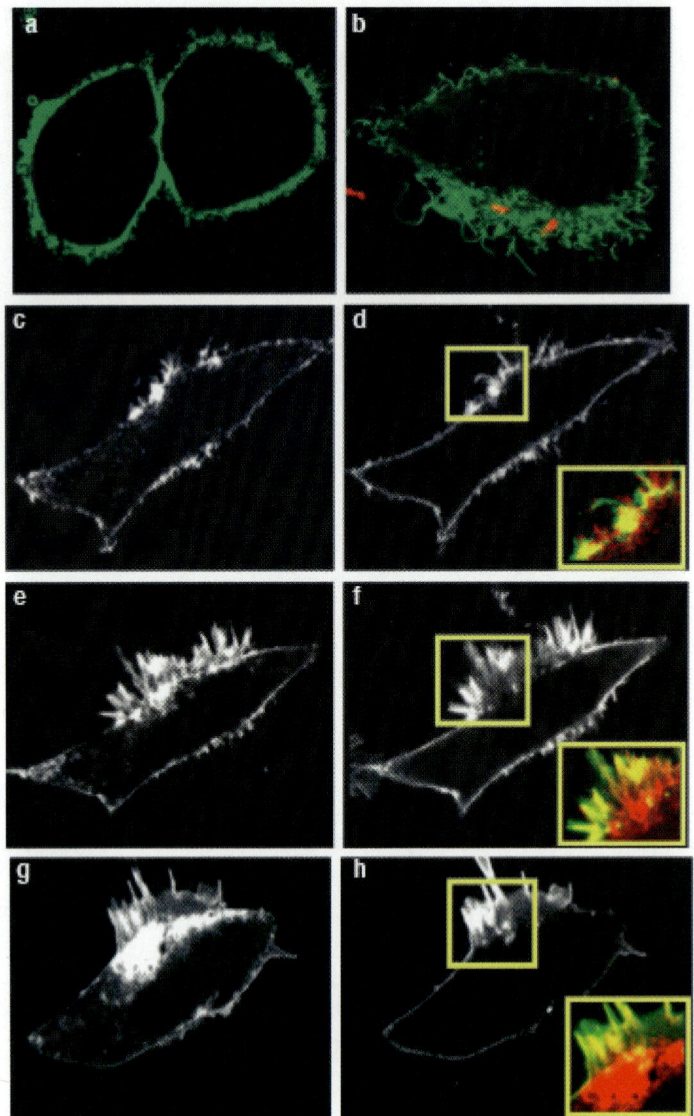

Plate 10 (Figure 9.1 on page 176 of this volume).

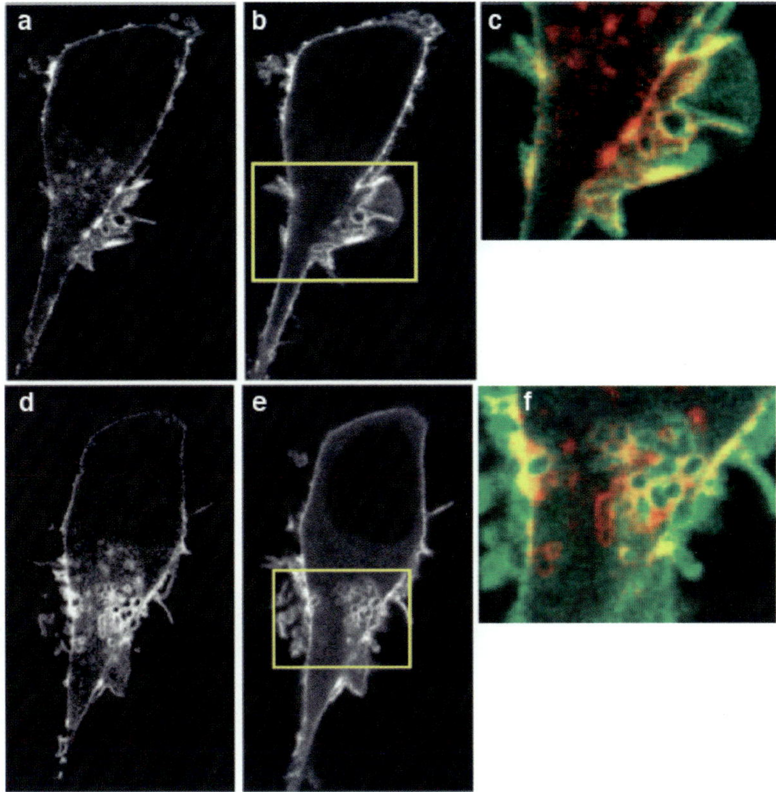

Plate 11 (Figure 9.2 on page 177 of this volume).

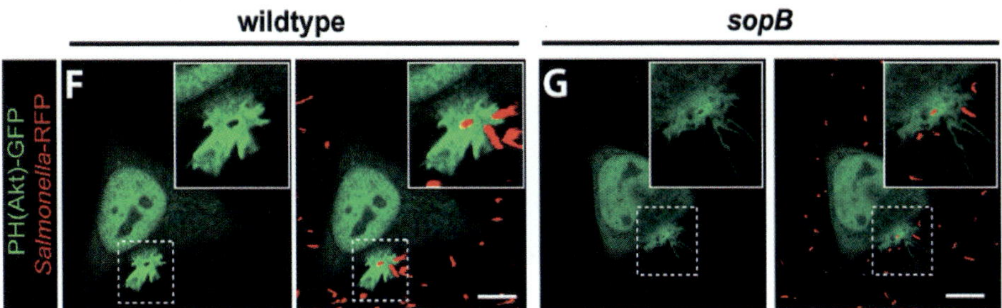

Plate 12 (Figure 9.3 on page 178 of this volume).

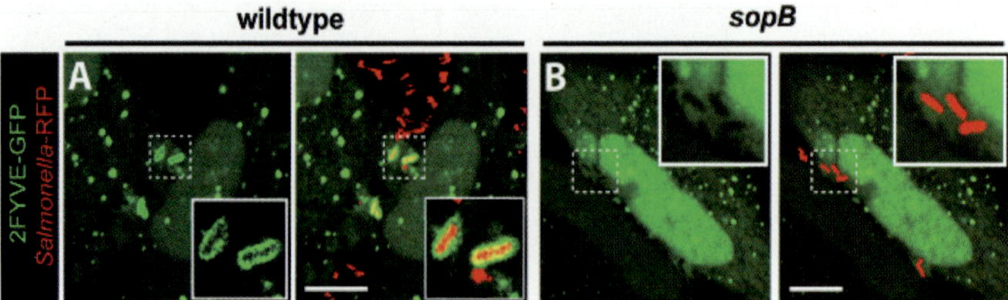

Plate 13 (Figure 9.4 on page 179 of this volume).

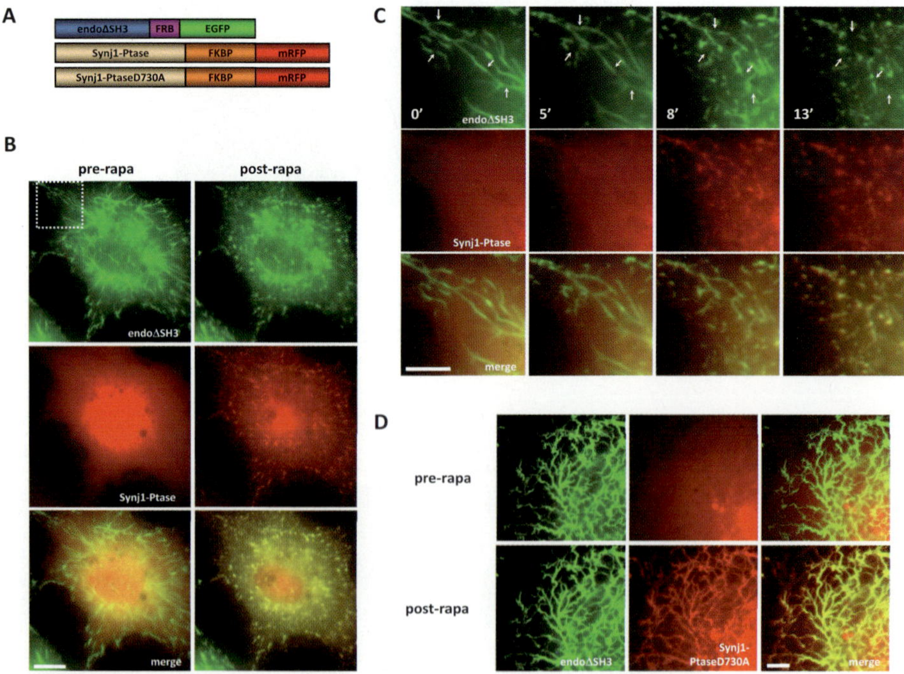

Plate 14 (Figure 10.3 on page 197 of this volume).

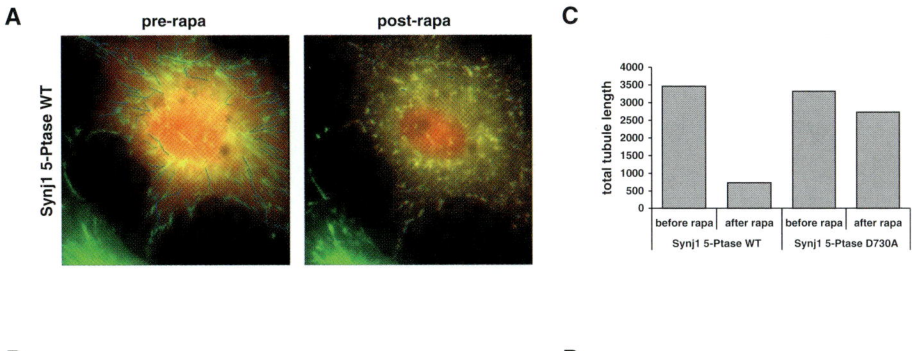

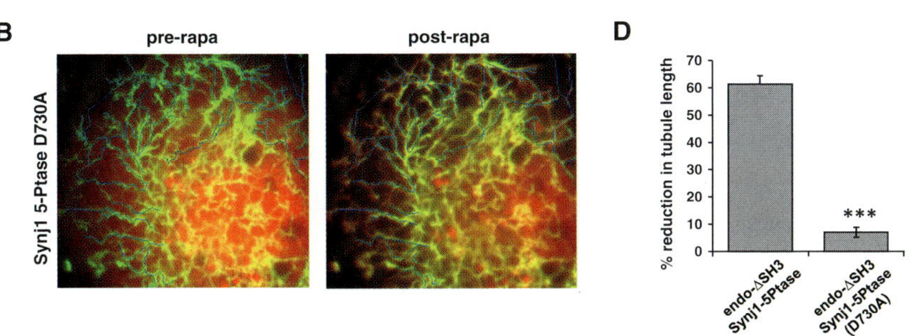

Plate 15 (Figure 10.4 on page 201 of this volume).

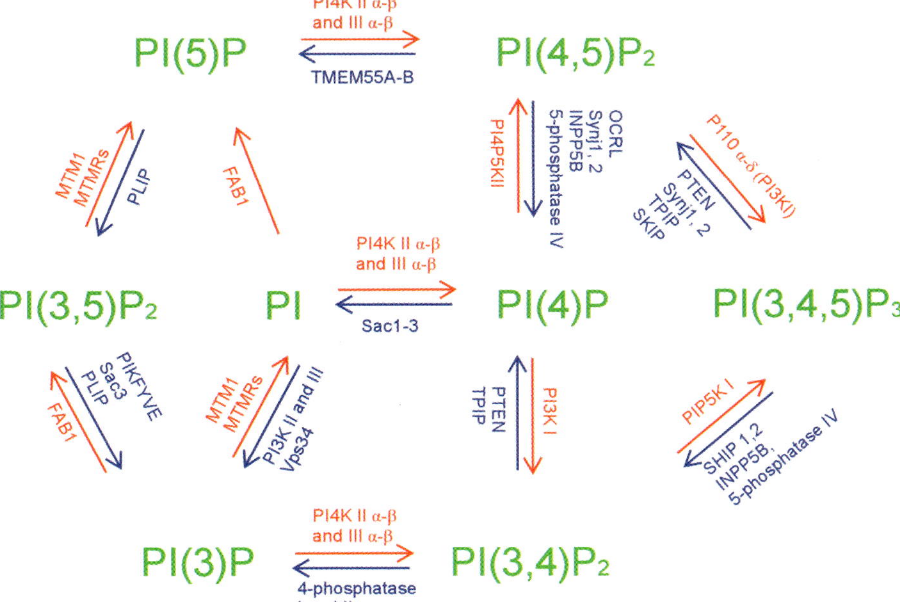

Plate 16 (Figure 12.1 on page 230 of this volume).

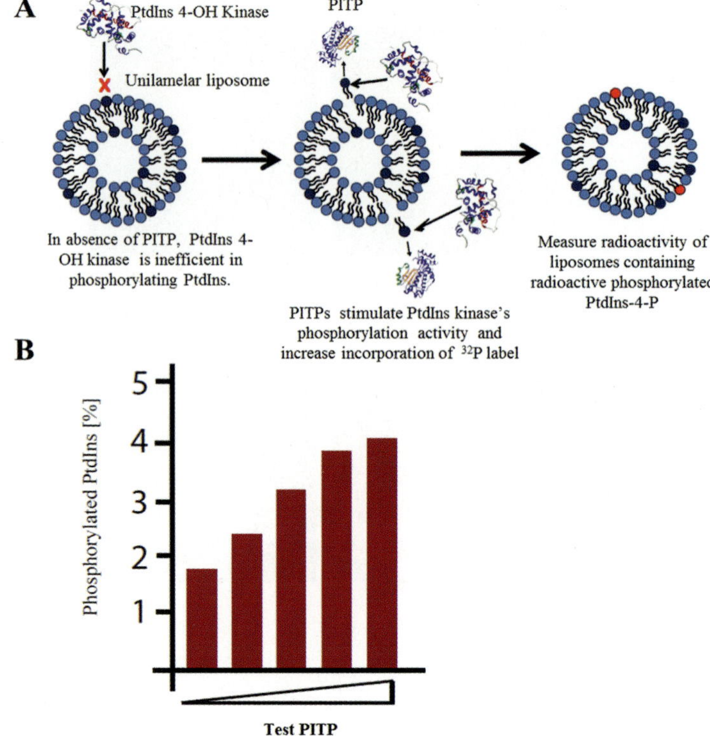

Plate 17 (Figure 13.3 on page 265 of this volume).

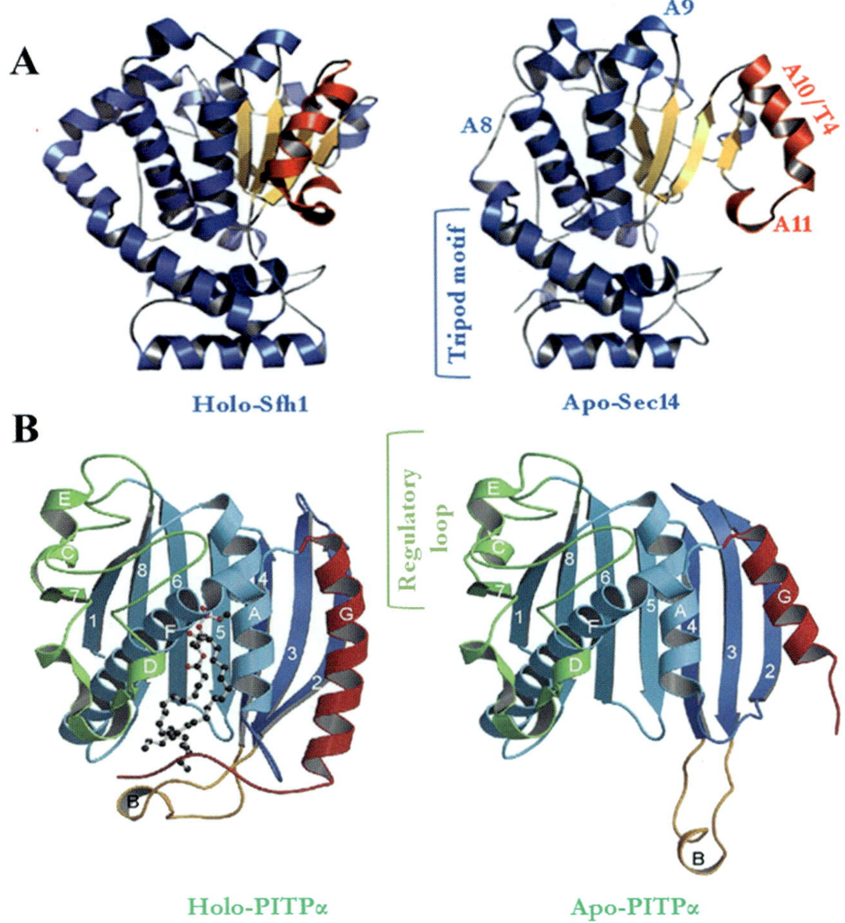

Plate 18 (Figure 13.8 on page 290 of this volume).

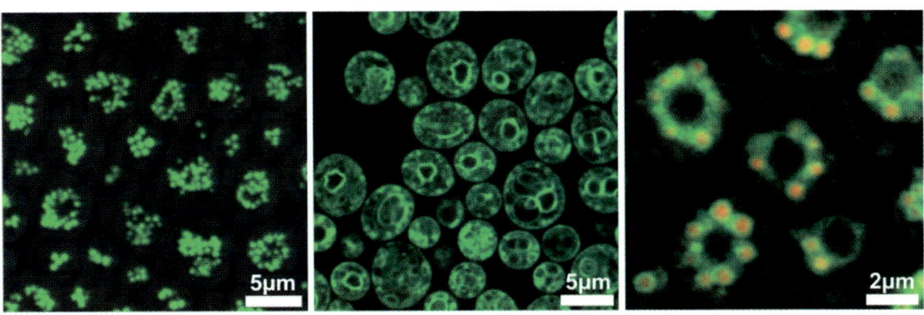

Plate 19 (Figure 16.1 on page 349 of this volume).

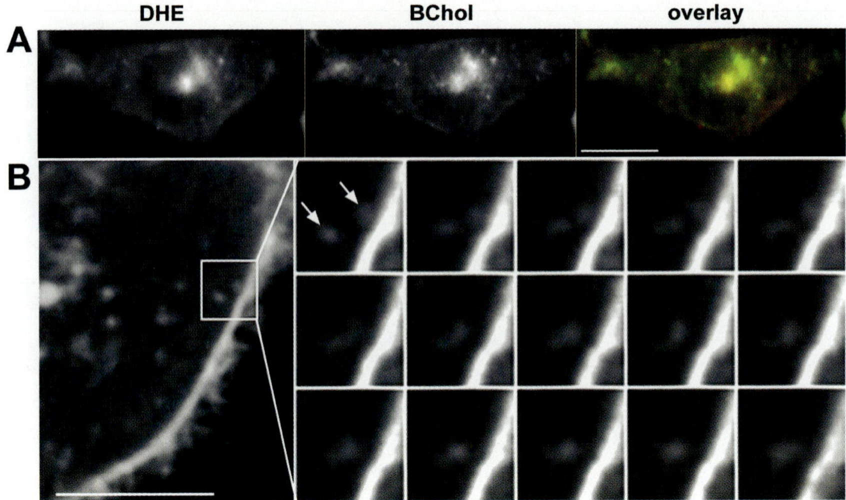

Plate 20 (Figure 17.3 on page 384 of this volume).

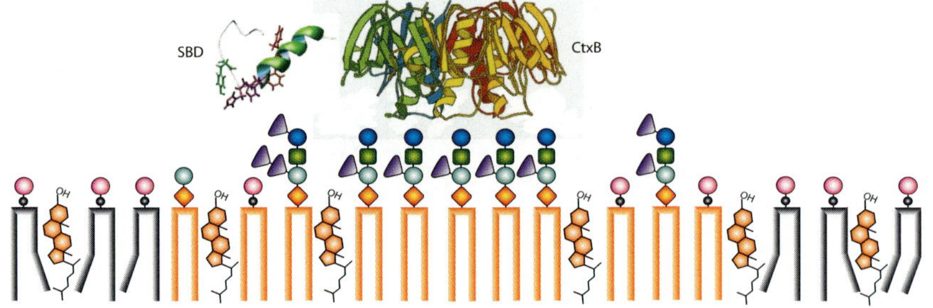

Plate 21 (Figure 18.1 on page 400 of this volume).

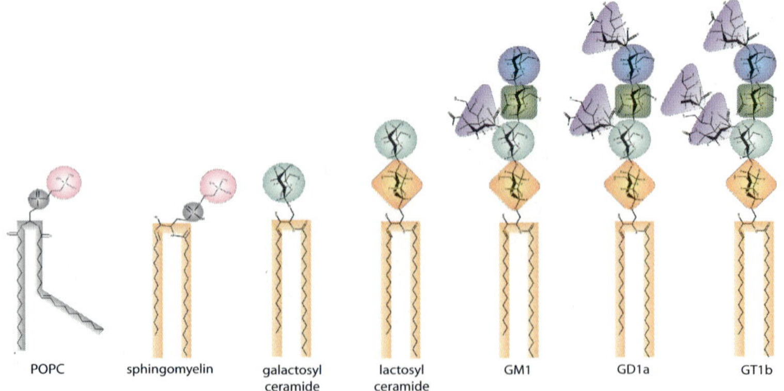

Plate 22 (Figure 18.2 on page 400 of this volume).

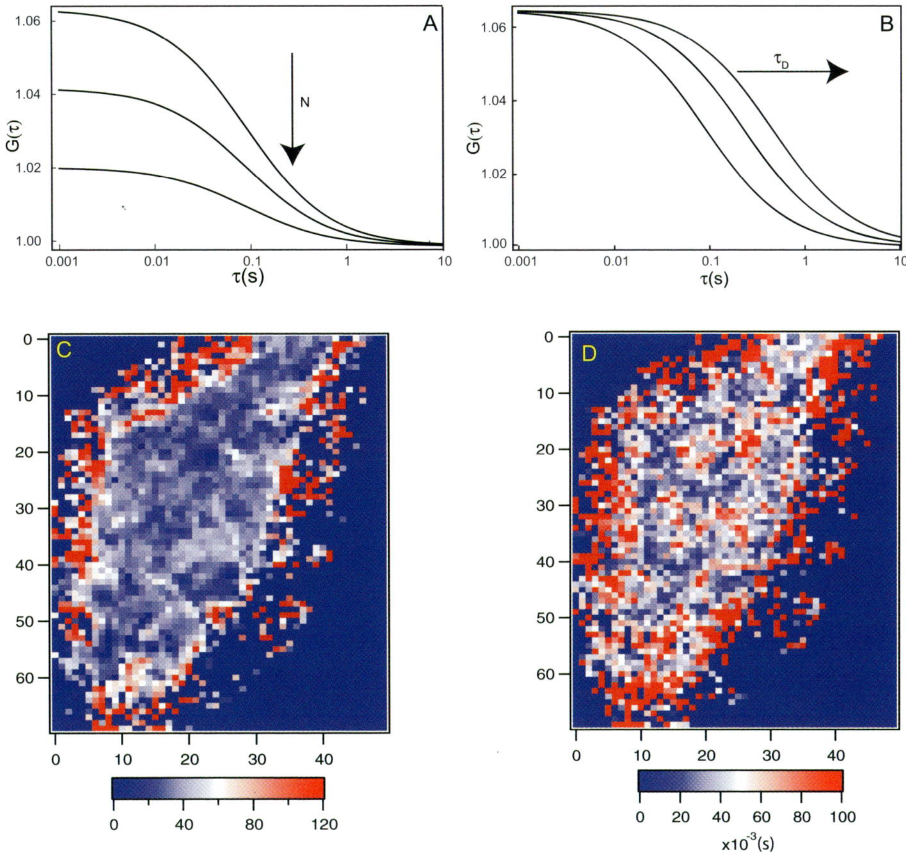

Plate 23 (Figure 18.4 on page 407 of this volume).

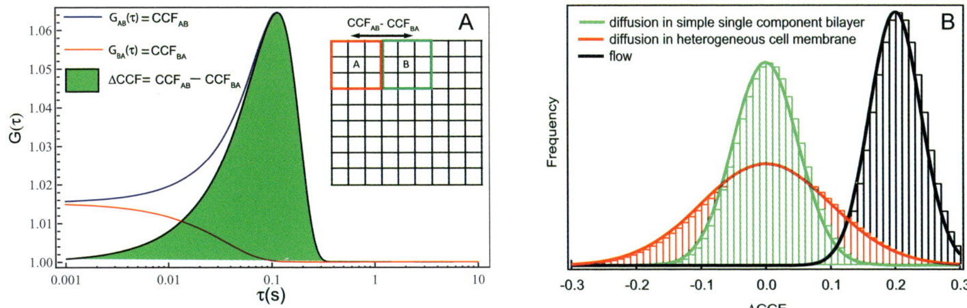

Plate 24 (Figure 18.5 on page 408 of this volume).

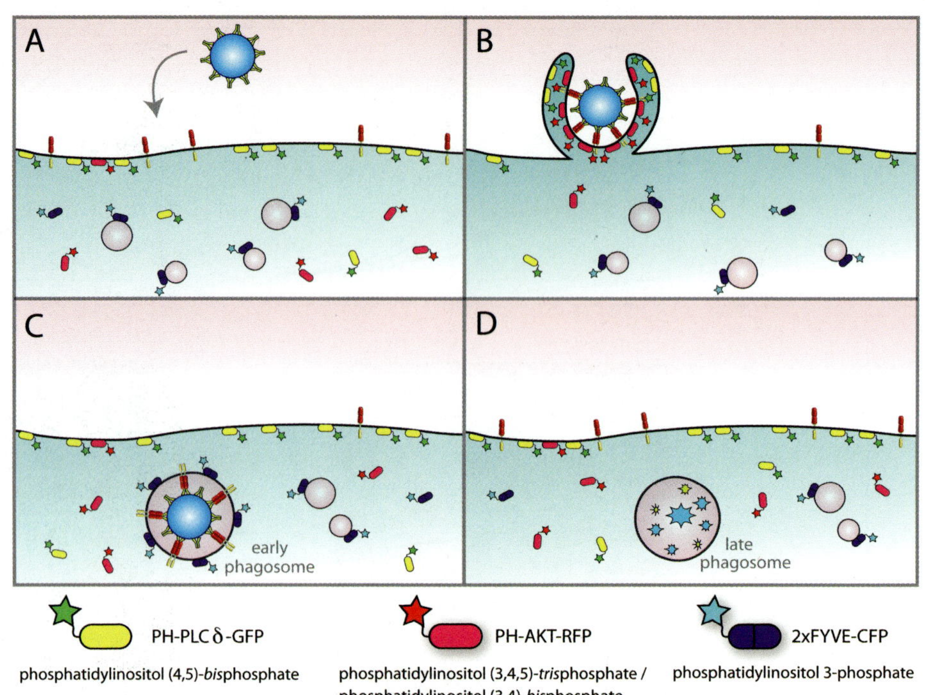

Plate 25 (Figure 19.1 on page 440 of this volume).

UNIVERSITY OF LINCOLN